T0222361

Lecture Notes in Computer Science 10931

Commenced Publication in 1973
Founding and Former Series Editors:
Gerhard Goos, Juris Hartmanis, and Jan van Leeuwen

More information about this series at http://www.springer.com/series/7407

James H. Davenport · Manuel Kauers
George Labahn · Josef Urban (Eds.)

Mathematical Software – ICMS 2018

6th International Conference
South Bend, IN, USA, July 24–27, 2018
Proceedings

 Springer

Editors
James H. Davenport 🆔
University of Bath
Bath
UK

Manuel Kauers
Johannes Kepler University
Linz
Austria

George Labahn
University of Waterloo
Waterloo, ON
Canada

Josef Urban
Czech Technical University in Prague
Prague 6
Czech Republic

ISSN 0302-9743 ISSN 1611-3349 (electronic)
Lecture Notes in Computer Science
ISBN 978-3-319-96417-1 ISBN 978-3-319-96418-8 (eBook)
https://doi.org/10.1007/978-3-319-96418-8

Library of Congress Control Number: 2018948224

LNCS Sublibrary: SL1 – Theoretical Computer Science and General Issues

This Springer imprint is published by the registered company Springer Nature Switzerland AG
The registered company address is: Gewerbestrasse 11, 6330 Cham, Switzerland

Preface

These are the proceedings of the 6th International Congress on Mathematical Software, which was held during July 24–27, 2018 at the Department of Applied and Computational Mathematics and Statistics at the University of Notre Dame.

The ICMS community believes that the appearance of mathematical software is one of the most important current developments in mathematics, and this phenomenon should be studied as a coherent whole. We hope this conference can serve as the main forum for mathematicians, scientists, and programmers who are interested in development of mathematical software.

The program of the 2018 meeting consisted of 20 topical sessions, each of which provided an overview of the challenges, achievements, and progress in a subfield of mathematical software research, development, and use. The topical sessions made up the core of the program, consisting of more than 150 contributed talks. Session contributors were given the option to submit their work for publication in these proceedings, and 59 papers were selected through a peer reviewing process.

The conference also featured three invited talks. Folkmar Bornemann spoke on "Short of Proof: How Many Digits Are Nonetheless Correct?"; Thomas C. Hales spoke on "Formal Abstracts in Mathematics"; and William Stein gave a talk about "CoCalc: Making Open Source Mathematical Software Collaborative and Easily Available on the Web." Short abstracts of these talks also appear in these proceedings. We thank the invited speakers for accepting our invitations to speak at ICMS 2018. We also thank all the contributors, session organizers, Program Committee members, as well the local arrangements team and the members of the advisory board for helping to make this conference a success. Finally, we thank our sponsors, listed on the following pages, for the financial support of the event.

June 2018

James H. Davenport
Manuel Kauers
George Labahn
Josef Urban

Organization

Program Committee

Erika Abraham	RWTH Aachen University, Germany
Dan Bates	Colorado State University, USA
Dani Brake	University of Wisconsin Eau Claire, USA
Bruno Buchberger	Johannes Kepler University Linz, Austria
Jin-San Cheng	Academia Sinica, Taiwan
James H. Davenport	University of Bath, UK
Yihe Dong	Wolfram Research, USA
Matthew England	Coventry University, UK
Jonathan Hauenstein	University of Notre Dame, USA
Patrick Ion	Mathematical Reviews/AMS
Mikolas Janota	University of Lisbon, Portugal
Tudor Jebelean	RISC-Linz, Austria
Christopher Jefferson	University of St. Andrews, UK
Michael Joswig	TU Berlin, Germany
Delaram Kahrobaei	City University of New York, USA
Masataka Kaneko	Toho University, Japan
Manuel Kauers	Johannes Kepler University, Austria
Michael Kohlhase	FAU Erlangen-Nürnberg, Germany
Christoph Koutschan	Johann Radon Institute for Computational and Applied Mathematics, Austria
Temur Kutsia	Johannes Kepler University Linz, Austria
George Labahn	University of Waterloo, Canada
Robert Lewis	Vrije Universiteit Amsterdam, The Netherlands
Alexander Maletzky	Johannes Kepler University Linz, Austria
Stephen Melczer	University of Waterloo, Canada and ENS Lyon, University of Lyon, Inria, CNRS, UCBL, France
Chenqi Mou	Beihang University, China and LIP6-UPMC, France
Yasuyuki Nakamura	Nagoya University, Japan
Markus Pfeiffer	University of St. Andrews, UK
Marco Pollanen	Trent University, Canada
Florian Rabe	FAU Erlangen-Nürnberg, Germany and LRI Paris, France
Yue Ren	MPI MIS Leipzig, Germany
Andrew Reynolds	University of Iowa, USA
Vikram Sharma	IMSc, Tamil Nadu, India
Vladimir Shpilrain	City University of New York, USA
Wolfram Sperber	FIZ Karlsruhe, Germany
Nicolas Thiery	Universite Paris Sud, Paris, France

Josef Urban   Czech Technical University in Prague, Czech Republic
Dongming Wang   CNRS, Paris, and Beijhang University, China
Wolfgang Windsteiger   Johannes Kepler University Linz, Austria
Chee Yap   New York University, USA

Abstracts of Invited Talks

Short of Proof: How Many Digits are Nonetheless Correct?

Folkmar Bornemann

Technische Universität München
bornemann@tum.de

Ever since my participation in Nick Trefethen's "SIAM 100-Digit Challenge" I have been obsessed by asking and (trying to) answering that question. Though most textbooks on numerical analysis address sources and propagation of errors, they shy away from providing tools to answer my question (short of proof, that is, when tools such as interval arithmetic and verification are not an option). Most numerical software leaves it to the discretion of the user to draw the line between the meaningful and the contingent in the output. What kind of habits developed by experienced users to stay on the safe side can be used to increase the accountability of numerical software? The question becomes pertinent when writing software for the elaborate numerical evaluation of a new class of special functions (such as higher-order gap probabilities in random matrix theory) where users implicitly expect all but the last digit given to be correct. And the question becomes a necessity when a community of users asks for tables of numbers instead of the software itself. I will report on the tools that I use in my software to put me at rest when providing such tables.

Formal Abstracts in Mathematics

Tom Hales

University of Pittsburg
hales@pitt.edu

A formal abstract is a statement of a mathematical theorem (and its accompanying definitions) that is represented in both a computer and human readable way. The computer representation of the theorem is required to be fully grounded in the foundations of mathematics, so that the theorem statement can be manipulated according to the rules of logic and mathematics. This talk will discuss an initiative to express large bodies of published mathematics as formal abstracts.

CoCalc: Making Open Source Mathematical Software Collaborative and Easily Available on the Web

William Stein

University of Washington
`wstein@uw.edu`

In 2013, I created https://CoCalc.com (then called "SageMathCloud"), as an easy way for students and instructors to streamline their use of open source mathematics software such as R, SageMath, Octave, Jupyter notebooks, and LaTeX. Everything in CoCalc now fully supports realtime synchronized editing, and there is a huge preinstalled software stack. In this talk, I will explain how you can ensure software you write is available in CoCalc and use CoCalc in teaching courses. I will also describe the current architecture of CoCalc, which has undergone many rewrites due to increased usage, and the introduction of major open source technologies, including Kubernetes and React.

Contents

Inferring Safe Maude Programs
with ÁTAME

María Alpuente[1], Demis Ballis[2]([⊠]), and Julia Sapiña[1]

[1] DSIC-ELP, Universitat Politècnica de València,
Camino de Vera s/n, 46022 Valencia, Spain
{alpuente,jsapina}@dsic.upv.es
[2] DMIF, University of Udine, Via delle Scienze, 206, 33100 Udine, Italy
demis.ballis@uniud.it

Abstract. In this paper, we present ÁTAME, an assertion-based program specialization tool for the multi-paradigm language Maude. The program specializer ÁTAME takes as input a set $\mathcal{A}$ of system assertions that model the expected program behavior plus a Maude program $\mathcal{R}$ to be specialized that might violate some of the assertions in $\mathcal{A}$. The outcome of the tool is a safe program refinement $\mathcal{R}'$ of $\mathcal{R}$ in which every computation is a good run, i.e., it satisfies the assertions in $\mathcal{A}$. The specialization technique encoded in ÁTAME is fully automatic and ensures that no good run of $\mathcal{R}$ is removed from $\mathcal{R}'$, while the number of bad runs is reduced to zero. We demonstrate the tool capabilities by specializing an overly general nondeterministic dam controller to fulfill a safety policy given by a set of system assertions.

Keywords: Program specialization · Program adaptability
Assertions · Maude · Rewriting logic

1 Introduction

Adaptability refers to the ability of a piece of software to satisfy requirements dedicated to the specific context in which it is used. In concurrent object-oriented software, adaptability is very fragile as the slightest attempt to modify the foundation of any program component may damage the whole system, ruining the effectiveness of standard reusing mechanisms.

Maude is a high-level programming language and system that supports functional, concurrent, logic, and object-oriented computations and provides equational reasoning modulo algebraic axioms such as associativity, commutativity, and identity. In this paper, we propose an adaptation technique for Maude programs that integrates system assertions and program specialization.

This work has been partially supported by the EU (FEDER) and the Spanish MINECO under grants TIN2015-69175-C4-1-R, and by Generalitat Valenciana ref. PROMETEOII/2015.013.

© Springer International Publishing AG, part of Springer Nature 2018
J. H. Davenport et al. (Eds.): ICMS 2018, LNCS 10931, pp. 1–10, 2018.
https://doi.org/10.1007/978-3-319-96418-8_1

In the literature, program specialization is often used to mean partial evaluation [5], which takes a program of n inputs and produces a simpler and usually faster version where some of the inputs are fixed to particular values. In this paper, we consider a somehow dual specialization transformation where we take a program of n outputs, or more generally, a program that explores n execution traces, and then we produce a more specific version of the original program where we disregard some of the output traces according to the assertional constraints being considered.

Our specialization technique works with Maude programs that are equipped with system assertions, with each assertion consisting of a pair $\Pi \mid \varphi$ where Π (the *state template*) is a term and φ (the *state invariant*) is a quantifier-free first-order formula with equality that defines a safety property φ which must be enforced on all the system states that match (modulo equations and axioms) the state template Π. In our technique, assertions take an active role since they are directly embedded into the specialized program to safely guide its execution. Given a set of system assertions $\mathcal{A}$ and an overly general Maude program $\mathcal{R} = (\Sigma, E, R)$ (i.e., a program that deploys all desired traces but may disprove some of the assertions), our transformation coerces $\mathcal{R}$ into a specialized program $\mathcal{R}'$ that enforces $\mathcal{A}$. This means that: (i) every execution of $\mathcal{R}'$ is an execution of $\mathcal{R}$ (i.e., no spurious computation states are produced); and (ii) every assertion in $\mathcal{A}$ is satisfied by all computation states in $\mathcal{R}'$. The program $\mathcal{R}'$ is obtained from $\mathcal{R}$ by inserting suitable conditions (abetted by the assertions of $\mathcal{A}$) in the rules of R and defining them by means of new equations that are added to E until a suitable adaptation of the original program is automatically inferred which satisfies all the assertions.

The advantage of this technique is that more refined versions of a program can be incrementally built without any programming effort by simply adding new logical constraints into the given assertion set. Specifically, this makes it possible to adapt existing Maude programs to predefined safety policies and allows the inexperienced user to largely forget about Maude syntax and semantics.

This paper is organized as follows. After some technical preliminaries in Sect. 2, we introduce a running example that we use to illustrate the kind of specialization that we aim to produce automatically. Section 3 shows how safety policies can actually be defined as system assertions in our rewriting setting, and then applied for program specialization. Section 4 shows how software adaptation can be performed efficiently in the ÁTAME system, which implements our specialization methodology. Section 5 concludes the paper.

2   Modeling Software Systems in Maude

Nondeterministic as well as concurrent software systems can be formalized through Maude programs. A Maude program essentially consists of two components, E and R, where E is a canonical (membership) equational theory that models system states as terms of an algebraic data type, and R is a set of rewrite rules that define transitions between states. Algebraic structures often involve axioms like associativity (assoc), commutativity (comm), and/or identity (also known as unity) (id) of function symbols, which cannot be handled by ordinary term rewriting but instead are handled implicitly by working with congruence classes of terms. More precisely, the membership equational theory E is decomposed into a disjoint union $E = \Delta \uplus Ax$, where the set Δ consists of (conditional) equations and membership axioms (i.e., axioms that assert the type or *sort* of some terms) that are implicitly oriented from left to right as rewrite rules (and operationally used as simplification rules), and Ax is a set of algebraic axioms that are implicitly expressed as function attributes and are only used for Ax-matching.

The system evolves by rewriting states using *equational rewriting*, i.e., rewriting with the rewrite rules in R *modulo* the equations and axioms in E [7]. Formally, system computations (also called execution traces) correspond to rewrite sequences $t_0 \xrightarrow{r_0}_E t_1 \xrightarrow{r_1}_E \ldots$, where $t \xrightarrow{r}_E t'$ denotes a transition (modulo E) from state t to t' via the rewrite rule of R that is uniquely labeled with label r. The transition space of all computations in $\mathcal{R}$ from the initial state t_0 can be represented as a *computation tree* whose branches specify all of the system computations in $\mathcal{R}$ that originate from t_0.

The following Maude program will be used as a running example throughout the paper.

Example 1. Consider a Maude program $\mathcal{R}_{\mathsf{DAM}}$ that models a simplified, non-deterministic dam controlling system to monitor and manage the water volume of a given basin[1]. In the program code, variable names are fully capitalized.

We assume that the dam is provided with three spillways called s1, s2, and s3 each of which has 4 possible aperture widths of increasing discharge capacity close, open1, open2, open3. Each spillway is formally specified by a term [S,O], where S ∈ {s1, s2, s3} and O ∈ {close, open1, open2, open3}. A global spillway configuration is a multiset [s1,O1] [s2,O2] [s3,O3] that groups together the three spillways by means of the usual associative and commutative infix, union operator __ (written in mixfix notation with empty syntax) whose identity is the constant empty. System states are defined by terms of the form { SC | V | T | AC } where SC is a global spillway configuration, V is a rational number that indicates the basin water volume (in m^3), T is a natural number that timestamps the current configuration, and the Boolean flag AC, called apertureCommand, enables changes of the spillway aperture widths only when its value is true.

Figure 1 shows the equational specification that formalizes basin water inflow and outflow. To keep the exposition simple, we assume that the basin water inflow is constant, while the basin outflow depends on the width of the spillway apertures and can be computed as the sum of the outflows of each spillway in the spillway configuration. Note that inflow and outflow values are measured in

---

[1] Maude's syntax is hopefully self-explanatory. Due to space limitations and for the sake of clarity, we only highlight those details of the system that are relevant to this work. A complete Maude specification of the dam controller is available at the ÁTAME website at http://safe-tools.dsic.upv.es/atame. For more information about the Maude language, see [4].

```
eq inflow = 3000 .                  --- Basin water inflow
eq aperture(close) = 0 .            --- Outflow for a closed spillway
eq aperture(open1) = 200 .          --- Outflow for aperture width open1
eq aperture(open2) = 400 .          --- Outflow for aperture width open2
eq aperture(open3) = 1200 .         --- Outflow for aperture width open3

--- Basin water outflow for a given spillway configuration
eq outflow(empty) = 0 .
eq outflow([S,O] SS) = aperture(O) + outflow(SS) .
```

Fig. 1. Equational definition of basin inflow and outflow.

m^3/min and are hard-coded into the dam controller. More realistic scenarios could be easily defined by sophisticating the basin inflow and outflow functions.

The system dynamics is specified by the eight rewrite rules in Fig. 2, which implement system state transitions. The openX-Y rewrite rules progressively increment the aperture width of a given spillway (e.g., the rule open1-2 increases the aperture of the spillway S from level open1 to level open2). Dually, closeX-Y rewrite rules progressively decrement the aperture width of a spillway. The rule nocmd specifies the empty command which basically states that no action is taken on the spillway configuration by the dam controller at time instant T. The rule is fired only when the AC flag is enabled, and its application disables the flag to allow a new basin water volume to be computed in the next time instant. These eight rules, called *aperture command* rules, implement instantaneous spillway modifications that do not change the time instant or the basin water volume.

The temporal evolution of the basin water volume is specified by the conditional rewrite rule volume that computes the volume V' at time T + deltaT, given the input volume V at time T. The parameter deltaT is measured in

```
rl [nocmd] : { SC | V | T | true } => { SC | V | T | false } .
rl [openC-1] :
    { [S,close] SS | V | T | true } => { [S,open1] SS | V | T | false } .
rl [open1-2] :
    { [S,open1] SS | V | T | true } => { [S,open2] SS | V | T | false } .
rl [open2-3] :
    { [S,open2] SS | V | T | true } => { [S,open3] SS | V | T | false } .
rl [close1-C] :
    { [S,open1] SS | V | T | true } => { [S,close] SS | V | T | false } .
rl [close2-1] :
    { [S,open2] SS | V | T | true } => { [S,open1] SS | V | T | false } .
rl [close3-2] :
    { [S,open3] SS | V | T | true } => { [S,open2] SS | V | T | false } .
crl [volume] : { SC | V | T | false } => { SC | V' | (T + deltaT) | true }
    if V' := (V + inflow * deltaT) - (outflow(SC) * deltaT) .
```

Fig. 2. (Conditional) rewrite rules for the dam controlling system.

minutes and can be set by the user. The volume computation changes the input volume V by adding the water inflow and subtracting the corresponding water outflow over the deltaT interval.

The use of the apertureCommand flag in the rule definitions guarantees a fair interleaving between the applications of the rule volume and the remaining aperture command rules. Specifically, this implies that a new basin water volume is computed after each spillway aperture width modification.

Note that computations in $\mathcal{R}_{\mathsf{DAM}}$ may reach potentially hazardous system states (e.g., an extremely high water volume). This is because $\mathcal{R}_{\mathsf{DAM}}$ does not implement any spillway management policy that safely restricts the applications of the aperture command rules.

3   Defining Safety Policies Through Assertions

A *safety policy* for a Maude program $\mathcal{R}$ is defined by means of a set $\mathcal{A}$ of system assertions, each assertion being of the form $\Pi \mid \varphi$, which $\mathcal{R}$ must satisfy. Intuitively, system assertions specify those computation states such that, for every subterm of a state that matches the algebraic structure of the state template Π with substitution (modulo the axioms) σ, the constraints given by the instantiated invariant $\varphi\sigma$ are satisfied. Besides the usual Boolean operators and Maude predefined predicates, the state invariant φ may include user-defined predicates as well as functions that can be specified via suitable equational definitions.

Example 2. Let us consider the user-defined function openSpillways(SC) that returns the number of open spillways in the spillway configuration SC, whose equational definition is

```
eq openSpillways(empty)    = 0 .
eq openSpillways([S,O] SC) = if (O =/= close)
                             then (1 + openSpillways(SC))
                             else openSpillways(SC)
                             fi .
```

and the safety policy $\mathcal{A}_{\mathsf{DAM}}$ of Fig. 3 for the dam controller of Example 1 that specifies some safety constraints to prevent basin critical situations.

More specifically, assertion a1 states that, in every system state, the basin water volume must be less than 50 million m^3 to avoid dam bursts and potentially disastrous floods. Assertion a2 specifies that, whenever the basin water volume is greater than 40 million m^3, all of the spillways must be open and the aperture width of at least one spillway must be maximal (level open3). Assertion a3 requires the closure of all the spillways when the basin water volume is particularly low (10 million m^3). Finally, assertion a4 specifies the spillway handling for an intermediate water volume (10 million $m^3 \leq$ V $\leq$ 40 million m^3); in this scenario we require that exactly two spillways be constantly open.

```
(a1) { SC | V | T | AC } | (V < 50000000)
(a2) { [ S1,O1 ] [ S2,O2 ] [ S3,O3 ] | V:Rat | T:TimeStamp | AC:Bool } |
                 (V:Rat > 40000000) implies (
                    (O1 == open3 and O2 =/= close and O3 =/= close) or
                    (O2 == open3 and O1 =/= close and O3 =/= close) or
                    (O3 == open3 and O1 =/= close and O2 =/= close))
(a3) { SC | V | T | AC } | (V < 10000000) implies
                               (openSpillways(SC) == 0)
(a4) { SC | V | T | AC } | ((V >= 10000000) and (V <= 40000000)) implies
                               (openSpillways(SC) == 2)
```

Fig. 3. Safety policy $\mathcal{A}_{\text{DAM}}$ for the dam controller $\mathcal{R}_{\text{DAM}}$.

4   Computing Safe Maude Programs with ÁTAME

Program specialization techniques make it possible to automatically transform a program into a specialized version, according to an execution context. In our approach, we use assertions to set the specialization scenario and guide a two-phase program specialization technique that allows a Maude program $\mathcal{R}$ to be refined into a program $\mathcal{R}'$ w.r.t. a safety policy $\mathcal{A}$ as follows.

The first phase translates the safety policy $\mathcal{A}$ to be fulfilled into an executable equational definition $Eq(\mathcal{A})$ that can be used to detect assertion violations within system states. Roughly speaking, given a system state t, a violation of some assertion in $\mathcal{A}$ is detected in t if t can be simplified into the special constant fail by using the equational theory E of $\mathcal{R}$ extended with $Eq(\mathcal{A})$.

The second phase transforms the original rewrite rules of $\mathcal{R}$ into guarded, conditional rewrite rules that can only be fired if no system assertion is violated. Intuitively, this is achieved by transforming each rewrite rule $r : (\lambda \Rightarrow \rho \text{ if } C)$ of $\mathcal{R}$ into a refined version $r' : (\lambda \Rightarrow \rho \text{ if } C \wedge \text{check}(\rho) =/= \text{fail})$ of r that contains the extra constraint $\text{check}(\rho) =/= \text{fail}$ that holds when (the instances of) the right-hand side ρ cannot by simplified to fail by using the extended equational theory $E \cup Eq(\mathcal{A})$. This ensures that any state transition $t_1 \xrightarrow{r'}_{E \cup Eq(\mathcal{A})} t_2$, that yields the system state t_2 by means of the application of the rule r', is enabled only if t_2 is a safe state, that is, a state that does not violate any assertion.

Computations in the resulting program $\mathcal{R}'$ are both reproducible in $\mathcal{R}$ and guaranteed to meet $\mathcal{A}$. In other words, for each computation $\mathcal{C}$ in $\mathcal{R}'$, (i) $\mathcal{C}$ is also a computation in $\mathcal{R}$, and (ii) there is no system state t in $\mathcal{C}$ that violates one or more system assertions of $\mathcal{A}$.

The proposed specialization technique has been efficiently implemented in a Maude tool called ÁTAME (*Assertion-based Theory Amendment in MaudE*) that has been implemented in Maude itself by using Maude's meta-level capabilities. ÁTAME integrates a RESTful Web service that is written in Java, and an intuitive Web user interface that is based on AJAX technology and is written in HTML5 and Javascript. The implementation contains about 600 lines of Maude source code, 600 lines of C++ code, 750 lines of Java code, and 700 lines of HTML5 and JavaScript code.

As an additional feature, ÁTAME provides the interconnection with the ANIMA Maude stepper [1], which integrates program animation capabilities into the ÁTAME system. Indeed, we can execute the computed specialization by incrementally building and exploring the computation tree of $\mathcal{R}'$ w.r.t. a given input initial state. The tool ÁTAME is publicly available together with a number of examples at http://safe-tools.dsic.upv.es/atame.

In order to demonstrate the tool capabilities, in the following we show the specialization of the dam controller $\mathcal{R}_{DAM}$ w.r.t. the safety policy $\mathcal{A}_{DAM}$ that can be achieved by ÁTAME.

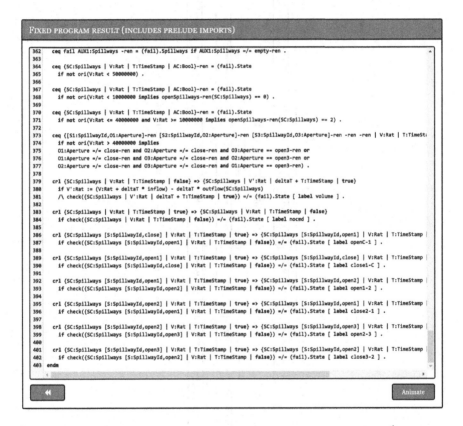

Fig. 4. A fragment of the safe specialization for $\mathcal{R}_{DAM}$ computed by ÁTAME.

Example 3. By feeding the ÁTAME system with the Maude program for the dam controller $\mathcal{R}_{DAM}$ of Example 1 and the safety policy $\mathcal{A}_{DAM}$ of Example 2, a program specialization $\mathcal{R}'_{DAM}$ for $\mathcal{R}_{DAM}$ is automatically computed. Figure 4 shows a fragment of such a specialization that includes $Eq(\mathcal{A}_{DAM})$ (i.e., the equations for detecting assertion violations) and the constrained, conditional versions of the

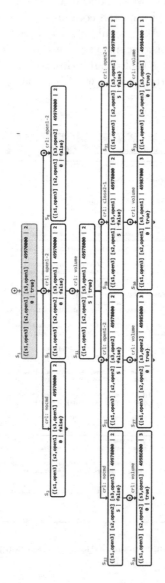

Fig. 5. A computation tree fragment for $\mathcal{R}'_{\mathsf{DAM}}$.

original rewrite rules. Note that all the operators in the equations of $Eq(\mathcal{A}_{\text{DAM}})$ are renamed by adding the textual suffix -ren. This guarantees that assertion checking is orthogonal to system computations, that is, there is no interference between the assertion checking mechanism and the applications of the rewrite rules that make the system evolve only through safe states that meet $\mathcal{A}_{\text{DAM}}$. A fragment of the computation tree that is deployed by the Maude stepper ANIMA for the initial state $s = \{$[s1,open3] [s2,open1] [s3,open1] | 49970000 | 20 | true$\}$ in $\mathcal{R}'_{\text{DAM}}$ is shown in Fig. 5. Note that all of the states in the considered tree fragment fulfill the system assertions formalized in $\mathcal{A}_{\text{DAM}}$.

In practice, the runtime cost of checking the assertions must be weighed against the saving gained from embedding them into the code and thus omitting the need for executing programs within a monitored runtime environment. The manual inclusion of safety policies as a piece of code is generally problematic, since such conditions may not be easily coded by non-specialists. Moreover, as shown in [2], the monitored runtime verification of external constraints generally incurs more cost than running the specialized program that is automatically inferred by our approach. In the case of the running example of this paper, as expected the specialized program $\mathcal{R}'_{\text{DAM}}$ is slightly slower than the original program $\mathcal{R}_{\text{DAM}}$. Nevertheless, running $\mathcal{R}'_{\text{DAM}}$ is 68% faster than running $\mathcal{R}_{\text{DAM}}$ within a runtime environment that supports dynamic assertion-checking. As for the time necessary for computing the program specializations, it is almost negligible (a few milliseconds). For a detailed empirical evaluation, we refer to [2].

5   Concluding Remarks

The technique described in this paper presents similarities with automated program correction and related problems such as code fixing and repair techniques. The discussion of these similarities is outside the scope of this paper; a detailed comparison can be found in [2]. Loosely related to this work is also the concept of program specialization of terminating programs based on output constraints (i.e., program post-conditions) [6]. This methodology translates the constraints into a characterization function for the program's input that is used to guide a partial evaluation process. In contrast, we deal with non-terminating concurrent programs and the specialization that we achieve cannot be produced by any (conventional or unconventional) partial evaluation techniques for Maude [3]. To our knowledge, the assertion-based functionality for molding programs supported by ÁTAME is beyond the capabilities of all existing Maude tools.

References

1. Alpuente, M., Ballis, D., Frechina, F., Sapiña, J.: Exploring conditional rewriting logic computations. J. Symbolic Comput. **69**, 3–39 (2015)
2. Alpuente, M., Ballis, D., Sapiña, J.: Static correction of maude programs with assertions. Technical report, Universitat Politècnica de València (2018). http://hdl.handle.net/10251/100268

3. Alpuente, M., Cuenca-Ortega, A., Escobar, S., Meseguer, J.: Partial evaluation of order-sorted equational programs modulo axioms. In: Hermenegildo, M.V., Lopez-Garcia, P. (eds.) LOPSTR 2016. LNCS, vol. 10184, pp. 3–20. Springer, Cham (2017). https://doi.org/10.1007/978-3-319-63139-4_1
4. Clavel, M., Durán, F., Eker, S., Escobar, S., Lincoln, P., Martí-Oliet, N., Meseguer, J., Talcott, C.: Maude Manual (Version 2.7.1). Technical report, SRI International (2016). http://maude.cs.uiuc.edu/maude2-manual/
5. Danvy, O., Glück, R., Thiemann, P. (eds.): Proceedings of the International Seminar on Partial Evaluation (Dagstuhl 1996). LNCS, vol. 1110. Springer, Heidelberg (1996). https://doi.org/10.1007/3-540-61580-6
6. Khoo, S.C., Shi, K.: Program adaptation via output-constraint specialization. Higher Order Symbolic Comput. **17**(1), 93–128 (2004)
7. Meseguer, J.: Conditional rewriting logic as a unified model of concurrency. Theor. Comput. Sci. **96**(1), 73–155 (1992)

Finding a Middle Ground
for Computer-Aided Cryptography

Evan Austin$^{(\boxtimes)}$, Scott Batson, Peter Curry, and Bryan Williams

SPAWAR Systems Center Atlantic,
P.O. Box 190022, North Charleston, SC 29419, USA
{evan.austin,scott.batson,peter.j.curry,bryan.l.williams1}@navy.mil

Abstract. Motivated by the ever-increasing difficulty of proofs of security and correctness, cryptographers have drawn inspiration from the more general software and hardware verification communities and integrated formal methods tools and techniques into their workflows. Though this practice of computer-aided cryptography is still comparatively young, it has spawned a number of automated cryptographic analysis tools. These tools can be categorized in one of two ways: tools focused on theoretical, or "provable," aspects of security; and tools focused on verifying more practical implementation details. This paper discusses our motivation for, and early work towards, finding an approachable middle ground of the current cryptographic tool spectrum.

1 Introduction

The looming threat of a quantum computing breakthrough has shifted researchers' focus away from cryptographic schemes founded on simple algebraic properties towards those based on more complex abstractions. In addition to not being as well studied as traditional cryptographic primitives, the underlying "hardness" assumptions of these new abstractions greatly challenge any formal verification of their associated security and correctness properties. This has exacerbated an already significant issue in that "[cryptographers] generate more proofs than [they] carefully verify" [10] due to the commonly shared opinion that "many proofs in cryptography have become essentially unverifiable" [3].

This problem is not unique to cryptography; complexity is the enemy of any verification effort. Drawing inspiration from the more general software and hardware verification communities, cryptographic researchers have begun employing a variety of formal methods tools and techniques in their work, giving birth to the comparatively new practice of computer-aided cryptography. The automated subset of cryptographic analysis tools, or AutoCrypto tools for short, typically fall into two schools of thought. First, there are proof assistants like EasyCrypt [1], which are specialized to reason about the theoretical security of abstract cryptographic schemes in the game-based style of Shoup [12] and the previously cited Bellare and Rogaway. The second mindset is more concerned

J. H. Davenport et al. (Eds.): ICMS 2018, LNCS 10931, pp. 11–18, 2018.
https://doi.org/10.1007/978-3-319-96418-8_2

with establishing the practical security and functional correctness of specific instantiations of a cryptographic scheme. The Cryptol language [9] perhaps best exemplifies this approach.

Regardless of where on the spectrum AutoCrypto tools lie, there is what we perceive to be a shared flaw among them in that their specification languages are logically and, to a lesser degree semantically, distant from the mathematical notations typically used to describe cryptographic algorithms. In our experience, this has challenged the adoption of these tools by otherwise knowledgeable users well beyond the usual growing pains that come with learning to apply formal methods. Furthermore, we found it extremely difficult to experiment with, or otherwise refine, a specification in environments where provable correctness was a point of constant emphasis. This pushed us to first develop cryptographic prototypes in an informal environment before attempting any notion of verification, leading to duplicated and wasted effort.

Our hope is that by documenting our trials and tribulations we can further motivate the need for more approachable formal methods tools. With that in mind, the content that follows is one part experience report and one part work in progress. Section 2 provides relevant background information and introduces a simple cryptographic scheme to be used as a motivating example. Section 3 compares and contrasts the two AutoCrypto tools cited above, EasyCrypt and Cryptol, and discusses our initial impressions of working with them. Finally, Sect. 4 discusses our ongoing work to develop an approachable methodology for the rapid prototyping of cryptographic systems that we envision sitting somewhere between these two tools – the titular "middle ground" that is being sought.

2   Background

Public-key encryption schemes are so named because secure communication between two parties can be conducted by having the receiving party openly share their encryption key with any sending party. Provided that the associated decryption key remains private, ciphertext messages can be publicly transmitted with a reasonable belief that adversaries can intercept them, but not recover the original plaintext message. This property is referred to as one-wayness. A stricter interpretation of privacy of communication would require that an adversary not be able to recover any part of a plaintext from its associated ciphertext. This property, semantic security, implies the indistinguishabilty of any two ciphertexts produced by the same cryptographic scheme. Proving that a scheme possesses these, and possibly other, security properties requires reasoning about the computational assumptions of the underlying algorithms.

The hardness assumption mentioned in the introduction is tied to the notion of a trap-door function; a mathematical problem that is computationally difficult to solve in general, but can be easily solved given additional knowledge. As an example, we take the discrete logarithm problem: Given a cyclic group $\mathbb{G}$ of order q that is generated by element g, such that $\mathbb{G} = \{g^0, g^1, ..., g^{(q-1)}\}$, and some $g^x \in \mathbb{G}$, where $x \in \mathbb{Z}_q$, compute the value of x. Depending on the structure of the group $\mathbb{G}$, the difficulty of this problem ranges from trivial to intractable.

keyGen():

    $privKey \Leftarrow \mathbb{Z}_q$

    $pubKey := g^{privKey} \, mod \, p$

    $return(pubKey, privKey)$

enc(pubKey, message):

    $r \Leftarrow \mathbb{Z}_q$

    $shared := g^r \, mod \, p$

    $cipher := (pubKey^r \, mod \, p) * message$

    $return(shared, cipher)$

dec(privKey, shared, cipher):

    $return(cipher * (shared^{privKey} \, mod \, p)^{-1} \, mod \, p)$

Fig. 1. The ElGamal Cryptosystem for $\mathbb{F}_p$

For the purposes of public-key encryption, we can reduce the discrete logarithm problem to a computationally equivalent problem: Given $g^x \in \mathbb{G}$ and $g^y \in \mathbb{G}$, again with $x, y \in \mathbb{Z}_q$, compute the value $g^{xy} \in \mathbb{G}$. This problem is the basis of the Diffie-Hellman key exchange protocol, with $\mathbb{G}$ typically being defined as a subset of some large prime field for prime p such that q divides $(p - 1)$ [7]. If two parties individually select x and y values and transmit the corresponding group values $g^x \, mod \, p$ and $g^y \, mod \, p$ then they can establish a shared secret $g^{xy} \, mod \, p$, as each can compute one side of the equation $(g^y \, mod \, p)^x \, mod \, p = (g^x \, mod \, p)^y \, mod \, p$. All variants of the ElGamal cryptosystem, regardless of the underlying group, use this shared secret to directly compute the ciphertext for a given plaintext [8]. For ease of reference, we provide an algorithmic definition of ElGamal over the prime field $\mathbb{F}_p$ in Fig. 1.

3   EasyCrypt and Cryptol

EasyCrypt is an interactive theorem prover whose design was inspired by the increasingly popular game-based reasoning style of cryptographic security proofs. Rather than providing the capability to reason exclusively about cryptographic systems, EasyCrypt has extended this approach to allow users to construct adversarial models for reasoning about the relations between more general procedures. In order to achieve this generality, EasyCrypt pairs an expression language reminiscent of a polymorphic lambda calculus with a stateful procedure language that includes primitives for probabilistic interactions with memory. Additionally, EasyCrypt has a robust module language that promotes a high level of specification and proof reuse.

    EasyCrypt's proof logic combines a higher-order logic that can be used to reason about expressions with deterministic, probabilistic, and relational variants of a Hoare-style logic capable of reasoning about procedures. The proof language itself is derived from earlier work on the CertiCrypt library [2], such that EasyCrypt proofs look very similar to those of the Coq proof system. Reasoning can be conducted in either a forward or backward manner with large proofs utilizing tactics to proceed in a subgoal-directed fashion, as is the style in most modern interactive proof systems. EasyCrypt also takes advantage of SMT solvers to quickly reduce or eliminate arithmetic and other simple expressions.

```
proc kg(): group * t = {
  var sk;
  sk = $dt;
  return (g ^ sk, sk);
}

proc enc(pk:group, m:group): group * group = {
  var y;
  y = $dt;
  return (g ^ y, pk^y * m);
}

proc dec(sk:t, c:group * group): group option = {
  var gy, gm;
  (gy, gm) = c;
  return Some (gm * gy^(-sk));
}
```

Fig. 2. The ElGamal cryptosystem in EasyCrypt

To continue with our motivating example, an EasyCrypt definition of the ElGamal cryptosystem is provided in Fig. 2. This definition matches quite closely with the specification from Fig. 1, with the exception that details of the modular arithmetic have been abstracted away. Note that no information is given about the construction of variable values in this specification beyond asserting that they are elements of a cyclic group, or its related base type, as indicated by the types group and t respectively. In theory, this abstract group type is intended to represent all cyclic groups, regardless of their underlying structure.

In practice, however, the definition of cyclic groups and their prerequisite operations in EasyCrypt are heavily axiomatized and tailored to the prime field $\mathbb{F}_p$. When we attempted to utilize EasyCrypt to reason about cryptographic schemes based on groups of different structures, e.g. the elliptic curve variant of the ElGamal system [11], we found its standard theories to be incompatible with other models of computation. The root cause of this incompatibility was not immediately obvious. Frequently we had constructed what we thought to be a correct proof only to have some invocation of the smt tactic greet us with the less than helpful error message "cannot prove goal (strict)."

This frustration aside, we were more concerned that EasyCrypt operated at an abstraction level that might prevent us from verifying, or possibly even ascribing, certain classes of cryptographic properties. For example, we previously noted that the computational difficulties of the discrete logarithm and related Diffie-Hellman problems are directly dependent on the structure of the cyclic group on which the problem is defined. For the specific subset of cyclic groups based on prime fields that we have discussed so far, this makes the selection of the parameters p and q absolutely critical. However, in EasyCrypt these parameters are largely ignored, with the exception of abstractly introducing the order of a prime field in a prelude theory to reason about the probability of sampling random elements.

Sitting on the other end of the abstraction spectrum for AutoCrypto tools is Cryptol. Cryptol's driving goal is to bridge the gap between mathematical specifications of cryptographic algorithms and their actual implementations. To this end, Cryptol represents an executable environment in which cryptographers can develop and refine a specification, verify its correctness, and then ultimately use it to generate code for an implementation that provably corresponds. In order to facilitate this generation, Cryptol requires complete, unambiguous specification of an algorithm, including statically inferable, concrete representations for all data values.

The majority of these constraints and implementation details can be conveniently encoded at the type level. Given that Cryptol is both implemented in and heavily inspired by the Haskell programming language, it should be of no surprise that its type system is in the style of Hindley-Milner. More specifically, Cryptol extends simple parametric polymorphism with support for fixed-size types, type-level arithmetic, and basic type predicates.

When considering how to implement an ElGamal system in Cryptol, one must move away from the general towards the specific. For the specific instance of ElGamal over the finite field $\mathbb{F}_p$ where the random exponents are selected from the subgroup $\mathbb{Z}_q$, one source recommends a 1024-bit prime for p and a 160-bit prime for q [4]. Thus, assuming that messages are elements of the primary group, we might end up with type signatures for algorithms of the scheme that look like the ones shown below. Of course, it should be noted that these signatures do not include any considerations about how to model random number generation or potential decryption failures; they account only for the bit widths of the inputs and outputs of the functions.

```
type pubKey = 1024
type privKey = 160
type plaintext = pubKey
type ciphertext = 2*plaintext

kg :: ([pubKey], [privKey])
enc :: [pubKey] -> [plaintext] -> [ciphertext]
dec :: [privKey] -> [ciphertext] -> [plaintext]
```

Similar to what you would expect to find in related systems with more robust refinement types, Cryptol utilizes SMT solvers to assist with type inference and reduction. For informal verification of specifications, Cryptol provides a capability for automated, random testing in the style of QuickCheck [5]. And, much like EasyCrypt, the same SMT solvers that drive Cryptol's type system can be used to formally prove the correctness of specifications.

We found the executable environment of Cryptol quite pleasant to work in, however, for us it had the opposite problem of EasyCrypt. We are not concerned with practical implementations of cryptographic systems at this point in time, such that it was frequently a burden to have to work with systems at a such a high level of specificity. As an example, the modular arithmetic employed by ElGamal is significantly easier to implement and reason about for arbitrary length integers compared to their fixed-size, bitstring equivalents.

4   The Middle Ground

To reiterate our concerns, we were of the opinion that the specification languages of EasyCrypt and Cryptol were too abstract and too specific, respectively, for our needs. Ultimately, what we desire is an environment for experimenting with novel cryptographic schemes where we can specify constructs using a syntax that more closely matches the mathematical notations we are familiar with. We envision this new environment as being an informal precursor to formal reasoning tools, such that we want to easily translate whatever specifications we develop to EasyCrypt or Cryptol for further verification. Eventually we would like this new system to exist as a standalone domain-specific language (DSL) that is purpose-built to be highly approachable; however, for now we are content working within the confines of Haskell.

```
class (CryptoScheme a, CyclicGroup (ValueSpace a),
       ValueSpace a ~ Message a) => ElGamal a where
  genVal :: a -> ValueSpace a
  pVal   :: a -> Integer
  qVal   :: a -> Integer

type PubKey a = ValueSpace a

encrypt :: ElGamal r => PubKey r -> Message r
        -> CryptoM r (PubKey r, Message r)
encrypt pub m =
  do y <- randInt pVal
     g <- asks genVal
     return (g .^ y, pub .^ y .* m)
```

Fig. 3. The ElGamal system in Haskell

Continuing with the motivating example of ElGamal, a subset of our Haskell implementation is shown in Fig. 3. We began this implementation by abstracting out commonly used structures to type classes, as is the standard approach in Haskell. Type classes support a notion of inheritance, such that our definition of the ElGamal class extends both the CryptoScheme class, where the ValueSpace and Message type families are defined, and the CyclicGroup class, where the modular arithmetic operators are defined.

We structured computation within our implementation using our CryptoM monad which is a simple stacking of the Reader and IO monads. This allows us to implicitly pass around a parameter set while providing access to effectful methods such as random number generation and exception handling. The resultant pairing of type classes and the monadic implementation style can be seen in the definition for our encrypt function. Our goal was to be able to write a definition that matched closely with what we have already seen in the algorithmic and

EasyCrypt specifications from Figs. 1 and 2, something that we feel we have achieved.

Much like the developers of Cryptol, we believe there is value in performing informal testing before undertaking a formal verification effort. As such, we have elected to follow their lead and test our cryptographic schemes using QuickCheck. As an example, the functional correctness property for our ElGamal implementation is shown below. Interestingly enough, given that we are working with monadic definitions, our correctness property appears visibly closer to an EasyCrypt procedure definition than a Cryptol property:

```
prop_elgamal :: (Eq (Message r), ElGamal r)
             => Message r -> CryptoM r Bool
prop_elgamal m =
   do (pk, sk) <- generateKeys
      c <- encrypt pk m
      m' <- decrypt sk c
      return (m == m')
```

Unlike EasyCrypt, however, we can easily instantiate and test this property for a variety of cyclic group structures without requiring any modifications to the scheme's definitions. In order to simplify testing, we require that users specify how to generate a random message for a given parameter set as part of providing an instantiation for the CryptoScheme class. Provided we select appropriate parameter values, the prop_elgamal property presents an accurate test of correctness. Shown below is an instance where a test failed because we tried to construct a finite field with a non-prime value for the modulus p:

```
*Crypto.ElGamal> checkCryptoProp prop_elgamal badParams
*** Failed! Assertion failed (after 1 test):
PrimeField {prime = 10, val = 2}
```

5 Future and Related Work

We are in the process of implementing a large number of cryptosystems in the same style as what was shown in Sect. 4. Our long-term goal is to use this collection of cryptographic implementations to influence the design of a DSL for the rapid prototyping of novel cryptographic systems. We are pursuing the DSL approach because the vast majority of our research team (everyone except the first author, in fact) are mathematicians, each with varying levels of comfort with programming languages and formal methods.

Our hope is that, by tailoring the syntax and semantics of this DSL to fit their academic strengths, we will end up with a computer-aided cryptographic tool that they find to be more approachable and usable. Additionally, if this DSL is embedded within Haskell, then the construction of parsers, interpreters, and other tools to analyze the cryptographic constructs we write is greatly simplified. The problem of moving to EasyCrypt or Cryptol, therefore, is reduced to figuring out a translation semantics.

Our ultimate goal is not unlike that of Crockett and Peikert with their $\Lambda \circ \lambda$ project [6]. Their software framework, however, is targeting a very specific subset of lattice-based cryptography, whereas we hope to provide a more general solution. We are both working with Haskell, though, so we are optimistic that we should be able to interoperate with their library should we get to the point of investigating that class of problems.

References

1. Barthe, G., Dupressoir, F., Grégoire, B., Kunz, C., Schmidt, B., Strub, P.-Y.: EasyCrypt: a tutorial. In: Aldini, A., Lopez, J., Martinelli, F. (eds.) FOSAD 2012-2013. LNCS, vol. 8604, pp. 146–166. Springer, Cham (2014). https://doi.org/10.1007/978-3-319-10082-1_6
2. Barthe, G., Grégoire, B., Zanella Béguelin, S.: Formal certification of code-based cryptographic proofs. SIGPLAN Not. **44**(1), 90–101 (2009)
3. Bellare, M., Rogaway, P.: Code-Based Game-Playing Proofs and the Security of Triple Encryption. Cryptology ePrint Archive, Report 2004/331 (2004)
4. Chevallier-Mames, B., Paillier, P., Pointcheval, D.: Encoding-free ElGamal encryption without random oracles. In: Yung, M., Dodis, Y., Kiayias, A., Malkin, T. (eds.) PKC 2006. LNCS, vol. 3958, pp. 91–104. Springer, Heidelberg (2006). https://doi.org/10.1007/11745853_7
5. Claessen, K., Hughes, J.: QuickCheck: a Lightweight Tool for Random Testing of Haskell Programs. In: Proceedings of the Fifth ACM SIGPLAN International Conference on Functional Programming, ICFP 2000, pp. 268–279. ACM, New York (2000)
6. Crockett, E., Peikert, C.: $\Lambda \circ \lambda$: functional lattice cryptography. In: Proceedings of the 2016 ACM SIGSAC Conference on Computer and Communications Security, CCS 2016, pp. 993–1005. ACM, New York (2016). http://doi.acm.org/10.1145/2976749.2978402
7. Diffie, W., Hellman, M.: New directions in cryptography. IEEE Trans. Inf. Theory **22**(6), 644–654 (1976)
8. Elgamal, T.: A public key cryptosystem and a signature scheme based on discrete logarithms. IEEE Trans. Inf. Theory **31**(4), 469–472 (1985)
9. Erkök, L., Matthews, J.: High assurance programming in cryptol. In: Proceedings of the 5th Annual Workshop on Cyber Security and Information Intelligence Research: Cyber Security and Information Intelligence Challenges and Strategies, CSIIRW 2009, pp. 60:1–60:2. ACM, New York (2009)
10. Halevi, S.: A plausible approach to computer-aided cryptographic proofs. Cryptology ePrint Archive, Report 2005/181 (2005)
11. Koblitz, N.: Elliptic curve cryptosystems. Math. Comput. **48**(177), 203–209 (1987)
12. Shoup, V.: Sequences of games: a tool for taming complexity in security proofs. Cryptology ePrint Archive, Report 2004/332 (2004)

Quadratic Time Algorithm for Inversion of Binary Permutation Polynomials

Lucas Barthelemy[1], Delaram Kahrobaei[2], Guénaël Renault[3,4], and Zoran Šunić[5(✉)]

[1] Quarkslab, Paris, France
lbarthelemy@quarkslab.com
[2] Graduate Center, CUNY, New York, NY, USA
dkahrobaei@gc.cuny.edu
[3] Agence Nationale de la Sécurité des Systèmes d'Information,
51 boulevard de La Tour-Maubourg, 75700 Paris 07 SP, France
guenael.renault@ssi.gouv.fr
[4] Sorbonne Université, UPMC, LIP6,
4 place Jussieu, 75252 Paris Cedex 5, France
[5] Department of Mathematics, Hofstra University, Hempstead, NY 11549, USA
zoran.sunic@hofstra.edu

Abstract. In this paper, we propose a new version of the Lagrange interpolation applied to binary permutation polynomials and, more generally, permutation polynomials over prime power modular rings. We discuss its application to obfuscation and reverse engineering.

Keywords: Permutation polynomial · Lagrange interpolation
Obfuscation

1 Motivation and Introduction

Permutation polynomials in the context of Galois' fields are very well studied in particular for their applications in cryptography. The study of binary polynomials (polynomials with coefficients in an integer ring modulo a power of 2) is less extensive, but it has been shown recently that they are important for computer security. As discussed in [1] and in [2], a straightforward application of binary permutation polynomials is obfuscation. Here we define obfuscation as a way to write computer programs that prevents reverse engineering of applications while minimizing the overhead in memory/computation cost.

In comparison with finite field permutation polynomials, the binary polynomials allow fast computation of bijective functions, since they can be directly implemented with low level arithmetic operations on computers. Moreover their

The second-named author was partially supported by a PSC-CUNY grant from the CUNY Research Foundation and by the ONR (Office of Naval Research) grant N000141512164.

use adds diversity to obfuscation techniques. The last point is of primary concern for obfuscation. Indeed, obfuscation usually does not rely on one overwhelming method, but on an aggregation of several layers of different techniques that aim to prevent automated attacks. For example, in [2] new classes of polynomials were considered and proved to be resistant to the attacks defined in [3].

In this application context, the study of permutation polynomials is purely algorithmic and a central operation is the computation of the inverse of such a polynomial.

Newton's method for inverting binary permutation polynomials is an effective algorithm, but we present in this paper a new technique based on Lagrange interpolation with two important properties:

– In a designer point of view, it is very important to measure the strength of any obfuscation technique based on binary permutation polynomials. The interpolation algorithm analyzed in this paper is proven to have a fixed complexity. This provides a more precise framework when measuring attack complexities with regard to computational overhead of using a binary permutation polynomial.
– From the reverse engineering point of view, this algorithm enables inversion techniques in a black-box context (i.e. when an encoding function is given as an evaluation function only). This is of importance when considering the reliance of encodings based on binary permutation polynomials since this algorithm can retrieve the explicit function through interpolation.

In addition, our version of Lagrange interpolation allows a better understanding on how to use binary permutation polynomials. This should prove useful for future work on the subject.

2   Interpolation of the Inverse Polynomial over $\mathbb{Z}_{2^n}$

2.1   Reduction of Integer Polynomials

Integer multiples of Newton polynomials may be used to reduce any integer polynomial to a polynomial of relatively small degree (no greater than $n + \log_2 n$) that induces the same function on $\mathbb{Z}_{2^n}$. The approach follows Mullen and Stevens [4] and has recently been used in [2] in the context of inversion of polynomials by using Largange interpolation and also Newton's method.

For $i \geq 0$, let t_i be the largest integer ℓ such that 2^ℓ divides $\ell!$, and let d_n be the largest integer i such that $n - t_i > 0$. Note that d_n is always odd and not greater than $n + \log_2 n$. Define

$$P_i(x) = 2^{n - t_i} \prod_{j=0}^{i-1} (x - j) \text{ for } i = 0, 1, \ldots, d_n, \text{ and } P_{d_n+1}(x) = \prod_{j=0}^{d_n} (x - j).$$

Each polynomial $P_i(x)$, for $i = 0, 1, \ldots, d_n + 1$, is an integer multiple of the Newton polynomial $\prod_{j=0}^{i-1}(x-j)$ of degree i, and only the last one, $P_{d_n+1}(x)$, is monic.

The ideal I of $\mathbb{Z}[x]$ generated by $P_0(x), \ldots, P_{d_n+1}(x)$ consists precisely of all integer polynomials that induce the zero function on $\mathbb{Z}_{2^n}$. Define the set of reduced polynomials $\mathcal{R}_n$ as the set of all integer polynomials $b_0 + b_1 x + \ldots + b_{d_n} x^{d_n}$ of degree at most d_n, such that, for $i = 0, \ldots, d_n$,

$$0 \leq b_i < 2^{n-t_i}.$$

For every integer polynomial $P(x)$ there exists a unique reduced polynomial $P_R(x)$ such that $P(x)$ and $P_R(x)$ induce the same function on $\mathbb{Z}_{2^n}$. The reduction is performed as follows. First, $P(x)$ is replaced by its reminder modulo the monic polynomial $P_{d_n+1}(x)$, and this yields a polynomial $R(x)$ of degree at most d_n. If $R(x)$ is reduced we are done. Otherwise, let i be the largest degree such that the i-coefficient of $R(x)$ is not in the range from 0 to $2^{n-t_i} - 1$ and let c_i be the value of this coefficient. Then there exists a nonzero q such that $c_i = 2^{n-t_i} q + r$, where $0 \leq r \leq 2^{n-t_i} - 1$. Thus, the i-coefficient of the polynomial $R(x) - qP_i(x)$ is equal to r, which is in the correct range. Continuing in the same fashion we may push all coefficients, one by one, in the order from highest to lowest degree, into the correct range and obtain a reduced polynomial.

2.2   Precise Description of the Inversion Problem

If the degree of the original integer polynomial $P(x)$ is high and/or if its coefficients are large integers, the reduction procedure may take a long time. We are not interested in this issue, our quadratic algorithm assumes that $P(x)$ is given either in reduced form or as a black box that can calculate the sequence of values $P(0), P(1), \ldots, P(d_n)$ in $\mathbb{Z}_{2^n}$ in $O(n^2)$ time. Note that if we are given a reduced polynomial $P(x)$, then we can calculate $P(0), P(1), \ldots, P(d_n)$ in $O(n^2)$ time.

We formulate precisely the input and output for our problem.

Let $P(x)$ be an integer polynomial that induces a permutation on $\mathbb{Z}_{2^n}$.
Input: the sequence of values $P(0), \ldots, P(d_n)$ in $\mathbb{Z}_{2^n}$.
Output: the sequence of coefficients $b_0, b_1, \ldots, b_{d_n}$ of the unique reduced polynomial $Q(x)$ that induces the inverse permutation to $P(x)$ on $\mathbb{Z}_{2^n}$.

We know with certainty that a polynomial solution exists, since $P(x)$ induces a permutation on a finite set, which implies that some iteration of $P(x)$, which is also a polynomial with integer coefficients, induces the inverse permutation.

Our quadratic "time" complexity actually refers to the number of multiplications and/or additions and/or inversions of units in $\mathbb{Z}_{2^n}$ necessary to calculate the sequence of coefficients of $Q(x)$. The numbers involved in these calculations have $O(n)$ digits, but each addition/multiplication/inversion is counted as being performed in unit time.

We state our main result.

Theorem 1. Let $P(x) \in \mathbb{Z}[x]$ be a polynomial that induces a permutation on $\mathbb{Z}_{2^n}$, given by its sequence of values $P(0), \ldots, P(d_n)$ in $\mathbb{Z}_{2^n}$. There exists an algorithm of time complexity $O(n^2)$ that determines the sequence of coefficients $b_0, b_1, \ldots, b_{d_n}$ of the unique reduced polynomial $Q(x)$ that induces the inverse permutation to $P(x)$ on $\mathbb{Z}_{2^n}$.

2.3 Binary Permutation Polynomials

There is a simple characterization of binary permutation polynomials in terms of the coefficients of the polynomial. Namely, a polynomial $P(x) = a_0 + \cdots + a_m x^m \in \mathbb{Z}[x]$ induces a permutation on $\mathbb{Z}_{2^n}$ if and only if (i) a_1 is odd, (ii) the sum $a_3 + a_5 + a_7 \ldots$ is even, and (iii) the sum $a_2 + a_4 + a_6 + \ldots$ is even.

The criterion is stated and proved in this form by Rivest [5], but he points out that it also follows easily from the following more general criterion: $P(x)$ induces a permutation on $\mathbb{Z}_{p^n}$, where p is a prime and $n \geq 2$, if and only if (i) $P(x)$ induces a permutation on $\mathbb{Z}_p$ and (ii) $P'(a) \not\equiv 0 \pmod{p}$ for $a \in \mathbb{Z}$. The last criterion is stated in the work of Mullen and Stevens [4], who consider it a direct corollary of Theorem 123. in the book by Hardy and Wright [6].

The following corollary is crucial for our purposes.

Corollary 1. *The polynomial $P(x) = a_0 + \cdots + a_m x^m \in \mathbb{Z}[x]$ induces a permutation on $\mathbb{Z}_{2^n}$ if and only if, for all $a, b \in R$, with $a \neq b$, the Newton quotient $k_{a,b} = \frac{P(a) - P(b)}{a - b}$ is an odd integer.*

Proof. Indeed, $k_{a,b} = a_1 A_1 + a_2 A_2 + \cdots + a_m A_m$, where $A_1 = 1$ and, for $i = 2, 3, \ldots, m$, $A_i = a^{i-1} + a^{i-2}b + \cdots + ab^{i-2} + b^{i-1}$. If both a and b are even then, modulo 2, $k_{a,b} \equiv a_1$, if they have different parity then $k_{a,b} \equiv a_1 + a_2 + \cdots + a_m$, and if they are both odd, $k_{a,b} \equiv a_1 + a_3 + a_5 + \ldots$ and the conclusion follows.

2.4 Solving the Associated Linear System

Fix n, and to simplify notation, set $d = d_n$.

For $i = 0, \ldots, d$, set $x_i = P(i)$ and $y_i = Q(x_i) = i$. We need to solve, over $\mathbb{Z}_{2^n}$, the linear system of equations

$$V[x_0, x_1, \ldots, x_d](b_0, b_1, \ldots, b_d)^T = (y_0, y_1, \ldots, y_d)^T,$$

where $V = V[x_0, x_1, \ldots, x_d] = [v_{i,j}]_{(d+1) \times (d+1)}$ is the $(d+1) \times (d+1)$ Vandermonde matrix in which $v_{i,j} = x_i^j$.

We will use the following two results by Oruç and Phillips.

Theorem 2 (Oruç-Phillips 2000 [7]). *Let $x_0, x_1, \ldots, x_m$ be distinct. An explicit LDU decomposition of the Vandermonde matrix $V = V[x_0, x_1, \ldots, x_m]$ is given by $V = LDU$, where D is the diagonal matrix*

$$Diag(1, \ x_1 - x_0, \ (x_2 - x_1)(x_2 - x_0), \ldots, \ (x_m - x_{m-1})(x_m - x_{m-2}) \ldots (x_m - x_0)),$$

L is the lower triangular matrix $L = [\ell_{i,j}]$ given by

$$\ell_{i,j} = \prod_{t=0}^{j-1} \frac{x_i - x_{j-1-t}}{x_j - x_{j-1-t}}, \qquad 0 \leq j \leq i \leq m,$$

and U is the upper triangular matrix $U = [u_{i,j}]$ given by

$$u_{i,j} = \tau_{j-i}(x_0, \ldots, x_i) \qquad 0 \leq i \leq j \leq m,$$

with the understanding that empty products are equal to 1 (thus all diagonal entries in both L and U are equal to 1), and $\tau_r(x_0, \ldots, x_i)$ is the complete symmetric function evaluated at $x_0, \ldots, x_i$, that is,

$$\tau_r(x_0, \ldots, x_i) = \sum_{\lambda_0 + \lambda_1 + \cdots + \lambda_i = r} x_0^{\lambda_0} x_1^{\lambda_1} \ldots x_i^{\lambda_i}.$$

Example 1. For $m = 4$, we have

$$L = \begin{bmatrix} 1 & 0 & 0 & 0 & 0 \\ 1 & 1 & 0 & 0 & 0 \\ 1 & \frac{x_2 - x_0}{x_1 - x_0} & 1 & 0 & 0 \\ 1 & \frac{x_3 - x_0}{x_1 - x_0} & \frac{(x_3 - x_1)(x_3 - x_0)}{(x_2 - x_1)(x_2 - x_0)} & 1 & 0 \\ 1 & \frac{x_4 - x_0}{x_1 - x_0} & \frac{(x_4 - x_1)(x_4 - x_0)}{(x_2 - x_1)(x_2 - x_0)} & \frac{(x_4 - x_2)(x_4 - x_1)(x_4 - x_0)}{(x_3 - x_2)(x_3 - x_1)(x_3 - x_0)} & 1 \end{bmatrix}$$

$$U = \begin{bmatrix} 1 & x_0 & x_0^2 & x_0^3 & x_0^4 \\ 0 & 1 & x_0 + x_1 & x_0^2 + x_0 x_1 + x_1^2 & x_0^3 + x_0^2 x_1 + x_0 x_1^2 + x_1^3 \\ 0 & 0 & 1 & x_0 + x_1 + x_2 & x_0^2 + x_0 x_1 + x_1^2 + x_0 x_2 + x_1 x_2 + x_2^2 \\ 0 & 0 & 0 & 1 & x_0 + x_1 + x_2 + x_3 \\ 0 & 0 & 0 & 0 & 1 \end{bmatrix}$$

The entries of U can be obtained recursively, by $u_{0,j} = x_0^j$, $u_{i,i} = 1$, and

$$u_{i,j} = u_{i-1,j-1} + u_{i,j-1} \cdot x_i, \qquad \text{for } 1 \le i < j. \tag{1}$$

Theorem 3 (Oruç-Phillips 2000 [7]). *Let $x_0, x_1, \ldots, x_m$ be distinct. The matrix L from the explicit LDU decomposition of the Vandermonde matrix $V = V[x_0, x_1, \ldots, x_m]$ given in Theorem 2 decomposes as the product*

$$L = L^{(1)} L^{(2)} \ldots L^{(m)}$$

of subdiagonal $(m + 1) \times (m + 1)$ matrices $L^{(k)} = [\ell_{i,j}^{(k)}]$ with 1s on the diagonal and the subdiagonal entries given, for $j = 0, \ldots, m - 1$, by

$$\ell_{j+1,j}^{(k)} = \begin{cases} 0, & 0 \le j < m - k, \\ \displaystyle\prod_{t=0}^{j-(m-k)-1} \frac{x_{j+1} - x_{j-t}}{x_j - x_{j-1-t}}, & m - k \le j \le m. \end{cases}$$

Example 2. For $m = 4$, the following table provides the subdiagonal entries:

j	$\ell_{j+1,j}^{(1)}$	$\ell_{j+1,j}^{(2)}$	$\ell_{j+1,j}^{(3)}$	$\ell_{j+1,j}^{(4)}$
0	0	0	0	1
1	0	0	1	$\frac{x_2 - x_1}{x_1 - x_0}$
2	0	1	$\frac{x_3 - x_2}{x_2 - x_1}$	$\frac{(x_3 - x_2)(x_3 - x_1)}{(x_2 - x_1)(x_2 - x_0)}$
3	1	$\frac{x_4 - x_3}{x_3 - x_2}$	$\frac{(x_4 - x_3)(x_4 - x_2)}{(x_3 - x_2)(x_3 - x_1)}$	$\frac{(x_4 - x_3)(x_4 - x_2)(x_4 - x_1)}{(x_3 - x_2)(x_3 - x_1)(x_3 - x_0)}$

The subdiagonal entries can be calculated recursively as follows. For fixed j and $k = m - j$, we have $\ell^{(k)}_{j+1,j} = 1$, and for $k > m - j + 1$,

$$\ell^{(k)}_{j+1,j} = \ell^{(k-1)}_{j+1,j} \cdot \frac{x_{j+1} - x_{m-k+1}}{x_j - x_{m-k}}. \tag{2}$$

Going back to our situation, we see that the entries of U and D are integers and, as such, are well defined over $\mathbb{Z}_{2^n}$. The entries of L and $L^{(k)}$ are not necessarily integers, but they are still well defined over $\mathbb{Z}_{2^n}$.

Proposition 1. *Let $P(x) \in \mathbb{Z}[x]$ induce a permutation on $\mathbb{Z}_{2^n}$.*

(a) *Each entry of L is has odd denominator in its simplest form.*
(b) *Each entry of $L^{(k)}$, for $k = 1, \ldots, d$ has odd numerator and odd denominator in its simplest form.*

Proof. (a) By Corollary 1, we have, for $0 \le j \le i \le d$,

$$\ell_{i,j} = \prod_{t=0}^{j-1} \frac{x_i - x_{j-1-t}}{x_j - x_{j-1-t}} = \prod_{t=0}^{j-1} \frac{P(y_i) - P(y_{j-1-t})}{P(y_j) - P(y_{j-1-t})} = \prod_{t=0}^{j-1} \frac{(y_i - y_{j-1-t})k_{i,j-1-t}}{(y_j - y_{j-1-t})k_{j,j-1-t}}$$

$$= \prod_{t=0}^{j-1} \frac{(i - j + 1 + t)k_{i,j-1-t}}{(1+t)k_{j,j-1-t}} = \binom{i}{j} \prod_{t=0}^{j-1} \frac{k_{i,j-1-t}}{k_{j,j-1-t}},$$

where each $k_{*,*}$ is an odd integer.

(b) Let $d - k \le j \le d$ and set $s = j - (d - k) - 1$. By Corollary 1, we have,

$$\ell^{(k)}_{j+1,j} = \prod_{t=0}^{s} \frac{x_{j+1} - x_{j-t}}{x_j - x_{j-1-t}} = \prod_{t=0}^{s} \frac{P(y_{j+1}) - P(y_{j-t})}{P(y_j) - P(y_{j-1-t})} = \prod_{t=0}^{s} \frac{(y_{j+1} - y_{j-t})k_{j+1,j-t}}{(y_j - y_{j-1-t})k_{j,j-1-t}}$$

$$= \prod_{t=0}^{s} \frac{(1+t)k_{j+1,j-t}}{(1+t)k_{j,j-1-t}} = \prod_{t=0}^{s} \frac{k_{j+1,j-t}}{k_{j,j-1-t}},$$

where each $k_{*,*}$ is an odd integer.

We are ready to prove the main result.

Proof (Proof of Theorem 1). Recall that, in our situation, $m = d = d_n < n + \log_2 n$ and we are solving the system $LDU\mathbf{b} = \mathbf{y}$. The recursive formulas (1) and (2) show that the entries of U and the subdiagonal entries in all $L^{(k)}$, $k = 1, \ldots, d$, can be calculated in $O(n^2)$ steps. The diagonal entries of D can also be calculated recursively in $O(n^2)$ steps. The inverse of $L^{(k)}$ is obtained by simply changing the sign in all subdiagonal entries. Therefore, we can calculate $\mathbf{y}' = L^{-1}\mathbf{y} = L^{(m)^{-1}} L^{(m-1)^{-1}} \ldots L^{(1)^{-1}} \mathbf{y}$ in $O(n^2)$ steps.

We then solve the system $DU\mathbf{b} = \mathbf{y}'$ by backward substitution in $O(n^2)$ steps. Note that the i-entry of D has the form $2^{t_i} f_i$, where f_i is odd. Because of our constraints on the coefficients of reduced polynomials, we are seeking only for solutions for b_i in the range $0 \le b_i < 2^{n-t_i}$, and a solution exists and is unique in this range. More precisely, once $b_{i+1}, \ldots, b_d$ are substituted in, we need to solve for b_i from an equation of the form $2^{t_i} f_i b_i = g_i \pmod{2^n}$, for some odd f_i and some $g_i \in \mathbb{Z}_{2^n}$. We already know that a solution exists, so it must be that $g_i = 2^{t_i} g_i'$ for some $g_i' \in \mathbb{Z}_{2^n}$. After canceling the term 2^{t_i} we solve for b_i from $f_i b_i = g_i' \pmod{2^{n-t_i}}$ by inverting f_i, and thus produce the unique solution in the range $0 \leq b_i < 2^{n-t_i}$.

2.5 Another Solution

If we are not interested in producing the coefficients of the inverse polynomial $Q(x)$, but rather just in calculating the values of $Q(x)$ at various points, a slightly different algorithm exists and we outline it here.

Let $\mathcal{U}_n$ be the ring of units of the ring $\mathbb{Z}_{2^n}$. Without loss of generality we may assume that $P(x)$ separately permutes $\mathcal{U}_n$, the odds, and its complement, the evens (if it does not, we may replace $P(x)$ by $P(x) + 1$).

The ideal I' of integer polynomials that induce the zero function on $\mathcal{U}_n$ is described in [8]. It is generated by $P_i(x) = 2^{n-i-t_i} \prod_{j=0}^{i-1}(x - (2j + 1))$, for $i = 0, 1, \ldots, d_n'$, and $P_{d_n'+1}(x) = \prod_{j=0}^{d_n'}(x - (2j + 1))$, where d_n' is the largest integer i such that $n - i - t_i > 0$. Every integer polynomial that permutes $\mathcal{U}_n$ has a unique representative modulo I', which is a polynomial of degree at most d_n' with the i-coefficient in the range from 0 to $2^{n-i-t_i} - 1$. The maximum degree d_n' is approximately half of d_n. We may calculate, by using the same approach as above (the Vandermonde matrix will have dimension $(d_n' + 1) \times (d_n' + 1)$ and the interpolation is preformed for $x_i = P(2i + 1)$, $i = 0, \ldots, d_n'$) the coefficients of the unique reduced polynomial $\overline{Q}(x)$ modulo I' that inverts the values of $P(x)$ on $\mathcal{U}_n$ (and not necessarily on its complement).

By a similar approach, the coefficients of another polynomial, $\overline{\overline{Q}}(x)$, of degree at most d_n' that inverts the values of $P(x)$ on the complement of $\mathcal{U}_n$ may be calculated. The two polynomials $\overline{Q}(x)$ and $\overline{\overline{Q}}(x)$ may then be used to calculate the values of $Q(x)$ (use the former for odd x and the latter for even). Since the degrees of $\overline{Q}(x)$ and $\overline{\overline{Q}}(x)$ are, in general, smaller than the degree of $Q(x)$, this approach may be faster if we need to calculate many values of $Q(x)$.

3 Interpolation of the Inverse Polynomial over $\mathbb{Z}_{p^n}$

Fix a prime p and $n > 1$.

We claim that the same inversion technique works equally well for permutation polynomials over the ring $\mathbb{Z}_{p^n}$.

The ideal of integer polynomials that induce the zero function on $\mathbb{Z}_{p^n}$ is generated by the polynomials

$$P_i(x) = p^{n-t_{p,i}} \prod_{j=0}^{i-1}(x - j) \text{ for } i = 0, 1, \ldots, d_{p,n}, \text{ and } P_{d_{p,n}+1}(x) = \prod_{j=0}^{d_{p,n}}(x - j).$$

where, for $i \geq 0$, $t_{p,i}$ is the largest integer ℓ such that p^ℓ divides $\ell!$, and $d_{p,n}$ is the the largest integer i such that $n - t_{p,i} > 0$. Each integer polynomial is equivalent, as a function over $\mathbb{Z}_{p^n}$, to a unique reduced polynomial, that is, polynomial $b_0 + b_1 x + \cdots + b_{d_{p,n}} x^{d_{p,n}}$ of degree at most $d_{p,n}$, such that, for $i = 0, \ldots, d_{p,n}$, we have $0 \leq b_i < p^{n - t_{p,i}}$ (see [4, Theorem 2.1]).

We prove an analog of Corollary 1.

Proposition 2. *The polynomial $P(x) = a_0 + \cdots + a_m x^m \in \mathbb{Z}[x]$ induces a permutation on $\mathbb{Z}_{p^n}$ if and only of, for all $a, b \in \mathbb{Z}$, with $a \neq b$, the Newton quotient $k_{a,b} = \frac{P(a) - P(b)}{a - b}$ is an integer that is not divisible by p.*

Proof. Recall that $P(x)$ induces a permutation on $\mathbb{Z}_{p^n}$ if and only if it induces a permutation on $\mathbb{Z}_p$ and $P'(a) \not\equiv 0 \pmod{p}$, for all a.

We work modulo p. Let $a \neq b$ and, moreover, $a - b \not\equiv 0$. Since $P(a) - P(b) \equiv (a - b)k_{a,b}$ and $a - b \not\equiv 0$, we have $P(a) - P(b) \equiv 0$ if and only if $k_{a,b} \equiv 0$. Thus, $P(x)$ induces a permutation on $\mathbb{Z}_p$ if and only if $k_{a,b} \not\equiv 0$, for all $a \neq b$. Let $a \neq b$, but $a \equiv b$. Then, for $i \geq 2$, we have $A_i = a^{i-1} + a^{i-2}b + \cdots + ab^{i-2} + b^{i-1} \equiv ia^{i-1}$ and $k_{a,b} \equiv a_1 + 2a_2 a + \cdots + m a_m a^{m-1} \equiv P'(a)$. Thus, $P'(a) \not\equiv 0$, for all a, if and only if $k_{a,b} \not\equiv 0$, for all $a \neq b$ with $a \equiv b$.

The rest of the proof is exactly the same as in the binary case, except, of course, that the analog of Proposition 1 should state that, in their simplest form, all denominators of the entries in L are integers not divisible by p, and all numerators and denominators of the entries in $L^{(k)}$, $k = 1, \ldots, d_{p,n}$, are integers not divisible by p. Thus, L and $L^{(k)}$ are well defined over $\mathbb{Z}_{p^n}$.

Thus we may state a more general version of our main result.

Theorem 4. *Let p be a prime and $P(x) \in \mathbb{Z}[x]$ a polynomial that induces a permutation on $\mathbb{Z}_{p^n}$, given by its sequence of values $P(a), P(a+1), \ldots, P(a+d_{p,n})$ in $\mathbb{Z}_{p^n}$ for some $a \in \mathbb{Z}_{p^n}$ (not necessarily 0). There exists an algorithm of time complexity $O(n^2)$ that determines the coefficients $b_0, b_1, \ldots, b_{d_{p,n}}$ of the unique reduced polynomial $Q(x)$ that induces the inverse permutation to $P(x)$ on $\mathbb{Z}_{p^n}$.*

References

1. Zhou, Y., Main, A., Gu, Y.X., Johnson, H.: Information hiding in software with mixed boolean-arithmetic transforms. In: Kim, S., Yung, M., Lee, H.-W. (eds.) WISA 2007. LNCS, vol. 4867, pp. 61–75. Springer, Heidelberg (2007). https://doi.org/10.1007/978-3-540-77535-5_5
2. Barthelemy, L., Eyrolles, N., Renault, G., Roblin, R.: Binary permutation polynomial inversion and application to obfuscation techniques. In: Proceedings of the 2nd International Workshop on Software Protection, Vienna, Austria. ACM, October 2016
3. Biondi, F., Josse, S., Legay, A., Sirvent, T.: Effectiveness of synthesis in concolic deobfuscation. Comput. Secur. **70**, 500–515 (2017)
4. Mullen, G., Stevens, H.: Polynomial functions (mod m). Acta Math. Hungar. **44**(3–4), 237–241 (1984)

5. Rivest, R.L.: Permutation polynomials modulo 2^w. Finite Fields Appl. **7**(2), 287–292 (2001)
6. Hardy, G.H., Wright, E.M.: An Introduction to the Theory of Numbers, 4th edn. Clarendon Press, Oxford (1960)
7. Oruç, H., Phillips, G.M.: Explicit factorization of the Vandermonde matrix. Linear Algebra Appl. **315**(1–3), 113–123 (2000)
8. Markovski, S., Šunić, Z., Gligoroski, D.: Polynomial functions on the units of Z_{2^n}. Quasigr. Relat. Syst. **18**(1), 59–82 (2010)

Paramotopy: Parameter Homotopies in Parallel

Dan Bates[1]([✉]), Danielle Brake[2], and Matt Niemerg[3]

[1] Colorado State University, Fort Collins, USA
bates@math.colostate.edu
[2] University of Wisconsin - Eau Claire, Eau Claire, USA
brakeda@uwec.edu
[3] Knoxville, USA
research@matthewniemerg.com
http://www.math.colostate.edu/~bates, http://danibrake.org,
http://www.matthewniemerg.com

Abstract. Numerical algebraic geometry provides tools for approximating solutions of polynomial systems. One such tool is the parameter homotopy, which can be an extremely efficient method to solve numerous polynomial systems that differ only in coefficients, not monomials. This technique is frequently used for solving a parameterized family of polynomial systems at multiple parameter values. This article describes **Paramotopy**, a parallel, optimized implementation of this technique, making use of the Bertini software package. The novel features of this implementation include allowing for the simultaneous solutions of arbitrary polynomial systems in a parameterized family on an automatically generated or manually provided mesh in the parameter space of coefficients, front ends and back ends that are easily specialized to particular classes of problems, and adaptive techniques for solving polynomial systems near singular points in the parameter space.

1 Introduction

The methods of numerical algebraic geometry provide a means for approximating the solutions of a system of polynomials $F : \mathbb{C}^N \to \mathbb{C}^n$, i.e., those points $z \in \mathbb{C}^N$ such that $F(z) = 0$. There are many variations on these methods, but the key point is that polynomial systems of moderate size can be solved efficiently via homotopy continuation-based methods. In the case of a parameterized family of polynomial systems $F : \mathbb{C}^N \times \mathcal{P} \to \mathbb{C}^N$, where the coefficients are polynomial in the parameters $p \in \mathcal{P} \subset \mathbb{C}^M$, a particularly efficient technique comes into play: the parameter homotopy [1][1].

The process of using a standard homotopy to solve a system F begins with the construction of a polynomial system G that is easily solved. Once the system

---

[1] In fact, this technique applies when the coefficients are *holomorphic functions* of the parameters [1], but we restrict to the case of polynomials for simplicity.

© Springer International Publishing AG, part of Springer Nature 2018
J. H. Davenport et al. (Eds.): ICMS 2018, LNCS 10931, pp. 28–35, 2018.
https://doi.org/10.1007/978-3-319-96418-8_4

G is solved, the solutions of G are tracked numerically by predictor-corrector methods as the polynomials of G are transformed into those of F. Thanks to the underlying geometry, discussed for example in [2] or [3], we are guaranteed to find a superset $\widehat{V}$ of the set V of isolated solutions of F. The set $\widehat{V}$ is easily trimmed down to V in a post-processing step [4].

Parameter homotopies are particularly powerful as the number of solutions to be followed is exactly equal to the number of isolated solutions of $F(z, p)$ for almost all values of $p \in \mathcal{P}$ (under the common assumption that $\mathcal{P}$ has positive volume in its ambient Euclidean space). Furthermore, the solution of a single G will work for almost all values of $p \in \mathcal{P}$, so only one round of precomputation is needed regardless of the number of polynomial systems to be solved.

Parameter homotopies are not new and have been used in several areas of application [5–8] and implemented in at least two software packages for solving polynomial systems: Bertini [9] and PHCpack [10]. These implementations allow the user to run a single parameter homotopy from one parameter value p_0 with known solutions to the desired parameter value, p_1, with the solutions at p_0 provided by the user. The software package that is the focus of this article differs from these other two implementations in the following ways:

1. Paramotopy accepts as input the general form of the parameterized family $F(z, p)$ (p given as indeterminates), chooses a random $p_0 \in \mathcal{P}$, and solves $F(z, p_0)$ via a Bertini run[2];
2. Paramotopy builds a mesh in the parameter space given simple instructions from the user (or uses a user-provided set of parameter values) and performs parameter homotopy runs from p_0 to each other p in the mesh;
3. Paramotopy carries out all of these runs in parallel, as available[3];
4. Paramotopy includes adaptive schemes to automatically attempt to find the solutions of $F(z, p)$ from starting points other than p_0 if ill-conditioning causes path failure in the initial attempt; and
5. Paramotopy is designed to simplify the creation of front ends and back ends specialized for particular applications.

The full version of this article [11] includes more background and examples.

2   Homotopies

2.1   Homotopy Continuation

Given a polynomial system $F : \mathbb{C}^N \to \mathbb{C}^N$ to be solved, standard homotopy continuation consists of three basic steps:

1. Choose a *start system* $G : \mathbb{C}^N \to \mathbb{C}^N$ similar in some way to $F(z)$ that is "easy" to solve;

---

[2] Bertini provides this functionality as well.
[3] Bertini and PHCpack both have parallel versions, but not for multiple parameter homotopy runs.

2. Find the solutions of $G(z)$ and form the new homotopy function $H : \mathbb{C}^N \times \mathbb{C} \to \mathbb{C}^N$ given by $H(z,t) = F(z) \cdot (1 - t) + G(z) \cdot t \cdot \gamma$, where $\gamma \in \mathbb{C}$ is randomly chosen; and
3. Using predictor-corrector methods (and various other numerical routines [2, 12–14]), track the solutions of G at $t = 1$ to those of F at $t = 0$.

There are many variations on this general theme, but we focus here on the basic ideas, leaving details and alternatives to the references. A discussion of the choice of an adequate start system G goes beyond the scope of this paper. It is enough to know that there are several such options [2,3,15,16].

Once $G(z)$ is solved and $H(z,t)$ is formed, the solutions of $H(z,t)$ for varying values of t may be visualized as curves. Indeed, as t varies continuously, the solutions of $H(z,t)$ will vary continuously, so each solution sweeps out a curve or path (also sometimes called a *solution curve* or *solution path*) as t moves from 1 to 0. A schematic of four such paths is given in Fig. 1. Predictor-corrector methods are used to follow the solutions of G to those of F along these paths. See [11] for more background.

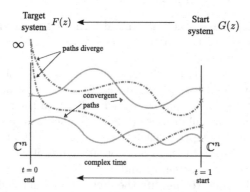

Fig. 1. A schematic depiction of a homotopy from system F to system G. There are four solutions of $G(z)$ at $t = 1$. Two solution paths diverge as $t \to 0$, while the other two lead to solutions of F at $t = 0$.

2.2   Parameter Homotopies

Suppose we wish to solve a parameterized polynomial system $F(z, p)$ in variables z and parameters p at a (possibly very large) number of points in parameter space, i.e., we want to find z such that $F(z, p') = 0$ for varying values $p = p'$. If we know all isolated, finite, complex solutions at some generic point $p = p_0$ in a *convex*[4] parameter space $\mathcal{P}$, the underlying theory allows us to make

---

[4] Handling non-convex parameter spaces is significantly more difficult and is described later.

use of a *parameter* or *coefficient-parameter homotopy* [1]. The usefulness of this software becomes readily apparent from the following proposition, proved in somewhat different language in [2]. The proposition guarantees that we can find the isolated, finite, complex solutions of $F(z, p')$ simply by following paths through the parameter space, $\mathcal{P} \subset \mathbb{C}^M$, from the solutions of $F(z, p_0)$.

Proposition 1. *The number of finite, isolated solutions of $F(z, p)$ is the same for all $p \in \mathcal{P}$ except for a measure zero, algebraic subset $\mathcal{B}$ of $\mathcal{P}$.*

This proposition gives us a probability one guarantee that a randomly chosen path through parameter space will avoid $\mathcal{B}$. Assuming further that $\mathcal{P}$ is convex, a straight line segment through parameter space from a randomly chosen $p_0 \in \mathcal{P}$ to a prespecified target $p_1 \in \mathcal{P}$ will, with probability one, not pass through the set $\mathcal{B}$. This immediately implies a (known) technique for solving many polynomial systems from the same parameterized family with parameter space $\mathcal{P}$. First, find all finite, isolated, complex solutions for some randomly chosen $p_0 \in \mathcal{P}$. We refer to this as Step 1. Second, for each parameter value of interest, $p_i \in \mathcal{P}$, simply follow the finite, isolated, complex solutions through the simple homotopy $H(z, t) = F(z, p_0) \cdot t + F(z, p_i) \cdot (1 - t)$. We refer to this as Step 2. Notice that the randomly chosen γ from standard homotopies can be neglected in this homotopy since p_0 is chosen randomly. We describe in Sect. 3.2 how we monitor these Step 2 runs in case paths fail and also how we handle such failures.

For the cost of a single Step 1 solve at some random point p_0 in the parameter space, we may rapidly solve many other polynomial systems in the same parameterized family. Indeed, there are a minimal number of paths to follow in each Step 2 run and there is no pre-computation cost beyond the initial solve.

3   Implementation

Paramotopy is a C++ implementation of parameter homotopies, relying heavily on Bertini [9]. In this section, we provide many details about this software.

3.1   Main Algorithm

We first present the main parameter homotopy algorithm that is implemented in Paramotopy. Note in particular the input value K and the while loop at the end, both included to help manage path failures during the Step 2 runs. Also, note that this algorithm assumes that $\mathcal{P} = \mathbb{C}^M$, for some M. The use of Paramotopy for other parameter spaces is described in Sect. 3.3.

Remark 1. To find all solutions for all $p \in L$, we must have that all solutions of $F(z, p_0)$ are nonsingular as we can only follow paths starting from nonsingular solutions during the parameter homotopies after the first run. Deflation [17,18] could be used to regularize singularities in Step 1 before beginning Step 2, but this is not currently implemented.

Input : $F(z; p)$, a set of polynomial equations, variables $z \in \mathbb{C}^N$, and parameters $p \in L \subset \mathcal{P} = \mathbb{C}^M$; $\ell = | L |$ parameter values at which the solutions of $F(z; p)$ are desired; bound K on the number of times to try to find solutions for any given $p \in L$, in the case of path failures.

Output: List of solutions of $F(z; p) = 0$ for each $p \in L$.

1 Choose random $p_0 \in \mathcal{P}$;
2 Solve $F(z; p_0) = 0$ with any standard homotopy. (Step 1);
3 Store all nonsingular finite solutions in set S;
4 Set $\mathcal{F} := \varnothing$. (Beginning of Step 2.);
5 **for** $i=1$ *to* ℓ **do**
6 | Construct parameter homotopy from $F(z; p_0)$ to $F(z; p_i)$;
7 | Track all $|S|$ paths starting from points in S;
8 | Set $\mathcal{F} := \mathcal{F} \cup \{i\}$ if any path fails;
9 Set $k := 0$. (Beginning of path failure mitigation.);
10 **while** $|\mathcal{F}| > 0$ *and* $k < K$ **do**
11 | Set $\mathcal{F}' = \varnothing$;
12 | Choose random $p' \in \mathcal{P}$;
13 | Solve $F(z; p') = 0$ with a parameter homotopy from p_0;
14 | **for** $m=1$ *to* $|\mathcal{F}|$ **do**
15 | | Solve $F(z; p_{\mathcal{F}[m]}) = 0$ with a parameter homotopy from p' to $p_{\mathcal{F}[m]}$;
16 | | Set $\mathcal{F}' := \mathcal{F}' \cup \{m\}$ if any path fails;
17 | Set $\mathcal{F} := \mathcal{F}'$ and increment k;

Algorithm 1: Paramotopy.

3.2 Handling Path Failures During Step 2

If a path fails during a Step 2 run for some parameter value $p \in L$, Paramotopy will automatically attempt to find the solutions at p by tracking from a different randomly chosen parameter value $p' \neq p_0 \in \mathcal{P}$. It will repeat this process K times, with K specified by the user. This is the content of the while loop at the end of the Main Algorithm.

The idea behind this is that paths often fail for one of two reasons, either the path seems to be diverging or the Jacobian matrix becomes so ill-conditioned that either the steplength drops below the minimum allowed or the precision needed rises above the maximum allowed. For parameter homotopies, a path failure of the first type is possible for either of two reasons: either the path really is diverging or the norm of the solution is above a particular threshold. In the former case, it can happen that the nature of the solution set at target value p differs from that at a generic point in the parameter space, e.g., there could be fewer finite solutions at p. Such path failures are captured and reported by Paramotopy, but there is simply no hope for "fixing" them as this result is a natural consequence of the geometry of the solution set, i.e., p is inherently different from other points in parameter space, so Paramotopy takes the correct action in reporting it. In the latter case, it can happen that the scaling of the problem results in solutions that are large in some norm, e.g., $|z|_\infty > 10^5$ as is the default in the current version of Bertini. If this is suspected, the user could rescale the system or adjust the threshold MaxNorm and run the problem again.

For the second type of path failure, the ill-conditioning is caused by the presence of a singularity $b \in \mathcal{B}$ near or on the path between p_0 and p. By choosing new starting point p' "adequately far" from p_0, it should be feasible to avoid the ill-conditioned zone around b unless b is near the target value p. In this last case, it is unlikely that choosing different starting points p' will have any value, which is why we have capped the number of new starting points allowed at K.

For now, the new point p' is chosen randomly in the unit hypercube. Future work will detect where in parameter space the failures have occurred and bound p' away from this region. Since it cannot easily be determined which paths from p' to p correspond to the failed paths from p_0 to p, there is no choice but to follow all paths from p' to p. To find all solutions at p', we simply use a parameter homotopy to move the solutions at p_0 to those at p'. Of course, if there are path failures, we must choose yet another p' and try again.

3.3 Handling Parameter Spaces Other Than $\mathbb{C}^M$

As described near the end of Sect. 2.2, Paramotopy may be used to handle parameter spaces other than the simplest parameter space, $\mathbb{C}^M$ for some M. However, some changes are needed in the algorithm.

If $\mathcal{P} \subset \mathbb{C}^M$ is a proper, convex subset of $\mathbb{C}^M$, Algorithm 1 needs only one change: p_0 must be somehow chosen within $\mathcal{P}$. To accommodate this, Paramotopy allows the user to specify p_0.

If $\mathcal{P}$ is a proper, non-convex set, more work is required. The Step 1 run would be the responsibility of the user, as in the previous paragraph, and it would be up to the user to string together subsequent Paramotopy runs to stay within $\mathcal{P}$.

3.4 Parallelization and Data Management

One of the features of Paramotopy that sets it apart from Bertini is the use of parallel computing for multiple parameter homotopies. Bertini includes parallel capabilities for a single homotopy run, but not for a sequence of runs. Parallelization was achieved using the head-worker paradigm, implemented with MPI. A single process controls the distribution of parameter points to the workers, which constitute the remainder of the processes. Workers are responsible for writing the necessary files for Bertini and for writing their own data to disk.

Bertini creates structures in memory by parsing an input file. As input is interpreted, several other files are created. These contain the straight line program, coefficient values, variable names, etc. Since the monomial structure of the polynomials in each Step 2 run is the same, almost all of these files are identical from one run to the next, so almost all this parsing is unnecessary. The only file that needs to be changed between runs is the file containing parameter values.

To prevent proliferation in the number of files needed to contain the data from the Paramotopy run, the Bertini output data is read back into memory, and dumped into a collective data file. The collective data files have a maximum buffer size, and once the buffer size is reached, the data in the buffer is written to the file, and the process repeats by storing the Bertini output data in memory until the buffer is full once more.

Repeated writing and reading is taxing on hard drives and clogs a LAN if the workers are using network drives. To free workers from having to physically write temporary files to electronic media storage, an option is provided to the user to exploit a shared memory location (or ramdisk), should it be available.

3.5   Front Ends and Back Ends

Real-world problems may involve many parameters. This could be problematic when one wants to discretize a parameter space into a uniform sample as the number of parameter points of interest can easily reach into the astronomical. Hence, Paramotopy contains support for both linear uniform meshes of parameters as well as user-defined sets of parameter values stored in a text file. A generic `Matlab` interface for gathering, saving, and plotting data from an arbitrary Paramotopy run is provided on the Paramotopy website. See [11] for further details.

4   Conclusions

Paramotopy can be used to solve parameterized polynomial systems efficiently for large numbers of parameter values. This extends the reach of numerical algebraic geometry in a new direction, particularly one that might be useful for mathematicians, scientists, and engineers who would like to rapidly test a hypothesis or would like to find regions of a parameter space over which the polynomial system has the same number of solutions. While Bertini and PHCpack have some parameter homotopy capabilities, Paramotopy has been optimized for the scenario of using many-processor computers to solve at many parameter values of interest.

Acknowledgements. The authors appreciate the useful comments from several anonymous referees and Andrew Sommese as these have greatly contributed to the quality of this paper. The first author would also like to recognize the hospitality of Institut Mittag-Leffler and the Mathematical Biosciences Institute, as well as partial support from the NSF via award DMS-1719658.

References

1. Sommese, A., Morgan, A.: Coefficient-parameter polynomial continuation. Appl. Math. Comp. **29**, 123 160 (1989)
2. Sommese, A.J., Wampler, C.W.: The Numerical Solution of Systems of Polynomials Arising in Engineering and Science. World Scientific Publishing, Singapore (2005)

3. Bates, D.J., Hauenstein, J.D., Sommese, A.J., Wampler, C.W.: Numerical Solution of Polynomial Systems Using the Software Package Bertini. SIAM, Philadelphia (2013)
4. Bates, D., Hauenstein, J., Peterson, C., Sommese, A.: A numerical local dimension test for points on the solution set of a system of polynomial equations. SIAM J. Numer. Anal. **47**(5), 3608–3623 (2009)
5. Brake, D.A., Bates, D.J., Putkaradze, V., Maciejewski, A.A.: Illustration of numerical algebraic methods for workspace estimation of cooperating robots after joint failure. In: 15th IASTED International Conference on Robotics and Applications, pp. 461–468 (2010)
6. He, Y.H., Mehta, D., Niemerg, M., Rummel, M., Valeanu, A.: Exploring the potential energy landscape over a large parameter-space. J. High Energy Phys. **2013**(7), 1–29 (2013)
7. Newell, A.J.: Transition to superparamagnetism in chains of magnetosome crystals. Geochem. Geophys. Geosy. **10**(11), Q11Z08 (2009)
8. Rostalski, P., Fotiou, I.A., Bates, D.J., Beccuti, A.G., Morari, M.: Numerical algebraic geometry for optimal control applications. SIAM J Optimiz. **21**(2), 417–437 (2011)
9. Bates, D.J., Hauenstein, J.D., Sommese, A.J., Wampler, C.: Bertini: software for numerical algebraic geometry (2006)
10. Verschelde, J.: Algorithm 795: PHCpack: a general-purpose solver for polynomial systems by homotopy continuation. ACM Trans. Math. Softw. (TOMS) **25**(2), 251–276 (1999)
11. Bates, D., Brake, D., Niemerg, M.: Paramotopy: parameter homotopies in parallel. arXiv.org/abs/1804.04183 (2018)
12. Bates, D., Hauenstein, J., Sommese, A., Wampler, C.: Adaptive multiprecision path tracking. SIAM J. Numer. Anal. **46**(2), 722–746 (2008)
13. Bates, D.J., Hauenstein, J.D., Sommese, A.J., Wampler, C.W.: Stepsize control for path tracking. Contemp. Math. **496**, 21–31 (2009)
14. Bates, D.J., Hauenstein, J.D., Sommese, A.J.: Efficient path tracking methods. Numer. Algorithms **58**(4), 451–459 (2011)
15. Wampler, C.W.: Bezout number calculations for multi-homogeneous polynomial systems. Appl. Math. Comput. **51**(2), 143–157 (1992)
16. Li, T.Y.: Numerical solution of polynomial systems by homotopy continuation methods. Handb. Numer. Anal. **11**, 209–304 (2003)
17. Leykin, A., Verschelde, J., Zhao, A.: Newton's method with deflation for isolated singularities of polynomial systems. Theoret. Comput. Sci. **359**, 111–122 (2006)
18. Hauenstein, J., Wampler, C.: Isosingular sets and deflation. Found. Comput. Math. **13**, 371–403 (2013)

DiscreteZOO: Towards a Fingerprint Database of Discrete Objects

Katja Berčič[1]([✉]) and Janoš Vidali[2]

[1] UNAM, Morelia, Mexico
`katja@matmor.unam.mx`
[2] University of Ljubljana, Ljubljana, Slovenia
`janos.vidali@fmf.uni-lj.si`
`http://katja.not.si`, `http://jaanos.github.io`

Abstract. There have been various efforts to collect certain mathematical results into searchable databases. In this paper, we present Discrete-ZOO: a repository and a fingerprint database for discrete mathematical objects. At the moment, it hosts collections of vertex-transitive graphs and maniplexes, which are a common generalisation of maps and abstract polytopes. The project encompasses a tool for handling and maintaining collections of objects, as well as a website and SageMath package for interacting with the database. The project aims to become a general platform to make collections of mathematical objects easier to publish and access.

Keywords: Fingerprint database · Vertex-transitive graphs
Maniplexes · SageMath package · Website

1 Introduction

Collections of mathematical results of various kinds are becoming more and more common, which is not at all surprising given the technological advances and their usefulness. Billey and Tenner [6] described an important concept in this context, that of a *fingerprint database*,

> a searchable, collaborative database of citable mathematical results indexed by small, language-independent, and canonical data.

The more of these properties a database of mathematical results satisfies, the more useful it is. Moreover, fingerprint databases also have the potential to be more than just an efficient way to look things up: they can help uncover connections between fields, provide a tool for mathematical experimentation. As such databases make it easier to uncover prior work, they can improve refereeing.

Perhaps the most famous example of a database of mathematical results is the On-Line Encyclopedia of Integer Sequences (OEIS) [16]. The OEIS is a searchable and collaborative database of integer sequences, which serve as fingerprints

© Springer International Publishing AG, part of Springer Nature 2018
J. H. Davenport et al. (Eds.): ICMS 2018, LNCS 10931, pp. 36–44, 2018.
https://doi.org/10.1007/978-3-319-96418-8_5

for their associated entries. The sequences are citable via their unique identifier (such as A000055) and the indexing is based on small, language-independent, and canonical data: the first few elements of a sequence. The usefulness of OEIS stems in part from the fact that the fingerprint is simple to search for.

Collections of mathematical objects such as the Foster census [7, 9, 10] of cubic symmetric graphs are also natural candidates for searchable databases. These collections are already commonly used to look up references and properties of mathematical objects or to browse for patterns and counterexamples. They are even more useful when they observe at least some of the fingerprint database principles, and most importantly, when they are computer searchable.

We have started collecting various partially overlapping censuses of vertex-transitive graphs [9, 17, 18] into a database with the intent to make the collections of graphs searchable. We soon realised that we need not restrict ourselves to graphs: we could store any discrete mathematical object, as long as there is an efficient way to compute a fingerprint of said object. Thus, we have started the DiscreteZOO project with the aim for it to become a repository and fingerprint database for discrete mathematical objects.

Our fingerprint database is somewhat different from the idea of Billey and Tenner. They require indexing mathematical statements by canonical data. We also are interested in storing mathematical results and objects without a canonical fingerprint. For these we use small and language-independent data which are not necessarily canonical, but provide an intuitive way to search for objects. The simplest example of such a "semantic fingerprint" that applies to nearly all object types is size. Other examples for graphs include valency, degree of symmetry, etc.

Any collection of mathematical results or objects beyond a certain size threshold presents the author with a choice on how to make it available to the research community. While a short collection can be simply included in a table in a paper, this approach does not work with larger collections. There are two related difficulties a researcher could face when publishing such a collection, and Discrete-ZOO [2] aims to address both.

1. A searchable database is often out of the scope of a typical researcher's work.
2. Most mathematical objects do not have a canonical fingerprint.

While every database of mathematical results necessarily has its own peculiarities, they have enough in common for a reusable infrastructure to make sense. This infrastructure can then provide common features to all databases while being flexible enough to accommodate any database-specific features.

DiscreteZOO offers a service both to the authors of collections and the mathematical community in general. For the authors, it is a platform for publishing a collection as a searchable database. For researchers, it provides interfaces for interacting with the databases.

2   Project Description

At the centre of DiscreteZOO is the core database and the connections to its derived databases. In addition to the objects and their properties, the DiscreteZOO core database keeps track of additions and changes as well as records of references for each object and object property. There are two main user interfaces: the website and the SageMath package. Each of the interfaces uses a simplified database optimised for its needs. In addition to the core and package databases, a user can download a subset of the core database for offline use with the SageMath package. This local database can store any properties that she computes. From here, she can submit the changes she makes to the core database.

A typical DiscreteZOO database entry describes a single discrete object and several of its properties, including description, identifiers in other databases, and related objects. We intend to maintain a core set of features for each object type supported by DiscreteZOO. For example, all objects included should have a citable unique identifier and have a shareable encoding consisting of printable characters. Such an encoding makes it possible to transfer an object between software tools. *Precomputed properties* are properties stored in the database for the purpose of searching and can be thought of as semantic fingerprints mentioned earlier.

Every object has a GUID (globally unique identifier), a citable unique identifier, and any number of human-readable aliases and descriptions. Furthermore, precomputed properties are stored with every object. For example, the object representing the Petersen graph is marked as a graph, and has all the precomputed properties relevant to graphs. Additionally, this object is also marked as being a vertex-transitive graph and a cubic vertex-transitive graph, and each of these types carries its own additional properties.

A canonical form or labelling [1] of a graph G is a labeled graph $\mathrm{Canon}(G) \simeq G$ with the additional property that for every other graph $H \simeq G$, H has the same canonical labelling $\mathrm{Canon}(H) = \mathrm{Canon}(G)$. DiscreteZOO uses software such as Nauty [14] and Bliss [11] to obtain canonical labellings of graphs, which are used to find a given graph or maniplex in the database.

2.1   GUIDs and Citable Unique Identifiers

In the OEIS [16], the sequence for the number of trees with n unlabeled nodes is identified by A000055. One can obtain more information about an integer sequence by typing the unique identifier after the domain in the URL, like so: https://oeis.org/A000055. Databases like The Database of Permutation Pattern Avoidance [20] and FindStat [19] use similar citable unique identifiers.

Each object in the DiscreteZOO repository has a unique identifier: the hexadecimal representation of the SHA-256 hash of some canonical string representation of the object. The use of a cryptographic hash function is a strong guarantee that no two objects will ever be found to have the same identifier. The reason for choosing such a hash-based identifier instead of a sequential one (as with other databases) is the desire for decentralisation. For instance, a researcher may encounter an object that is not yet in the database; another researcher may then reproduce her work and easily verify that they have obtained the same object by comparing the identifier. Once the object is in the database, the identifier may then be used to quickly access the object and its properties.

The resulting 64-character strings are infeasibly long for reproduction in a text intended to be read by humans. We chose a standard abbreviation technique. In the Git versioning system [8], objects are identified by 40-character hashes, but are usually referred to by simply taking the first 7 characters. In DiscreteZOO, the citable unique identifier is obtained by taking the first 12 hexadecimal digits of the hash and using further characters when necessary to avoid conflicts. The DiscreteZOO citable identifier is described in more detail in the project documentation [4]. For readability, the characters are split into groups of 4 characters and the letter Z is prepended:

$$123456789\texttt{abcdef}\ldots \quad \rightarrow \quad \texttt{Z1234-5678-9abc}. \tag{1}$$

2.2 Collaborative Aspects

The databases have a journaling system that keeps track of the tables, rows, and columns changed, as well as of who introduced the changes. The data repository [3] is intended solely for the ease of adding information to the database and exporting the database into usable forms. It is composed of the following three parts.

– **Contributions**: user contributions to the database to be merged into the main database.
– **Datasets**: specifications of datasets to be exported from the database for local use.
– **Objects**: specifications of database objects.

To submit a contribution, an author can make a pull request to the data repository. The request is then checked. If accepted, the contribution is merged into the database, the database downloads are updated, and other users can choose to update their local databases. A dataset is simply a collection of objects in the database, identified by one of their identifiers, together with the specification of the types of objects it describes. Datasets may also be nested – i.e., a parent dataset will contain all objects in the child dataset (but will not necessarily describe the same object types).

3 User Interfaces

Both the website and the SageMath package make it possible for users to search for objects and filter object sets. The SageMath package supports adding new objects and properties into the local database. If a researcher wants to submit some of them to the core database, the package helps with preparing the changes file.

In preparing the DiscreteZOO interfaces, we took advantage of the design process as described in, for example, [12,15] from the beginning. The project stemmed from the observation that at least some researchers wished for a tool to work with collections of mathematical objects. We are constructing the interface to support the features that are simplest to implement and that have the greatest usability and we hope to improve the platform as researchers start to use it.

3.1 Website

The DiscreteZOO website is dedicated to simple searches, downloads and displaying encyclopaedic information. For example, it is possible to download graph search results in the `sparse6` [13] format. For a search of symmetric objects, it is possible to export the corresponding list of automorphism groups for certain computer algebra systems like GAP, Magma and SageMath. The website also provides a list of references (online resources, authors, papers) relevant to the search result. These references can be conveniently downloaded in BibTeX format. The website also supports various other downloads, including code snippets for datasets.

On the first page, a visitor is presented with the search box. Each result in the search results display has options to copy data to the clipboard (such as a link to the description page, references, formats for computer algebra systems, etc.). In the following list, we describe the functionality of the search box shown in Fig. 1.

1. The user selects the type of objects (graphs, maniplexes, etc.).
2. Corresponding contextual filters are shown in the adjacent area.
3. The number of matches found in the database is displayed in real time to give instant feedback on the search.
4. The user can modify the search results display by choosing the properties that get shown.
5. To optimise responsiveness, results are not displayed until the user presses the "display results" button.
6. Search result downloads are supported for various formats and do not require displaying the search results first.

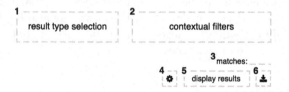

Fig. 1. Search box

The contextual filters area shown in Fig. 2 contains all properties stored in the database for the chosen object type and supports the functionality in list below. Numeric properties can be filtered with simple equations.

1. When there are many filtering options, the interface shows a filter search.
2. The filter list shows all filters available for the chosen type of objects.
3. On mouse over (or tap), the filter line shows further information. The definition of the filter is available as a tooltip on the question mark icon and the arrow activates the filter.
4. The right hand side of the contextual filters area displays currently active filters.
5. On mouse over, the filter line shows action options for an active filter: edit and remove.
6. Filter edit mode.

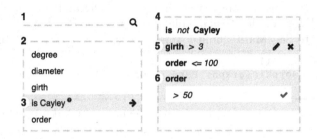

Fig. 2. Sample contextual filters for graphs

Example 1. Alice wants to find more information about her favourite abstract polytope. She does not remember its name, so she uses its properties to find it in the database. She filters for self-dual and self-Petrial abstract polytopes with the Schläfli symbol $\{12, 12\}$ and is then able to access the description page of the polytope.

Example 2. Bob wants to test a subgroup condition for automorphism groups of certain maps. Since maps are exactly maniplexes of rank 3, he chooses these as his object type, along with any other filters necessary to get the ones he wants. Bob has GAP code that tests his subgroup condition. He downloads GAP code with a list of the automorphism groups of the maps that he is interested in from DiscreteZOO. Bob then runs his code against the list from GAP.

3.2 SageMath Interface

The objects in the database can also be accessed using the SageMath interface [5]. SageMath is an open source computer algebra system based on the Python programming language, and it already provides many structures for representing various mathematical objects. The DiscreteZOO SageMath interface defines its own structures that inherit and override SageMath's structures, thus allowing a user to utilise the full potential of SageMath while adding the functionality of accessing precomputed properties in the database, as well as storing newly computed properties back to the database for later reuse.

After installing the interface and the database, the user can either import the entire discretezoo package, or load submodules as needed. In the following examples, we will use the submodules for the census of connected cubic vertex-transitive graphs by Potočnik et al. [17] (also known as the CVT census). We only need to import the class CVTGraph and the object info from the discretezoo.entities.cvt submodule as well as the objects from the discretezoo.entities.cvt.fields submodule. The first submodule is intended for cubic vertex-transitive graphs, while the latter contains the objects representing the precomputed properties that a user can use in search queries.

The CVTGraph class extends the ZooGraph class representing general graphs in the database, and the latter in turn extends SageMath's Graph class. It is possible to construct a CVTGraph instance by specifying the order and index as given in the CVT census. For example, it is possible to compare the Petersen graph obtained in such a way to SageMath's builtin version using the is_isomorphic method.

```
sage: G = CVTGraph(10, 3)
sage: G.is_isomorphic(graphs.PetersenGraph())
True
```

Note that the object G already contains the precomputed properties from the database. It is possible to use the usual SageMath methods to access them.

```
sage: G.girth()
5
sage: G.is_cayley()
False
```

A graph may also be constructed manually – if the obtained graph is in the database, it will be recognized and its precomputed properties will be loaded.

```
sage: CVTGraph([[(u, i) for u in GF(7) for i in (-1, 1)],
....:      lambda (u, i), (v, j): i != j and u*i + v*j in (1, 2, 4)])
Heawood graph: cubic vertex-transitive graph
              on 14 vertices, number 1
```

The info object is used to make queries to the database. The user may restrict the queries by specifying conditions using those field objects, which share names with the methods used to access the corresponding properties.

The simplest type of query is a counting query – how many objects satisfy the given conditions. The user may also request that the counts be broken down by the values of one or more properties.

```
sage: info.count(diameter == 5) # CVT graphs with diameter 5
37
sage: info.count(diameter == 5, groupby = girth) # break down by girth
{3: 1, 4: 7, 5: 2, 6: 18, 7: 3, 8: 3, 9: 3}
```

Alternatively, the user may want to list graphs satisfying the specified properties. The info.all method returns a generator yielding the requested graphs. It is thus possible to use the usual Python methods to either generate the graphs one by one or to obtain all of them.

```
sage: gen = info.all(girth == 5, orderby = order)  # order by the
sage: next(gen)                                     # number of vertices
Petersen graph: cubic vertex—transitive graph
                on 10 vertices, number 3
sage: next(gen)
Dodecahedron: cubic vertex—transitive graph on 20 vertices, number 6
sage: list(info.all(is_partial_cube, ~is_prism))
[Desargues graph: cubic vertex—transitive graph
                on 20 vertices, number 7,
 Truncated Octahedron: cubic vertex—transitive graph
                on 24 vertices, number 11,
 Truncated Cuboctahedron: cubic vertex—transitive graph
                on 48 vertices, number 29,
 Truncated Icosidodecahedron: cubic vertex—transitive graph
                on 120 vertices, number 60]
sage: for G in info.all(is_connected, ~is_hamiltonian): print(G)
Truncated Petersen graph
Truncated Coxeter graph
Petersen graph
Coxeter graph
```

Sometimes, we are only interested in a single graph with given properties – this can be achieved using the info.one method.

```
sage: info.one(girth >= 7, diameter == 4, orderby = order)
Generalized Petersen graph (13, 5): cubic vertex—transitive graph
                on 26 vertices, number 5
```

4 Future Work

DiscreteZOO started out as a database for symmetric graphs. We introduced some large improvements with the new version and added maniplexes, which are a common generalisation of maps and abstract polytopes. We are currently working to add finite automata and plan to add other kinds of objects as the project grows. We are also working to implement more search features, including providing information on the completeness of search results. For example, we want to be able to say up to which order the search results show all graphs satisfying the search conditions.

DiscreteZOO will be used in a planned classification of maniplexes of small ranks with a small number of orbits. We hope that it will be helpful with other classification attempts as well. We would be delighted if authors of any of the existing databases were interested in using DiscreteZOO, in which case we would implement any missing features.

Our main priority is for DiscreteZOO to be useful to the mathematical community, which is why we are planning to develop more features or change existing ones as the need arises. We are confident that the project will be a valuable tool for many mathematicians in near future.

References

1. Babai, L., Luks, E.M.: Canonical labeling of graphs. In: Proceedings of the Fifteenth Annual ACM Symposium on Theory of Computing. pp. 171–183. STOC 1983. ACM, New York (1983). http://doi.acm.org/10.1145/800061.808746
2. Berčič, K., Vidali, J.: DiscreteZOO. http://discretezoo.xyz/
3. Berčič, K., Vidali, J.: DiscreteZOO data repository. https://github.com/DiscreteZOO/DiscreteZOO-data
4. Berčič, K., Vidali, J.: DiscreteZOO documentation. https://github.com/DiscreteZOO/DiscreteZOO-docs
5. Berčič, K., Vidali, J.: DiscreteZOO SageMath interface repository. https://github.com/DiscreteZOO/DiscreteZOO-sage
6. Billey, S.C., Tenner, B.E.: Fingerprint databases for theorems. Not. Am. Math. Soc. **60**(8), 1034–1039 (2013). https://doi.org/10.1090/noti1029
7. Bouwer, I. (ed.): The Foster census. R. M. Foster's census of connected symmetric trivalent graphs. Co-editors: Chernoff, W.W., Monson, B., Star, Z.: With a foreword by H. S. M. Coxeter and a biographical preface by S. Schuster. Charles Babbage Research Centre, Winnipeg (1988)
8. Chacon, S., Straub, B.: Pro Git (2014). https://git-scm.com/
9. Conder, M., Dobcsányi, P., McKay, B., Royle, G.: The Extended Foster Census. http://www.cs.uwa.edu.au/gordon/remote/foster
10. Foster, R.M.: Geometrical circuits of electrical networks. Trans. Am. Inst. Electr. Eng. **51**(2), 309–317 (1932). https://doi.org/10.1109/T-AIEE.1932.5056068
11. Junttila, T., Kaski, P.: Bliss: A Tool for Computing Automorphism Groups and Canonical Labelings of Graphs. http://www.tcs.hut.fi/Software/bliss/
12. Kholmatova, A.: Design Systems. Smashing Media AG (2017)
13. McKay, B.: Description of graph6, sparse6 and digraph6 encodings. http://users.cecs.anu.edu.au/~bdm/data/formats.txt
14. McKay, B.D., Piperno, A.: Practical graph isomorphism II. J. Symbolic Comput. **60**, 94–112 (2014)
15. Norman, D.: The Design of Everyday Things. Revised and Expanded edn. Basic Books (2013)
16. OEIS Foundation Inc. The On-Line Encyclopedia of Integer Sequences (2018). http://oeis.org
17. Potočnik, P., Spiga, P., Verret, G.: Cubic vertex-transitive graphs on up to 1280 vertices. J. Symbolic Comput. **50**, 465–477 (2013). https://doi.org/10.1016/j.jsc.2012.09.002
18. Royle, G.: Transitive graphs. http://staffhome.ecm.uwa.edu.au/~00013890/remote/trans/index.html
19. Rubey, M., Stump, C., et al.: FindStat - The combinatorial statistics database (2017). http://www.FindStat.org
20. Tenner, B.E.: Database of permutation pattern avoidance. http://math.depaul.edu/bridget/patterns.html

A Framework for Unconditionally Secure Public-Key Encryption (with Possible Decryption Errors)

Mariya Bessonov[1], Dima Grigoriev[2], and Vladimir Shpilrain[3(✉)]

[1] New York City College of Technology, Brooklyn, USA
mbessonov@citytech.cuny.edu
[2] Université de Lille, Lille, France
dmitry.grigoryev@math.univ-lille1.fr
[3] The City College of New York, New York, USA
shpil@groups.sci.ccny.cuny.edu

Abstract. We offer a public-key encryption protocol where decryption of a single bit by a legitimate party is correct with probability p that is greater than $1/2$ but less than 1. At the same time, a computationally unbounded (passive) adversary correctly recovers the transmitted bit with probability exactly $1/2$.

1 Preface

It is well known (and easy to show) that unconditionally secure (i.e., secure without any computational assumptions) public-key encryption is impossible if the legitimate receiver decrypts correctly with probability exactly 1. The question is: what if this probability is less than 1? More precisely, what if the sender transmits a single encrypted bit and the legitimate receiver decrypts it correctly with probability P greater than $1/2$ but less than 1?

One can say "since the legitimate receiver has the same information about the secret bit as the eavesdropper does, he cannot have any advantage over a computationally unbounded eavesdropper, so the latter will decrypt correctly with probability at least P". This is, indeed, correct. Note however that if decryption is not necessarily accurate (i.e., if decryption errors are possible), then the legitimate *sender* has an advantage over the eavesdropper since the sender, unlike the eavesdropper, knows *exactly* what the transmitted secret bit is. Therefore, if instead of making the receiver guess the transmitted bit we make the sender guess the receiver's decryption key, we may get some advantage. Thus, what we do in our scheme is:

Research of Mariya Bessonov was partially supported by the NSF grant DMS-1515800. Research of Vladimir Shpilrain was partially supported by the ONR (Office of Naval Research) grant N000141512164. Research of Dima Grigoriev was partially supported by the RSF grant 16-11-10075.

J. H. Davenport et al. (Eds.): ICMS 2018, LNCS 10931, pp. 45–54, 2018.
https://doi.org/10.1007/978-3-319-96418-8_6

We make the adversary compete with the sender, not with the receiver, in contrast with the existing encryption schemes.

Competing with the sender is dramatically different from competing with the receiver because the adversary and the sender have *different goals*:

The goal of the sender is to guess the receiver's decryption key to have him decrypt her secret bit correctly, whereas the goal of the adversary is to guess the sender's secret bit correctly.

Thus, the adversary and the sender may have *different probability spaces* for making their guess and therefore it is not surprising that their probabilities of success may be different. Note that *the adversary's guess of the receiver's decryption key is at least as good as that of the sender* (for information-theoretical reasons), but again – the goal of the adversary is to guess the sender's bit, not the receiver's decryption key.

We will show that it is, in fact, not too hard to arrange for the sender to have a higher probability of success (in her guessing) compared to that of the adversary, see Proposition 2 in our Sect. 4.2. What is nontrivial is to have the adversary's probability of success in such a scenario to be equal to exactly $1/2$, which is what we claim in our scheme.

Finally, we note that in [4], the authors offered a simple public key encryption scheme where a computationally unbounded adversary cannot recover a secret bit with probability higher than 0.75 if she uses an encryption emulation attack. At the same time, the legitimate party recovers a secret bit with probability very close to 1. However, in that scheme the receiver's private key can be uniquely recovered from the public key, and therefore the private key is not secure against a computationally unbounded adversary. This is not the case with the scheme in the present paper; in fact, given a public key, any private key from the set of all possible private keys can be associated to it with nonzero probability.

2   Introduction

We consider a scenario where one party, Alice, wants to transmit a secret bit to another party, Bob, in the presence of a computationally unbounded (passive) adversary, Eve. We allow the legitimate parties, Alice and Bob, to fail with some controlled probability.

The way it works is roughly as follows. Bob applies a randomized (public) function F to his private decryption key b and obtains the result $B = F(b)$ that he makes public. Based on B, Alice tries to guess b. The probability to guess b is the same for Alice and Eve since they both have the same information about b in that case. However, what Eve really wants is not to recover b, but to recover Alice's bit, which means she needs to recover not the actual b, but rather what Alice thinks b is. (Think about a scenario where a customer Alice wants to transmit her credit card number to an Internet retailer Bob. Then what Eve really wants is Alice's credit card number, not Bob's decryption key.)

Therefore, probability spaces for Alice and Eve are different in general, and by (privately) manipulating her probability space Alice can get advantage over Eve as far as their probabilities of success are concerned. Once again, *success for Alice (the sender) is to guess b while success for Eve is to guess the bit Alice wants to transmit to Bob.* Note that success for Alice is the same as success for Bob in the sense that Bob decrypts Alice's bit correctly if and only if Alice is successful in our terminology.

Computing exact probabilities of success for Alice and Eve theoretically can be tedious in general; we denote these probabilities by P_A and P_E, respectively. We use the following trick to make computation of P_E easy. Alice will select, with equal probability, between two mutually exclusive strategies for guessing b, thus making P_E equal to exactly $\frac{1}{2}$.

Computing P_A precisely remains a difficult theoretical task. However, in Sect. 4 we give an "existence-type" argument showing that there exists a choice of parameters that makes P_A strictly greater than $\frac{1}{2}$, see Proposition 1. Experimentally, the best we could do for P_A is about 0.55, see our Sect. 6. It remains an interesting theoretical question what the maximum possible value of P_A (as a function of n, the interval length) in our protocol in Sect. 3 is.

Finally, we mention that it is not immediately clear whether our protocol in Sect. 3 has any practical significance; we discuss this in Sect. 5.

3   Basic Protocol

The protocol below is for transmitting a single secret bit from Alice to Bob.

There are (private or public) functions $f(n)$ and $g(n)$ and a public function $h(n)$ in the protocol below that have to be selected to maximize P_A, Alice's probability of guessing Bob's decryption key b. Parameters are discussed in our Sect. 6.

1. Bob selects, uniformly at random on integers from the interval $[0, n-1]$, a starting point b of his random walk. This b will be his private decryption key. Bob then does a simple symmetric random walk with $h(n)$ steps. Let B be the end point of Bob's random walk. If $B \geq n-1$, then Bob starts over. Otherwise, he publishes B. Bob publishes B.

2. Step 2 is repeated by Alice m times, for a sufficiently large m.
   Alice selects, uniformly at random on integers from the interval $[B, n-1]$, a starting point a of her random walk. She then selects, with probability $\frac{1}{2}$, between $f(n)$ steps and $g(n)$ steps. Alice then does a random walk starting at the point a with the number of steps selected. Denote by A the end point of Alice's random walk. After she does her random walk, Alice moves the end point A either $\frac{1}{2}$ left or $\frac{1}{2}$ right, with probability $\frac{1}{2}$. She then moves the point a $\frac{1}{2}$ in the same direction. (This is needed to avoid situations where $a = b$ or $A = B$.)

3. Alice arranges all her m random walks at Step 2 in two groups: in one group there are walks satisfying the condition $A < B$, while in the other group there

are walks satisfying the condition $B < A$. Then she selects between the two groups, with probability $\frac{1}{2}$.

4. Alice then splits all walks in the group selected at Step 3 in two groups again: in one group there are walks with $f(n)$ steps, in the other group there are walks with $g(n)$ steps. Then she selects between the two groups, with probability $\frac{1}{2}$. If the selected group turns out to be empty, Alice starts over from Step 2. If the selected group is not empty, then from this group, Alice selects one random walk uniformly at random. Let a_0 be the starting point of that selected random walk.

5. If the random walk selected by Alice at Step 4 has $f(n)$ steps and satisfies $A < B$, she chooses the interval $\{x < a_0\}$. If it has $g(n)$ steps and satisfies $A < B$, she chooses the interval $\{x > a_0\}$. If it has $f(n)$ steps and satisfies $A > B$, she chooses the interval $\{x > a_0\}$. If it has $g(n)$ steps and satisfies $A > B$, she chooses the interval $\{x < a_0\}$.

6. Alice assumes that Bob's decryption key b is in the interval she selected at Step 5 of the protocol and encrypts her bit accordingly, i.e., by labeling the selected interval with her secret bit c and the other interval with the bit $1 - c$. She then sends the point a_0 and the above interval labeling to Bob.

7. Bob recovers the bit corresponding to the label of the interval where his b is.

Remark 1. At Step 2 of the above protocol Alice selects a starting point a uniformly at random on integers from the interval $[B, n - 1]$. We note that, in fact, the distribution of a on $[B, n - 1]$ does not have to be uniform. It can be closer to geometric, say (with points closer to B more likely to be selected). This will not affect security, but can increase the probability of correct decryption by legitimate party.

Below we summarize public as well as private information relevant to this protocol.

Private information consists of:
Alice's choices between the options at Steps 2, 3, 4.
Alice's private key: point A (the end point of Alice's random walk).
Bob's private key: point b (the starting point of Bob's random walk).
Functions $f(n)$ and $g(n)$ can be private but they do not have to be.

Public information consists of:
Public parameters: interval $[0, n - 1]$; the number $h(n)$ of steps in Bob's random walk; the number m of Alice's random walks at Step 2.
Transmitted information: point a_0 (the starting point of Alice's selected random walk) and labeling of the interval $\{x > a_0\}$ by a bit.
Bob's public key: point B (the end point of Bob's random walk).

3.1   Informal Explanation

We think it will be helpful to the reader if we give an informal explanation of what is actually going on in the above protocol. The core of the whole thing is the following non-obvious fact: if Alice and Bob do independent random walks starting at two random points, a and b, respectively, then the conditional probability $P(b < a \mid A < B < a)$ is higher when the number of steps in Alice's random walk is larger (with the number of steps in Bob's random walk fixed). Refer to our Appendix to see how to explain and theoretically quantify this statement.

Now suppose that the number $f(n)$ of steps is large while $g(n)$ is small. Then, to increase her probability of success P_A, Alice could have just done $f(n)$ steps and guess that $b < a$, conditioned on $A < B < a$. This guess would be correct with high probability. However, what Alice tries to do in our protocol is confuse Eve and make sure that Eve is unable to guess Alice's transmitted bit with probability greater than $\frac{1}{2}$. This is why Alice deliberately decreases her probability of success by selecting, in case she does $g(n)$ steps, the interval $\{x > a\}$ where the point b belongs with probability less than $\frac{1}{2}$, in the hope that her total probability of success will still be greater than $\frac{1}{2}$. This is indeed the case under appropriate choice of parameters, see our Sect. 6.

At the same time, the conditional probability $P(b < a \mid B < A < a \text{ or } B < a < A) = P(b < a \mid B < A \text{ and } B < a)$ "almost" does not depend on the number of steps in Alice's walk, so here Alice can have her probability of success only slightly above $\frac{1}{2}$. Nevertheless, we need to include the walks satisfying this condition in Alice's probability space to make it "symmetric" since otherwise, if we just use the walks satisfying $A < B < a$, Eve might get some idea about the number of steps in Alice's walk. Specifically, if the points B and a are far apart, then the condition $A < B < a$ makes it appear likely that the number of steps in Alice's walk was rather large than small. Symmetrizing Alice's probability space by adding walks with $B < A$ eliminates this problem, but there is a price to pay for that: the difference $P_A - \frac{1}{2}$ gets cut in half.

Finally, we note that the fact that $P(b < a \mid B < A \text{ and } B < a)$ "almost" does not depend on the number of steps in Alice's walk is in sharp contrast with the fact that $P(b < a \mid B < A)$ does strongly depend on the number of steps, see [1].

4 Probabilities of Success

Recall that success for Alice (the sender) in our scenario is, given two intervals $\{x > a\}$ and $\{x < a\}$, to guess the interval where Bob's private number b is. On the other hand, success for Eve (the passive adversary) is to guess the bit Alice wants to transmit to Bob, i.e., to "guess Alice's guess" of the interval where b is. We denote by P_A and P_E the probabilities of success for Alice and Eve, respectively.

4.1 Alice's Probability P_A to Guess the Interval Where b Is

Proposition 1. *There exists a choice of parameters that makes P_A strictly greater than $\frac{1}{2}$.*

Proof. Recall that, while executing the protocol in Sect. 3 (cf. Steps 3, 4), Alice selects between two mutually exclusive options (selecting the interval $\{x < a\}$ or $\{x > a\}$) with probability $\frac{1}{2}$. Denote her probability of success if she uses the option 1 by p, and her probability of success if she uses the option 2 by q. Then $P_A = \frac{1}{2}(p + q)$. If it happens so that $p + q < 1$, then $(1 - p) + (1 - q) > 1$. This means that if Alice switches the interval assignments between the options, then $P_A = \frac{1}{2}((1 - p) + (1 - q)) > \frac{1}{2}$. This shows that there is a choice of interval assignments that gives $P_A > \frac{1}{2}$, unless $p + q = 1$ for any choice of parameters. The latter however is impossible because by varying the number of steps in a random walk for one of the two possible options, one varies the probability of guessing in this option only, see our Sect. 3.1 and Appendix.

4.2   Eve's Probability P_E to Guess Alice's Bit

The following follows directly from the protocol description.

Proposition 2. $P_E = \frac{1}{2}$.

Proof. As follows from the protocol description (Steps 3, 4, 5), Alice selects, with equal probability $\frac{1}{4}$, between 4 possibilities. Two of these possibilities result in selecting the interval $\{x > a_0\}$, while the other two result in selecting the interval $\{x < a_0\}$. Thus, any third party cannot guess Alice's selection with probability greater than $\frac{1}{2}$.

4.3   If $P_E = \frac{1}{2}$, How Is It Possible That $P_A > \frac{1}{2}$?

Note that, given Alice's probability space $\{B < a\}$, the point B always belongs to the interval $\{x < a\}$. This implies, in particular, that if Eve selects the interval $\{x < a_0\}$ where the point B is, she will guess Alice's bit with probability $\frac{1}{2}$ because for any given point a, Alice selects between the intervals $\{x < a\}$ or $\{x > a\}$ with probability $\frac{1}{2}$.

One might ask here: why is then $P_A > \frac{1}{2}$? Since the point b is either left or right of a and Alice selects between left and right with probability $\frac{1}{2}$, then should not P_A be equal to $\frac{1}{2}$, too? An informal explanation is: the probability spaces for Eve and Alice are different. Eve only sees one point a_0, whereas Alice selects this a_0 from a pool with different points a. The point b may be left of some of these points a but right of the others, so there is no contradiction between $P_A > \frac{1}{2}$ and $P_E = \frac{1}{2}$.

The same kind of reasoning applies if Eve tries to emulate Bob: her probability space will include a *fixed* point B, which is not the case for Bob.

We note in passing that selecting the interval $\{x < a_0\}$ where the point B is will let Eve guess, with significant probability (still less than 1), the interval where the point b is. However, we remind the reader that success for Eve is not to guess b but to guess Alice's bit.

5 How to Use This in Real Life

Given that, according to experimental results, the probability of successfully transmitting a single bit from Alice to Bob is as low as 0.55 (see our Sect. 6.1), a natural question now is: how can our scheme be used in real life? It is possible to transform an encryption scheme susceptible to decryption errors into one that is immune to these errors by using techniques from [2] or [3]. This, however, can increase Eve's probability of success as well.

We mention that the most straightforward way to boost the probability of success is to run the protocol from our Sect. 3 k times (every time with fresh randomness), every time transmitting the same bit c. If, say, $k = 1000$, then the probability that there will be less than 501 occurrences of c out of 1000 is $\sum_{i=0}^{501} \binom{k}{i} (0.55)^i (0.45)^{k-i} \approx 0.000846$. (This was computed using the normal approximation of the binomial distribution.) This means that if Bob goes with the bit that has more occurrences out of k than the other bit does, he will recover Alice's bit correctly with probability at least 0.99915 if $k = 1000$.

However, different runs of the protocol are not independent in this case since Alice is transmitting the same bit every time. Therefore, we cannot claim that Eve's probability of success will stay at $\frac{1}{2}$. In other words, statistical attacks on multiple runs of the protocol for transmitting the same bit may be possible. These statistical attacks would be based on the fact that for some points B (specifically, those that are farther from Alice's points a) Alice's success rate will be higher than with others. To counter these attacks, Bob will have to be more proactive with his public key, e.g. make the correspondence between his points B and b such that sometimes points closer to a make Alice more successful and sometimes not. This could mean, in particular, fluctuating parameters of his random walk, e.g. using random walks in random environment. This suggestion, of course, is very informal; more precise proposals should be based on more serious probability theory, so we leave this for a future work. Here we offer an example of how Alice's probability of success can be somewhat amplified if we use two independent runs of the protocol. Note that Eve's probability of success, too, is amplified in this case.

Example 1. Instead of transmitting a bit, Alice can use the protocol in Sect. 3 to transmit an integer. Thus, in a single run of the protocol she transmits labels m_{11} and m_{12} of two subintervals, where, say, m_{12} is the integer Alice wants Bob to receive.

In the second run of the protocol, Alice transmits labels m_{21} and m_{22}, and let m_{22} be the integer Alice wants Bob to receive.

Thus, Bob receives m_{12} with probability (approximately) 0.55 and m_{22} also with probability (approximately) 0.55. Therefore, Bob receives at least one of these two numbers with probability $1 - (1 - 0.55)^2 \approx 0.8$. To capitalize on that, Alice now sends a polynomial $P(x, y) + M$ to Bob, where $P(x, y)$ is a polynomial such that $P(m_{12}, m_{21}) = P(m_{12}, m_{22}) = P(m_{11}, m_{22}) = 0$, and M is a secret number. Bob plugs in for x one of the two numbers he received from Alice, for y the other number and recovers the secret number M with probability approximately 0.8.

Eve recovers the secret number M here with probability $\frac{3}{4} = 0.75$.

6 Parameters and Computer Experiment Results

Suggested parameter values for the protocol in Sect. 3 are: $n = 256$, $h(n) = 2000$, $g(n) = 2000$, $f(n) = 100,000$.

6.1 Computer Simulation Results

With $f(n) = 100,000$ steps for Alice, success rate in a single run of the protocol was 76%. With $g(n) = 2000$ steps for Alice, success rate in a single run of the protocol was 34%. Thus, $P_A = \frac{1}{2}(0.76 + 0.34) = 0.55$ for a single run.

7 Conclusions

- We offered a public-key encryption scheme where decryption of a single bit by a legitimate party is correct with probability p that is strictly greater than $1/2$. With suggested parameters, $p \approx 0.55$.
- In this scheme, even a computationally unbounded (passive) adversary cannot recover the transmitted bit correctly with probability greater than $1/2$.

Appendix

How $P(b < a | A < B < a)$ Depends on the Number of Steps

Let $\alpha > 0$ and n^α be the number of steps in Alice's walk and suppose initially that this number is odd (to avoid parity issues, although the conclusion that $P(b < a | A < B < a)$ depends on α still holds when the number of steps is even). Let n^β be the fixed number of steps in Bob's walk with $0 < \beta < 2$. Then $P(b < a | A < B < a)$ depends on α as follows:

- When α is very small, $P(b < a | A < B < a)$ is very close to $1/2$.
- As α increases, $P(b < a | A < B < a)$ tends to $P(b < a | B < a)$, which tends to 1 as $n \to \infty$.

Suppose that $n^\alpha = 1$. Then $P(b < a | A < B < a)$ is the probability that $b < a$, given that Alice's one step was to the left and Bob's final location happens to be between A and a, for which there is only one possibility $B = a - 0.5$ and $A = a - 1$. In this case,

$$P(b < a | A < B < a) = P(b < B) = \frac{1}{2} - O\left(n^{-\beta/2}\right),$$

or, if we remove the possibility that $B = b$, by shifting Bob's end point by adding or subtracting 0.5 with equal probability, then

$$P(b < a | A < B < a) = P(b < B) \xrightarrow{n \to \infty} \frac{1}{2}$$

The probability is not exactly equal to $1/2$ due to the restriction that $0 \leq b \leq n - 1$ and $B < n - 1$. However, as $n \xrightarrow{\infty}$, the probability that b is close to 1 or n goes to zero. As α increases, given that $B < a$, B is more likely to be farther from a, and when B is farther from and to the left of a, b is more likely to be less than a. This is because the number of steps in Bob's walk remains fixed, and Bob is (almost) equally likely have started to be to the left or to the right of B. If $b < B$, certainly $b < a$. If $B > b$, the fact that $A - a$ can be larger, increases the probability that $B < b < a$. "Almost" because of the restriction on b and B mentioned above.

Now, as α increases, the condition $A < B < a$ implies that A will be farther from a. Eventually, for α large enough, A will be outside of the interval $\{0, 1, \ldots, n - 1\}$ with probability close to 1. The probability of A being in the interval will be exponentially small in α. If A is outside of this interval, then $P(b < a | A < B < a) = P(b < a | B < a$, Alice's walk ends to the left of her starting point$) = P(b < a | B < a)$.

Lemma 1. $P(b < a | B < a) \to 1$ as $n \to \infty$.

To see that this is true, consider

$$P(b < a | B < a) = P(b < B < a | B < a) + P(B < b < a | B < a) \tag{1}$$

The first term, $P(b < B < a | B < a) = P(b < B) \to 1/2$ as $n \to \infty$. If B is distance $O(n^{\beta/2+\epsilon})$ for small $\epsilon > 0$, $P(b < B)$ goes to $1/2$ as $n \to \infty$, as it is just the probability that the endpoint of the walk is to the right of the starting point. If B is close to 0, the probability is under $1/2$ since Bob's starting point b is restricted to $\{0, 1, \ldots, n - 1\}$. As $n \to \infty$, the probability that B is close to 0 goes to zero. If B is close to $n - 1$, $P(b < B)$ is actually close to 1, but the probability that B is close to $n - 1$ also goes to zero.

The second term, $P(B < b < a | B < a) \xrightarrow{n \to \infty} 1/2$ as well. Here, we consider two possibilities:

- $P(B < b < a | B < a, a - B \geq n^{\beta/2+\epsilon}) \xrightarrow{n \to \infty} 1/2$, since the probability of the displacement being greater than $O(n^{\beta/2})$ is exponentially small.
- $P(B < b < a | B < a, a - B < n^{\beta/2+\epsilon})$ is not close to 1, however,

$$P\left(a - B < n^{\beta/2+\epsilon}\right) \xrightarrow{n \to \infty} 0$$

From this,

$$P(B < b < a | B < a) \xrightarrow{n \to \infty} P(B < b < a | B < a, a - B \geq n^{\beta/2+\epsilon}) \xrightarrow{n \to \infty} 1/2.$$

Why $P(b < a | B < A < a$ or $B < a < A)$ does not depend greatly on the number of steps

Consider the two events in the condition separately and note that they are disjoint.

- If $B < a < A$, then the probability that $b < a$ does not depend on Alice's walk, and thus on α, at all, since the condition is that Alice ended to the right of her starting point a (the probability of which is the same as Alice ending to the left of a) and B is always to the left of a in our setup. Note also that in the sample space consisting of the events $\{B < a < A\} \cup \{B < A < a\}$, the event $\{B < a < A\}$ has probability greater than $1/2$ since Alice is more likely to end to the right of A with no other restriction than to the left of a but to the right of B.
- If $B < A < a$, the probability that $b < a$ does depend on the number of steps in both walks, however, if $\beta < \alpha$ are fixed, the probability will approach 1 as $n \to \infty$. Thus, the dependence on α is weak, so long as $\beta < \alpha$. We have here that under this condition, Alice ended her walk to the left of where she started, and Bob ended to the left of Alice's endpoint. If Alice performed a greater number of steps than Bob, to not have $b < a$, Bob's displacement would have to be greater than Alice's.

References

1. Bessonov, M., Grigoriev, D., Shpilrain, V.: Probabilistic solution of Yao's millionaires' problem, preprint. https://eprint.iacr.org/2017/1129
2. Dwork, C., Naor, M., Reingold, O.: Immunizing encryption schemes from decryption errors. In: Cachin, C., Camenisch, J.L. (eds.) EUROCRYPT 2004. LNCS, vol. 3027, pp. 342–360. Springer, Heidelberg (2004). https://doi.org/10.1007/978-3-540-24676-3_21
3. Holenstein, T., Renner, R.: One-way secret-key agreement and applications to circuit polarization and immunization of public-key encryption. In: Shoup, V. (ed.) CRYPTO 2005. LNCS, vol. 3621, pp. 478–493. Springer, Heidelberg (2005). https://doi.org/10.1007/11535218_29
4. Osin, D., Shpilrain, V.: Public key encryption and encryption emulation attacks. In: Hirsch, E.A., Razborov, A.A., Semenov, A., Slissenko, A. (eds.) CSR 2008. LNCS, vol. 5010, pp. 252–260. Springer, Heidelberg (2008). https://doi.org/10.1007/978-3-540-79709-8_26

Classifying Cubic Surfaces over Finite Fields Using Orbiter

Anton Betten[(✉)]

Colorado State University, Fort Collins, USA
betten@math.colostate.edu
http://www.math.colostate.edu/~betten/

Abstract. We present two algorithms to classify cubic surfaces over a finite fields. An implementation in the programming system Orbiter will be described.

Keywords: Cubic surface · Clebsch · Algebra · Geometry
Classification · Finite field

1 Introduction

The classification of cubic surfaces is a long-standing problem, whose roots trace back to the 19th century, with groundbreaking work done by Cayley, Salmon, Clebsch, Schlaefli and many more. It is well known that a smooth cubic surface has 27 lines. In Fig. 1, the Clebsch surface with its 27 lines is shown.

The (affine) equation of the surface shown is

$$0 = -3x^3 + 7x^2y + 7x^2z + 1x^2 + 7xy^2 - 2xyz - 14xy + 7xz^2$$
$$-14xz + 3x - 3y^3 + 7y^2z + y^2 + 7yz^2 - 14yz + 3y - 3z^3 + z^2 + 3z - 1$$

where x, y, z are the coordinates of affine 3-space. The canonical (projective) equation of the Clebsch surface is

$$x_0^3 + x_1^3 + x_2^3 + x_3^3 - (x_0 + x_1 + x_2 + x_3)^3 = 0.$$

The somewhat complicated affine equation was chosen to give a pleasant picture and to make sure that all 27 lines are real and hence visible.

What makes cubic surfaces very interesting is the existence of a special kind of mapping which takes the points of a surface to points of a plane. Because of [6] we call them Clebsch maps. This map proceeds in an almost one-to-one way. These kinds of maps arise naturally, and they can be described algebraically and geometrically. Objects which admit such a map are called rational. In the algebraic description, there are polynomial equations which describe the maps in both ways. In the geometric description, one argues that an arbitrary point in $PG(3, q)$ determines a unique line which intersects two given skew lines. This line intersects the surface in three points, and a given plane in one (cf. Fig. 2).

© Springer International Publishing AG, part of Springer Nature 2018
J. H. Davenport et al. (Eds.): ICMS 2018, LNCS 10931, pp. 55–61, 2018.
https://doi.org/10.1007/978-3-319-96418-8_7

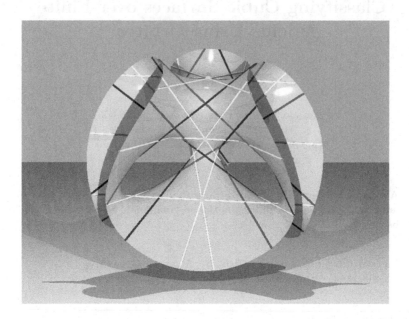

Fig. 1. The Clebsch surface with 27 lines

Fig. 2. The Clebsch map

There is an exceptional locus where the map is either undefined or many-to-one. On a cubic surface with 27 lines, six lines are mapped to points each. In the plane, the reverse map is undefined on these six points. The isomorphism type of the surface is determined by the position of the six points in the plane which are associated to the exceptional locus of the map. Figure 3 is a picture of a cubic (the Clebsch cubic), a planar circle and its image under the Clebsch map, which is a certain curve lying on the surface. It is the same view as in Fig. 2, just with the lines removed.

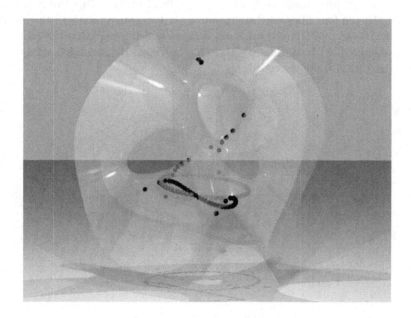

Fig. 3. The Clebsch map illustrated

2   Underlying Theory

The theory of cubic surfaces is very rich. For an account of the geometry over finite fields, see [8]. Some notions which are relevant for this work are the following:

A double six is a set of twelve lines a_i $i = 1, \ldots, 6$ and b_i $i = 1, \ldots, 6$ such that the six a_i are pairwise skew, the six b_i are pairwise skew and a_i intersects b_j if and only if $i \neq 6$. Often, a double six is denoted in the array

$$a_1 \; a_2 \; a_3 \; a_4 \; a_5 \; a_6$$
$$b_1 \; b_2 \; b_3 \; b_4 \; b_5 \; b_6$$

Once the double sixes in $PG(3, q)$ are classified, the cubic surfaces in $PG(3, q)$ are classified. In order to classify the double sixes in $PG(3, q)$, the algorithm first classifies the ways in which 5 pairwise skew lines with a common transversal can be chosen. We can think of these configurations as the beginning of a double six, such as

$$a_1\ a_2\ a_3\ a_4\ a_5\ \cdot$$
$$\cdot\quad\cdot\quad\cdot\quad\cdot\quad\cdot\quad b_6$$

Some additional testing is needed if a double six can be completed from the set of 5 lines $a_1, \ldots, a_5$ with the common transversal b_6. The five lines have to be sufficiently general. Specifically, any three of the a_i determine a hyperboloid and the other two a_i lines must be bisecants to it (cf. Fig. 4). In light of Schlaefli's theorem [11], the four lines need to have exactly two transversals. In the picture on the left, the second transversal is not unique, which means that the four lines cannot be embedded into a double six. On the right, the second transversal is unique and this is what is needed.

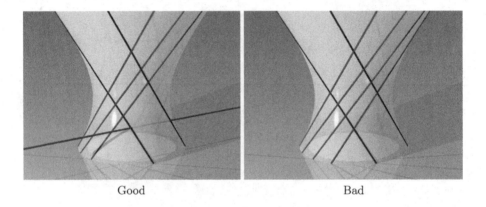

Good   Bad

Fig. 4. The condition on 4 lines

3   Functionality

Orbiter is an open source library of C++ classes for algebraic computations. It offers functionality for finite groups such as permutation groups or matrix groups over finite fields. It has geometry over finite fields, such as points lines, planes and general linear subspaces of projective space. Besides that, Orbiter offers a suite of algorithms to compute orbits. The main tool is a bosed based classification algorithm which builds up orbits on bigger sets from orbits on smaller sets, using a relation between the two sets. The main idea is outlined in [3]. This can be used to classify combinatorial objects of many different kinds. Classification means that a system of representatives is chosen, so that each isomorphism type of object is represented exactly once. The stabilizers of the orbit representatives are readily available. After the classification of a type of objects is complete, the system can perform isomorphism tests for objects of this type. More specifically, the system can map objects to the chosen orbit representatives. If two objects are given, both objects can be mapped to the orbit representative. If the orbit representative is the same, an isomorphism between the two given objects can be computed. Otherwise, proof that the objects are nonisomorphic is obtained. A critical aspect in all of this is speed. Backtracking is avoided at the expense of memory.

In this note, we focus on two applications of classifying cubic surfaces. In one program, six-arcs not on a conic are classified in a plane $PG(2, q)$. The classification of six-arcs in a plane is then used to obtain a classification of the cubic surfaces with 27 lines in $PG(3, q)$. This algorithm has been described in [4] and improved in [9]. In another program, double sixes of lines in $PG(3, q)$ are classified first, as described in Sect. 2.

4   Application

The classification of cubic surfaces over small finite fields is helpful in understanding better the theory of these objects. Determining the surfaces up to isomorphism is very useful to any further investigation. Surfaces which are isomorphic have the same properties and hence can be studied all at once. Surfaces which are not isomorphic show different behavior and need to be studied separately. The goal is to form parametrized families of cubic surfaces which behave in a certain way. The members of a family can be described in a unified fashion. Known families of cubic surfaces are the Clebsch surface and the Fermat surface. Another family of cubic surfaces was recently found based on the data of the classification described in this paper. This new family is described in [5]. It exists for every finite field $\mathbb{F}_q$ whose order q is odd. The equation of this surface is

$$x_3^3 - b^2(x_0^2 + x_1^2 + x_2^2)x_3 + \frac{b^3}{a}(a^2 + 1)x_0x_1x_2 = 0.$$

where a, b are elements of $\mathbb{F}_q$ and $a \notin \{0, \pm 1\}$, $a^2 \neq \pm 1$, and $b \neq 0$. This surface has an automorphism group isomorphic to Sym_4 of order 24. The projectivities associated with the matrices

$$S_1 = \begin{bmatrix} 1 & 0 & 0 & 0 \\ 0 & 0 & 1 & 0 \\ 0 & 1 & 0 & 0 \\ 0 & 0 & 0 & 1 \end{bmatrix}, \quad S_2 = \begin{bmatrix} 0 & 1 & 0 & 0 \\ 1 & 0 & 0 & 0 \\ 0 & 0 & 1 & 0 \\ 0 & 0 & 0 & 1 \end{bmatrix}, \quad S_3 = \begin{bmatrix} -1 & 0 & 0 & 0 \\ 0 & 0 & 1 & 0 \\ 0 & 1 & 0 & 0 \\ 0 & 0 & 0 & -1 \end{bmatrix}$$

satisfy the Coxeter relations for Sym_4 and generate this group. For certain congruences on q, the surface is isomorphic to the Clebsch (or diagonal) surface and for certain other congruences on q, the surface is isomorphic to the Fermat (or equianharmonic) surface. The presence of the parameter b can be ignored as all choices for b lead to isomorphic surfaces (for this reason, we can safely put $b = 1$). The presence of the parameter a has an interesting consequence: For a given q, the construction can give several nonisomorphic surfaces.

Table 1 is taken from [5] and shows the number of isomorphism types of cubic surfaces over all finite fields $\mathbb{F}_q$ for $q \le 97$. This table was computed using the algorithm which classifies double sixes in $PG(3, q)$. The table confirms all previously known classification results. The numbers for $q \le 9$ are due to Hirschfeld. The number for $q = 11$ is due to Sadeh. The number for $q = 13$ is from [4], and the numbers for $q = 17$ and $q = 19$ have been confirmed in [9] using the method of lifting six-arcs. The computing time for $q = 97$ was about two to three weeks on a machine using a single CPU.

Table 1. Isomorphism types of cubic surfaces with 27 lines in $PG(3, q)$

q	#	q	#	q	#	q	#
2	0	16	5	37	77	67	595
3	0	17	7	41	107	71	731
4	1	19	10	43	126	73	813
5	0	23	16	47	169	79	1081
7	1	25	18	49	121	81	331
8	1	27	11	53	258	83	1292
9	2	29	34	59	376	89	1673
11	2	31	43	61	427	97	2304
13	4	32	11	64	101		

5   Technical Contribution

Orbiter is an open source project, announced in [1] and available on github [2]. The main issue with Orbiter is the lack of a user-interface. At present, Orbiter is best used using Makefiles, using command line interface to specify options. This is not optimal for many users, but it has some advantages. The commands that one issues are automatically saved in a file (the makefile), and computations can be repeated. Also, all data that is computed by orbiter is stored in files. That way, data is available for later re-use. Finally, the data from classification jobs can be exported in C++ source code, which can then be compiled into Orbiter. That way, a classification becomes mathematical knowledge that is available for other users. Tables are hard-coded in the source files and are distributed in later versions of Orbiter. For instance, the almost 10,000 cubic surfaces from the classification described in [5] have been integrated into the Orbiter source code as a catalogue of cubic surfaces over small finite fields ($q \le 97$). The equations of those surfaces as well as the automorphism groups are readily available. A program is distributed with Orbiter which recreates any of the surfaces from the catalogue.

Acknowledgements. The author thanks Alain Esculier [7] for providing the original figure of the Clebsch surface with the accompanying Povray [10] source code. He also thanks Professor Hirschfeld and Fatma Karaoglu for stimulating discussions during a Sabbatical stay at the University of Sussex in the Fall of 2017.

References

1. Betten, A.: Classifying discrete objects with orbiter. ACM Commun. Comput. Algebra **47**(3/4), 183–186 (2014). https://doi.org/10.1145/2576802.2576832
2. Betten, A.: Orbiter - a program to classify discrete objects, 2016–2018. https://github.com/abetten/orbiter
3. Betten, A.:. Rainbow cliques and the classification of small BLT-Sets. In: Kauers, M. (ed.) ISSAC 2013, 26–29 June 2013, Boston, Massachusetts, pp. 53–60 (2013)
4. Betten, A., Hirschfeld, J.W.P., Karaoglu, F.: Classification of cubic surfaces with twenty-seven lines over the finite field of order thirteen. Eur. J. Math. **4**(1), 37–50 (2018)
5. Betten, A., Karaoglu, F.: Cubic surfaces over small finite fields. Submitted to Designs, Codes and Cryptography
6. Clebsch, A.: Die Geometrie auf den Flächen dritter Ordnung. J. Reine Angew. Math. **65**, 359–380 (1866)
7. Esculier, A.: Clebsch, ses 27 droites, les points de Eckardt. http://aesculier.fr/fichiersMaple/ClebschDroites/ClebschDroites.html. Accessed 15 Apr 2018
8. Hirschfeld, J.W.P.: Finite Projective Spaces of Three Dimensions, x+316 pp. Oxford University Press, Oxford (1985)
9. Karaoglu, F.: The cubic surfaces with twenty-seven lines over finite fields, Thesis, University of Sussex (2018). Submitted
10. Povray. Persistence of vision raytracer. http://www.povray.org
11. Schläfli, L.: An attempt to determine the twenty-seven lines upon a surface of the third order and to divide such surfaces into species in reference to the reality of the lines upon the surface. Quart. J. Math. **2**, 55–110 (1858)

How Fast Can We Compute Orbits of Groups?

Anton Betten[(✉)]

Colorado State University, Fort Collins, USA
betten@math.colostate.edu
http://www.math.colostate.edu/~betten/

Abstract. Many problems in Combinatorics and related fields reduce
to the problem of computing orbits of groups acting on finite sets. One
of the techniques is known under the name Snakes and Ladders. We offer
the alternate name poset classification algorithm. We will describe this
technique and compare the performance on example problems.

Keywords: Group orbits · Classification · Combinatorial object

1   Introduction

The classification problem for combinatorial objects can be expressed as a problem of computing orbits of groups. Let X be a finite set whose elements are the
instances of a certain kind of combinatorial structure. There is a group G acting
on X. Two objects in X are isomorphic if they belong to the same G-orbit.
The problem of classifying the combinatorial object is finding a transversal for
the orbits of G on X. Once a transversal has been computed, the recognition
problem is that of identifying the element in a transversal whose orbit contains
a given element x of X. The constructive recognition problem is finding a group
element $g \in G$ such that xg is the unique element in the transversal representing
the orbit of x.

The groups and the associated sets whose orbits we need to compute can
get quite large. For this reason, it is required that we design efficient algorithms.
Two classes of algorithms have emerged. In the first class, the notion of canonical
orbit representative is important. This notion means that we are able to map
an arbitrary element of a group orbit to a specific element that is determined
through some algorithm. This specific element could be the lexicographically
least element in the orbit. Other variants distinguish this element by means of
an algorithm which computes the canonical form. The unique property that this
algorithm has is that two input elements have the same output if and only if the
two input elements are in the same group orbit.

There are two main issues with the canonical augmentation method. First,
the need to perform backtracking to compute the canonical form. Secondly, there
is only a limited number of implementations of the canonical form algorithm.

© Springer International Publishing AG, part of Springer Nature 2018
J. H. Davenport et al. (Eds.): ICMS 2018, LNCS 10931, pp. 62–70, 2018.
https://doi.org/10.1007/978-3-319-96418-8_8

The present scarcity of implementations means that researchers interested in specific combinatorial objects have to devise clever ways of reducing the classification problem to one for which a canonical form algorithm has been implemented. The reduction from one combinatorial structure to another may be inefficient.

The algorithm described in this work avoids canonical forms and expensive backtracking as much as possible. One of the main characteristics is that it uses the symmetry in the most natural way, namely in the way that it appears in the defining group action. The algorithm is available in the open-source software package Orbiter [3,4]. Perhaps one of the most striking benefits of using Orbiter is the way in which group actions can be set up. Starting from a set of atomic groups which are builtin (symmetric group, semilinear and linear groups, various subgroups thereof), induced group actions can be created. This way, combinatorial objects can be classified using the defining group actions. There is no need to reduce the classification problem to a related classification problem of graphs.

2 Some Theory

Let G be a group and let $\mathcal{A}$ and $\mathcal{B}$ be sets on which G acts. Suppose that $\mathcal{R}$ is a relation between $\mathcal{A}$ and $\mathcal{B}$. Thus, $\mathcal{R} \subseteq \mathcal{A} \times \mathcal{B}$. The elements of $\mathcal{R}$ are called flags. The relation $\mathcal{R}$ is assumed to be G-invariant, so $(a, b) \in \mathcal{R}$ implies that $(ag, bg) \in \mathcal{R}$ also for all $g \in G$. Let Π_1 and Π_2 be the projection maps from $\mathcal{A} \times \mathcal{B}$ onto the first and onto the second component. Thus, for $(a, b) \in \mathcal{A} \times \mathcal{B}$, we have $\Pi_1((a, b)) = a$ and $\Pi_2((a, b)) = b$. For $a \in \mathcal{A}$, let

$$\mathrm{Up}(a) = \mathcal{R} \cap \Pi_1^{-1}(a) = \{(a, b) \in \{a\} \times \mathcal{B} \mid (a, b) \in \mathcal{R}\}.$$

Likewise, for $b \in \mathcal{B}$, let

$$\mathrm{Down}(b) = \mathcal{R} \cap \Pi_2^{-1}(b) = \{(a, b) \in \mathcal{A} \times \{b\} \mid (a, b) \in \mathcal{R}\}.$$

From now on, assume that $\mathcal{A}$ and $\mathcal{B}$ are both finite. Suppose that $P_1, \dots, P_m$ is a transversal for the orbits of G on $\mathcal{A}$. Likewise, suppose that $Q_1, \dots, Q_n$ is a transversal for the orbits of G on $\mathcal{B}$. Thus

$$\mathcal{A} = \bigcup_{i=1}^{m} \mathrm{Orb}_G(P_i) \quad \text{and} \quad \mathcal{B} = \bigcup_{j=1}^{n} \mathrm{Orb}_G(Q_j).$$

For $i = 1, \dots, m$, let $G_i := \mathrm{Stab}_G(P_i)$ be the stabilizer of P_i in G. For $j = 1, \dots, n$, let $H_j := \mathrm{Stab}_G(Q_j)$ be the stabilizer of Q_j in G. Notice that G_i acts on $\mathrm{Up}(P_i)$ and that H_j acts on $\mathrm{Down}(Q_j)$. Let $t_{i,r}, r = 1, \dots, R_i$ be a set of orbit representatives for the orbits of G_i on $\mathrm{Up}(P_i)$. Likewise, let $s_{j,k}, k = 1, \dots, K_j$ be a set of orbit representatives for the orbits of H_j on $\mathrm{Down}(Q_j)$. Thus,

$$\mathrm{Up}(P_i) = \bigcup_{r=1}^{R_i} \mathrm{Orb}_{G_i}(t_{i,r}) \quad \text{and} \quad \mathrm{Down}(Q_j) = \bigcup_{k=1}^{K_j} \mathrm{Orb}_{H_j}(s_{j,k}).$$

We define
$$\mathcal{T}_{i,r} = \mathrm{Orb}_{G_i}(t_{i,r}), \quad i = 1, \ldots, m, \; r = 1, \ldots, R_i$$
and
$$\mathcal{S}_{j,k} = \mathrm{Orb}_{H_j}(s_{j,k}), \quad j = 1, \ldots, n, \; k = 1, \ldots, K_j.$$

In this situation, we have the following Lemma. Part *1.* is from [6]. Parts *2.* and *3.* are immediate.

Lemma 1. *1. There is a canonical bijection between the set of orbits $\{\mathcal{T}_{i,r} \mid i = 1, \ldots, m, \; r = 1, \ldots, R_i\}$ and the set of orbits $\{\mathcal{S}_{j,k} \mid j = 1, \ldots, n, \; k = 1, \ldots, K_j\}$.*

2.
$$\sum_{i=1}^{m} R_i = \sum_{j=1}^{n} K_j.$$

3. If $\mathcal{T}_{i,r}$ and $\mathcal{S}_{j,k}$ correspond under the canonical bijection from 1., then
$$|\mathcal{T}_{i,r}| \cdot |H_j| = |\mathcal{S}_{j,k}| \cdot |G_i|.$$

3   A Lifting Algorithm

Suppose that the orbits of G on $\mathcal{A}$ are known, and that $P_1, \ldots, P_m$ have been determined, together with the associated stabilizer subgroups G_i. Suppose further that for an arbitray element $a \in \mathcal{A}$ we can determine the orbit representative P_i such that $P_i \sim_G a$. Suppose moreover that we can determine an element $g \in G$ such that $ag = P_i$. Suppose furthermore that $\mathrm{Down}(b) \neq \emptyset$ for all $b \in \mathcal{B}$. Then we can employ the following algorithm to classify the orbits of G on $\mathcal{B}$ by listing orbit representatives $Q_1, \ldots, Q_n$. We can provide the stabilizers $H_j = \mathrm{Stab}_G(Q_j)$, and provide an algorithm which for arbitrary $b \in \mathcal{B}$ determines a group element $g \in G$ such that $bg = Q_j$ for some j. The algorithm selects elements $\Delta_{i,r} \in G$ for $i = 1, \ldots, m$ and $r = 1, \ldots, K_i$ which need to be stored to make the recognition procedure work. Here is the algorithm:

1. For $i = 1, \ldots, m$.
   (a) Compute $\mathrm{Up}(P_i)$. Moreover, compute a transversal $t_{i,r}$ $r = 1, \ldots, K_i$ of the orbits of G_i on $\mathrm{Up}(P_i)$. Let $\mathcal{T}_{i,r} = \mathrm{Orb}_{G_i}(t_{i,r})$.
2. Mark all $\mathcal{T}_{i,r}$ as unprocessed. Let $j = 1$.
3. While there is still an unprocessed $\mathcal{T}_{i,r}$, do the following:
   (a) Let $Q_j := \Pi_2(t_{i,r})$. Let $\Delta_{i,r} = 1_G$ be the identity element of G. Initialize the group H_j with the group $\mathrm{Stab}_{G_i}(t_{i,r})$.
   (b) Compute $\mathcal{D} = \mathrm{Down}(Q_j)$.
   (c) Loop over all elements in $\mathcal{D}$. Let d be the next unprocessed element in $\mathcal{D}$.
      i. Let $a := \Pi_1(d)$. Using the recognition algorithm for $\mathcal{A}$, find $\alpha \in G$ with $a\alpha = P_{i'}$ for some $i' \leq m$.
      ii. Find an element $\beta \in G_{i'}$ such that $d\alpha\beta = t_{i',r'}$ for some $r' \leq K_{i'}$.
      iii. If $i' = i$ and $r' = r$ then extend the group H_j by the element $\alpha\beta$.

  iv. Otherwise, mark the orbit $\mathcal{T}_{i',r'}$ as processed and set $\Delta_{i',r'} := (\alpha\beta)^{-1}$.
  v. Continue with the next element $d \in D$ in (c) until done.
(d) At this point, the group H_j is equal to $\mathrm{Stab}_G(Q_j)$. Increment j and continue with the next unprocessed orbit $\mathcal{T}_{i,r}$ in 3 until done.
4. Let $n = j$. The transversal $Q_1, \ldots, Q_n$ has been computed.

Once the procedure is finished, the group elements $\Delta_{i,r}$ can be used to provide constructive recognition. Given an element $b \in \mathcal{B}$, the following algorithm produces an element $g \in G$ such that $bg = Q_j$ for some $j \leq n$. We assume that we have constructive recognition for $\mathcal{A}$.

1. Compute $\mathcal{D} = \mathrm{Down}(b)$.
2. Pick one element $d \in \mathcal{D}$.
3. Let $a := \Pi_1(d)$.
4. Using the recognition algorithm for $\mathcal{A}$, find $\alpha \in G$ with $a\alpha = P_i$ for some $i \leq m$.
5. Find an element $\beta \in G_i$ such that $da\beta = t_{i,r}$ for some $r \leq K_i$.
6. Let $g := \alpha\beta\Delta_{i,r}$ and determine $j \leq n$ with $\Pi_2(d)g = P_j$.

4   The Poset Classification Algorithm

In order to classify combinatorial objects, it is often helpful to utilize an ordered structure, whose elements are the elements of X together with certain objects which are "smaller" and serve as a stepping stone. The collection of all objects forms a partially ordered set, and the group G acts on it. The lifting algorithm described in the previous section forms the basis for a larger algorithm which classifies the G-orbits on the poset and hence also on X.

The algorithm is based on a data structure which keeps track of all orbits of the group on a poset associated with the set X. The classification algorithm builds up this data structure. Once the data structure is established, the recognition problem can be solved constructively. The requirement on the poset is that it should be ranked, so that layers are defined. A layer is the set of elements of equal rank.

There are many situations where a combinatorial object either has a substructure or is related to a smaller combinatorial object. One can build a poset structure from these substructures or related objects and apply the poset classification algorithm. Here are some of the most common examples:

In the set-situation, the objects are embedded in the lattice of subsets of a set. The group acts on the poset. The rank function is the size of a subset. The lifting algorithm proceeds from one layer to the next, by increasing the size of the sets. In the subspace-situation, the objects are embedded in the lattice of subspaces of a finite vector space. The rank function is the dimension of a subspace. The lifting algorithm proceeds from one layer to the next, by increasing the dimension of the subspaces. The orbit classification proceeds bottom up in a breadth first manner. In the subset lattice, the level is the size of the subset.

In the subspace lattice, the level is the dimension of the subspace. Other types of posets can be considered also.

The presence of the relation in Lemma 1 is very important for practical applications. It means that custom designed conditions can be imposed on the partially ordered set and that the classification algorithm will work within the scope of these constraints. By restricting the search, the algorithm can progress faster and with less memory. The conditions on the poset is that it is invariant under the group action and that it is hereditary. If an element is part of the poset, all elements below it must be in the poset as well. In practical implementations, one may define a test-function on the elements of the power set lattice (or of the lattice of subspaces). The purpose of the test-function is to tell whether an element needs to be considered. This kind of restricted search is very important for practical computations, in order to keep the size of the poset under control, and hence make the search feasible.

5   Applications

Let us look at some example problems to illustrate the poset classification algorithm.

5.1   Graphs on n Vertices

In order to classify simple graphs on n vertices (no loops, no multi edges), we consider the group $G = \mathrm{Sym}_n$ in its action on unordered pairs. Let $V = \{0, \ldots, n-1\}$. Let X be the set of two-subsets of X. The graphs on n vertices are identified with subsets of X. Two graphs are isomorphic if they belong to the same orbit under the action of G. The poset classification starts from the empty graph and uses the lifting algorithm to inductively classify graphs with $i + 1$ edges from the known classification of graphs with i edges. It is possible to adapt to certain specific classes of graphs if desired. For instance, it is possible to classify graphs on n vertices which are regular of degree r for some r. To do so, we consider the poset of those subsets of X where each element of V is incident with *at most* r elements from the chosen set. Likewise, it would be possible to ask for a girth of the graph by testing if the subset is free of cycles whose size is less than the value of the girth.

5.2   Orbits of Linear Groups

Let G be a subgroup of $\mathrm{P\Gamma L}(n, q)$ for some n and q. In order to classify the orbits of G on subspaces, one starts from the zero subspace, and then uses the lifting algorithm to classify the $i + 1$ subspaces from the previously computed classification of i-subspaces.

There are too many applications to list here. One case that we will study below is the problem of computing what is known as the Kramer-Mesner matrix of t-subspaces versus k-subspaces. This matrix describes how the orbits of a group on t-dimensional subspaces relate to the orbits of the same group on k-dimensional subspaces. An interesting group to consider is the group generated by the Singer cycle, which is transitive on the points of the vector space. The Kramer-Mesner matrix is interesting for the construction of geometric designs (also known as q-analogs of t-designs).

5.3   Applications in Finite Geometry

Many problems in finite geometry reduce to problems involving orbits of a certain linear group on objects inside a projective space. For instance, in [1], the orbits of an orthogonal group on subsets of the quadric $Q(4, q)$ of size $q+1$ called BLT-sets are important. The BLT-condition is just a condition which restricts the way in which the set can be chosen. We call the objects satisfying the condition partial BLT-sets, and they form a poset which is invariant under the group and hereditary. The most successful approach in this case is to use the poset classification to classify the partial BLT-sets of $Q(4, q)$ of size 5 and then proceed with a different technique to lift these sets to sets of size $q + 1$. The final classification is obtained by applying Lemma 1 for the set $\mathcal{A}$ which is the partial BLT-sets of size 5 and the set $\mathcal{B}$ which is the BLT-sets in $Q(4, q)$.

Other objects that are often studied in finite geometry are arcs in projective planes. These are sets of points where no more than d lie on a line. This problem reduces to a poset classification problem if V is the set of points in the projective plane under consideration and X is the set of subsets of V with no more than d collinear. The poset classification plays an important role in the classification of cubic surfaces with 27 lines as reported in a series of papers [7–9].

5.4   Applications in Coding Theory and Design Theory

A lot of work has been done in coding theory and design theory to utilize the computer to perform searches and classifications. Some of it has been based on the poset classification algorithm, and some other work has been done using the canonical augmentation (see below). We refer to [5] for an application of the poset classification algorithm and to [13] for applications of the canonical augmentation procedure.

6   Time and Space Issues

The poset classification algorithm stores a lot of data. There are the orbit representatives at every level in the search. In addition there are the associated flag orbits $\mathcal{T}_{i,r}$ and the group elements $\Delta_{i,r}$ (the orbits $\mathcal{S}_{j,k}$ do not have to be stored). The cost in memory is perhaps offset by the advantage in speed. The algorithm which perform constructive recognition depends linearly on the rank of the element times the cost to perform constructive recognition for the $\mathcal{T}_{i,r}$. The time to classify a class of combinatorial objects is the time to build up the data structure. This depends on the number of orbits times the time for computing the relevant up and down sets and the time to compute the orbits $\mathcal{T}_{i,r}$ on flags. In addition, for each node, the number of constructive recognition steps is proportional to the size of the down set of that node. Since too many of these quantities depend on the nature of the poset and the group acting on it, an explicit analysis of the time and space complexity seems difficult at this point.

7   Implementations

The poset classification algorithm has been described by several authors. The original implementation is [18]. Orbiter [3,4] is an open-source software package devoted to the poset classification algorithm and applications in combinatorics, algebra and finite geometry. The poset classification algorithm for subspaces has recently been implemented in GAP [12], using the package FinInG [2] by the author together with Michel Lavrauw. Further implementations of the poset classification algorithm have been reported by Koch [14] (subset situation) and by Braun [11] (subspace situation).

8   Comparisons

The main competitors for the poset classification algorithm are based on the notion of canonical form. These algorithms work quite differently, using a back-tracking procedure to compute a canonical representative of each G-orbit. The most popular implementation of the canonical representative procedures are McKay's program Nauty which relates to [16] and Jeff Leon's partition back-track algorithm [15]. A new development is described in [17]. Many of these algorithms are available in Magma [10] or GAP [12], for instance. The main advantage of these algorithms is that they classify the poset depth first, thereby requiring minimal storage. On the other hand, the analysis of the backtrack algorithm is difficult.

It may be illustrative to run some example problems on some of the available implementations. We choose benchmark problems from combinatorics, finite geometry, and algebra. We measure execution time and peak memory usage. An entry N/A means that no data is available. We measure the performances in terms of time and memory. The notation is as follows: t_X stands for measured CPU time using software X, and m_X stands for measured peak memory usage under software X. The keys for the software systems are: O is orbiter [3], K is Koch [14], N is Nauty (version 2.6) [17], and G is the author's implementation in GAP [12]. The data for t_K and m_K is taken from [14] (the Koch implementation is not published). All other measurements were taken from actual runs.

The first benchmark is the problem of classifying graphs on n vertices (cf. Table 1). We classify the complete poset, not taking into account the fact that there is an obvious isomorphism between graphs with e edges and graphs with $\binom{n}{2} - e$ edges (complementation). The next problem is that of computing the orbits of the irreducible Singer group acting on the subspaces of $\mathbb{F}_q^n$ (cf. Table 2).

We compute the orbits on the whole poset, disregarding the fact that under duality, the orbits on subspaces of dimension t are in canonical correspondence to the orbits on subspaces of dimension $n - t$. The next problem is that of classifying arcs in $PG(2, q)$ under the group $P\Gamma L(3, q)$ (cf. Table 3). The number of orbits is the overall number of orbits at all levels in the poset.

Table 1. Classifying graphs with n vertices

n	# orbits	t_O	m_O	t_K	m_K	t_N
8	12346	2 s	34 MB	1 min 47 s	11 MB	0 s
9	274668	1 min 14 s	224 MB	1 h 40 min	11 MB	0.18 s
10	12005168	3 h 38 min	51 MB	N/A	N/A	3.77 s
11	1018997864	N/A	N/A	N/A	N/A	4 min 58 s

Table 2. Classifying the orbits on subspaces of $\mathbb{F}_q^n$ under the irreducible Singer group

(n, q)	# orbits	t_O	m_O	t_G	m_G
(6,2)	49	0 s	561 MB	11 s	8503 MB
(7,2)	232	1 s	561 MB	1 min 40 s	8493 MB
(8,2)	1643	33 s	567 MB	49 min 40 s	76 GB
(9,2)	16214	22 min 27 s	634 MB	N/A	N/A
(10,2)	224617	17 h 21 min	1912 MB	N/A	N/A

Table 3. Classifying arcs in $PG(2, q)$

q	# orbits	t_O	m_O
11	73	0 s	562 MB
13	438	0 s	563 MB
16	4214	2 s	570 MB
17	52420	23 s	631 MB
19	711709	6 min 17 s	1556 MB

The benchmarks allow for some interesting comparisons. The differences in the running times can be striking. Classifying graphs is clearly best done with Nauty, which is a factor of 2200 times faster than Orbiter, which is a factor of 100 faster than the Koch program. For geometric objects, Orbiter is a factor of 100 faster than the implementation in GAP. As pointed out before, for problems in geometry, Nauty can only be used indirectly. For this reason, Nauty does not appear in the benchmarks which involve problems from geometry. It is questionable if a system like GAP should be used for classification at all. Nauty's use for geometric problems is unsettled.

References

1. Al-Azemi, A., Betten, A., Chowdhury, S.R.: A rainbow clique search algorithm for BLT-sets. In: Davenport, J.H., Kauers, M., Labahn, G., Urban, J. (eds.) ICMS 2018. LNCS, vol. 10931, pp. 71–79. Springer, Cham (2018)
2. Bamberg, J., Betten, A., Cara, Ph., De Beule, J., Lavrauw, M., Neunhöffer, M.: Finite Incidence Geometry. FinInG - a GAP package, version 1.4 (2017)
3. Betten, A.: Classifying discrete objects with orbiter. ACM Commun. Comput. Algebra **47**(3/4), 183–186 (2014). https://doi.org/10.1145/2576802.2576832
4. Betten, A.: Orbiter - a program to classify discrete objects (2016–2018). https:// github.com/abetten/orbiter
5. Betten, A., Braun, M., Fripertinger, H., Kerber, A., Kohnert, A., Wassermann, A.: Error-Correcting Linear Codes, Classification by Isometry and Applications. Algorithms and Computation in Mathematics, vol. 18. Springer, Heidelberg (2006). https://doi.org/10.1007/3-540-31703-1
6. Betten, A.: Rainbow cliques and the classification of small BLT-sets. In: Kauers, M. (ed.) ISSAC 2013, 26–29 June 2013, Boston, Massachusetts, pp. 53–60 (2013)
7. Betten, A., Hirschfeld, J.W.P., Karaoglu, F.: Classification of cubic surfaces with twenty-seven lines over the finite field of order thirteen. Eur. J. Math. **4**(1), 37–50 (2018)
8. Betten, A., Karaoglu, F.: Cubic surfaces over small finite fields. Submitted to Designs, Codes and Cryptography
9. Betten, A.: Classifying cubic surfaces over finite fields using orbiter. In: Davenport, J.H., Kauers, M., Labahn, G., Urban, J. (eds.) ICMS 2018. LNCS, vol. 10931, pp. 55–61. Springer, Cham (2018)
10. Bosma, W., Cannon, J., Playoust, C.: The Magma algebra system. I. The user language. J. Symbolic Comput. **24**, 235–265 (1997)
11. Braun, M.: Some new designs over finite fields. Bayreuth. Math. Schr. **74**, 58–68 (2005)
12. The GAP Group, GAP - Groups, Algorithms, and Programming, Version 4.8.10 (2018). https://www.gap-system.org
13. Kaski, P., Östergård, P.: Classification Algorithms for Codes and Designs. Algorithms and Computation in Mathematics, vol. 15. Springer, Heidelberg (2006). https://doi.org/10.1007/3-540-28991-7
14. Koch, M.: Neue Strategien zur Lösung von Isomorphieproblemen. (German) [New strategies for the solution of isomorphism problems] Ph.D. thesis. University of Bayreuth (2015)
15. Leon, J.S.: Partitions, refinements, and permutation group computation. In: Groups and Computation, II (New Brunswick, NJ, 1995), vol. 28. DIMACS Series Discrete Mathematics Theoretical Computer Science, pp. 123–158. American Mathematical Society, Providence (1997)
16. McKay, B.D.: Isomorph-free exhaustive generation. J. Algorithms **26**(2), 306–324 (1998)
17. McKay, B.D., Piperno, A.: Practical graph isomorphism II. J. Symbolic Comput. **60**, 94–112 (2014). https://doi.org/10.1016/j.jsc.2013.09.003
18. Schmalz, B.: Verwendung von Untergruppenleitern zur Bestimmung von Doppelnebenklassen. (German) [Use of subgroup ladders for the determination of double cosets]. Bayreuth. Math. Schr. **31**, 109–143 (1990)

A Rainbow Clique Search Algorithm
for BLT-Sets

Abdullah Al-Azemi[1], Anton Betten[2($\boxtimes$)], and Sajeeb Roy Chowdhury[2]

[1] Kuwait University, Kuwait City, Kuwait
alazmi95@gmail.com
[2] Colorado State University, Fort Collins, USA
betten@math.colostate.edu, srchowdh@rams.colostate.edu
http://www.math.colostate.edu/~betten/

Abstract. We discuss an algorithm to search for rainbow cliques in vertex-colored graphs. This algorithm is a generalization of the Bron-Kerbosch algorithm to search for maximal cliques in graphs. As an application, we describe a larger algorithm to classify a certain type of geometric-combinatorial objects called BLT-sets. We report on the classification of BLT-sets of order 71.

Keywords: Classification · Rainbow clique · Graph · BLT-set
Finite geometry

1 Introduction

Let $\mathrm{PG}(n, q)$ denote the n-dimensional projective space over the field $\mathbb{F}_q$. A $Q(4, q)$ space is a projective space $\mathrm{PG}(4, q)$ equipped with a quadratic form, such as $x_0^2 + x_1 x_2 + x_3 x_4 = 0$. There is an associated bilinear form $\beta(\mathbf{x}, \mathbf{y})$ and a group which preserves the quadratic form (for more details, see [13]). By considering the points and lines on the quadric with respect to inclusion, an incidence structure of points and lines is obtained. There are $(q^2 + 1)(q + 1)$ points and equally many lines. Each point is on $q + 1$ lines and each line has $q + 1$ points. It is an example of a finite generalized quadrangle [12]. This means that if a point is not on a line, the point is collinear to exactly one point on the line. The symmetry group of the incidence structure is the orthogonal group $\mathrm{P\Gamma O}(5, q)$.

A BLT-set (named after the initials of the authors of [2]) of $Q(4, q)$ is a set of $q + 1$ points such that no point on the quadric is collinear to three points in the set. In terms of the collinearity graph associated to $Q(4, q)$, a BLT-set induces an equitable partition. BLT-sets are interesting because they give rise to quadratic flocks, which in turn give rise to translation planes. In a curious twist, BLT-sets predate themselves, having been introduced in [8] under the name $(0, 2)$-sets (by the same author). BLT-sets give rise to generalized quadrangles and other objects of interest to the finite geometry community. Two BLT-sets over the same field $\mathbb{F}_q$ are equivalent if there is an orthogonal transformation

© Springer International Publishing AG, part of Springer Nature 2018
J. H. Davenport et al. (Eds.): ICMS 2018, LNCS 10931, pp. 71–79, 2018.
https://doi.org/10.1007/978-3-319-96418-8_9

which takes one to the other. We consider the problem of classifying BLT-sets up to equivalence. It is known that BLT sets in $Q(4, q)$ exist if and only if q is odd. A list of the 40 points P_i of $Q(4, 3)$ is given in Table 1.

Table 1. The points of $Q(4, 3)$

0: 0, 1, 0, 0, 0	10: 0, 0, 1, 0, 1	20: 1, 2, 1, 1, 0	30: 1, 0, 2, 2, 1
1: 0, 0, 1, 0, 0	11: 0, 0, 1, 0, 2	21: 1, 1, 2, 1, 0	31: 1, 0, 0, 1, 2
2: 0, 0, 0, 1, 0	12: 0, 1, 2, 2, 2	22: 1, 2, 1, 0, 1	32: 1, 2, 0, 1, 2
3: 0, 1, 0, 1, 0	13: 0, 1, 2, 1, 1	23: 1, 1, 2, 0, 1	33: 1, 1, 0, 1, 2
4: 0, 1, 0, 2, 0	14: 0, 1, 1, 2, 1	24: 1, 2, 1, 0, 2	34: 1, 0, 1, 1, 2
5: 0, 0, 1, 1, 0	15: 0, 1, 1, 1, 2	25: 1, 1, 2, 0, 2	35: 1, 0, 2, 1, 2
6: 0, 0, 1, 2, 0	16: 1, 2, 1, 0, 0	26: 1, 0, 0, 2, 1	36: 1, 1, 1, 1, 1
7: 0, 0, 0, 0, 1	17: 1, 1, 2, 0, 0	27: 1, 2, 0, 2, 1	37: 1, 2, 2, 1, 1
8: 0, 1, 0, 0, 1	18: 1, 2, 1, 2, 0	28: 1, 1, 0, 2, 1	38: 1, 1, 1, 2, 2
9: 0, 1, 0, 0, 2	19: 1, 1, 2, 2, 0	29: 1, 0, 1, 2, 1	39: 1, 2, 2, 2, 2

Based on this labeling of points, Table 2 lists the 40 lines. Each line is represented as the set of points incident with it.

Table 2. The lines of $Q(4, 3)$

0: 0, 31, 32, 33	10: 8, 18, 29, 38	20: 5, 22, 27, 37	30: 13, 20, 25, 26
1: 0, 26, 27, 28	11: 5, 9, 12, 15	21: 4, 22, 29, 36	31: 4, 25, 35, 39
2: 1, 26, 29, 30	12: 9, 19, 30, 39	22: 13, 19, 22, 31	32: 0, 7, 8, 9
3: 1, 31, 34, 35	13: 9, 20, 34, 36	23: 6, 24, 32, 39	33: 1, 7, 10, 11
4: 15, 16, 28, 35	14: 10, 20, 32, 37	24: 3, 24, 34, 38	34: 7, 16, 22, 24
5: 14, 16, 30, 33	15: 10, 19, 28, 38	25: 12, 21, 24, 26	35: 7, 17, 23, 25
6: 14, 17, 27, 34	16: 4, 10, 12, 14	26: 6, 23, 28, 36	36: 0, 2, 3, 4
7: 15, 17, 29, 32	17: 11, 18, 27, 39	27: 12, 18, 23, 31	37: 1, 2, 5, 6
8: 6, 8, 13, 14	18: 11, 21, 33, 36	28: 3, 23, 30, 37	38: 2, 16, 18, 20
9: 8, 21, 35, 37	19: 3, 11, 13, 15	29: 5, 25, 33, 38	39: 2, 17, 19, 21

An example of a BLT-set in $Q(4, 3)$ is the set $\{P_0, P_1, P_{14}, P_{15}\}$: Each point of $Q(4, 3)$ other than the four points in the BLT-set lies on either 0 or 2 lines in pencils through these points (cf. Table 3).

The points which lie on two pencils are $\{3, 4, 5, 6, 7, 8, 9, 10, 11, 12, 13, 16, 17, 26, 27, 28, 29, 30, 31, 32, 33, 34, 35\}$. The points which are not collinear to points of the BLT-set are $\{2, 19, 19, 20, 21, 22, 23, 24, 25, 36, 37, 38, 39\}$. In terms of the

Table 3. The pencils of lines through the points in the BLT-set

$0, 31, 32, 33$	$1, 26, 29, 30$	$14, 16, 30, 33$	$15, 16, 28, 35$
$0, 26, 27, 28$	$1, 31, 34, 35$	$14, 17, 27, 34$	$15, 17, 29, 32$
$0, 7, 8, 9$	$1, 7, 10, 11$	$14, 6, 8, 13$	$15, 5, 9, 12$
$0, 2, 3, 4$	$1, 2, 5, 6$	$14, 4, 10, 12$	$15, 3, 11, 13$

collinearity graph of $Q(4,3)$, we have an equitable partition with three classes of size $4, 23$ and 13, respectively.

Several infinite families of BLT-sets exist. However, many examples are known that are currently sporadic. It is very likely that new infinite families can be constructed from the known examples, though this might be a difficult problem. Besides that, there is interest in resolving the problem of classifying BLT-sets for additional finite fields. At present, all BLT-sets in $Q(4, q)$ for $q \leq 67$ have been classified [6]. In this note, the case $q = 71$ will be settled. Our computations are facilitated using the computer algebra package Orbiter [4]. Orbiter is a library of C++ classes, devoted to the problem of classifying combinatorial objects.

2   Underlying Theory

One of the centerpieces in the classification of BLT-sets is an algorithm to find all rainbow cliques in a vertex colored graph. The reduction of the problem of classifying BLT-sets to the problem of searching for rainbow cliques in graphs is interesting. In this section, we wish to outline this reduction briefly.

Let $\Gamma = (V, E)$ be a finite graph. For two vertices $x, y \in V$, we say that x and y are adjacent in Γ, if $\{x, y\}$ is an edge in E, and we express this by writing $x \sim y$. Also, for a vertex $x \in V$, let $N(x)$ be the set of neighbors of x in Γ, i.e. $N(x) = \{y \in V \mid x \sim y\}$. A clique in Γ is a set $T \subseteq V$ of vertices of Γ such that $x \sim y$ for any two vertices of T. The Bron-Kerbosch algorithm [7] is a systematic way to find all maximal cliques in a graph. A clique is called maximal if there is no clique which strictly contains it. The automorphism group of a graph $\Gamma = (V, E)$ is the set of permutations of V which preserve the adjacency relation. This group is denoted by $\mathrm{Aut}(\Gamma)$. Two subsets X and Y of V are called equivalent if there is an automorphism $\alpha \in \mathrm{Aut}(\Gamma)$ with $X\alpha = Y$.

Let C be a finite set, whose elements are called colors. A colored graph is a tuple $\Gamma = (V, E, C, c)$ where (V, E) is a graph, C is a set of colors, and $c : V \rightarrow C$ is a function which assigns exactly one color to each vertex. We assume that $c(x) \neq c(y)$ whenever x and y are adjacent. For $k \in C$, let C_k be the set of vertices of Γ of color k, i.e. $C_k = \{y \in V \mid c(y) = k\}$. The set C_k is called a color-class. A rainbow clique in Γ is a clique which intersects each color class in exactly one element. Rainbow cliques are maximal but not every maximal clique is a rainbow clique. In this paper, we will utilize rainbow cliques to facilitate the classification of BLT-sets. This will be explained next.

A partial BLT-set is a set S of points of $Q(4,q)$ with the property that no point of $Q(4,q)$ is collinear with more than two points of S. A BLT-set is a partial BLT-set of size $q+1$. An efficient way to construct and classify BLT-sets is this:

- (Problem 1) Classify the partial BLT-sets of some size s, say. The output from this step is a list $S_1, \ldots, S_N$ of representatives of the orbits of $P\Gamma O(5,q)$ on the set of partial BLT-sets of size s.
- (Problem 2) Given a partial BLT-set of size s, find all BLT-sets containing it.
- (Problem 3) Given the data created when solving Problems 1 and 2, produce the classification of BLT-sets of $Q(4,q)$ under the action of $P\Gamma O(5,q)$.

For Problem 1, the poset classification algorithm described in [5] can be used. Regarding Problem 2, the rainbow clique techniques described in this paper are useful. For Problem 3, the algorithm of [5] is again useful.

The choice of the parameter s is important. If s is too large, we may not be able to classify the partial BLT-sets of that size, so we fail with Problem 1. If s is too small, we may either be unable to solve Problem 2 or we may end up with too many solutions. Based on practical experiments, we find that $s = 5$ seems to work well.

Let us now focus on solving Problem 2. Suppose we have a partial BLT-set set S of size $s > 0$. The goal is to find all BLT-sets which contain S. To this end, we define a graph Γ_S. We say that a point P of $Q(4,q)$ is alive with respect to S if $S \cup \{P\}$ is a partial BLT-set of size $s+1$. The vertices of the graph Γ_S are the live points. Two vertices of Γ_S corresponding to points P_1 and P_2 are adjacent in Γ_S if $S \cup \{P_1, P_2\}$ is a partial BLT-set of size $s+2$. Now, pick a line through any point of S. Any line will do, but it is important that we fix this line. So, suppose that $S = \{Q_1, \ldots, Q_s\}$. Let ℓ be a line through Q_1, say. By the generalized quadrangle property, the points of $S \setminus \{Q_1\} = \{Q_2, \ldots, Q_s\}$ are collinear to certain points on ℓ. By the partial BLT-set property, these points are all distinct. Let $\mathcal{C} = \{C_1, \ldots, C_{q+1-s}\}$ be the set of points of $\ell \setminus \{Q_1\}$ which are not collinear to any of the points in $\{Q_2, \ldots, Q_s\}$. The set $C = \{1, \ldots, q+1-s\}$ will serve as the set of colors of Γ_S. A live point P is collinear to exactly one point of $\mathcal{C}$. If P is collinear to the point C_k, then we set $c(P) = k$. Assume that P_1 and P_2 are live points collinear to the same point Q on ℓ. Then Q is collinear to P_1, P_2 and Q_1, which contradicts the BLT-property. Hence $S \cup \{P_1, P_2\}$ is not a partial BLT-set, and so $P_1 \not\sim P_2$ in Γ_S. This shows that the function $c : V \to C$ is a coloring of Γ_S.

The following lemma from [3] is helpful to create the adjacency relation E of Γ_S. Let N_q be the set of non-squares in $\mathbb{F}_q$, i.e.

$$N_q = \{x \in \mathbb{F}_q \setminus \{0\} \mid y^2 \neq x \text{ for all } y \in \mathbb{F}_q\}.$$

Lemma 1. *Let S be a non-empty partial BLT-set and let $a \in S$ arbitrary. Let $x, y \in Q(4,q)$ such that both $S \cup \{x\}$ and $S \cup \{y\}$ are partial BLT-sets. Then $S \cup \{x, y\}$ is partial BLT-set if and only if*

$$-\beta(a,x)\beta(a,y)\beta(x,y) \in N_q.$$

Every BLT-set containing S corresponds to a rainbow clique in Γ_S. Interestingly, the converse is true as well, though this is by no means obvious (the proof depends on a specific property of BLT-sets, see [3]). This connection is the basis of our classification algorithm. It reduces Problem 2 to the problem of finding all rainbow cliques in all graphs Γ_S associated to partial BLT-sets S from a transversal of the equivalence classes found in Problem 1.

Consider the example of BLT-sets in $Q(4, 3)$. Suppose we pick $s = 1$ in the classification algorithm. Suppose we consider the partial BLT-set $S = \{P_0\}$, and we wish to find all BLT-sets containing it. We may pick the line $\ell_0 = \{0, 31, 32, 33\}$ through P_0. The pencil of lines through P_0 is the four lines

$$\{0, 31, 32, 33\}, \ \{0, 26, 27, 28\}, \ \{0, 7, 8, 9\}, \ \{0, 2, 3, 4\}.$$

Based on this list, the set of points which are alive is $\{1, 5, 6, 10, 11, 12, 13, 14, 15, 16, 17, 18, 19, 20, 21, 22, 23, 24, 25, 29, 30, 34, 35, 36, 37, 38, 39\}$, which is all points not contained in the pencil. The colored graph $\Gamma_S = (V, E, C, c)$ can now be defined. The vertex set is the set V of live points just listed. The colors are defined using the points 31, 32 and 33 on the line that we picked. By looking at the pencil of lines through each of these points, we find that the color classes are

$$C_0 = \{1, 12, 13, 18, 19, 22, 23, 34, 35\},$$
$$C_1 = \{6, 10, 15, 17, 20, 24, 29, 37, 39\},$$
$$C_2 = \{5, 11, 14, 16, 21, 25, 30, 36, 38\}.$$

Thus we have a graph Γ_S with 27 vertices and 3 colors. Each color class has size 9. For the adjacency relation, we note that 2 is the only non-square in $\mathbb{F}_3$. The neighbors of 1 turn out to be $\{14, 15, 36, 37, 38, 39\}$. For instance, since

$$-\beta(P_0, P_1)\beta(P_0, P_{14})\beta(P_1, P_{14}) = 2$$

is a nonsquare, 1 is adjacent to 14. Similarly, the other adjacencies can be checked. The full adajcency list is shown in Table 4.

The search for rainbow cliques in this graph will be continued below.

A few words regarding the symmetry group of $Q(4, q)$ are in order. The orthogonal group $O(n, q)$ is generated by elements known as Siegel transformations and orthogonal reflections. A vector v is singular if $Q(v) = 0$. The singular vectors give rise to the points on the quadric. All $q - 1$ multiples of one singular vector give rise to one projective point on the quadric. The bilinear form associated to Q gives rise to a polarity $\perp$. Let u be a singular vector and let $v \in \langle u \rangle^\perp$. The linear map

$$\rho_{u,v}(x) = x + \beta(x, v)u - \beta(x, u)v - Q(v)\beta(x, u)u$$

is known as Siegel transformation. It belongs to the simple subgroup $\Omega(n, q)$ of $O(n, q)$. For a non-singular vector v, the linear map

$$t_v(x) = x - Q(v)^{-1}\beta(x, v)v$$

Table 4. Adjacency in Γ_S

35: 30, 11, 5, 20, 17, 24	30: 35, 22, 18, 6, 17, 10	20: 35, 22, 23, 38, 11, 14
22: 30, 38, 21, 20, 6, 15	36: 1, 13, 18, 37, 17, 24	39: 1, 13, 23, 38, 16, 21
1: 36, 38, 14, 39, 37, 15	38: 22, 1, 12, 20, 39, 17	37: 1, 19, 12, 36, 16, 25
13: 36, 16, 5, 39, 17, 10	16: 13, 34, 12, 39, 37, 29	6: 22, 34, 12, 30, 11, 25
18: 30, 36, 25, 24, 15, 10	21: 22, 34, 23, 39, 15, 10	17: 35, 13, 12, 30, 36, 38
34: 16, 21, 25, 6, 29, 10	11: 35, 19, 12, 20, 6, 29	29: 34, 19, 23, 16, 11, 5
19: 11, 25, 14, 37, 29, 24	25: 18, 34, 19, 37, 6, 15	24: 35, 18, 19, 36, 5, 14
12: 38, 16, 11, 37, 6, 17	5: 35, 13, 23, 29, 24, 10	15: 22, 1, 18, 21, 25, 14
23: 21, 5, 14, 20, 39, 29	14: 1, 19, 23, 20, 24, 15	10: 13, 18, 34, 30, 21, 5

is called orthogonal reflection and belongs to $O(n, q)$ as well. Both types of elements suffice to generate the orthogonal group $O(n, q)$. The projective orthogonal group $\mathrm{PGO}(n, q)$ is the factor group $O(n, q)/\pm I$, where I is the identity transformation. For q odd, the orthogonal group $O(2m + 1, q)$ has order

$$2q^{m^2} \prod_{i=1}^{m} \left(q^{2i} - 1\right).$$

The order of the projective group is half of this number. It is possible to create the orthogonal group by creating generators which are either Siegel transformations or orthogonal reflections at random. By using the order formula as a stopping rule, the algorithm will produce the full group.

Let us consider an example. The orthogonal group $\mathrm{PGO}(5, 3)$ has order 51840. Random generators are the six group elements

$$g_1 = \rho_{(0,0,1,1,0),(0,1,2,0,2)}, \quad g_4 = \rho_{(1,0,0,2,1),(2,0,2,0,1)},$$
$$g_2 = \rho_{(1,2,0,2,1),(2,0,1,2,2)}, \quad g_5 = \rho_{(1,1,2,0,2),(0,2,2,0,2)},$$
$$g_3 = \rho_{(0,0,0,1,0),(1,0,0,0,0)}, \quad g_6 = t_{(1,1,1,1,0)}.$$

This can be verified using the following piece of Magma [11] code, which confirms the group order:

```
K  := FiniteField(3);
GL53 := GeneralLinearGroup(5, K);
g1 := elt<GL53 | 1,0,0,0,0, 0,0,1,0,1, 0,0,2,1,0, 0,0,2,0,0, 0,2,2,1,2 >;
g2 := elt<GL53 | 0,0,1,2,2, 1,0,0,2,1, 1,1,2,0,1, 1,0,2,2,1, 0,2,1,1,1 >;
g3 := elt<GL53 | 1,0,0,2,0, 0,1,0,0,0, 0,0,1,0,0, 0,0,0,1,0, 2,0,0,2,1 >;
g4 := elt<GL53 | 2,0,2,1,0, 2,1,0,1,2, 0,0,1,0,0, 1,0,1,1,2, 0,0,2,2,0 >;
g5 := elt<GL53 | 2,0,1,0,1, 0,0,2,0,2, 1,2,1,0,0, 0,2,2,1,2, 0,0,0,0,1 >;
g6 := elt<GL53 | 0,2,2,2,0, 1,2,1,1,0, 1,1,2,1,0, 0,0,0,1,0, 1,1,1,1,1 >;
L := [g1,g2,g3,g4,g5,g6];
O := sub< GL53 | L >;
```

The output of this code is 51840, the order of PGO(5, 3). For more background on the orthogonal group, we refer to [13].

3   Technical Contribution

The basis of this algorithm is the Bron-Kerbosch algorithm [7]. A modified version for rainbow cliques is described next. Let $\Gamma = (V, E, C, c)$ be a colored graph. A partial rainbow clique is a clique which intersects each color class in at most one element. We say that a color has been satisfied by a set R if there is an element in R of that color. The algorithm considers a set R which is a partial rainbow clique. The set R is implemented as a stack, so we can tell in what order the elements have been added to R. The algorithm tries to extend R in all possible ways that are reasonable. A set L of live points holds all vertices under consideration at the current state. A set $M \subseteq L$ is determined to be the minimal color class. Vertices from M are used to extend R in the next step. Here is the recursive algorithm:

1. Input: a partial rainbow clique R, a set L of live points.
2. Test if $|R| = h$. In this case, R is a rainbow clique and will be printed. Return to the previous level in the search.
3. Let r be the element that was added last to R.
4. Let $L = L \cap N(r)$.
5. Determine the set M which is the smallest among the sets $C_i \cap L$ where C_i varies over the color classes which have not yet been satisfied by R. If there are several choices for M, pick one arbitrarily.
6. Loop over all elements $m \in M$.
7. Form $R \cup \{m\}$ and recurse.
8. Move to the next element $m \in M$ and repeat, until M has been exhausted.
9. Return to the previous level of the backtrack search.

Initially, the algorithm is started with $R = \emptyset$ (the empty set) and $L = \{0, \ldots, n-1\}$ (the set of vertices of Γ).

In order to illustrate the algorithm, we continue the example of BLT-sets in $Q(4, 3)$. The search tree of the rainbow clique algorithm is illustrated in Fig. 1.

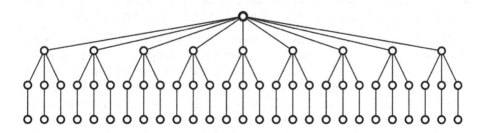

Fig. 1. The search tree

The algorithm starts with the empty set, which is a partial rainbow clique. This is the root node in the figure. At this point, the live point set is all of the vertices. The algorithm then proceeds to pick a color class, say C_0. Within C_0, there are nine possible vertices that can be picked. These possibilities are represented as the nine nodes which descend from the root node. For each of the possibilities, the neighbor set is computed, and the new live point set is formed. As we have seen, there are three neighbors in each of the remaining two color classes. The algorithm picks one of the remaining color classes, and then loops over all three vertices. This amounts to the $9 \times 3 = 27$ vertices at depth two in the search tree. For each choice, the new live point set has size one. The algorithm picks the remaining color class and adds the unique live point to form a rainbow clique. This amounts to the 27 nodes at the bottom of the search tree. The resulting 27 rainbow cliques are listed in Table 5. The associated BLT-sets are constructed by adding S, which amounts to adding P_0.

Table 5. Rainbow cliques in Γ_S

$35, 30, 17$	$22, 21, 15$	$13, 5, 10$	$34, 25, 6$	$19, 11, 29$	$23, 21, 39$
$35, 11, 20$	$1, 38, 39$	$13, 16, 39$	$34, 16, 29$	$12, 11, 6$	$23, 14, 20$
$35, 5, 24$	$1, 36, 37$	$18, 36, 24$	$34, 21, 10$	$12, 16, 37$	
$22, 30, 6$	$1, 14, 15$	$18, 25, 15$	$19, 25, 37$	$12, 38, 17$	
$22, 38, 20$	$13, 36, 17$	$18, 30, 10$	$19, 14, 24$	$23, 5, 29$	

4   Application

We will now describe an application of this algorithm. The problem is that of classifying BLT-sets in $Q(4, 71)$. This is the next open case following the results published in [6]. This in turn improves on earlier results published in [10].

Theorem 1. *The number of BLT-sets in $Q(4, 71)$ up to equivalence is 8. The BLT-sets are the Linear, the Fisher, the Penttila, and the Fisher-Thas, together with four more, which do not fall into any of the known families. The four sporadic sets have automorphism groups of order $3, 8, 24$ and 24. One of the sets with automorphism group of order 24 was described in Law [9]. The two BLT-sets with groups of order 24 can be distinguished by the number of planes which intersect them in 4 points. One has 810 and the other has 876 such planes.*

The computations leading to this result were performed on the Summit compute cluster [1]. We used Orbiter [4] to perform all computations. The number of partial BLT-sets of size 5 in $Q(4, 71)$ is 196891. This is also the number of instances of rainbow clique finding problems that had to be solved. Each graph has 67 color classes, so each rainbow clique has size 67. In the hardest instance, the graph had 18,211 vertices. The search for this instance took about 53 min. On average, the graphs were smaller and the search was shorter, in part because we used the lexicographic ordering reduction described in [6].

Acknowledgements. The authors acknowledge generous access to the HPC resource Summit [1], which was essential to perform the computations necessary for this work.

References

1. Anderson, J., Burns, P.J., Milroy, D., Ruprecht, P., Hauser, T., Siegel, H.J.: Deploying RMACCS summit: an HPC resource for the rocky mountain region. In: Proceedings of PEARC 2017, New Orleans, LA, USA, 09–13 July 2017, 7 pages (2017)
2. Bader, L., Lunardon, G., Thas, J.A.: Derivation of flocks of quadratic cones. Forum Math. **2**(2), 163–174 (1990)
3. Bader, L., O'Keefe, C.M., Penttila, T.: Some remarks on flocks. J. Aust. Math. Soc. **76**(3), 329–343 (2004)
4. Betten, A.: Orbiter - a program to classify discrete objects (2016). https://github.com/abetten/orbiter
5. Betten, A.: How fast can we compute orbits of groups? In: Davenport, J.H., Kauers, M., Labahn, G., Urban, J. (eds.) ICMS 2018. LNCS, vol. 10931, pp. 62–70. Springer, Cham (2018)
6. Betten, A.: Rainbow cliques and the classification of small BLT-sets. In: ISSAC 2013–Proceedings of the 38th International Symposium on Symbolic and Algebraic Computation, pp. 53–60. ACM, New York (2013)
7. Bron, C., Kerbosch, J.: Algorithm 457: finding all cliques of an undirected graph. Commun. ACM **16**(9), 575–577 (1973)
8. De Soete, M., Thas, J.A.: A characterization theorem for the generalized quadrangle $T_2^*(O)$ of order $(s, s + 2)$. Ars Comb. **17**, 225–242 (1984)
9. Law, M.: Flocks, generalised quadrangles and translation planes from BLT-sets. Thesis presented to the Department of Mathematics and Statistics, The University of Western Australia, March 2003
10. Law, M., Penttila, T.: Classification of flocks of the quadratic cone over fields of order at most 29. Adv. Geom. (suppl.), S232–S244 (2003)
11. Magma: The Computational Algebra Group within the School of Mathematics and Statistics of the University of Sydney (2004)
12. Payne, S.E., Thas, J.A.: Finite Generalized Quadrangles. EMS Series of Lectures in Mathematics, 2nd edn. European Mathematical Society (EMS), Zürich (2009)
13. Taylor, D.E.: The Geometry of the Classical Groups. Sigma Series in Pure Mathematics, vol. 9. Heldermann Verlag, Berlin (1992)

Numerical Software to Compute Newton Polytopes

Taylor Brysiewicz[✉]

Texas A&M University, College Station, USA
tbrysiewicz@math.tamu.edu
http://www.math.tamu.edu/~tbrysiewicz/

Abstract. We present our implementation of an algorithm which functions as a numerical oracle for the Newton polytope of a hypersurface in the **Macaulay2** package **NumericalNP.m2**. To showcase this software, we investigate the Newton polytope of both a hypersurface coming from algebraic vision and the classical Lüroth invariant.

1 Introduction

Often hypersurfaces are presented as the image of a variety under some map. Determining the defining equation $f \in \mathbb{C}[x_1, \ldots, x_n]$ of such a hypersurface $\mathcal{H} \subseteq \mathbb{C}^n$ is computationally difficult and often infeasible using symbolic methods such as Gröbner bases. Moreover, many times the defining equation is so large that it is not human-readable and so one naturally desires a coarser description of the polynomial, such as the Newton polytope. The Newton polytope of f, or equivalently that of $\mathcal{H}$, is the convex hull of the exponent vectors appearing in the support of f and provides a large amount of information about the hypersurface. Newton polytopes are necessary to compute the BKK bound on the number of solutions to a polynomial system [4] and can also provide topological information such as the Euler characteristic of the hypersurface [10]. Knowing $\mathrm{New}(f)$ also reduces the computational difficulty of finding f via interpolation: the size of the linear system one must solve is $|\mathrm{New}(f) \cap \mathbb{Z}^n|$, which is usually much smaller than the naïve bound of $\binom{n+d-1}{d}$ where $d = \deg(f)$.

In 2012 Hauenstein and Sottile [9] proposed an algorithm we call the HS-algorithm and showed that this algorithm functions as a vertex oracle for linear programming on $\mathrm{New}(\mathcal{H})$. This algorithm requires that the hypersurface is represented numerically by a witness set. Because a witness set is the only requirement, the HS-algorithm applies to hypersurfaces which arise as images of maps, such as rational varieties. We observe that the HS-algorithm is stronger than a vertex oracle and so we introduce the notion of a numerical oracle which returns some information when the linear program is not solved by a vertex.

We implemented the IIS-algorithm in the **Macaulay2** [7] package **NumericalNP.m2**. It uses the package **Bertini.m2** [1] to call **Bertini** [2] to perform numerical path tracking. Section 2 contains background on polytopes, numerical algebraic geometry, and a brief description of the HS-algorithm. Section 3

© Springer International Publishing AG, part of Springer Nature 2018
J. H. Davenport et al. (Eds.): ICMS 2018, LNCS 10931, pp. 80–88, 2018.
https://doi.org/10.1007/978-3-319-96418-8_10

outlines the three main user functions in the package and Sect. 4 advertises its strength on much larger examples.

2 Underlying Theory

2.1 Polytopes

A *polytope* $P \subseteq \mathbb{R}^n$ is the convex hull of finitely many points $V \subseteq \mathbb{R}^n$. Equivalently, P is the bounded intersection of finitely many halfspaces. The former presentation is a *V-representation* of P while the later is an *H-representation* of P. Given $\omega \in \mathbb{R}^n$ the set $P_\omega := \{x \in P \mid \langle x, \omega \rangle$ is maximized$\}$ is called the *face* of P *exposed* by ω and the function $h_P(\omega) = \max_{x \in P}\langle x, \omega \rangle$ is the *support function* of P. We define a *numerical oracle* to be the function

$$\mathcal{O}_P : \mathbb{R}^n \to \mathbb{N}^n \cup \{\text{EEP}\}$$

$$\omega \mapsto \begin{cases} P_\omega & \dim(P_\omega) = 0 \\ \min(P_\omega) & 0 < \dim(P_\omega) < \dim(P) \\ \text{EEP} & P_\omega = P \end{cases}$$

where $\min(P_\omega)$ is the coordinate-wise minimum of all points in P_ω and EEP abbreviates Exposes Entire Polytope. We remark that when a numerical oracle returns a vertex $v = \mathcal{O}_P(\omega)$, it also reveals that $\{x \in \mathbb{R}^n | \langle x, \omega \rangle \leq \langle v, \omega \rangle\}$ is a halfspace containing P. Finding a *V-representation* given an oracle is difficult but possible [6].

Given a polynomial

$$f = \sum_{\beta \in \mathcal{A}} c_\beta x_1^{\beta_1} \cdots x_n^{\beta_n} \in \mathbb{C}[x_1, \ldots, x_n] \qquad c_\beta \neq 0, \mathcal{A} \subseteq \mathbb{N}^n, |\mathcal{A}| < \infty$$

its *Newton polytope* $\mathrm{New}(f)$ is the convex hull of $\mathcal{A}$. Motivated by language for polynomials, we say that P is *homogeneous* whenever $\mathcal{O}_P(1, 1, \ldots, 1) = \text{EEP}$ and define $\deg(P) := h_P(1, 1, \ldots, 1)$. The *homogenization* of P denoted $\tilde{P}$ is the convex hull of $\{(x, \deg(P) - |x|) | x \in P\}$ where $|x| := \sum_{i=1}^n x_i$.

2.2 Numerical Algebraic Geometry

Let $X \subseteq \mathbb{C}^N$ be an algebraic variety of dimension k and degree d appearing as an irreducible component of the zero set of a collection of polynomials $\mathcal{F}_X \subseteq \mathbb{C}[x_1, \ldots, x_n]$. For a generic $N - k$ dimensional linear space $\mathcal{L} \subseteq \mathbb{C}^N$, the intersection $S = X \cap \mathcal{L}$ is zero dimensional and consists of d points. The triple $(\mathcal{F}_X, \mathcal{L}, S)$ is called a *witness set* for X and is the fundamental data type in numerical algebraic geometry. The standard numerical method of *homotopy continuation* quickly computes any witness set $(\mathcal{F}_X, \mathcal{L}', S')$ for X from a pre-computed witness set $(\mathcal{F}_X, \mathcal{L}, S)$ by numerically tracking the solutions S along a homotopy from S to S' [3]. A major feature of numerical algebraic geometry is that we can compute witness sets for varieties without access to their equations. Let $X \subseteq \mathbb{C}^N$ be an irreducible and reduced component of a variety, $\pi : X \to \mathbb{C}^n$ a finite projection, and $Y := \overline{\pi(X)}$ the Zariski closure of its image. A witness set for Y is encoded as a quadruple $(\mathcal{F}_X, \pi, \mathcal{L}', S')$ where $S' = \mathcal{L}' \cap Y$. Given a witness set $(\mathcal{F}_X, \mathcal{L}, S)$, we produce a witness set for Y by performing a linear homotopy from the points in S to the points $\pi^{-1}(S') = \pi^{-1}(\mathcal{L}') \cap X$ [8]. Since every map is an embedding followed by a projection, this means we can compute a witness set for the image of any map (Fig. 1).

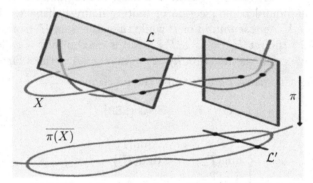

Fig. 1. Computing a witness set for a projection

2.3   The HS-Algorithm

Let $\mathcal{H} \subseteq \mathbb{C}^n$ be a degree d hypersurface defined by

$$f = \sum_{\alpha \in \mathcal{A}} c_\alpha x_1^{\alpha_1} \cdots x_n^{\alpha_n} \in \mathbb{C}[x_1, \ldots, x_n] \qquad c_\alpha \neq 0, \mathcal{A} \subseteq \mathbb{N}^n, |\mathcal{A}| < \infty$$

so that $\mathrm{New}(f)$ is the convex hull of the points in $\mathcal{A}$. Let $\omega \in \mathbb{R}^n$ be a direction, $a, b \in (\mathbb{C}^*)^n$, and consider the family of lines $\mathcal{L}_t$ parametrized by

$$s \xrightarrow{\mathbf{L}_t} (\ell_{(t,1)}(s), \ldots, \ell_{(t,n)}(s))$$

where $\ell_{(t,i)}(s) = t^{\omega_i}(a_i s - b_i)$. For any fixed t value, $f(\mathbf{L}_t)$ is a univariate polynomial in s whose solutions $p(t) := \{p_1(t), \ldots, p_d(t)\}$ correspond to intersection points of $\mathcal{H}$ and $\mathcal{L}_t$. We may write $f(\mathbf{L}_t)$ as

$$f(\ell_{(t,1)}, \ldots, \ell_{(t,n)}) = \sum_{\alpha \in \mathcal{A}} c_\alpha [t^{\omega_1}(a_1 s - b_1)]^{\alpha_1} \cdots [t^{\omega_n}(a_n s - b_n)]^{\alpha_n}$$

$$= \sum_{\alpha \in \mathcal{A}} t^{\langle \omega, \alpha \rangle}(a_1 s - b_1)^{\alpha_1} \cdots (a_n s - b_n)^{\alpha_n}$$

As $t \to \infty$, the terms $\mathcal{A}_\omega$ corresponding to points of $\mathcal{A}$ which maximize $\langle \omega, \alpha \rangle$ will dominate the behavior of the zeros and so the solutions $p(t)$ will converge to those of

$$f_\omega(\ell_{(t,1)}, \ldots, \ell_{(t,n)}) := \sum_{\alpha \in \mathcal{A}_\omega} c_\alpha (a_1 s - b_1)^{\alpha_1} \cdots (a_n s - b_n)^{\alpha_n}.$$

If $\mathcal{A}_\omega = \{\beta\}$ then f_ω is a monomial and so $f_\omega(\ell)$ has roots $\gamma_i := b_i/a_i$ where γ_i occurs with multiplicity β_i. If $|\beta| := \sum \beta_i$ is less than d, then there are $\beta_\infty := d - |\beta|$ points which have diverged towards infinity. One can see this by observing that if we began with the homogenization F, this would be the exponent of homogenizing variable in the term F_ω.

If ω exposes the entire polytope defined by $\mathcal{A}$, then the roots $p(t)$ remain constant as t varies since all $f(\mathbf{L}_t)$ are all scalar multiples of each other.

If ω exposes a proper non-trivial subset of $\mathcal{A}$, then there is more than one term in f_ω, but these terms will have a common factor of $\prod_{i=1}^n (a_i s - b_i)^{m_i}$ where the vector m is the coordinate-wise minimum of the points $\mathcal{A}_\omega$. Therefore, m_i roots will converge to γ_i and $m_\infty := \min_{\beta \in \mathcal{A}_\omega} (d - |\beta|)$ points will diverge to infinity. All other roots will converge to generic points in $\mathbb{C}$. These observations give rise to the HS-algorithm.

Algorithm 1 HS-Algorithm

Input:

- A witness set W for a hypersurface $\mathcal{H} \subseteq \mathbb{C}^n$
- A direction $\omega \in \mathbb{R}^n$

Output:

- $\mathcal{O}_{\widetilde{\mathrm{New}(\mathcal{H})}}(\omega)$

Steps:

1. Pick random $a, b \in \mathbb{C}^n$ and construct $\{\ell_{(t,i)}\}_{i=1}^n$ described above
2. Track the witness points in W to the intersection $\mathcal{H} \cap \mathcal{L}_1$
3. Initialize vector $\beta = \mathbf{0} \in \mathbb{N}^{n+1}$
4. Track the witness points in $\mathcal{H} \cap \mathcal{L}_t$ from $t = 1$ toward ∞
5. If none of the solutions move, return **EEP**
6. If a solution has converged, stop tracking it
   - If it has converged to some γ_i increment β_i by one
7. If a solution has diverged increment β_∞ by one
8. If all solutions have converged or diverged, return $\beta = (\beta_1, \ldots, \beta_n, \beta_\infty)$

---

3 Functionality

We have implemented the HS-algorithm in **NumericalNP.m2** via three main functions. Function 1, computes a witness set for the image of an irreducible and reduced variety $X \subseteq \mathbb{C}^N$ under a projection $\pi : \mathbb{C}^N \to \mathbb{C}^n$.

> **Function 1** `witnessForProjection`
> **Input:**
>
> - I: Ideal defining $X \subseteq \mathbb{C}^N$
> - ProjCoord: List of coordinates which are forgotten by π
> - OracleLocation (option): Path in which to create witness files
>
> **Output:** A subdirectory `/OracleLocation/WitnessSet` containing
>
> - witnessPointsForProj: Preimages of witness points of $\overline{\pi(X)}$
> - projectionFile: List of coordinates in ProjCoord
> - equations: List of equations defining $X' \subseteq X$ such that $\pi|_{X'}$ is finite and $\overline{\pi(X')} = \overline{\pi(X)}$

Function 2, `witnessToOracle`, creates all necessary **Bertini** files to track the witness set $\mathcal{H} \cap \mathcal{L}_t$ as $t \to \infty$ for any $\omega \in \mathbb{R}^n$. These files treat ω as a parameter so that the user only needs to produce these files once.

> **Function 2** `witnessToOracle`
> **Input:**
>
> - OracleLocation: Path containing the directory `/WitnessSet`
>
> **Optional Input:**
>
> - `PointChoice`: Prescribes a and b explicitly (see Algorithm 1)
> - `TargetChoice`: Prescribes targets b_i/a_i
> - `NPConfigs`: List of **Bertini** path tracking configurations
>
> **Output:**
>
> - A subdirectory `/OracleLocation/Oracle` containing all necessary files to run the homotopy described in Algorithm 1.

Function 2 by default chooses $a, b \in \mathbb{C}^n$ such that $\gamma_i := a_i/b_i$ are n-th roots of unity. One may choose to either specify a and b (`PointChoice`), or $\gamma_i := a_i/b_i$ (`TargetChoice`) or request that these choices are random. When random, the function ensures that the points γ_i are far from each other so that convergence to γ_i is easily distinguished from convergence to γ_j. **Bertini** is called to track the solutions in `/OracleLocation/WitnessSet` to points $\overline{\pi(X)} \cap \mathcal{L}_1$. These become start solutions to the homotopy described in Algorithm 1 with parameters ω and t. There are many numerical choices for **Bertini**'s native pathtracking algorithms which can be specified via `NPConfigs`.

Function 3 `oracleQuery`
Input:

- OracleLocation (Option): Location containing the directory `/Oracle`
- ω: A vector in $\mathbb{R}^n$

Optional Input:
- `Certainty` − `Epsilon` − `UseCauchy` − `MinTracks` − `MaxTracks`
- `StepResolution` − `MakeSageFile`
Output:

- $\mathcal{O}_{\widetilde{\mathrm{New}(\mathcal{H})}}(\omega)$ or `Reached MaxTracks`
- A subdirectory `/OracleLocation/OracleCalls/Call#` containing
  - `SageFile`: Sage code animating the paths $p(t)$
  - `OracleCallSummary`: a human-readable file summarizing the results

The fundamental function, `oracleQuery`, runs the homotopy in the HS-algorithm, monitors convergence, and outputs the result of the numerical oracle.

To monitor convergence of solutions $p(t)$ we track $t \to \infty$ in discrete steps. The option `StepResolution` specifies these t-step sizes. In each step, for each path $p_i(t)$, a numerical derivative is computed to determine convergence or divergence of the solution. If the solution is large and the numerical derivative exceeds $10^{\text{Certainty}}$ in two consecutive steps the path is declared to diverge, and if the numerical derivative is below $10^{-\text{Certainty}}$ in two consecutive steps the point is declared to converge. If `UseCauchy` is set to `true` a Cauchy loop is performed to corroborate convergence. If a converged point is at most `Epsilon` from some γ_i, then the software deems that it has converged to γ_i. When a point is declared to converge or diverge, it is not tracked further. The option `MaxTracks` allows the user to specify how long to wait for convergence of the paths $p(t)$. The rate of convergence is slower when ω is close to exposing a positive dimensional face of $\mathrm{New}(\mathcal{H})$, as illustrated by Fig. 2. This figure shows the Newton polytope of a plane sextic (see Example) and a graphic describing convergence rate of the algorithm on different directions $\omega \in S^1$: the black rays indicate that the algorithm returned `Reached MaxTracks` and the length of the green (grey) rays is proportional to the number of steps it took the algorithm to finish. One may also specify `MinTracks` which indicates the time step convergence should begin to be monitored. The option to create a Sage [12] animation (see Fig. 3) of the solution paths helps the user recognize pathological behavior in the numerical computations and fine tune parameters such as `Certainty` or `Epsilon` accordingly so that the software is more careful.

Example: Consider the curve in $X \subseteq \mathbb{C}^3$ defined by

$$I = \langle xyt - (x - y - t)^2 + 3x + t, x + y^2 + t^2 \rangle \subseteq \mathbb{C}[x, y, t]$$

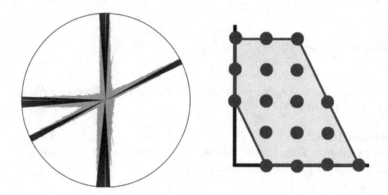

Fig. 2. (Left) Convergence of different ω for queryOracle on Newton polytope (Right) (Color figure online)

and let π be the projection forgetting the t coordinate. The following **Macaulay2** code computes a witness set for $\mathcal{C} := \overline{\pi(X)}$, prepares oracle files for the HS-algorithm and then runs the HS-algorithm in the direction $(3,2)$. The software returns $\{2,4,0\}$ indicating that $\mathrm{New}(\overline{\pi(X)})_{(3,2)} = (2,4)$.

```
i1: loadPackage("NumericalNP");
i2: R=CC[x,y,t];
i3: I=ideal(x*y*t-(x-y-t)^2+3*x+t,x+y^2+t^2);
i4: witnessForProjection(I,{2},OracleLocation=>"Example");
i5: witnessToOracle("Example") ;
i6: time oracleQuery({3,2},OracleLocation=>"Example",MakeSageFile=>true)
    -- used 0.178448 s
o6: {2,4,0}
```

The full Newton polytope of $\overline{\pi(X)}$ is displayed in Fig. 2 and snapshots of the Sage animation created by `queryOracle` are shown in Fig. 3. There, the red circles are centered at $\gamma_1 = 1$ and $\gamma_2 = -1$ and have radius `epsilon`.

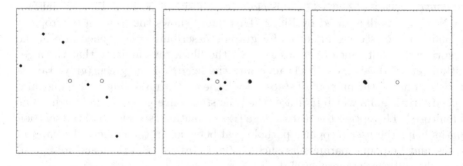

Fig. 3. Three snapshots of Sage animation from example with viewing window $[-4, 4]^2$ (Color figure online)

4   Applications

4.1   Algebraic Vision Tensor

The multiview variety X of a pinhole camera and a two slit camera is a hypersurface in the space of $3 \times 2 \times 2$ tensors given by the image of twelve particular minors of

$$[A \ B \ C] = \begin{bmatrix} a_{1,1} & a_{1,2} & a_{1,3} & b_{1,1} & b_{1,2} & c_{1,1} & c_{1,2} \\ a_{2,1} & a_{2,2} & a_{2,3} & b_{2,1} & b_{2,2} & c_{2,1} & c_{2,2} \\ a_{3,1} & a_{3,2} & a_{3,3} & b_{3,1} & b_{3,2} & c_{3,1} & c_{3,2} \\ a_{4,1} & a_{4,2} & a_{4,3} & b_{4,1} & b_{4,2} & c_{4,1} & c_{4,2} \end{bmatrix}.$$

We consider X as a subvariety of $\mathbb{P}^{11}$ given by the image of

$$F : \mathbb{C}^{28} \to \mathbb{C}^{12} \qquad\qquad [ABC] \xrightarrow{F} \{f_{i,j,k}\}_{i \in \{1,2,3\}, j,k \in \{1,2\}}$$

where $f_{i,j,k}$ is the minor not involving columns a_i, b_j, and c_k. This map has 17 dimensional fibres so `witnessForProjection` automatically slices $\mathbb{C}^{28}$ with 17 hyperplanes to compute a witness set for X which shows that $\deg(X) = 6$. Therefore, its defining polynomial has an *a priori* upper bound of 12376 terms. There is a group action of $G \cong S_3 \times S_2 \times S_2$ on $[ABC]$ permuting the a, b, and c columns appropriately. This extends to a transitive action on the coordinates of the Newton polytope. A few oracle calls quickly determine that New(X) is contained in a 7-dimensional subspace of $\mathbb{R}^{12}$ and only has 4 vertices and 2 facets up to the G-action. In total, New(X) has 60 vertices and 6 interior points. With only 66 possible terms, interpolation recovers the polynomial found in Proposition 7.5 of [11].

4.2   The Lüroth Polytope

A *Lüroth quartic* is a plane quartic which interpolates the ten intersection points of a configuration of five lines. The set of all Lüroth quartics is a hypersurface $\mathcal{H}$ of degree 54 in the 15 coefficients of a plane quartic called the *Lüroth hypersurface*. The group $G = S_3$ acts on the vertices of New$(\mathcal{H})$ by permuting the three indeterminants of a homogeneous quartic. A face of New$(\mathcal{H})$ was found in [9]. Using our software, we have rediscovered that New$(\mathcal{H})$ is 12-dimensional and have, so far, found 1743 vertices, belonging to $1, 1, 28,$ and 276 orbits of sizes $1, 2, 3,$ and 6 respectively.

Up-to-date computations regarding the Lüroth invariant as well as the package **NumericalNP.m2** can be found at the authors webpage [5].

References

1. Bates, D.J., Gross, E., Leykin, A., Israel Rodriguez, J.: Bertini for Macaulay2 (2013)
2. Bates, D.J., Hauenstein, J.D., Sommese, A.J., Wampler, C.W.: Bertini: software for numerical algebraic geometry. Available at bertini.nd.edu with permanent https://doi.org/10.7274/R0H41PB5
3. Bates, D.J., Hauenstein, J.D., Sommese, A.J., Wampler, C.W.: Numerically Solving Polynomial Systems with Bertini. SIAM, Philadelphia (2013)
4. Bernstein, D.: The number of roots of a system of equations. Funct. Anal. Appl. **9**, 183–185 (1975)
5. Brysiewicz, T.: Numerical computations of newton polytopes (2018). http://www.math.tamu.edu/~tbrysiewicz/NumericalNP
6. Emiris, I.A., Fisikopoulos, V., Konaxis, C., Penaranda, L.: An oracle-based, output-sensitive algorithm for projections of resultant polytopes. Int. J. Comput. Geom. Appl. **23**(04n05), 397 (2013)
7. Grayson, D.R., Stillman, M.E.: Macaulay2, a software system for research in algebraic geometry. http://www.math.uiuc.edu/Macaulay2/
8. Hauenstein, J.D., Sommese, A.J.: Witness sets of projections. Appl. Math. Comput. **217**(7), 3349–3354 (2010)
9. Hauenstein, J.D., Sottile, F.: Newton polytopes and witness sets. Math. Comput. Sci. **8**(2), 235–251 (2012)
10. Khovanskii, A.: Newton polyhedra (algebra and geometry). Amer. Math. Soc. Transl. **153**(2) (1992)
11. Ponce, J., Sturmfels, B., Trager, M.: Congruences and concurrent lines in multiview geometry. Adv. Appl. Math. **88**, 62–91 (2017)
12. Stein, W., et al.: Sage Mathematics Software (Version x.y.z). The Sage Development Team (2017). http://www.sagemath.org

On the Interference Problem for Ellipsoids: Experiments and Applications

Jorge Caravantes[1] and Laureano Gonzalez-Vega[2]([⊠])

[1] Universidad Complutense de Madrid, Madrid, Spain
jcaravan@mat.ucm.es
[2] Universidad de Cantabria, Santander, Spain
laureano.gonzalez@unican.es

Abstract. The problem of detecting when two moving ellipsoids overlap is of interest to robotics, CAD/CAM, computer animation, etc. By analysing symbolically the sign of the real roots of the characteristic polynomial of the pencil defined by two ellipsoids $\mathcal{A}$ and $\mathcal{B}$ we use and analyse the new closed formulae introduced in [9] characterising when $\mathcal{A}$ and $\mathcal{B}$ overlap, are separate and touch each other externally for determining the interference of two moving ellipsoids. These formulae involves a minimal set of polynomial inequalities depending only on the entries of the matrices A and B (defining the ellipsoids $\mathcal{A}$ and $\mathcal{B}$), need only to compute the characteristic polynomial of the pencil defined by A and B and do not require the computation of the intersection points between them. This characterisation provides a new approach for exact collision detection of two moving ellipsoids since the analysis of the univariate polynomials (depending on the time) in the previously mentioned formulae provides the collision events between them.

Keywords: Ellipsoids separation problem
Real quantifier elimination · Closed form solutions

1   Introduction

The problem of detecting the collisions or overlap of two ellipsoids is of interest to robotics, CAD/CAM, computer animation, etc., where ellipsoids are often used for modelling (or enclosing) the shape of the objects under consideration (see for example [6, 10–12, 14, 15]). The problem to be considered here is obtaining closed formulae characterising the separation by a plane of two ellipsoids in the three dimensional real affine space by using tools coming from Real Algebraic Geometry and Computer Algebra. Moreover this characterisation should

Both authors are partially supported by the Spanish Ministerio de Economía y Competividad and by the European Regional Development Fund (ERDF), under the project MTM2017-88796-P.

© Springer International Publishing AG, part of Springer Nature 2018
J. H. Davenport et al. (Eds.): ICMS 2018, LNCS 10931, pp. 89–97, 2018.
https://doi.org/10.1007/978-3-319-96418-8_11

provide easily the manipulation of the formulae for exact collision detection of two ellipsoids under rational motions (see [2, 3, 16]).

Note that the problem considered in this paper is not the computation of the intersection points between the two considered ellipsoids. This intersection problem can be solved by any numerical nonlinear solver or by "ad–hoc" methods. Nevertheless, the results later described can be used as a preprocessing step since any intersection problem is highly simplified if the structure of the intersection set is known in advance: i.e. nothing to compute if it is known that the considered ellipsoids are separated by a plane.

Our approach is based on the characterisation presented in [17] where the separation of two ellipsoids is determined by the sign of the real roots of the characteristic polynomial of the matrix pencil defined by the two considered ellipsoids.

The main result of this paper is to use and analyse the new formula, introduced in [9], characterizing when two ellipsoids are separate, overlap or touch each other externally improving the best existing one introduced in [13]. Compared with this solution, this approach presents, at least, three clear improvements:

1. Less polynomials are involved: the solution in [13] requires to deal with 5 polynomials and our solution does require only 4 polynomials.
2. Less arithmetic operations are required: 23 multiplications and 12 additions (against 28 multiplications and 12 additions in [13]).
3. The way the relative positions are characterized is simpler than in [13] (less sign conditions are involved).

These formulae generalise to the ellipsoids separation problem the approach introduced in [1, 7] to characterise in a similar way the relative positive positions of two ellipses in the plane. Moreover the approach presented in this paper is specially well suited for analysing the relative position of two ellipsoids depending on a parameter t (or two moving ellipsoids; see [4] for the similar problem for two moving ellipses).

This paper is divided into three sections. First one is devoted to show the approach to be used for solving the interference problem for ellipsoids. Second one shows how to compute the intervals where two moving ellipsoids do not overlap. Last one introduces some examples where the introduced method is applied.

2   Underlying Theory: Formulae for the Separation, Overlapping and External Touching of Ellipsoids

The equation of any quadric $\mathcal{A}$ in $\mathbb{R}^3$ can be written as

$$a_{11}x^2 + a_{22}y^2 + a_{33}z^2 + 2a_{12}xy + 2a_{13}xz + 2a_{23}yz$$
$$+2a_{14}x + 2a_{24}y + 2a_{34}z + a_{44} = 0$$

or in matricial form

$$\begin{bmatrix} x & y & z & 1 \end{bmatrix} A \begin{bmatrix} x \\ y \\ z \\ 1 \end{bmatrix} = 0$$

where A is the symmetric matrix:

$$A = \begin{bmatrix} a_{11} & a_{12} & a_{13} & a_{14} \\ a_{12} & a_{22} & a_{23} & a_{24} \\ a_{13} & a_{23} & a_{33} & a_{34} \\ a_{14} & a_{24} & a_{34} & a_{44} \end{bmatrix}.$$

Moreover, if the quadric is an ellipsoid then $\det(A) < 0$. We assume that the interior of the ellipsoid $\mathcal{A}$ is defined by $X^T A X < 0$.

We are going to consider three possible configurations for two ellipsoids when regarded as a solid bounded by the boundary surface $X^T A X = 0$: separation, overlapping and external touching. Two ellipsoids are separate if they are separated by a plane not touching the ellipsoids, are overlapping if their interiors share a common point and are touching each other externally if they are separated by a plane tangent to both ellipsoids.

Given two ellipsoids $\mathcal{A} : X^T A X = 0$ and $\mathcal{B} : X^T B X = 0$, their characteristic equation (or polynomial) is defined as

$$f(\lambda) = \det(\lambda A + B) = \det(A)\lambda^4 + \ldots + \det(B)$$

which is a quartic polynomial in λ with real coefficients.

The characterization of the relative position of two ellipsoids in terms of the sign of the real roots of their characteristic equation was introduced by [17].

Theorem 1. *Let $\mathcal{A}$ and $\mathcal{B}$ be two ellipsoids with the characteristic equation $f(\lambda)$. Then:*

1. *The characteristic equation $f(\lambda)$ always has at least two negative roots.*
2. *$\mathcal{A}$ and $\mathcal{B}$ are separated if and only if $f(\lambda)$ has two distinct positive roots.*
3. *$\mathcal{A}$ and $\mathcal{B}$ touch each other externally if and only if $f(\lambda)$ has a positive double root.*

In what follows, if $\mathcal{A}$ and $\mathcal{B}$ are two ellipsoids, we will turn monic its characteristic polynomial and will be denoted by

$$f(\lambda) = \lambda^4 + a\lambda^3 + b\lambda^2 + c\lambda + d.$$

Since $\mathcal{A}$ and $\mathcal{B}$ are ellipsoids and $d = \det(B)/\det(A)$, we have $d > 0$.

This theorem moves to a simpler one the quantifier elimination problem over the reals we want to solve: looking for the conditions ellipsoids $\mathcal{A}$ and $\mathcal{B}$ must verify in order they do not share an interior common point. The problem we will solve in the most efficient possible way will be the determination of the conditions the coefficients of $f(\lambda)$ must verify in order to have two different positive real roots (and knowing in advance that it is the characteristic equation of two ellipsoids implying that has, at least, two negative real roots and that $d > 0$).

Next theorems, proven in [9], present the formulae to be used for characterising when two ellipsoids are separate, overlap or touch each other externally. The characterisation uses the subresultant polynomials of $f(\lambda)$ and $f'(\lambda)$ (index 0 and 1) $\Delta(f)$, $\Delta_1(f)$ and $\Delta_{10}(f)$:

$$\Delta(f) = -27d^2a^4 - 4a^3c^3 + 18a^3dcb + a^2c^2b^2 - 6a^2c^2d - 4a^2b^3d$$
$$+144a^2bd^2 - 192ad^2c - 80ab^2cd,$$
$$\Delta_1(f) = -6a^3c + 2a^2b^2 - 12a^2d + 28abc - 8b^3 + 32bd - 36c^2,$$
$$\Delta_{10}(f) = -9a^3d + a^2bc + 32abd + 3ac^2 - 4b^2c - 48cd.$$

Together with Descartes' law of signs, they characterise the sign of the real roots of $f(\lambda)$ according to the conditions in Theorem 1. Subresultants and Descartes' law of signs were also used in [13] to solve the problem at hand but the derived formulae are different from those introduced in [9] and used here.

For a sequence of real numbers $b_0, b_1, \ldots, b_n$, $\mathbf{Var}(b_0, b_1, \ldots, b_n)$ will denote the number of sign changes in $b_0, b_1, \ldots, b_n$ after dropping the zeros in the sequence.

Theorem 2. *Let A and B be two ellipsoids with the characteristic equation*

$$f(\lambda) = \lambda^4 + a\lambda^3 + b\lambda^2 + c\lambda + d.$$

A and B are separate if and only if

1. $\Delta(f) > 0$, $\mathbf{Var}(1, a, b, c, 1) = 2$, *or*
2. $\Delta(f) = 0$, $\Delta_1(f) > 0$, $\Delta_{10}(f) > 0$, $a\Delta_1(f) - 2\Delta_{10}(f) < 0$.

Theorem 3. *Let A and B be two ellipsoids with the characteristic equation*

$$f(\lambda) = \lambda^4 + a\lambda^3 + b\lambda^2 + c\lambda + d.$$

A and B touch each other externally if and only if

1. $\Delta(f) = 0$, $\Delta_{10}(f)\Delta_1(f) < 0$, *or*
2. $\Delta(f) = 0$, $\Delta_1(f) = 0$, $\mathbf{V}(1, a, b, c, 1) = 2$.

Theorem 4. *Let A and B be two ellipsoids with the characteristic equation*

$$f(\lambda) = \lambda^4 + a\lambda^3 + b\lambda^2 + c\lambda + d.$$

A and B overlap if and only if

1. $\Delta(f) < 0$, *or*
2. $\Delta(f) > 0$, $\mathbf{Var}(1, a, b, c, 1) = 0$, *or*
3. $\Delta(f) = 0$, $\Delta_1(f) = 0$, $\mathbf{V}(1, a, b, c, 1) = 0$, *or*

4. $\Delta(f) = 0$, $\Delta_1(f) < 0$, $\Delta_{10}(f) < 0$, or
5. $\Delta(f) = 0$, $\Delta_1(f) > 0$, $\Delta_{10}(f) > 0$, $a\Delta_1(f) - 2\Delta_{10}(f) > 0$.

Compared with the best solution dealing with this problem (see [13]), our approach presents, at least, two clear improvements:

1. Less polynomials are involved: we need to deal only with $\Delta(f)$, $\Delta_1(f)$, $\Delta_{10}(f)$ and $a\Delta_1(f) - 2\Delta_{10}(f)$, while the solution in [13] requires to use two additional polynomials ($\Delta_2(f) = 3a^2 - 8b$ and $\Delta_{20}(f) = ac - 16d$).
2. The way the relative position is determined is simpler than in [13]: starting with the (strict) sign of $\Delta(f)$ and generically ending with the computation of $\mathbf{V}(1, a, b, c, 1)$. When $\Delta(f) = 0$, we continue by analysing the (strict) signs of $\Delta_1(f)$, $\Delta_{10}(f)$ and $a\Delta_1(f) - 2\Delta_{10}(f)$. And when $\Delta_1(f) = 0$, we end by computing $\mathbf{Var}(1, a, b, c, 1)$.

3 Application: Moving Ellipsoids

The formulae presented in the previous section can be applied to study the case of two ellipsoids depending on one parameter. For example, given two moving ellipsoids $\mathcal{A}(t) : X^T A(t)X = 0$ and $\mathcal{B}(t) : X^T B(t)X = 0$, respectively, $\mathcal{A}(t)$ and $\mathcal{B}(t)$ are said to be collision-free if $\mathcal{A}(t)$ and $\mathcal{B}(t)$ are separate for all t in a given interval; otherwise $\mathcal{A}(t)$ and $\mathcal{B}(t)$ collide.

The characteristic equation of $\mathcal{A}(t)$ and $\mathcal{B}(t)$,

$$f(\lambda; t) := \det\left(\lambda A(t) + B(t)\right) = 0$$

is a degree four polynomial in λ with real coefficients depending on the parameter t. At any time t_0, if $\mathcal{A}(t_0)$ and $\mathcal{B}(t_0)$ are separate then $f(\lambda; t_0)$ has two distinct positive roots; otherwise $\mathcal{A}(t_0)$ and $\mathcal{B}(t_0)$ are either touching externally or overlapping, and $f(\lambda; t_0)$ has a double positive root or no two positive roots, respectively.

In order to determine the relative position of the considered ellipsoids, the study of the sign behaviour of the roots of the characteristic polynomial for all the possible values of the parameter t is required. This is accomplished by using the formulae introduced in the previous section producing in an automatic manner (and in terms of t) the behaviour of the sign of the real roots of $f(t; \lambda)$.

We start by computing $\Delta(f)$ which is a polynomial in t. If it is not identically zero then we determine the real roots $\gamma_1 < \ldots < \gamma_s$ of the polynomials $a(t)$, $b(t)$, $c(t)$ and $\Delta(f)$. Next, we analyse the partition of $\mathbb{R}$ provided by them and keep those points and intervals where

1. $\Delta(f) > 0$, $\mathbf{Var}(1, a, b, c, 1) = 2$, or
2. $\Delta(f) = 0$, $\Delta_1(f) > 0$, $\Delta_{10}(f) > 0$, $a\Delta_1(f) - 2\Delta_{10}(f) < 0$.

If $\Delta(f)$ is identically zero and $\Delta_1(f)$ is not identically zero then we determine its real roots $\tau_1 < \ldots < \tau_q$ and keep those intervals where $\Delta_1(f) > 0$, $\Delta_{10}(f) > 0$

and $a\Delta_1(f) - 2\Delta_{10}(f) < 0$. In this way we obtain those values of t such that $\mathcal{A}(t)$ and $\mathcal{B}(t)$ are separate.

If $\Delta_1(f)$ is identically zero then, for every t, $\mathcal{A}(t)$ and $\mathcal{B}(t)$ touch each other externally or overlap. The first case arises when $\mathbf{Var}(1, a, b, c, 1) = 2$ and the second one when $\mathbf{Var}(1, a, b, c, 1) = 0$. In this situation it is enough to analyse the partition of $\mathbb{R}$ provided by the real roots of $a(t)$, $b(t)$ and $c(t)$.

4   Functionality

The formulae presented in the previous section allows to consider in a very efficient way the problem of determining the non interference intervals for two moving ellipsoids (see [2,3]). First experiments in Maple shows a very good practical behaviour. According to Theorem 2, Theorem 3 and Sect. 3, we need to compute the following polynomials:

$$\Delta(f),\ \Delta_1(f),\ \Delta_{10}(f),\ a\Delta_1(f) - 2\Delta_{10}(f)\ .$$

Following [13] (and [5]) we proceed in the following way by denoting first:

$$\overline{b} = -\frac{a}{4},\ \overline{c} = \frac{b}{6},\ \overline{d} = -\frac{c}{4},\ \overline{e} = d$$

and determining:

$$D_2 = \overline{b}^2 - \overline{c} \quad W_1 = \overline{d} - \overline{b}\,\overline{c} \quad T = -9W_1^2 + 27D_2D_3 - 3W_3D_2$$
$$D_3 = \overline{c}^2 - \overline{b}\,\overline{d} \quad W_3 = \overline{e} - \overline{b}\,\overline{d} \quad A = W_3 + 3D_3$$
$$B = -\overline{d}W_1 - \overline{e}D_2 - \overline{c}D_3$$
$$T_2 = AW_1 - 3\overline{b}B$$
$$D_1 = A^3 - 27B^2$$

Finally the searched polynomials are given by

$$\Delta(f) = D_1,\ \Delta_1(f) = T,\ \Delta_{10}(f) = T_2,\ a\Delta_1(f) - 2\Delta_{10}(f) = aT - 2T_2.$$

The above expressions take 23 multiplications and 12 additions (agains the 28 multiplications and the 12 additions in [13]).

Example 1. Let $\mathcal{A}(t)$ and $\mathcal{B}(t)$ be two moving ellipsoids defined by

$$t^4 - 2t^2x + 2t^2 - 2ty - 2tz + x^2 + y^2 + z^2 = 1$$

and

$$x^2\left(t^4 + 2t^2 + 1\right) + t^2y^2\left(t^4 + 2t^2 + 1\right) + t^6 - 2zt^4 + t^2z^2 = t^2$$

respectively, where $t \in \mathbb{R}$. In this case $\Delta(f)$ is a degree 76 polynomial in t with only three different real roots, $\gamma_1 = -0.8195884928$, $\gamma_2 = 0$ and $\gamma_3 = 1.035723920$. The real roots of $a(t)$, $b(t)$ and $c(t)$ are:

$$a(t) : \gamma_4 = -0.9389188732,\ \gamma_5 = 1.072516819$$
$$b(t) : \gamma_6 = -0.7654533193,\ \gamma_2 = 0,\ \gamma_7 = 0.9435132848$$
$$c(t) : \gamma_8 = -0.7536913873,\ \gamma_2 = 0,\ \gamma_9 = 1.070639969$$

Analysing the partition of $\mathbb{R}$ given by the γ_i and the τ_j, we conclude that $\Delta f > 0$ and $\mathbf{Var}(1, a, b, c, 1) = 2$ in $(-\infty, \gamma_1)$ and $(\gamma_3, +\infty)$. By using Theorem 3, this implies that $\mathcal{A}(t)$ and $\mathcal{B}(t)$ are separate when t belongs to $(-\infty, \gamma_1)$ or $(\gamma_3, +\infty)$. Since $\Delta(f) < 0$ in (γ_1, γ_3) we conclude that $\mathcal{A}(t)$ and $\mathcal{B}(t)$ overlap when $t \in (\gamma_1, \gamma_3)$. By continuity, we conclude that $\mathcal{A}(\gamma_i)$ and $\mathcal{B}(\gamma_i)$ $(i = 1, 3)$ touch each other externally. Figure 1 shows images of $\mathcal{A}(t)$ and $\mathcal{B}(t)$ for different values of t.

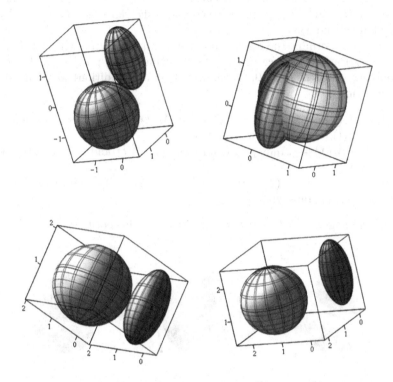

Fig. 1. Upper left: $\mathcal{A}(\gamma_1)$ and $\mathcal{B}(\gamma_1)$. Upper right: $\mathcal{A}(\frac{1}{2})$ and $\mathcal{B}(\frac{1}{2})$. Down left: $\mathcal{A}(\gamma_3)$ and $\mathcal{B}(\gamma_3)$. Down right: $\mathcal{A}(\frac{5}{4})$ and $\mathcal{B}(\frac{5}{4})$.

Example 2. Let $\mathcal{A}(t)$ and $\mathcal{B}(t)$ be two moving ellipsoids defined by

$$\frac{(x + 12t - 11)^2}{4} + y^2 + z^2 = 1$$

and

$$\frac{(x - 3)^2}{4} + (y - 4t + 2)^2 + (z - 4t + 4)^2 = 1,$$

respectively, where $t \in [0, 1]$ (example extracted from [13]). In this case

$$\begin{aligned} f(\lambda; t) &= \lambda^4 + \left(-68t^2 + 96t - 32\right)\lambda^3 + \left(-136t^2 + 192t - 66\right)\lambda^2 \\ &\quad + \left(-68t^2 + 96t - 32\right)\lambda + 1 \\ &= (\lambda + 1)^2(\lambda^2 + (-68t^2 + 96t - 34)\lambda + 1), \end{aligned}$$

$\Delta(f) \equiv 0$ and $\Delta_1(f) = \Delta_{10}(f)$ is a degree 8 polynomial in t with only two different real roots, $\gamma_1 = 0.5395042868$ and $\gamma_2 = 0.8722604191$. This implies, by using Theorem 3, that $\mathcal{A}(\gamma_i)$ and $\mathcal{B}(\gamma_i)$ $(i = 1, 2)$ touch each other externally because

$$\mathbf{Var}(1, a(\gamma_i), b(\gamma_i), c(\gamma_i), 1) = \mathbf{Var}(1, 0, -2, 0, 1) = 2.$$

In this particular case, using Theorem 3.10 in [13] to determine the relative position of $\mathcal{A}(t_i)$ and $\mathcal{B}(t_i)$ requires to compute the signs of $\Delta_2(f)$ and $\Delta_{20}(f)$ evaluated at γ_1 and γ_2.

In order to determine the relative position of $\mathcal{A}(t)$ and $\mathcal{B}(t)$ when t belongs to $[0, \gamma_1)$, (γ_1, γ_2) and $(\gamma_2, 1]$ we proceed by selecting a test point at each interval and applying, since $\Delta(f) = 0$, Theorem 2 (second condition) and Theorem 4 (forth condition):

- $t = 1/4$: $\Delta_1(f) = \Delta_{10}(f) > 0$ and $a(1/4)\Delta_1(f) - 2\Delta_{10}(f) < 0$ implies that $\mathcal{A}(t)$ and $\mathcal{B}(t)$ are separate when $t \in [0, \gamma_1)$.
- $t = 7/10$: $\Delta_1(f) = \Delta_{10}(f) < 0$ implies that $\mathcal{A}(t)$ and $\mathcal{B}(t)$ overlap when $t \in (\gamma_1, \gamma_2)$.
- $t = 9/10$: $\Delta_1(f) = \Delta_{10}(f) > 0$ and $a\Delta_1(f) - 2\Delta_{10}(f) < 0$ implies that $\mathcal{A}(t)$ and $\mathcal{B}(t)$ are separate when $t \in (\gamma_2, 1]$.

Fig. 2 shows images of $\mathcal{A}(t)$ and $\mathcal{B}(t)$ for different values of t.

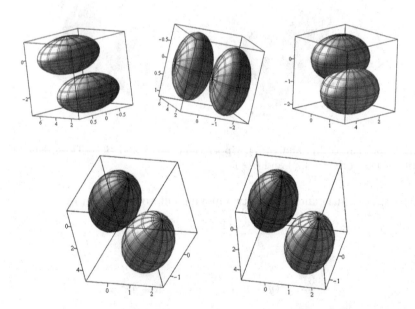

Fig. 2. Upper left: $\mathcal{A}(\frac{1}{4})$ and $\mathcal{B}(\frac{1}{4})$. Upper center: $\mathcal{A}(\gamma_1)$ and $\mathcal{B}(\gamma_1)$. Upper right: $\mathcal{A}(\frac{7}{10})$ and $\mathcal{B}(\frac{7}{10})$. Down left: $\mathcal{A}(\gamma_2)$ and $\mathcal{B}(\gamma_2)$. Down right: $\mathcal{A}(\frac{9}{10})$ and $\mathcal{B}(\frac{9}{10})$.

The main difficulty of this approach relies on the degree of $\Delta(f)$ but there are very efficient software packages computing very fast the real roots of the polynomial arising as the discriminant of $f(t; \lambda) = \det(\lambda A(t) + B(t))$.

References

1. Alberich-Carramiñana, M., Elizalde, B., Thomas, F.: New algebraic conditions for the identification of the relative position of two coplanar ellipses. Comput. Aided Geom. Des. **54**, 35–48 (2017)
2. Choi, Y.-K., Chang, J.-W., Wang, W., Kim, M.-S., Elber, G.: Continuous collision detection for ellipsoids. IEEE Trans. Visual. Comput. Graphics **15**, 311–324 (2009)
3. Choi, Y.-K., Kim, M.-S., Wang, W.: Exact collision detection of two moving ellipsoids under rational motions. In: 2003 IEEE International Conference on Robotics & Automation, pp. 349–354. IEEE Press, New York (2003)
4. Choi, Y.-K., Wang, W., Liu, Y., Kim, M.-S.: Continuous collision detection for two moving elliptic disks. IEEE Trans. Rob. **22**, 213–224 (2006)
5. Emiris, I.Z., Tsigaridas, E.P.: Real algebraic numbers and polynomial systems of small degree. Theor. Comput. Sci. **409**, 186–199 (2008)
6. Englert, C., Ferrando, J., Nordström, K.: Constraining new resonant physics with top spin polarisation information. Eur. Phys. J. C **77**, 407 (2017)
7. Etayo, F., Gonzalez-Vega, L., del Rio, N.: A new approach to characterizing the relative position of two ellipses depending on one parameter. Comput. Aided Geom. Des. **23**, 324–350 (2006)
8. Gonzalez-Vega, L., Mainar, E.: Solving the separation problem for two ellipsoids involving only the evaluation of six polynomials. In: Milestones in Computer Algebra-MICA (2008)
9. Caravantes, J., Gonzalez-Vega, L.: On the interference problem for ellipsoids: new closed form solutions. In: preparation (2018)
10. Gupta, A., Udupa, D.V., Topkar, A., Sahoo, N.K.: Astigmatic multipass cell with cylindrical lens. J. Opt. **46**, 324–330 (2017)
11. Hughes, G.B., Chraibi, M.: Calculating ellipse overlap areas. Comput. Vis. Sci. **15**, 291–301 (2012)
12. Hyun, D.-E., Yoon, S.-H., Kim, M.-S., Jüttler, B.: Modeling and deformation of arms and legs based on ellipsoidal sweeping. In: 11th Pacific Conference on Computer Graphics and Applications, pp. 204–212. IEEE Press, New York (2003)
13. Jia, X., Choi, Y.-K., Mourrain, B., Wang, W.: An algebraic approach to continuous collision detection for ellipsoids. Comput. Aided Geom. Des. **28**, 164–176 (2011)
14. Lee, B.H., Jeon, J.D., Oh, J.H.: Velocity obstacle based local collision avoidance for a holonomic elliptic robot. Auton. Rob. **41**, 1347–1363 (2017)
15. Lester, C.G., Nachman, B.: Bisection-based asymmetric M_{T2} computation: a higher precision calculator than existing symmetric methods. J. High Energy Phys. **3**, 100 (2015)
16. Wang, W., Choi, Y.-K., Chan, B., Kim, M.-S., Wang, J.: Efficient collision detection for moving ellipsoids using separating planes. Computing **72**, 235–246 (2004)
17. Wang, W., Wang, J., Kim, M.-S.: An algebraic condition for the separation of two ellipsoids. Comput. Aided Geom. Des. **18**, 531–539 (2001)

Efficient Computation of Squarefree Separator Polynomials

Michela Ceria[1(⊠)], Teo Mora[2], and Andrea Visconti[1]

[1] Department of Computer Science, Università degli Studi di Milano, Milan, Italy
michela.ceria@gmail.com, andrea.visconti@unimi.it
[2] Department of Mathematics, University of Genoa, Genoa, Italy
theomora@disi.unige.it

Abstract. Given a finite set of distinct points, a separator family is a set of polynomials, each one corresponding to a point of the given set, such that each of them takes value one at the corresponding point, whereas it vanishes at any other point of the set. Separator polynomials are fundamental building blocks for polynomial interpolation and they can be employed in several practical applications. Ceria and Mora recently developed a new algorithm for squarefree separator polynomials. The algorithm employs as a tool the point trie structure, first defined by Felszeghy-Ráth-Rónyai in their Lex game algorithm, which gives a compact representation of the relations among the points' coordinates. In this paper, we propose a fast implementation in C of the aforementioned algorithm, based on an efficient storing and visiting of the point trie. We complete the implementation with tests on some sets of points, giving different configurations of the corresponding tries.

Keywords: Separator polynomials · Point trie

1 Introduction

Given a finite set of distinct points $\mathbf{X} := \{P_1, ..., P_N\} \subset \mathbf{k}^n$, separator polynomials for $\mathbf{X}$ are polynomials $Q_1, ..., Q_N \in \mathbf{k}[x_1, ..., x_n]$ such that $\forall 1 \leq i, j \leq n$, $Q_i(P_j) = \delta_{i,j}$.

They have many applications in all fields of science, since they are the building blocks for polynomial interpolation. They are usually computed by means of some version Moeller algorithm [5,6], which gives also the whole Groebner basis for the ideal $I(\mathbf{X})$ of the points. The currently available implementations of Moeller algorithm have complexity $\mathcal{O}(n^2 N^3)$ (see [7, Vol. 2, 29.4.2]); if the improvement by Lundqvist would have been implemented (it is still not available) we would have complexity $\mathcal{O}(min(N, n)N^3 + nN^2)$. There are also some formulas to compute such polynomials [2,4], but, as remarked in [4], they add redundancy to the polynomials, which can be removed after computing them.

In [1], the authors developed an algorithm, based on Felszeghy-Ráth-Rónyai's point trie, which directly computes the separator polynomials, avoiding the

© Springer International Publishing AG, part of Springer Nature 2018
J. H. Davenport et al. (Eds.): ICMS 2018, LNCS 10931, pp. 98–104, 2018.
https://doi.org/10.1007/978-3-319-96418-8_12

redundancy so not needing to prune it afterwards. The complexity of the algorithm is $\mathcal{O}(N^2 log(N)n + N \min(N, nr))$.

The aim of this paper is to describe an efficient implementation of the algorithm in [1] which leans on an efficient storing and visiting of the point trie. We complete the implementation with tests on some sets of points, giving different configurations of the corresponding tries.

2   Notation

Throughout this paper we mainly follow the notation of [7]. We denote by $\mathcal{P} := \mathbf{k}[x_1, ..., x_n]$ the ring of polynomials in n variables with coefficients in the field $\mathbf{k}$.

Let $\mathbf{X} = \{P_1, ..., P_N\} \subset \mathbf{k}^n$ be a finite set of distinct points

$$P_i := (a_{1,i}, ..., a_{n,i}), \, i = 1, ..., N.$$

We call

$$I(\mathbf{X}) := \{f \in \mathcal{P} : \, f(P_i) = 0, \, \forall i\},$$

the *ideal of points* of $\mathbf{X}$.

Finally we recall some definitions from Graph Theory, following the notation of [2].

Definition 1. *We call* tree *a connected acyclic graph. A* rooted tree *is a tree where a special* vertex *(or node) called* root *is singled out.*

We say that a vertex is on the h-th *level* of the tree if its distance from the root is h, i.e. we have to walk on h edges to come from the root to the given vertex. If v is a vertex different from the root, and u is the vertex preceding v on the path from the root, then u is the *parent* of v and v is a *child* of u. Two vertices with the same parent are called *siblings*. If v is a vertex different from the root and u is on the path from v to the root, then u is an ancestor of v and v is a descendant of u. Clearly the root has no parent. We call *leaves* all the vertices having no children and we say that a *branch* is a path from the root to a leaf.

We consider always trees where all branches have the same length. The vertices lying in the last level of the tree coincide with the *leaves*; there are no vertices of the tree under them.

3   Separator Polynomials

In this section, following the notation of [4], we define separator polynomials.

Definition 2. *A family of separators for a finite set of distinct points* $\mathbf{X} = \{P_1, ..., P_N\}$ *is a set* $Q = \{Q_1, ..., Q_N\}$ *s.t.* $Q_i(P_j) = \delta_{ij}, \, 1 \leq i, j \leq N,$ *where* δ_{ij} *denotes the Kronecker delta.*

Separators are useful building blocks for polynomial interpolation, in the sense that, every time one has to find a polynomial $p \in \mathbf{k}[x_1, .., x_n]$ such that $p(P_i) = b_i$ for $b_i \in \mathbf{k}$, $1 \leq i \leq N$, it is possible to find it by computing a separator family for $\mathbf{X}$ and setting

$$p(x_1, ..., x_n) = \sum_{i=1}^{N} b_i Q_i(x_1, ..., x_n).$$

We denote the points in $\mathbf{X}$, as in [1], so by $P_i := (a_{1,i}, ..., a_{n,i})$, $i = 1, ..., N$, and we define the witness matrix $C = (c_{i,j})$ [4], as the symmetric matrix s.t., for $i, j \in \{1, ..., N\}$, $c_{i,j} = 0$ if $i = j$ and if $i \neq j$, $c_{i,j} = \min\{h : 1 \leq h \leq n$ s.t. $a_{h,i} \neq a_{h,j}\}$. In other words, the witness matrix represents the minimal index h such that two points share the first $1, ..., h-1$ coordinates, but they have different h-coordinate. Using this matrix and the coordinates of the points, we can compute the polynomials we will use as constituting factors for our separator polynomials:

$$p_{i,j}^{[c_{i,j}]} = \frac{x_{c_{i,j}} - a_{c_{i,j},j}}{a_{c_{i,j},i} - a_{c_{i,j},j}}$$

In the paper [4], separator polynomials are built with the following variation of Lagrange's formula:

$$R_i = \prod_{i \neq j} \frac{x_{c_{i,j}} - a_{c_{i,j},j}}{a_{c_{i,j},i} - a_{c_{i,j},j}} = \prod_{j \neq i} p_{i,j}^{[c_{i,j}]}.$$

Then, it is observed that repeated factors do not affect the values taken by the polynomials R_i on the points of $\mathbf{X}$, and so the repeated factors are indicated as useless.

In the next section, we show how to compute *directly* the squarefree versions of these polynomials.

4 An Algorithm for Computing Separator Polynomials

In this section, following [1], we show how it is possible to compute directly squarefree separator polynomials, via a purely combinatorial algorithm. Our tool is the *point trie*, defined in [2] and presented in details also in [4].

Definition 3. *A trie is a rooted tree s.t. there is a symbol from a fixed alphabet, written on each edge.*

We use a trie of this kind, called *point trie*, to represent the points of the set $\mathbf{X}$ and the reciprocal relations among their coordinates. In particular, we label both the nodes and the edges:

- each edge is labelled by a coordinate; in particular the i-th coordinates are those labelling edges connecting nodes at levels $i - 1$ and i;

– the nodes, denoted by $v_{i,u}$, contain as label sets $V_{i,u}$ of indices, identifying the points whose $1...i$-th coordinates coincide (at least until level i). If for some i, u, $|V_{i,u}| \geq 2$ we call its elements *twin points*.

The trie is constructed iteratively on the points, appending to the trie the branches corresponding to the points one by one, as shown in the Fig. 1, that refers to the set $\mathbf{X} = \{P_1 = (1,0,0), P_2 = (0,1,0), P_3 = (1,1,2), P_4 = (1,0,3)\}$ of [1].

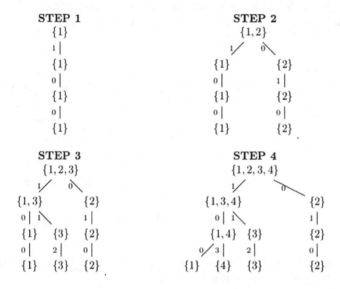

Fig. 1. The trie construction

Let now give a description of the algorithm, following [1]; in the next sections, we will give a closer look to the implementation.

If the given set $\mathbf{X}$ is composed by only one point $\mathbf{X} = \{P_1\}$, then the separator polynomial is $Q_1 = 1$. Suppose now to know the separator family $\{Q_1, ..., Q_{N-1}\}$ for $\{P_1, ..., P_{N-1}\}$ and to add the point P_N getting $\mathbf{X} = \{P_1, ..., P_N\}$. We compute the separator family $\{Q'_1, ..., Q'_N\}$ for $\mathbf{X}$, by computing the new polynomial Q_N, associated to P_N and by updating $Q_1, ..., Q_{N-1}$, making them fulfill Definition 2 for the whole $\mathbf{X}$:

1. set $Q'_N = 1$;
2. $\forall j = 1, ..., n$ (the index j represent a level of the trie, i.e. a variable, so we are actually performing a pre-order walk on the trie), consider the (unique) node $v_{j,u}$ with $N \in V_{j,u}$.
3. $\forall v_{j,u'}$, sibling of $v_{j,u}$, pick some $\bar{i} \in V_{j,u'}$ and set $Q'_N = Q'_N p^{[j]}_{N,\bar{i}}$;
4. if, at level j, N has no twin points, i.e. if $|V_{j,u}| = 1$, then for each sibling $v_{j,u'}$, for each $i \in V_{j,u'}$, we set $Q'_i = Q_i p^{[j]}_{i,N}$.

Once concluded the above procedure, if, for some $1 \leq h \leq N$ a separator polynomial Q_h, has *not* been modified by the above steps, we set $Q'_h = Q_h$, getting a separator family $\{Q'_1, ..., Q'_N\}$ for $\mathbf{X} = \{P_1, ..., P_N\}$.

5   How to Implement the Algorithm

In this section, we give some concrete details on the implementation and we provide some results of our testing activities.

First of all, Fig. 2 shows a toy example and represents both the trie construction and the resulting separator polynomials for the set $\mathbf{X} = \{P_1 = (0, 0), P_2 = (1, 2), P_3 = (4, 2), P_4 = (1, 3), P_5 = (7, 4)\}$:

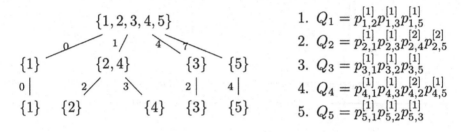

1. $Q_1 = p_{1,2}^{[1]} p_{1,3}^{[1]} p_{1,5}^{[1]}$
2. $Q_2 = p_{2,1}^{[1]} p_{2,3}^{[1]} p_{2,4}^{[2]} p_{2,5}^{[2]}$
3. $Q_3 = p_{3,1}^{[1]} p_{3,2}^{[1]} p_{3,5}^{[1]}$
4. $Q_4 = p_{4,1}^{[1]} p_{4,3}^{[1]} p_{4,2}^{[2]} p_{4,5}^{[1]}$
5. $Q_5 = p_{5,1}^{[1]} p_{5,2}^{[1]} p_{5,3}^{[1]}$

Fig. 2. A toy example: trie and separator polynomials

This trie is implemented in C using structs and pointers [8]. A graphical representation of its memory allocation is shown in Fig. 3

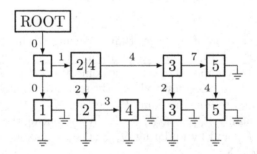

Fig. 3. Representation of the implemented trie for the toy example

We choose this approach because it minimizes the number of pointers allocated for each node and it also provides the possibility to add nodes dynamically at runtime. Moreover, the approach adopted avoids to store useless nodes, keeping, for example, the twin nodes as a part of a single node. The advantages of this approach can be appreciated as soon as the number of points grows consistently.

Our testing activities has been executed on point sets over some fields of the form $\mathbf{k} = \mathbb{F}_{2^m}$, $m \in \mathbb{N}$. For simplicity, our implementation treats the elements of $\mathbb{F}_{2^m}$ as *positive integers*. In particular we

- fix a primitive element $\alpha \in \mathbb{F}_{2^m}$;
- set $\mathbb{F}_{2^m} = \{0, \alpha, \alpha^2, ..., \alpha^{2^m-1} = 1\}$;
- identify α^i with i.

Notice that the element 0 is actually identified with 0, whereas since $1 = \alpha^0 = \alpha^{2^m-1}$, we associate $2^m - 1$ to the element $1 \in \mathbb{F}_{2^m}$.

In order to evaluate the performance of the algorithm, we run our code on a laptop equipped with an Intel Core i7-7700HQ processor — cache 6 MB, base frequency 2.8 GHz, maximum frequency 3.8 GHz, 4 cores, 8 threads — and 32 GB of RAM. The operative system installed is Kubuntu 16.04.

In our testing activities, we generate points with three and four coordinates, which give different configurations of the trie, but it is trivial to extend it to more coordinates. The computational time spent to construct the trie and to compute the separator polynomials runs between 0.01 s — best case, 1,024 points, three coordinates — and 6 min. — worse case, 65,536 points, four coordinates (see Table 1 for more details).

Table 1. Time spent to compute the separator polynomials

Number of points	Number of coordinates	Time spent (seconds)
1,024	3	0.01
4,096	3	1.02
12,341	3	2.68
16,384	3	8.41
65,536	3	240.52
512	4	0.01
65,536	4	373.37

6   Conclusions

In this paper, after recalling the definition an the importance of separator polynomials for interpolation, we have shown how to implement the algorithm introduced in [1] for computing them directly in a squarefree and redundancy-free way, which does not require pruning useless multiplicative factors.

Our testing activities suggest that the implementation of the algorithm does not use a large amount of memory and it runs quite fast enough, providing us the possibility to run the code with a high number of nodes that is relevant w.r.t. the numbers found in literature — see [3] for example. Notice that in [3] the number of variables is bigger than the number of points, but this is not relevant for the resources and time employed by our implementation. Anyway, it is possible to improve the performances of our implementation by keeping track of the last sibling for each node and of the last twin stored in a node. Of course, at a cost of increasing memory consumption, one can speed up the code, since insertion of new points would not require reading all the siblings/twins anymore.

References

1. Ceria, M., Mora, T.: Combinatorics of ideals of points: a cerlienco-mureddu-like approach for an iterative lex game. Preprint (2018)
2. Felszeghy, B., Ráth, B., Rónyai, L.: The lex game and some applications. J. Symbolic Comput. **41**(6), 663–681 (2006)
3. Laubenbacher, R., Stigler, B.: A computational algebra approach to the reverse engineering of gene regulatory networks. J. Theor. Biol. **229**(4), 523–537 (2004)
4. Lundqvist, S.: Vector space bases associated to vanishing ideals of points. J. Pure Appl. Algebra **214**(4), 309–321 (2010)
5. Marinari, M.G., Moeller, H.M., Mora, T.: Gröbner bases of ideals defined by functionals with an application to ideals of projective points. Appl. Algebra Eng. Commun. Comput. **4**(2), 103–145 (1993)
6. Möller, H.M., Buchberger, B.: The construction of multivariate polynomials with preassigned zeros. In: Calmet, J. (ed.) EUROCAM 1982. LNCS, vol. 144, pp. 24–31. Springer, Heidelberg (1982). https://doi.org/10.1007/3-540-11607-9_3
7. Mora, T.: Solving Polynomial Equation Systems, 4 Vols. Cambridge University Press, Cambridge, I (2003), II (2005), III (2015), IV (2016)
8. Ritchie, D.M., Kernighan, B.W., Lesk, M.E.: The C Programming Language. Prentice Hall, Englewood Cliffs (1988)

libtropicon: A Scalable Library for Computing Intersection Points of Generic Tropical Hyper-surfaces

Tianran Chen[(⊠)]

Auburn University at Montgomery, Montgomery, USA
ti@nranchen.org
http://www.tianranchen.org

Abstract. The computation of intersection points of generic tropical hyper-surfaces is a fundamental problem in computational algebraic geometry. An efficient algorithm for solving this problem will be a basic building block in many higher level algorithms for studying tropical varieties, computing mixed volume, enumerating mixed cells, constructing polyhedral homotopies, etc. libtropicon is a library for computing intersection points of generic tropical hyper-surfaces that provides a unified framework where the several conceptually opposite approaches coexist and complement one another. In particular, great efficiency is achieve by the data cross-feeding of the "pivoting" and the "elimination" step — data by-product generated by the pivoting step is selectively saved to bootstrap the elimination step, and vice versa. The core algorithm is designed to be naturally parallel and highly scalable, and the implementation directly supports multi-core architectures, computer clusters, and GPUs based on CUDA or ROCm/OpenCL technology. Many-core architectures such as Intel Xeon Phi are also partially supported. This library also includes interface layers that allows it to be tightly integrated into the existing ecosystem of software in computational algebraic geometry.

Keywords: Tropical hypersurfaces · Mixed volume · Mixed cells
BKK bound · Polyhedral homotopy

1 Introduction

Finding common solutions to a multivariate nonlinear polynomial system is a fundamental problem in computational mathematics with a great variety of important applications. While the most meaningful space for searching for such common solutions is the field of complex numbers $\mathbb{C}$ due to the rather special feature of being *algebraic closed*, the vastness of $\mathbb{C}$, however, poses challenges for computer programs. For instance, from a measure theory point of view, almost no complex number can be represented exactly using floating point numbers. In this light, the "tropical intersection problem" can be viewed as a particularly attractive discretization of this problem. In this tropical version, $\mathbb{C}$ is replaced

© Springer International Publishing AG, part of Springer Nature 2018
J. H. Davenport et al. (Eds.): ICMS 2018, LNCS 10931, pp. 105–112, 2018.
https://doi.org/10.1007/978-3-319-96418-8_13

by a *tropical semiring* [8] which is the extended real numbers with the addition and multiplication operations defined as minimum and addition respectively. The notion of polynomial functions and their zero sets can be extended to this semiring, and we thus have the "tropicalized" problem of finding intersections of zero sets for multivariate nonlinear polynomial systems in this context. Among several different computational aspects in this rich framework, libtropicon focuses on the problem of computing "generic intersections". The finite and combinatorial nature of the generic intersection problem make it particularly suitable for numerical computation (using floating point arithmetic). Yet great deal of information of about the original algebraic problems can be uncovered from these tropical intersections. For instance, generic root count (intersection number) of a given polynomial system can be directly extracted from tropical intersections. More importantly, from a computational view point, the rich data structures produced by such tropical intersection points are also the necessary ingredient for constructing a polyhedral homotopy method [3] for numerically locating all complex roots of the original polynomial system.

2   Problem Statement

The mathematical problem that libtropicon solves is the computation of intersection points of generic tropical hypersurfaces. Interestingly, this problem coincide with several different problems with seemingly independent origins. We briefly describe some of the key formulations.

2.1   Tropical Formulation

Consider the *tropical semiring* $\mathbb{T} = (\mathbb{R} \cup \{\infty\}, \oplus, \odot)$ with $a \oplus b = \min(a, b)$ and $a \odot b = a + b$. Given a Laurent polynomial (a polynomial with potentially negative exponents) $p(\mathbf{x}) = \sum_{\mathbf{a} \in S} c_{\mathbf{a}} \mathbf{x}^{\mathbf{a}}$ where $\mathbf{x}^{\mathbf{a}} = (x_1, \ldots, x_n)^{(a_1, \ldots, a_n)^\top} = x_1^{a_1} \cdots x_n^{a_n}$, it gives rise to the piecewise linear function

$$L(\mathbf{x}) = \min \{c_{\mathbf{a}} + \langle \mathbf{a}, \mathbf{x} \rangle \mid \mathbf{a} \in S\}, \tag{1}$$

if addition and multiplication are to be interpreted as the tropical operations $\oplus$ and $\odot$ respectively. The "zero set" $\mathcal{V}(p)$ of p over $\mathbb{T}$ is then defined to be the set of points where L is not differentiable, i.e., points where two or more linear pieces of $L(\mathbf{x})$ meet. Following the terminology from algebraic geometry (over $\mathbb{C}$), such a zero set will be known as a *tropical hypersurface* when it is of codimension one.

Problem 1. Given sets of monomials $M_1, \ldots, M_r \subset \mathbb{R}[x_1^\pm, \ldots, x_n^\pm]$, we consider generic Laurent polynomials $p_1, \ldots, p_r$ in these sets of monomials, i.e., each p_k is a linear combination of monomials in M_k with generic (nonzero) real coefficients. We want to find the intersection of the tropical hypersurface $\bigcap_{k=1}^r \mathcal{V}(p_k)$.

2.2   Mixed Cells Formulation

The above problem has a nearly equivalent formulation rooted from generalized coherent subdivision problem for point configurations: Given finite sets $S_1, \ldots, S_r \subset \mathbb{R}^n$, an r-tuple $(t_1, \ldots, t_r) \in (\mathbb{Z}^+)^r$ with $\sum_{k=1}^r t_k = n$, and functions $\omega_1, \ldots, \omega_r$ with each $\omega_k : S_k \to \mathbb{R}$ having generic images, we define $\hat{S}_k = \{(\mathbf{a}, \omega_k(\mathbf{a})) \mid \mathbf{a} \in S_k\} \subset \mathbb{R}^{n+1}$ for $k = 1, \ldots, r$. The main problem is to find all the r-tuple of faces of $\text{conv}(\hat{S}_1), \ldots, \text{conv}(\hat{S}_r)$ matching the dimensions given by $(t_1, \ldots, t_r)$ that can share the same "upward pointing" inner normal vector:

Problem 2. Given $S_1, \ldots, S_r$ and $\omega_1, \ldots, \omega_r$ described above, we want to find *all* possible r-tuples $(\{\mathbf{a}_{i_1,0}, \ldots, \mathbf{a}_{i_1,t_1}\}, \ldots, \{\mathbf{a}_{i_r,0}, \ldots, \mathbf{a}_{i_r,t_r}\})$ of subsets of $\hat{S}_1, \ldots, \hat{S}_r$ respectively for which there exists an $\alpha \in \mathbb{R}^n$ such that for each $k = 1, \ldots, r$

$$\langle \mathbf{a}_{i_k,0}, \alpha + \omega_k(\mathbf{a}_{i_k,0}) \rangle = \langle \mathbf{a}_{i_k,j}, \alpha + \omega_k(\mathbf{a}_{i_k,j}) \rangle \quad \text{for } j = 1, \ldots, t_k$$

$$\langle \mathbf{a}_{i_k,1}, \alpha + \omega_k(\mathbf{a}_{i_k,1}) \rangle < \langle \mathbf{a}, \alpha + \omega_k(\mathbf{a}) \rangle \quad \text{for } \mathbf{a} \in S_k \setminus \{\mathbf{a}_{i_k,0}, \ldots, \mathbf{a}_{i_k,t_k}\}.$$

It can be verified that in the $r = n$ case, Problems 1 and 2 are equivalent despite the rather different presentations. The r-tuples of points are known as *mixed cells* [3] of type $(t_1, \ldots, t_r)$, and they play a crucial role in the construction of polyhedral homotopies for solving a Laurent polynomial system.

2.3   Incremental Cayley's Trick Formulation

Yet another formulation of this problem is connected to the well known *Cayley's trick* and the *phase one problem* in linear programming. By introducing a new set of variables $h_k := \langle \mathbf{a}_{i_k,0}, \alpha \rangle + \omega_k(\mathbf{a}_{i_k,0})$ for $k = 1, \ldots, r$ as in Problem 2, we obtain the equivalent system

$$\langle \mathbf{a}_{i_k,j}, \alpha \rangle - h_k = -\omega_k(\mathbf{a}_{i_k,j}) \quad \text{for } j = 1, \ldots, t_k$$

$$\langle \mathbf{a}, \alpha \rangle - h_k > -\omega_k(\mathbf{a}) \quad \text{for } \mathbf{a} \in S_k \setminus \{\mathbf{a}_{i_k,1}, \mathbf{a}_{i_k,2}\},$$

With this, we get a reformulation of Problem 2 that resembles a generalized "Phase One" problem in linear programming:

Problem 3. Given $S_1, \ldots, S_r$, $(t_1, \ldots, t_r)$, and $\omega_1, \ldots, \omega_r$ described above, let $\check{A}_k$ to be the matrix whose rows are $(\mathbf{a}, -1)$ for points $\mathbf{a} \in S_k$, and let $\mathbf{c}_k$ be the column vector with corresponding entries of $-\omega_k(\mathbf{a})$ for $\mathbf{a} \in S_k$. We want to find *all* possible $(\alpha, h_1, \ldots, h_r) \in \mathbb{R}^{n+r}$ such that

$$\check{A}_k \begin{bmatrix} \alpha \\ h_k \end{bmatrix} \geq \mathbf{c}_k \text{ with } t_k \text{ equalities hold, for each } k = 1, \ldots, r. \qquad (2)$$

Here, the term "incremental" Cayley's trick refers to the fact that each group of inequality in (2) is embedded into $\mathbb{R}^{n+1}$ separately. Note that for each k, (2) is a generalized version of the Phase-One problem in linear programming. Therefore Problem 3 is a problem of simultaneous Phase-One problem: it requires the solutions $(\alpha, h_1), \ldots, (\alpha, h_r)$ to the r different generalized Phase-One problem to share the same projection onto the first n coordinates.

3   Applications

While finding the generic intersection points of tropical hypersurfaces is an interesting problem in its own right [5,9], the result also has a variety of important applications in computational mathematics which we shall outline below.

3.1   Root Counting Problem and the BKK Bound

One direct application of the generic tropical intersection points is the root counting problem. For each intersection point (component) in Problem 1, an intersection number (multiplicity) can be defined. The generic intersection number of the tropical hypersurfaces is defined to be the sum of all the intersection numbers at all the intersection points.

Theorem 1 (Huber and Sturmfels [3]). *Given a system of n Laurent polynomials $f_1, \ldots, f_n$ in n variables with generic coefficients, the total number of isolated common zeros in $(\mathbb{C}^*)^n$ equals the generic intersection number of the n tropical hypersurfaces defined by $f_1, \ldots, f_n$.*

Here $(\mathbb{C}^*)^n = (\mathbb{C} \setminus \{0\})^n$ is known as the "algebraic torus". This generic root count has since been known as the *BKK bound* or the *mixed volume bound*. This restriction on the domain of the root counting problem is minor and can be removed using the more general version of this theorem developed in [4,7,12].

3.2   Polyhedral Homotopy

The constructive proof of Theorem 1 gives rise to a numerical homotopy method for solving polynomial systems. Given a system of (Laurent) polynomials $F = (f_1, \ldots, f_n)$ in the n variables $\mathbf{x} = (x_1, \ldots, x_n)$ with generic coefficients, finding all isolated complex solutions of $F(\mathbf{x}) = \mathbf{0}$, i.e., $f_k(x_1, \ldots, x_n) = 0$ for $k = 1, \ldots, n$, is a fundamental problem in computational mathematics. If we write the i-th polynomial in F as $f_i = \sum_{\mathbf{a} \in S_i} c_{\mathbf{a}} \mathbf{x}^{\mathbf{a}}$ using the multi-index notation as before, we could consider the homotopy $H = (h_1, \ldots, h_n)$ given by

$$h_i(\mathbf{x}, t) = \sum_{\mathbf{a} \in S_i} c_{i,\mathbf{a}} \mathbf{x}^{\mathbf{a}} \, t^{\omega_i(\mathbf{a})} \quad \text{for } i = 1, \ldots, n \tag{3}$$

where $\omega_i : S_i \to \mathbb{R}$ are functions with generic images that play the same roles as the lifting functions in Problem 2. Clearly, H is continuous in $\mathbf{x}$ and t for $t > 0$, and $H(\mathbf{x}, 1) = F(\mathbf{x})$. Moreover as t varies between 0 and 1, the isolated solutions of $H(\mathbf{x}, t) = \mathbf{0}$ in $(\mathbb{C}^*)^n$ also move smoothly and form solution paths reaching the solutions of $F(\mathbf{x}) = H(\mathbf{x}, 1) = \mathbf{0}$ at $t = 1$. The end points of these paths include *all* solutions of the original system $F(\mathbf{x}) = \mathbf{0}$ in $(\mathbb{C}^*)^n$. Numerical *continuation method* can therefore be applied to trace these paths to reach the solutions if their starting points are known.

The difficulty, however, is that the starting points of these solution paths cannot be identified directly since at $t = 0$, $H(\mathbf{x}, 0)$ becomes identically zero.

An ingenious observation in [4] is that the tropical intersection points are exactly the right tool to resolve this difficulty: Under the genericity assumption of the functions ω_i, all solution paths escape $(\mathbb{C}^*)^n$ as $t \to 0$, and the asymptotic behavior of each path is characterized by $(y_1 t^{\alpha_1}, \ldots, y_n t^{\alpha_n})$ for some $\mathbf{y} = (y_1, \ldots, y_n) \in (\mathbb{C}^*)^n$ and a tropical intersection point $\alpha = (\alpha_1, \ldots, \alpha_n)$ in Problem 1. Finding the tropical intersection points is therefore the key step in bootstrapping the polyhedral homotopy construction.

3.3 Regular Triangulation of High Dimensional Polytopes

Finally, as a special case, libtropicon can be used to produce a regular triangulation for any convex polytope of any dimension. Indeed, given a finite set of vertices $S_1 \subset \mathbb{R}^n$, we consider the special case of Problem 2 with $r = 1$ and type $(t_1) = (n)$. It can be verified that in this case, the mixed cells of type (n) are in one-to-one correspondence with the cells in a regular triangulation of the convex polytope $\text{conv}(S)$.

4 Underlying Theory

To briefly outline the underlying theory behind libtropicon, we shall focus on the formulation given in Problem 3. At a solution $\mathbf{x}$ to the system $A\mathbf{x} \geq \mathbf{c}$ with rank $A = N = n + 1$, a row $\mathbf{a}$ of A is said to be *active* if $\langle \mathbf{a}, \mathbf{x} \rangle = c_k$ where c_k is the corresponding entry in $\mathbf{c}$. If there are N linear independent active rows, $\mathbf{x}$ is known as a *basic feasible solution*.

A *level-k basic feasible solution* is simply a basic feasible solution of the combined system

$$\check{A}_{i_j} \begin{bmatrix} \alpha \\ h_{i_j} \end{bmatrix} \geq \mathbf{c}_{i_j} \text{ with at least } t_{i_j} \text{ equalities hold for each } j = 1, \ldots, k \quad (4)$$

for a subset $\{i_1, \ldots, i_k\}$ of the indices[1] $1, \ldots, r$.

4.1 The Intersection-Elimination-Pivot Scheme

A series of successful software packages [1,6,10,11] for computing intersections of tropical hypersurfaces share a common basic incremental scheme that, in hindsight, could be described as an "intersection-elimination-pivot" scheme. This scheme starts with all the basic feasible solutions of $\check{A}_i \mathbf{x} \geq \mathbf{c}_i$ for certain $i \in \{1, \ldots, n\}$ and attempt to extend each into level-k extended basic feasible solutions for increasingly higher values of k until reaching all the level-r extended basic feasible solutions.

---

[1] Note that the level-k basic feasible solution defined here is but a simplified prototype of the much more technical concept of "level-k subfaces" that is actually used in family of algorithms [1,2,6,10] whence libtropicon inherits much of the core ideas.

This scheme consists of several complicated algorithms. The organization and detail of each algorithm will be outside the scope of this extended abstract. We only outline the general mathematical problem behind each step and highlight some of the new improvements over existing implementations. We refer to [1] for the complex organization of these steps.

Intersection. Using the information from two level-1 basic feasible solutions of $\check{A}_i \mathbf{x} \geq \mathbf{c}_i$ and $\check{A}_j \mathbf{y} \geq \mathbf{c}_j$ respectively, the intersection step seeks to construct a point $\alpha \in \mathbb{R}^n$ together with h_i and h_j such that

$$\check{A}_i \begin{bmatrix} \alpha \\ h_i \end{bmatrix} \geq \mathbf{c}_i \quad \text{and} \quad \check{A}_j \begin{bmatrix} \alpha \\ h_j \end{bmatrix} \geq \mathbf{c}_j. \tag{5}$$

This step also generalizes to the problem of finding a level-$(k_1 + k_2)$ basic feasible solution using a level-k_1 and level-k_2 basic feasible solutions. This question can be loosely interpreted as a local version of the tropical intersection problem.

Elimination. Using information produced from the previous step, the goal of the elimination step is to eliminate the possibility of certain level-k basic feasible solutions with minimum computational cost. This step closely resembles some of the key ideas in integer programming.

The *relation table* proposed in [2] is one of the first data structure designed for fast elimination. This simple idea has been proved to be extremely effective. It also sparked a series of related works on this idea. The technique developed in [11] is a major improvement on the relation table based method. libtropicon adopts a far generalization of this general idea to drastically improve the effectiveness of the elimination step with little computational cost. We will simply state the main theorem behind this idea in Sect. 4.2.

Pivot. Finally the pivot step walks from one level-k basic feasible solution to another. For a fixed k, the level-k basic feasible solutions are organized into a (not necessarily connected) graph. The pivot step is therefore constructed as a graph walking algorithm.

4.2   The Conic Elimination Method

Compare to other existing implementations, one distinguishing feature of libtropicon is the adoption of the "conic elimination method" which brings substantial improvement in the effectiveness of the elimination step.

Conic elimination is a series of tests that can quickly eliminate many candidates for level-$(k + 1)$ basic feasible solutions using only the local geometric information encoded in a level-k basic feasible solution. For brevity, we only state one version of such conic elimination tests. Recall that in linear programming, the set of indices of active constraints at a basic feasible solution is called *basic indices*, and the sub-matrix formed by active constraints is the *basic matrix*. Naturally, the inverse of the basic matrix is known as the *basic inverse*.

Theorem 2. *Let* $\mathbf{x}$ *and* $\mathbf{y}$ *be two level-1 basic feasible solutions to* $\check{A}_i\mathbf{x} \geq \mathbf{c}_i$ *and* $\check{A}_j\mathbf{y} \geq \mathbf{c}_j$ *respectively. Let* $\Delta_\mathbf{x}$ *and* $\Delta_\mathbf{y}$ *be the set of basic indices at* $\mathbf{x}$ *and* $\mathbf{y}$ *respectively. Also let* D_i *and* D_j *be the corresponding basic inverse matrix with columns of* D_j *denoted by* $\mathbf{d}_\ell$ *for* $\ell \in \Delta_j$. *For two sets* $F_i \subseteq \Delta_\mathbf{x}$ *and* $F_j \subseteq \Delta_\mathbf{y}$, *if there exists some* $b_1 \in F_i$ *and* $b_2 \in \Delta_\mathbf{x}$ *such that*

$$\langle \mathbf{d}_\ell, \check{\mathbf{a}}_{b_2} - \check{\mathbf{a}}_{b_1} \rangle \leq 0 \qquad \text{for each } \ell \in \Delta_\mathbf{y} \setminus F_j$$

$$\langle \mathbf{y} - \mathbf{x}, \check{\mathbf{a}}_{b_2} - \check{\mathbf{a}}_{b_1} \rangle < 0$$

then the constraints indexed by F_i *and* F_j *cannot belong to the same level-2 basic feasible solutions.*

This test can be easily extended to be used for eliminating general level-k basic feasible solutions.

5   Technical Contribution

5.1   Data Cross-Feeding

One notable technical feature of libtropicon, compared to previous implementations, is that data from different algorithms are shared and reused. In particular, the data by-product of the "intersection" step are selectively saved and used in the elimination step and vice versa. Memory occupied by data that are no longer needed can also be detected and freed immediately resulting in much more efficient memory usage.

5.2   Parallelization on Shared-Memory Architectures via Task Graphs

Today, parallel computation is an integral part of any high performance software for numerical computation. Parallel computation on shared-memory architectures such as modern multi-core systems are directly supported via a task-based model (independent calculations are organized into "tasks" that can be scheduled to run in parallel). Task-based models are generally considered to be much more scalable and flexible than thread-based models. While libtropicon inherited the basic "task pool" framework adopted in Hom4PS-3 [1], this framework is refined using the more flexible "task graphs" which organize the tasks into a dependency graph. This change allows libtropicon to be scaled to systems with more processor cores.

5.3   Low Latency GPU Implementations

An increasingly important trend in high performance computing is the use of GPU (graphics processing units) devices in general purpose computing tasks. GPU based parallel computation is also supported by libtropicon through

NVidia's CUDA and AMD's ROCm OpenCL framework[2]. In the previous preliminary GPU-based implementation [1] the data structure that completely describes the level-k basic feasible solutions are transferred back and forth between CPU and GPU devices. Such data structures contains several $N \times N$ matrices in double precision floating point numbers. Therefore, in hindsight, this is clearly the main cause of the rather high latency. In libtropicon, only the very short *hash keys* (256 bits by default) that identifies the basic feasible solutions are passed between CPU and GPU devices while the actual data remains in the GPU devices. This change greatly reduced the CPU-to-GPU latency.

References

1. Chen, T., Lee, T.L., Li, T.Y.: Mixed cell computation in Hom4PS-3. J. Symbolic Comput. **79**, 516–534 (2017)
2. Gao, T., Li, T.Y., Wu, M.: Algorithm 846. ACM Trans. Math. Softw. **31**(4), 555–560 (2005)
3. Huber, B., Sturmfels, B.: A polyhedral method for solving sparse polynomial systems. Math. Comput. **64**(212), 1541–1555 (1995)
4. Huber, B., Sturmfels, B.: Bernsteins theorem in affine space. Discrete Comput. Geom. **17**(2), 137–141 (1997)
5. Jensen, A.N.: A presentation of the Gfan software. In: Iglesias, A., Takayama, N. (eds.) ICMS 2006. LNCS, vol. 4151, pp. 222–224. Springer, Heidelberg (2006). https://doi.org/10.1007/11832225_21
6. Lee, T.L., Li, T.Y., Tsai, C.H.: HOM4PS-2.0: a software package for solving polynomial systems by the polyhedral homotopy continuation method. Computing, **83**(2–3), 109–133 (2008)
7. Li, T., Wang, X.: The BKK root count in $\mathbb{C}^n$. Math. Comput. Am. Math. Soc. **65**(216), 1477–1484 (1996)
8. Maclagan, D., Sturmfels, B.: Introduction to Tropical Geometry, vol. 161. American Mathematical Society, Providence (2015)
9. Malajovich, G.: Computing mixed volume and all mixed cells in quermassintegral time. Found. Comput. Math. **17**(5), 1–42 (2016)
10. Mizutani, T., Takeda, A.: DEMiCs: a software package for computing the mixed volume via dynamic enumeration of all mixed cells. In: Stillman, M., Verschelde, J., Takayama, N. (eds.) Software for Algebraic Geometry, vol. 148. The IMA Volumes in Mathematics and its Applications, pp. 59–79. Springer, New York (2008)
11. Mizutani, T., Takeda, A., Kojima, M.: Dynamic enumeration of all mixed cells. Discrete Comput. Geom. **37**(3), 351–367 (2007)
12. Rojas, J.M., Wang, X.: Counting affine roots of polynomial systems via pointed Newton polytopes. J. Complex. **12**(2), 116–133 (1996)

---

[2] ROCm is AMD's latest implementation of the OpenCL standard, an open standard for general purpose GPU computation. Currently, only ROCm have been tested. Support for other implementations of OpenCL could be added in the future with minimum changes to the code.

Plotting Planar Implicit Curves
and Its Applications

Jin-San Cheng[1,2]($\boxtimes$), Junyi Wen[1,2], and Wenjian Zhang[1,2]

[1] Key Lab of Mathematics Mechanization, Institute of Systems Science,
Academy of Mathematics and Systems Science, CAS, Beijing, China
jcheng@amss.ac.cn
[2] School of Mathematical Sciences,
University of Chinese Academy of Sciences, Beijing, China
{wenjunyi15,zhangwenjian16}@mails.ucas.ac.cn

Abstract. We present a new method to plot planar implicit curve in a given box $B \in \mathbb{R}^2$. Based on analyzing the geometry of the level sets of the given function, following the points with local maximal (or minimal) curvatures on the level sets, we compute points on each components of the given function in box B and trace each component to plot the curve. We also used this method to find real zeros of bivariate function systems in a given box. The experiments shows that our implementation works well. It works for polynomials with degrees more than 10,000. It also works for non-polynomial case.

Keywords: Plotting · Planar implicit curve · Level sets · Curvature
Real solving · Bivariate function systems

1 Introduction

Plotting planar implicit curves is a basic topic in computer aided geometric design and computer graphics. There are some methods that plot implicit curves with guaranteed topology, for example [9]. These methods require the topology of the curves is known. Computing the topology of the curves mainly use symbolic computation. One classic method is cylindrical algebraic decomposition and relevant modified methods [1,5,10]. Subdivision methods are well studied in plotting implicit curves, for example, marching cube, PV algorithm and the related modified algorithms [13–18]. Continuation methods need to find sample points on each component and then trace the curve components. The most difficult part is to find sample points on the components. There already exists some work [7,12]. There are also some related softwares [2,3,8].

In this paper, we present a new method, which we call it level set sweeping method. Before we present our new method, we introduce some notations.

Denote $C^i(\Omega)$ as a class of all i-order continuous differentiable functions defined in Ω, where $\Omega \subset \mathbb{R}^n$. Let $\Sigma = \{f_1,\ldots,f_m\} \subset C^2(\Omega), \Omega \subset \mathbb{R}^n$.

The work is partially supported by NSFC Grants 11471327.

We denote all the real zeros of $\Sigma = 0$ as $\mathbb{V}(\Sigma)$ and the gradient of f as $\nabla(f) = (\frac{\partial f}{\partial x_1}, \ldots, \frac{\partial f}{\partial x_n})$. In the following, we always assume that $f \in C^2(\Omega)$.

A point $p \in \Omega$ is called a **stationary point** or **critical point** of f, if $\nabla(f)(p) = 0$. Let $p_0 \in \Omega$. If $\exists \delta > 0$ s.t. $U = \{p \mid \|p - p_0\| < \delta\} \subset \Omega$, and $f(p) \leq f(p_0)$ ($f(p) \geq f(p_0)$), $\forall p \in U$, then we call p_0 is a **maximum point** (**minimum point**) of f. The minimum and maximum points are both called **extreme points** of f. It is clear that an extreme point of f is a stationary point of f.

A point $p \in \Omega$ is called a **saddle point** of f, if p is a stationary point but not an extreme point of f.

Definition 1. *Let $g \in C^k(\Omega)$ with $\Omega \subset \mathbb{R}^n$. We call $g = 0$ **singular** if the variety $\mathbb{V}(\nabla(g), g) \subset \mathbb{R}^n$ is non-empty.*

From the definitions, we can directly infer that $g = 0$ is singular if and only if there exists at least one stationary point on the curve $g = 0$.

Definition 2. *A **level set** of a real-valued function h of n real variables is a set of the form $L_r(h) = \{(x_1, \cdots, x_n) \mid h(x_1, \cdots, x_n) = r, r \in \mathbb{R}, (x_1, \cdots, x_n) \in \mathbb{R}^n\}$.*

Definition 3. *We call a zero set $W \subset \mathbb{V}(\Sigma) \subset \mathbb{R}^n$ as **a real connected component** (simply **component** without misunderstanding) if W is an isolated real zero of the function system Σ, or there exists a connected real path on W for any two distinct points on W.*

The level set sweeping method works as below. The algorithm is contained in [6] and the theoretical analysis is there.

Given a planar implicit curve $f(x, y) = 0$, we will plot the curve inside a real box B. We use the level set $f - r = 0$ of f to sweep throughout B. We follow the points on the level sets with local maximum (or minimum) curvature to trace the level sets during we sweep in the region B with them. When one real connected component splits into two or more real connected components during r varying, there exists $r' \in \mathbb{R}$ such that $f - r' = 0$ is a singular curve. We compute on each new component at least one sample point. Thus during the level set sweeping and splitting, we get points on each component of the input curve inside B when r is zero. In the end, we trace each component of the curve, which we get the plotting of the curve. Some of the components of the curve may be isolated singularities of the curve in B. We use the gradient method and Newton's method to get the points of the curve. We implement our algorithm in Maple.

2   Outline of Level Set Sweeping Algorithm

In this section, we will show how to sweep the whole given region B with level sets of f. We also need to find at least one point on each new component of the level sets of f when one component splits into two or more components during the value r changes.

Consider a given bounded box $B \subset \mathbb{R}^2$ and a real function $f \in C^2(B)$. It is obvious that f is bounded in B, i.e. $\exists r_0, r_0'$, s.t. $r_0' \leq f(p) \leq r_0, \forall p \in B$. Based on the fact, if we regard r as a variable and vary r from r_0' to r_0 continuously, the level set $L_r(f)$ will sweep the whole box B continuously. When r varies from r_0' to r_0, some two or more components will combine as one component (or one component splits into two or more components), that is to say, $\exists r^* \in (r_0', r_0)$, s.t. $f - r^*$ is singular in the process. Thus, if we trace the level set $L_r(f)$ when r varies from r_0' to r_0, we can get the real components of $f = 0$ finally.

But it is not easy to get the maximal value r_0 in B and it is not necessary. We can sweep B from some of its boundaries or endpoints from one side to the other side. For example, we choose one endpoint p of B as a start point. It is on the level set, say $L_p : f - f(p)$. We choose the (negative) gradient direction such that it points inside B. Choose a point, say q, on the line passing through p on the (negative) gradient direction. It determines another level set of f, say $L_q : f - f(q)$. Tracing L_q inside B, we search some interesting points on it. We trace the level set with a given step length, say Δ_h. From the points on L_q, we find next level set close to it. It works as below.

For the points $\mathbf{p}_i (i = 1, .., k)$ on L_q, we assume that the (negative) unit gradient directions are $\{\mathbf{v}_i, i = 1, \ldots, k\}$, respectively. For a given vertical step length, say Δ_v, we define the constant value of the next level set r as below:

$$r = \min_{i=1}^{k} f(\mathbf{p}_i - \mathbf{v}_i \times \Delta_v)(\text{ or } r = \max_{i=1}^{k} f(\mathbf{p}_i + \mathbf{v}_i \times \Delta_v)).$$

We call the tracing of the points on one level set **horizontal tracing** (see [4]) and the tracing from one level set to another level set **vertical tracing**. By horizontal tracing and vertical tracing (see Figs. 1 and 2 for an illustration), we can sweep B by the level sets of f. During the tracing, we can compute the singular points of the level sets which are related to some critical points of the curve. Also, there are some critical points between two level sets, we compute it by gradient method and Newton's method with the points with maximal (minimal) local curvature on the level set nearby. These critical points are always related to the position where one component splits into two or more components when the value of the level set varies. Thus we can get sample points on each

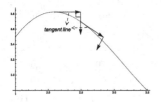

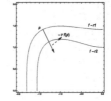

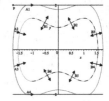

Fig. 1. Horizontal tracing   **Fig. 2.** Vertical tracing   **Fig. 3.** Tracing on boundary.

split components (see Fig. 4 for an illustration). When tracing near the boundary of the box, for the points on the boundary, we may require to do vertical tracing in two directions, see $A1, A4$ in Fig. 3 for example. For some level set, we need to do vertical tracing on two sides, see $B1, B2, B3$ in Fig. 3 for example.

3   The Main Functions of the Software

We implement the algorithm in Maple. Here are the main functions of our implementation.

HorizontalTracing is to trace the curve component of the level set of f inside B with a given step length Δ_h. The output is some points on the curve segment including the points with maximal (minimal) local curvatures (see [11] for details).

VerticalTracing is to trace the level set of f from $L_{r_1}(f)$ to $L_{r_2}(f)$ such that the Hausdorff distance between two level sets inside B is bounded by a step length Δ_v. The output is a value r_2 related to the next level set $L_{r_2}(f)$ and some point(s) on it.

TracingNearCriticalPoint is to trace the level set of f near a critical point of f. The output is at least one sample point on each component of the level sets of f near the critical point. The sample point(s) on the level set which has (have) already been traced is (are) ignored.

4   Examples on Plotting Curves

We test some examples with our code.

Examples 1, 2, 3 are defined by the following polynomials:

$$Ex1: \quad (x^2 + y^2 - 1) * (y - x^2),$$
$$Ex2: \quad (x^2 - 1)^2 + (y^2 - 1)^2 - 1,$$
$$Ex3: (x^2 + y^2 - 1) * (2 * x^2 + 2 * y^2 - 1).$$

Their plotting curves are in Figs. 5, 6 and 7. Their computing information are in Table 1.

Example 4 is defined by Ex4: $g * (h^2 + 1)$, where $g := 2 * y^2 - x^3 - x^2$ and $h := x^{4490}y^{550} + 2\,x^{1052}y^{3946} - 2\,x^{1287}y^{3697} - 2\,x^{689}y^{4290} - x^{4963}y^2 - x^{3471}y^{1395} + x^{3385}y^{1447} - 2\,x^{1766}y^{3011} + 2\,x^{3743}y^{967} - 2\,x^{4545}y^{137} - 2\,x^{4418}y^{203} - 2\,x^{2919}y^{1671} + x^{3258}y^{1256} - x^{1525}y^{2912} - x^{456}y^{3915} - 2\,x^{143}y^{4164} + 2\,x^{1351}y^{2686} + 2\,x^{3253}y^{748} + x^{1802}y^{1754} + 2\,x^{2728}y^{724} + 2\,x^{2137}y^{1253} - x^{2480}y^{874} + 2\,x^{1201}y^{2017} - x^{560}y^{2582} - x^{1844}y^{1284} + x^{942}y^{2117} - 2\,x^{1779}y^{1192} + x^{1609}y^{1347} + 2\,x^{1467}y^{1370} + 2\,x^{2073}y^{650} + x^{1041}y^{1563} + 2\,x^{1191}y^{1366} - x^{48}y^{2470} - 2\,x^{1782}y^{425} - 2\,x^{1607}y^{546} - 2\,x^{1073}y^{1036} - 2\,x^{300}y^{1714} + x^{377}y^{1600} + 2\,x^{635}y^{1023} - 2\,x^{1045}y^{550} - x^{1421}y^{92} + x^{266}y^{158}$. Its plotting curve is in Fig. 8 and related information is in Table 1.

Example 5 is formed as Example 4. Ex5: $g_1 * (h_1^2 + 1)$, where $g_1 = 4 * x^4 * y + x^3 * y^2 - 5 * x^2 * y^3 - 5 * x * y^4 - 4 * y^5 - 4 * x^4 + 2 * x^3 * y - x^2 * y^2 - 4 * x * y^3 - y^4 - 4 * x^3 - 4 * x * y^2 - y^3 - 3 * x^2 - 4 * x * y + y^2 - 4 * x - 2 * y - 4$ and $h_1 = 2 x^{4579} y^{498} + 2 x^{3949} y^{1125} - 2 x^{2612} y^{2421} + 2 x^{3215} y^{1768} + 2 x^{2600} y^{2272} - 2 x^{3448} y^{1231} + 2 x^{2315} y^{2227} - 2 x^{3584} y^{947} - x^{3888} y^{366} + x^{1386} y^{2798} + 2 x^{3920} y^{236} + 2 x^{2999} y^{1107} - 2 x^{2264} y^{1723} + x^{2776} y^{1208} - x^{567} y^{3310} - 2 x^{784} y^{2867} + x^{2496} y^{1154} + 2 x^{368} y^{3182} - 2 x^{2513} y^{896} + x^{584} y^{2806} - x^{2644} y^{717} + 2 x^{1242} y^{1906} + 2 x^{105} y^{2978} - 2 x^{1936} y^{1044} - 2 x^{1855} y^{1071} - 2 x^{2827} y^{32} + x^{15} y^{2486} + x^{861} y^{1615} + x^{357} y^{2062} + 2 x^{1206} y^{1030} - x^{477} y^{1606} + x^{1489} y^{554} + 2 x^{54} y^{1944} + x^{885} y^{1100} + x^{498} y^{1471} + 2 x^{542} y^{1066} + x^{272} y^{1324} + x^{1235} y^{330} + x^{180} y^{866} + 2 x^{124} y^{826}$.

We plot it twice. The first time we plot with the given form $g_1 * (h_1^2 + 1)$. The second time we plot with its expanded form. We can find that the expanded version takes more time than the given one in Table 1. The plotting curve is shown in Fig. 9.

Example 6 is defined as Ex6: $g_2 * (h_2^2 + 1)$, where $g_2 = -2 * x^5 + 5 * x^4 * y + 2 * x^3 * y^2 + 5 * x^2 * y^3 - 2 * x * y^4 - 2 * y^5 + 4 * x^3 * y + 5 * x^2 * y^2 + 2 * x * y^3 + 4 * y^4 - 4 * x^3 + 5 * x * y^2 - 3 * y^3 + 2 * x^2 + x * y - 4 * y^2$ and $h_2 = -3 x^{4619} y^{426} + 2 x^{1535} y^{3209} + x^{1793} y^{2862} - 3 x^{175} y^{4456} - 3 x^{4485} y^{75} + 3 x^{2710} y^{1846} + 3 x^{1671} y^{2882} + x^{1499} y^{3002} + 2 x^{2874} y^{1367} + 2 x^{550} y^{3653} - 3 x^{3046} y^{1125} - 3 x^{3834} y^{153} - 2 x^{3068} y^{889} + x^{2275} y^{1594} + 3 x^{3326} y^{511} + x^{851} y^{2964} - 3 x^{1020} y^{2727} - 2 x^{1042} y^{2662} - 3 x^{766} y^{2931} + x^{754} y^{2936} + x^{3187} y^{449} - 2 x^{1057} y^{2575} - 3 x^{2268} y^{1218} + x^{1925} y^{1436} + 3 x^{1940} y^{1345} + 2 x^{1253} y^{2026} + 2 x^{88} y^{3121} - 3 x^{2670} y^{520} - x^{567} y^{2416} - x^{2050} y^{880} - 3 x^{1985} y^{758} - 3 x^{1508} y^{837} - x^{2015} y^{211} - 3 x^{670} y^{1227} + 2 x^{83} y^{1808} - 3 x^{1786} y^{49} - 3 x^{66} y^{1732} + x^{813} y^{983} - 3 x^{193} y^{832} - 2 x^{280} y^{637} - 3 x^{519} y^{23} + 3 x^{308} y^{179}$. Its plotting curve is in Fig. 10 and related information is in Table 1.

We also test some non-polynomial implicit curves.

Example 7 is defined by Ex7: $\sin(x + y)^2 + \cos(x - y)^2 - 1/3$. Its plotting curve is in Fig. 11 and related information is in Table 1.

The last two examples can be found in Subsect. 5.2. They are two transcendental functions in the system $\Sigma_G = \{g_1, g_2\}$. The red curve is g_1 and the green one is g_2. Their plotting curves are in Fig. 12 and related computing times are in Table 1.

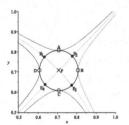

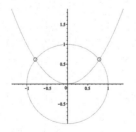

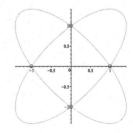

Fig. 4. Tracing near a critical point

Fig. 5. Plotting of Example 1

Fig. 6. Plotting of Example 2

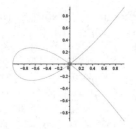

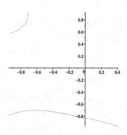

Fig. 7. Plotting of Example 3   **Fig. 8.**   Plotting   of Example 4   **Fig. 9.**   Plotting   of Example 5

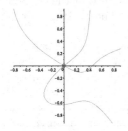

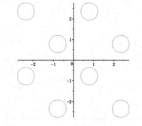

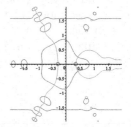

Fig. 10.   Plotting   of Example 6   **Fig. 11.**   Plotting   of Example 7   **Fig. 12.** Plotting $g_1 = 0$ (red), $g_2 = 0$ (green). (Color figure online)

Table 1. Plotting of bivariate functions

Case	d	t	$|c| \leq$	Box	Δ_v	Δ_h	Times
Ex1	4	5	1	$[-2,2] \times [-2,2]$	0.1	0.1	32.510 s
Ex2	4	5	2	$[-2,2] \times [-2,2]$	0.1	0.1	25.725 s
Ex3	4	6	4	$[-2,2] \times [-2,2]$	0.1	0.1	41.372 s
Ex4	10083	46	16	$[-1,1] \times [-1,1]$	0.05	0.05	239.602 s
Ex5	10159	60	42	$[-1,1] \times [-1,1]$	0.05	0.05	372.967 s
Ex5 (expanded)	10159	15590	42	$[-1,1] \times [-1,1]$	0.05	0.05	21407.143 s
Ex6	10095	59	90	$[-1,1] \times [-1,1]$	0.05	0.05	627.529 s
Ex6 (expanded)	10095	14447	90	$[-1,1] \times [-1,1]$	0.05	0.05	38550.624 s
Ex7	\	\	\	$[-3,3] \times [-3,3]$	0.1	0.1	181.928 s
$\Sigma_G : g_1$	\	\	\	$[-2,2] \times [-2,2]$	0.03	0.03	2223.093 s
$\Sigma_G : g_2$	\	\	\	$[-2,2] \times [-2,2]$	0.03	0.03	15887.017 s

5   Real Solving of Bivariate Function Systems

We denote $\Sigma = \{f_1, \ldots, f_m\} \subset C^2(\Omega), \Omega \subset \mathbb{R}^2, f = \sum_{i=1}^{m} f_i^2 \in C^2(\Omega)$. We regard f as a planar implicit curve inside a box B. It has only isolated singularities. We get all the singularities of the curve inside B, which means we get the real zeroes of the system Σ.

We do some experiments in the following.

5.1   Polynomial Case

Firstly, we consider the polynomial case. We randomly generate polynomial systems with different degrees, terms and coefficients.

Let $\Sigma = \{f_i, i = 1, \ldots, m\}$. Denote d as the maximal total degree of f_i, t as the maximal number of terms of f_i and $|c|$ as the maximal coefficients of f_i. $\#\mathbb{V}(\Sigma)$ is the number of real zeros of $\Sigma = 0$ and $\#P$ is the number of real zeros computed by our method of $\Sigma = 0$. Set $B = x_0 \times y_0 = [-3, 3] \times [-3, 3]$. The horizontal step length and the vertical step length both are $\Delta = 0.1$. The results are in Table 2.

Table 2. Real solving of polynomial systems

Examples	d	t	$|c| \leq$	$\#\mathbb{V}(\Sigma)$	$\#P$	Times
poly1	2	2	10	4	4	41.652 s
poly2	3	2	10	2	2	85.317 s
poly3	25	50	10	10	10	298.398 s
poly4	25	dense	10	7	7	760.926 s
poly5	25	50	100	9	9	625.424 s
poly6	25	dense	100	7	7	1102.927 s
poly7	10001	1075	100	9	9	1041.728 s
poly8	20003	2392	100	6	6	3069.866 s

Remark 1. The poly 1, 2, 3, 4, 5, 6 are examples which are expanded. Here poly1: $f_1 = x^2 - 1, f_2 = y^2 - 1$, poly2: $f_1 = x^2 y + 1, f_2 = x^2 y + x^2$.

Remark 2. The poly 3, 4, 5, 6 are generated randomly in Maple. For large systems poly 7, 8. We firstly generate two univariate polynomials $g_1(x)$ and $g_2(y)$ with low degrees, we can compute their real roots. Then we generate two bivariate polynomials with high degrees: h_1, h_2. Let $\Sigma = \{f_1, f_2\} = \{g_1(h_1^2 + 1), g_2(h_2^2 + 1)\}$. Assume that $\{g_1, g_2\}$ have m real zeros, then $\Sigma = 0$ also have m real zeros. We do not display these systems since they are very big.

5.2   Non-polynomial Case

Next, we do experiments on non-polynomial case. In [19], Strzebonski considers solving exp-log-arctan type function (equations). We can also solve this kind of equations and not limited to this kind. We also set $B = x_0 \times y_0 = [-3, 3] \times [-3, 3]$. The horizontal step length and the vertical step length both are $\Delta = 0.1$. The results are in Table 3. The related function systems are as below.

Table 3. Real solving of non-polynomial function systems

Examples	$\#\mathbb{V}(\Sigma)$	$\#P$	Times
$\{\sin(x+y), x^2 + y^2 - 1\}$	2	2	53.836 s
$\{x\sin(y) - y^2, \cos(x+y^2) + e^x - 3\}$	2	2	232.504 s
$\{e^{x^2 + 2y^2 - 1} - 1, \cos(xy) - 2x^2\}$	4	4	75.458 s
$\{\sin(x+y), \cos(x-y)\}$	8	8	92.103s

The below two examples are obtained from [20]:

(1) Let $F(z) = \tan(z) - \ln(z+3) - z^2$. $F(z) = 0$ has 4 complex roots in the box $-2 \le Re(z) \le 2$ and $-2 \le Im(z) \le 2$. Let $z = x + i * y$, and rewrite $F(z)$ to $f_1(x, y) + i * f_2(x, y)$, we have:

$$f_1 = \tan(x) - \frac{1}{2}\ln((x+3)^2 + y^2) - 2\tanh(y)\tan(x)xy$$
$$-\tanh(y)\tan(x)\arctan(\frac{y}{x+3}) - x^2 + y^2,$$
$$f_2 = \tanh(y) - \arctan(\frac{y}{x+3}) - 2xy - \tanh(y)\tan(x)y^2$$
$$+\frac{1}{2}\tanh(y)\tan(x)\ln((x+3)^2 + y^2) + \tanh(y)\tan(x)x^2.$$

Then, the complex roots of $F(z) = 0$ are the real roots of $\Sigma_F = \{f_1, f_2\} = 0$.

(2) Let $G(z) = \tan(z^3 + 1) - e^z$. $G(z) = 0$ has 23 complex roots in the box $-2 \le Re(z) \le 2$ and $-2 \le Im(z) \le 2$. Let $z = x + i * y$, and rewrite $G(z)$ to $g_1(x, y) + i * g_2(x, y)$, we have:

$$g_1 = \tan(x^3 - 3xy^2 + 1) - e^x\cos(y) - \tanh(3x^2y - y^3)^2\tan(x^3 - 3xy^2 + 1)$$
$$-\tanh(3x^2y - y^3)^2\tan(x^3 - 3xy^2 + 1)^2 e^x\cos(y),$$
$$g_2 = \tanh(3x^2y - y^3) - e^x\sin(y) + \tanh(3x^2y - y^3)\tan(x^3 - 3xy^2 + 1)^2$$
$$-\tanh(3x^2y - y^3)^2\tan(x^3 - 3xy^2 + 1)^2 e^x\sin(y).$$

Then, the complex roots of $G(z) = 0$ are the real roots of $\Sigma_G = \{g_1, g_2\} = 0$.

We compute the real roots of $\Sigma_F = 0$ and $\Sigma_G = 0$. We set $B = [-2, 2] \times [-2, 2]$. The result is in Table 4.

Table 4. Real solving of transcendental function systems

Examples	Δ_v	Δ_h	$\#\mathbb{V}(\Sigma)$	$\#P$	Times
Σ_F	0.1	0.1	4	4	164.362 s
Σ_G	0.1	0.1	23	19	781.394 s
Σ_G	0.03	0.03	23	21	4745.160 s
Σ_G	0.015	0.015	23	23	27260.598 s

References

1. Arnon, D.S., Collins, G., McCallum, S.: Cylindrical algebraic decomposition, II: an adjacency algorithm for plane. SIAM J. Comput. **13**(4), 878–889 (1984)
2. Berberich, E., et al.: EXACUS: efficient and exact algorithms for curves and surfaces. In: Brodal, G.S., Leonardi, S. (eds.) ESA 2005. LNCS, vol. 3669, pp. 155–166. Springer, Heidelberg (2005). https://doi.org/10.1007/11561071_16
3. Brake, D.A., Bates, D.J., Hao, W., Hauenstein, J.D., Sommese, A.J., Wampler, C.W.: Algorithm 976: bertini real: numerical decomposition of real algebraic curves and surfaces. ACM Trans. Math. Softw. **44**(1), 10 (2017)
4. Chen, F.L., Feng, Y., Kozak, J.: Tracing a plane algebraic curve. Appl. Math. J. Chin. Univ. **12**(1), 15–24 (1997)
5. Cheng, J.-S., Lazard, S., Peñaranda, L., Pouget, M., Rouillier, F., Tsigaridas, E.: On the topology of the real algebraic plane curves. Math. Comput. Sci. **4**, 113–117 (2010)
6. Cheng, J.-S., Wen, J., Zhang, W.: Level set sweeping method for bivariate function(s) and its applications, manuscript (2018)
7. Chandler, R.E.: A tracking algorithm for implicitly defined curves. IEEE Comput. Graphics Appl. **8**(2), 83–89 (1988)
8. Christoforou, E., Mantzaflaris, A., Mourrain, B., Wintz, J.: Axl, a geometric modeler for semi-algebraic shapes. In: Proceeding of ICMS 2018 (2018)
9. Gao, X.S., Li, M.: Rational quadratic approximation to real algebraic curves. Comput. Aided Geom. Des. **21**, 805–828 (2004)
10. González-Vega, L., Necula, I.: Efficient topology determination of implicitly defined algebraic plane curves. Comput. Aided Geom. Des. **19**, 719–743 (2002)
11. Goldman, R.: Curvature formulas for implicit curves and surfaces. Comput. Aided Geom. Des. **22**(7), 632–658 (2005)
12. Gomes, A.J.P.: A continuation algorithm for planar implicit curves with singularities, Special Section on CAD/Graphics 2013. Comput. Graph. **38**, 365–373 (2014)
13. Lien, J.-M., Sharma, V., Vegter, G., Yap, C.: Isotopic arrangement of simple curves: an exact numerical approach based on subdivision. In: Proceeding of International Congress on Mathematical Softwares (ICMS), Seoul (2014)
14. Lorensen, W.E., Cline, H.E.: Marching cubes: a high resolution 3D surface construction algorithm. In: Proceedings of SIGGRAPH 1987. ACM Press (1987)
15. Liang, C., Mourrain, B., Pavone, J.P.: Subdivision methods for the topology of 2D and 3D implicit curves. In: Jüttler, B., Piene, R. (eds.) Geometric Modeling and Algebraic Geometry. Springer, Heidelberg (2008). https://doi.org/10.1007/978-3-540-72185-7_11

16. Lin, L., Yap, C.: Adaptive isotopic approximation of nonsingular curves: the parameterizability and nonlocal isotopy approach. Discrete Comput. Geom. **45**(4), 760–795 (2011). Special Issue: 25th Annual Symposium on Computational Geometry SOCG 2009
17. Martin, R., Shou, H., Voiculescu, I., Bowyer, A., Wang, G.: Comparison of interval methods for plotting algebraic curves. Comput. Aided Geom. Des. **19**, 553–587 (2002)
18. Plantinga, S., Vegter, G.: Isotopic meshing of implicit surfaces. Vis. Comput. **23**, 45–58 (2007)
19. Strzebonski, A.: Real root isolation for exp-log-arctan functions. J. Symbolic Comput. **47**, 282–314 (2012)
20. https://exploration.open.wolframcloud.com/objects/exploration/TranscendentalEquations.nb

Software Products, Software Versions, Archiving of Software, and swMATH

Hagen Chrapary[1] and Wolfgang Dalitz[2(✉)]

[1] FIZ Karlsruhe/Zentralblatt MATH, Karlsruhe, Germany
hagen@zentralblatt-math.org
[2] Zuse Institute Berlin (ZIB), Berlin, Germany
dalitz@zib.de
http://www.zib.de/dalitz

Abstract. Management of software information is difficult for various reasons. First, software typically cannot be reduced to a single object: information about software is an aggregate of software code, APIs, documentation, installations guides, tutorials, user interfaces, test data, dependencies on hardware and other software, etc. Moreover, secondary information about software, especially use cases and experience with employing the software, is important to communicate. Second, typically named software, which we term here a 'software product', is taken to stand for all versions of the software which can have different features and properties and may produced different results from the same input data.

Software production is a dynamic process and software development is, increasingly, widely distributed. Therefore GitHub, GitLab, Bitbucket and other platforms for sharing are used. Information about software is alos provided in different locations, on websites, repositories, portals, etc. Each resource provides information about software from a particular point of view, but the information is often not linked together. Therefore swMATH has developed a conception which covers portals and a search engines for mathematical software, persistent and citable landing pages for specific software, and a method for software archiving. Based on the publication-based approach, swMATH collects and analyses semi-automatically the existing information about mathematical software found on the Web and makes it available in a user-oriented way. In the talk, we discuss recent extensions of the swMATH conception. We focus on the connection between the swMATH landing pages and different repositories for software.

Keywords: Knowlegde management · Digital preservation
Software · swMATH

1 Introduction

The discovery, description and long-term preservation of software in a scientific environment is more and more becoming the focus of the sciences themselves.

© Springer International Publishing AG, part of Springer Nature 2018
J. H. Davenport et al. (Eds.): ICMS 2018, LNCS 10931, pp. 123–127, 2018.
https://doi.org/10.1007/978-3-319-96418-8_15

Traditionally, scientists publish in journals. These serve as a reference and secure the first-time publication rights which may be important for the authors career. Increasingly, however, results are created using software. It is well-known that the solution of the four-color problem [1] was one of the first mathematical proofs, which was essentially written down with the help of a computer program (and lead to long discussions whether this is acceptable as a proof). Today, at least in applied science, the use of software to identify new results does not need an explanation.

The publications themselves mention the software used, but it is difficult to verify and reproduce the results obtained with the software. On the one hand, it needs access to the software and the input data itself, on the other hand, it must be known with which version of the software is required to achieve the results. In addition, there are the difficulties of properly embedding software systems in the respective computing environment, such as the libraries used, the version of the operating system and possibly also the existing (e.g. special) hardware. Even the user interfaces are not self-explanatory, it requires handbooks, technical manuals, descriptions of interfaces, just to name a few. Obviously, there are many dependencies to consider when using software in a scientific environment. To make matters worse, software development is a highly dynamic process. Software is often created by several developers, there are different versions and different releases. Existing interfaces are extended or changed, the documentation must be adapted accordingly. Ultimately, a long-term availability of software should be sought, which includes the individual versions and releases. In essence, it shows that the software used is an integral part of publications that does not yet find adequate consideration in these. But in a scientific context this is of particular interest.

2   The swMATH Approach

The connection of scientific publications in mathematics and used software is at the heart of the swMATH project. swMATH [2] wants to be a bridge between mathematics and applications in other sciences. A major goal of swMATH is to describe, index and present the content and other important features of the software packages in the context of the publications. The search for suitable mathematical software for research or applications should be supported. A connection with the published literature clarifies possible fields of application and application scenarios. Software is often no longer designed just for one application.

A special feature of swMATH is the publication-based approach. This means that bibliographic information from publications related to mathematical software is analyzed and used to describe the software. For each software, the list of articles that relate to it is given. In this way the embedding of the software in its mathematical context is achieved. The list of linked publications is also an indicator of the relevance of software in various fields. For a detailed discussion of this approach see the article of Wolfram Sperber' Mathematical Research

Data, Software, Models, and the Publication-based Approach' in the conference proceedings.

Originally, the database of the Zentralblatt für Mathematik, zbMATH [3], was used as a starting point for the publications. zbMATH is one of the world's most comprehensive and longest-running abstract and review service in pure and applied mathematics. The collection of publications in zbMATH is almost complete. Newer data is automatically integrated into the swMATH database. It makes sense to consider other sources established in the sciences such as arXiv.org. Conversely, software development today increasingly takes place via software platforms such as github [5], gitlab and bitbucket or software products are made accessible in institutional web servers or user-specific repositories such as CRAN [4]. These are also regularly inspected by swMATH in order to capture the most important software products in the field of scientific applications in mathematics and its environment.

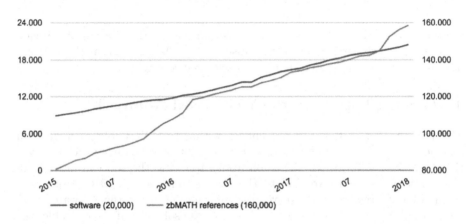

Fig. 1. Software and zbMATH references (2015–2018)

As of beginning of 2018, swMATH has identified around 20,000 software packages with more than 250,000 links generated from 160,000 scientific publications. Anyone can suggest new software or publications. The acquisition of software and the related references within swMATH are shown in Fig. 1 above.

The number of accesses to swMATH are shown in Fig. 2 below. They show a steady acceptance and a continuous growth. In three years traffic has tripled.

3   Citation Standard for Software

When analyzing scientific publications, the inconsistent style of software quotes is an obstacle. It makes automatic detection difficult and hard to find out which software was used in which version. Meanwhile, it has been recognized that there is a high need for action. Within the Software Citation Implementation Working

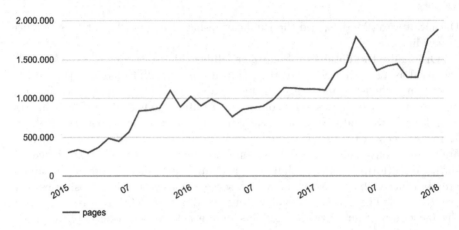

Fig. 2. Usage Statistic 2015–2018 per month including robots

Group [6] of the FORCE11 initiative [7], standards for the citation of software are developed, we participate in these discussions. The goal is that the software used needs to be clearly identifiable in a scientific context.

4   Landing Pages and the Web-Based Approach

The presentation of the software entries within swMATH follows a uniform scheme, which can be understood as the concept of a landing page. In addition to a brief description, the publications that use the software are listed. These references then also lead to the original entries of zbMATH or arXiv [8] or other sources used. Moreover the information about the software existing in the Web is used, especially websites of the software. If recognizable, the authors of the software are named and additional information is listed, such as homepages, license terms, versions, dependencies on other software and references to possible archives, where one can find this software. Within swMATH, each software receives a unique, persistent ID. This is included in the URL, so that any software can be quoted and referenced via swMATH. The scheme is www.swmath.org/software/ID.

5   Archiving

One of the biggest challenges today is finding and retrieving the software used in the specific version. Commercially-operated software usually relies on the archiving by the producers, the repositories and archives in the open-source context are usually not permanently accessible and secured. Often one finds copies of the website in the Internet Archive [9], which also reflects the timing of its collection. However, the completeness of the archived pages is quite different. swMATH links the landing page of the software with the Internet Archive wherever there are entries.

For evaluation and reproducing results which were achieved by software the access to the software code plays a central role which are especially provided on developer platforms. Moreover the Software Heritage Project [10] is paving the way to fully capture all versions of software code created through github or similar distributed development platforms. swMATH is strongly interested to extend the swMATH landing pages to the software archive run by the Software Heritage Project. This extension is currently under work.

6   Summary

In this talk we discuss the need for a powerful information infrastructure for mathematical software. The main concepts of swMATH are presented. Based on the publication-oriented approach, the web-oriented approach is pursued, which provides additional important information about the software products on landing pages. The current development of swMATH focuses on the linking of landing pages with software archives available on the Internet, which also takes into account different versions.

References

1. Appel, K., Haken, W.: Every Planar Map Is Four Colorable. Contemporary Mathematics, vol. 98. American Mathematical Society, Providence (1989)
2. swMATH. http://www.swmath.org
3. zbMATH. https://www.zbmath.org
4. CRAN. https://cran.r-project.org/
5. Github. https://github.com/
6. FORCE11 Software Citation Group. https://www.force11.org/group/software-citation-working-group
7. FORCE11 Initiative. https://www.force11.org/
8. arXiv. https://arxiv.org/
9. Internet Archive. https://archive.org/
10. Software Heritage. https://www.softwareheritage.org/

Axl, a Geometric Modeler
for Semi-algebraic Shapes

Emmanouil Christoforou[1,2]($\boxtimes$), Angelos Mantzaflaris[3], Bernard Mourrain[1],
and Julien Wintz[1]

[1] Inria Sophia Antipolis - Méditerranée, Sophia Antipolis, France
{emmanouil.christoforou,bernard.mourrain,julien.wintz}@inria.fr
[2] National and Kapodistrian University of Athens, Athens, Greece
echristo@di.uoa.gr
[3] Johannes Kepler University, Linz, Austria
angelos.mantzaflaris@oeaw.ac.at
http://team.inria.fr/aromath/

Abstract. We describe the algebraic-geometric modeling platform AXL,
which provides tools for the manipulation, computation and visualisation
of semi-algebraic models. This includes meshes, basic geometric objects
such as spheres, cylinders, cones, ellipsoids, torus, piecewise polynomial
parameterisations of curves, surfaces or volumes such as b-spline parameterisations, as well as algebraic curves and surfaces defined by polynomial equations. Moreover, AXL provides algorithms for processing these
geometric representations, such as computing intersection loci (points,
curves) of parametric models, singularities of algebraic curves or surfaces, certified topology of curves and surfaces, etc.

We present its main features and describe its generic extension mechanism, which allows one to define new data types and new processes on the
data, which benefit from automatic visualisation and interaction facilities. The application capacities of the software are illustrated by short
descriptions of plugins on algebraic curves and surfaces and on splines
for Isogeometric Analysis.

Keywords: Semi-algebraic model · Isogeometric analysis · b-splines
Algebraic surface · Algebraic-geometric computation
Generic programming

1 Introduction

Geometric modeling aims at providing shape descriptions and at developing computational tools for processing the models. It has strong interactions with other
application domains such as graphical rendering and visualisation, Computer
Aided Design and Computer Aided Manufacturing, numerical simulation, etc.

Many of the models which are used are semi-algebraic sets. Meshes, classically
used to approximate shapes, are piecewise linear models. b-splines or NURBS

© Springer International Publishing AG, part of Springer Nature 2018
J. H. Davenport et al. (Eds.): ICMS 2018, LNCS 10931, pp. 128–136, 2018.
https://doi.org/10.1007/978-3-319-96418-8_16

curves and surfaces used in CAD-CAM are the images of piecewise polynomial or rational parametrisation maps. Natural quadrics such as spheres, ellipsoids, cylinders, cones or higher order surfaces such as torus are algebraic surfaces defined by polynomial equations. Semi-algebraic models of order higher than one have interesting properties of approximation and regularity, allowing to construct high quality shape representations.

However, currently very few software are able to manipulate these different types of semi-algebraic sets. Software like MESHLAB are PARAVIEW, propose tools for visualization and computations with meshes. BLENDER or RHINO allow one to manipulate b-spline parametric objects. Software like SURF are able to render algebraic surfaces, but does not provide facilities to compute with them.

The goal of the AXL development project (axl.inria.fr) is to provide tools for the manipulation, computation and visualisation of semi-algebraic models of higher order. This includes meshes, basic algebraic objects, b-spline parameterisations of curves, surfaces or volumes and semi-algebraic sets defined by polynomial equations. Additionally, AXL provides algorithms to process these geometric representations such as computing intersection points or curves of parametric models, singularities of algebraic curves or surfaces, certified topology of curves and surfaces, etc.

To cope with the versatility of shape representations, AXL integrates a generic extension mechanism, which allows one to define new data types and new processes on this data. As soon as these new instances are constructed, visualisation and interaction facilities are provided essentially automatically. Via the production of dedicated plugins, external tools can be easily embedded, tested and demonstrated in this framework.

In Sect. 2, we describe the main feature of AXL platform. In Sect. 3, we describe the design of the code and its extension mechanism. In Sect. 4, we present applications, with a short description of plugins, respectively, on algebraic curves and surfaces and on splines for Isogeometric Analysis.

2   Functionalities

AXL software provides different types of semi-algebraic representations of shapes used in geometric modeling, such as the image of piecewise polynomial or rational maps from a bounded parameter domain into $\mathbb{R}^2$ or $\mathbb{R}^3$ or solutions of polynomial equations or geometric constructions on these objects.

Basic geometric objects such as points, segments, circular arcs, planes, spheres, cylinders, cones, ellipsoids or tori are available in the AXL library. These types correspond to specific classes with compact representations (`axlPoint`, `axlLine`, `axlSphere`, ...). For instance, an ellipsoid is represented by a center point and 3 orthogonal vectors defining its principal axis. The coordinates of these objects are stored as floating point numbers (`double` precision in the IEEE 754 standard). These objects can be edited interactively in two ways: either graphically via widget actors in the view window or through the object inspector panel of the application by changing directly the value of the numerical data representing the object. More complex objects such as meshes (`axlMesh`) are also available; they are represented as arrays of points, edges and faces with an arbitrary number of vertices. Normals or other attributes can be attached to the points.

b-spline and NURBS (Non-Uniform Rational B-Spline) parametrisations of curves, surfaces and volumes are provided by a specialized plugin `bsplinetools`, which is based on the library GoTools[1] developed by SINTEF. Such parametrisation maps are represented by arrays of control points, and knot sequences. Abstract classes (`axlAbstractCurveBSpline`, `axlAbstractSurfaceBSpline`, ...) specify the available methods, independently of the internal representation of the data. Dedicated widget actors allow one to edit dynamically these objects by changing graphically the control points, using the methods of the abstract interface classes.

Algebraic curves and surfaces defined by polynomial equations are implemented in the plugin `semialgebraictools`. They are represented by arrays of multivariate polynomials, with exact or approximate coefficients. They are embedded in the AXL framework, through abstract interface classes (`axlAbstractCurveAlgebraic`, `axlAbstractSurfaceAlgebraic`).

Geometric types of the AXL library derive from the generic class `axlAbstractData` and share color, transparency and shader attributes. Additionally, field attributes can be attached to the geometric objects. They can be scalar fields, visualized by a color map or vector fields visualized by small arrows. Their representations can depend on the type of the supporting geometric object. They can be discrete values at the vertices of a mesh, functions of the parameters on a parametric curve or surface or functions of the spatial coordinates of the points on the geometric object (Fig. 1).

(a) A collection of b-spline surfaces with different color attributes.   (b) A mesh with a spatial scalar field visualized by a color map.

Fig. 1. Visualization of different types of geometric objects

---

[1] https://www.sintef.no/projectweb/geometry-toolkits/gotools/.

To be able to exchange data, each data type in AXL is equipped with a reader and writer class. The writer class writes the information that define the objects in a XML-like format, which is simple to access and exploit. The reader class reads the XML text and constructs the corresponding AXL object. Here is an example of format for an ellipsoid, with name and color attributes:

```
<ellipsoid name="E0" color="160 0 32 1">
  <center>0.417312 0.466944 0.729041</center>
  <semix>0.603441 0.0735849 -0.425506</semix>
  <semiy>-0.00639521 0.395524 0.0593304</semiy>
  <semiz>0.116345 -0.022291 0.161143</semiz>
</ellipsoid>
```

The library AXL also provides facilities to run computation on geometric objects. This is implemented via the concept of process, corresponding to the abstract class `axlAbstractProcess`. A process class has input data, a `run` method and output data. This construction allows one to run interactively specific computations, implemented in the `run` command, on selected data and to view the result of the computation. Processes can be selected in the tool inspector panel of the application user interface and executed on the data selected from the user interface. Processes can also be stored in XML-format via their writer class and executed on data via their reader class.

A special type of process allows one to update interactively the result. It involves the, so called, dynamic data class (`axlDataDynamic`), consisting of input data, a process and output data. When the input data is modified, the output of the dynamic data is recomputed. Output data can be the input data of other dynamic objects. This allows one to develop complex constructions of geometric objects, which are updated interactively when some of their components are modified.

Another type of process allows one to create animations in AXL. These consist of input data and a `run` command, which transforms the input data according to a time parameter and visualizes the transformed data according to this time parameter.

3   A Generic Platform for Geometric Computation

The AXL application is developed in C++, depending on the *Qt*, *VTK* [12] and *dtk* [13] libraries. *Qt* is a cross-platform application development framework for applications that is compatible with various software and hardware platforms and is used for the graphical user interface of AXL. The *dtk* library is a meta-platform for modular scientific platform development, which provides generic tools for data, processes and views. The visualization plugin of AXL, `axlVtkView`, is based on *VTK*, which is an open source software system for 3D computer graphics, image processing, and visualization.

Using polymorphism, which can be either static or dynamic, AXL is a modular platform, in the sense that it formalizes abstract concepts such as data, processes or views. The latter are then virtual abstractions which can be specialized through plugins, dynamic libraries loaded at runtime fulfilling an abstraction specification. The extensive use of design patterns, such as factory, template method etc. makes it easy to specify the actual behavior of an algorithm by selecting a combination of processes acting on various data representations via their abstraction.

As a matter of fact, AXL provides a generic interface for geometric concepts that which curves, surfaces and volumes with different representations (that can be implicit, explicit or piecewise linear) and processes such as differentiation, intersection, arrangements, singular points computation etc. Instances of these concepts are commonly implemented by third-party libraries, using diverse algorithms under the hood, that are, in principle, conflicting one another, making their combination problematic. This problem is tackled by the abstraction level of AXL, ensuring the consistency of the different implementations.

Let us describe some of them starting with the virtual hierarchy of data, then some process and how they are combined in order to implement an algorithm. We will not focus on the view concept implemented in the plugin `axlVtkView` based on VTK, which basically renders the meshes output by the converters of the semi-algebraic models and instantiates the graphical actors.

Starting from the virtual hierarchy of data in AXL, in a simple case they inherit from the class `axlAbstractData`, described in previous section, which inherit from `dtkAbstractData`. Figures 2a, b show the inheritance of `axlPoint` and `axlLine` with more complex hierarchy. Likewise, AXL processes inherit from class `axlAbstractProcess`, also described previously, that inherit from `dtkAbstractProcess` (Fig. 2c, `axlIntersection` example). The use of the abstract classes of data as the processes default input and output, allows processes to handle multiple data-types when possible. In particular, the different data-types and processes acquire a common input and output, thus they can be easily combined or changed in their algorithmic implementation.

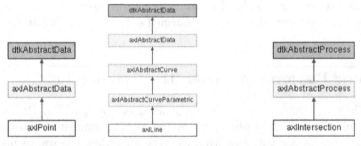

(a) Inheritance of data. `axlPoint` is a point defined by 3 coordinates.

(b) Inheritance of data. `axlLine` is a line (or segment) defined by a starting and an ending point (`axlPoints`) deriving from the interface abstract class of parametric curve.

(c) Inheritance of processes. `axlIntersection` intersects two lines and returns the point of intersection.

Fig. 2. AXL inheritance

AXL provides also the tools to extend data-types and processes. New data-types, which inherit from `axlAbstractData`, are created having their own reader, writer, creator, converter etc., by inheriting the corresponding abstract class (`axlAbstractDataReader`, `axlAbstractDataWriter`, etc.), in order to be properly integrated and functional in AXL. Also, new processes can be implemented using their abstraction (`axlAbstractProcess`) by defining the input data, a run method and the output data. The new data-types and processes can be used in AXL as plugins, by creating new packages.

4 Applications

In this section we describe two application plugins, that tightly integrate into AXL tools for real algebraic curves and surfaces and for isogeometric analysis.

4.1 Topology of Real Algebraic Sets

The AXL framework has been used to develop a plugin called `semialgebraictools`, dedicated to the topology analysis of algebraic curves and surfaces and to the computation of arrangements of such objects. Algebraic curves and surfaces are defined as the real solutions of polynomial equations. In this implementation, planar algebraic curves defined by one equation, curves in $\mathbb{R}^3$ defined by two equations and surfaces in $\mathbb{R}^3$ defined by one equation are considered. The polynomials are represented in the Bernstein basis associated to a given domain (that is, an axis aligned box of dimension 2 or 3) by a matrix or a tensor of control coefficients. The topology of the algebraic objects is analyzed in this region. Subdivision methods are used to compute a mesh approximation of the algebraic set, which is topologically certified. The subdivision steps consists in splitting the domain in one direction and in computing the Bernstein basis representation on each subdomain, using de Casteljau algorithm. Regularity criteria are used to determine whether the topology of the algebraic object can be determined from its intersection points with the edges (or faces) of the box [1].

The computation is performed on the tensor representation in the Bernstein bases with lower and upper approximate coefficient bounds (`double` type of the IEEE 754 standard). By choosing adequately the rounding mode during the computation, the exact value of the coefficients is guaranteed to stay between the computed lower and upper bounds [11].

New data types encoding the bounding box domain and the polynomial equations have been implemented in this plugin. The visualization of the algebraic sets is performed by a converter class, which computes a mesh from the polynomial equations by subdivision methods.

These subdivision methods have been used to compute the topology of algebraic curves [2,9] and algebraic surfaces [3], arrangements of curves [4], semi-algebraic sets [10] and Voronoï diagrams of curved objects [5], see Fig. 3.

 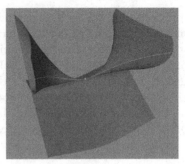

(a) The regions defined by an algebraic curve with singular points of degree 28, which contains the medial axis of two ellipses.

(b) An algebraic surface with a singular curve (in red) and a special singular point (in orange) defining its Whitney stratification and the apparent contour (in light blue).

Fig. 3. Views of an algebraic curve and an algebraic surface. (Color figure online)

4.2 Modeling and Simulation

Isogeometric Analysis is a new, innovative numerical technique that generalizes the Finite Element Method and uses splines or NURBS, normally used in Computer Aided Design, for both representing the geometry of the computational (physical) domain and for approximating the solution of the considered partial differential equation. In this paragraph we present the related AXL plugin based on the G+SMO library (http://www.gs.jku.at/gismo).

G+SMO is an open-source, object-oriented C++ library for isogeometric analysis [7,8]. The library makes use of object polymorphism and inheritance techniques in order to support a variety of different discretisation bases, namely b-spline, Bernstein, NURBS bases, hierarchical and truncated hierarchical b-spline bases of arbitrary polynomial order. The implementation of basis functions and geometries is dimension-independent, that is, curves, surfaces, volumes, bulks (in 4D) and other high–dimensional objects are instances of code templated with respect to the parameter domain dimension.

Three general guidelines have been set for the development process. Firstly, we promote both efficiency and ease of use; secondly, we focus on code quality and cross-platform compatibility and, thirdly, we encourage the exploration of new strategies, better suited for isogeometric analysis before adopting existing finite element practices.

The library is partitioned into modules that implement different functionalities. A basic module that is available is the *NURBS module*, which provides a dimension independent implementation of classical tensor-product b-splines and their rational counterpart. On top of the NURBS module we implemented the *hierarchical splines module* [6]. The functionalities can be used seamlessly in AXL via our plugin; Fig. 4 shows two instances of its use.

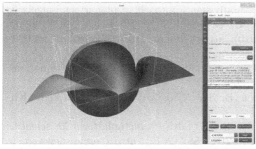 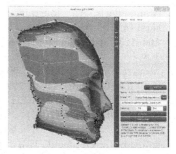

(a) Peeling of a NURBS sphere interactively in Axl. The original sphere is represented by a bi-quadratic tensor-product NURBS surface. Editing triggers evaluation of both the surface and the scalar field on a grid of points in real-time.

(b) A THB-spline model in Axl; note the accumulation of control point near the mouth region. A shader using isophotes is applied.

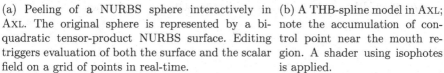

Fig. 4. Two snapshots of the G+Smo plugin.

References

1. Alberti, L., Mourrain, B.: Regularity criteria for the topology of algebraic curves and surfaces. In: Martin, R., Sabin, M., Winkler, J. (eds.) Mathematics of Surfaces XII. LNCS, vol. 4647, pp. 1–28. Springer, Heidelberg (2007). https://doi.org/10.1007/978-3-540-73843-5_1

2. Alberti, L., Mourrain, B.: Visualisation of implicit algebraic curves. In: Pacific Conference on Computer Graphics and Applications, Lahaina, Maui, Hawaii, United States, pp. 303–312. IEEE Computer Society, October 2007

3. Alberti, L., Mourrain, B., Técourt, J.P.: Isotopic triangulation of a real algebraic surface. J. Symb. Comput. **44**(9), 1291–1310 (2009)

4. Alberti, L., Mourrain, B., Wintz, J.: Topology and arrangement computation of semi-algebraic planar curves. Comput. Aided Geom. Des. **25**(8), 631–651 (2008)

5. Emiris, I., Mantzaflaris, A., Mourrain, B.: Voronoi diagrams of algebraic distance fields. Comput. Aided Des. **45**(2), 511–516 (2013)

6. Giannelli, C., Juettler, B., Kleiss, S.K., Mantzaflaris, A., Simeon, B., Speh, J.: THB-splines: an effective mathematical technology for adaptive refinement in geometric design and isogeometric analysis. Comput. Methods Appl. Mech. Eng. **299**, 337–365 (2016)

7. Juettler, B., Langer, U., Mantzaflaris, A., Moore, S., Zulehner, W.: Geometry + simulation modules: implementing isogeometric analysis. Proc. Appl. Math. Mech. **14**(1), 961–962 (2014)

8. Langer, U., Mantzaflaris, A., Moore, S.E., Toulopoulos, I.: Multipatch discontinuous galerkin isogeometric analysis. In: Jüttler, B., Simeon, B. (eds.) Isogeometric Analysis and Applications 2014. LNCSE, vol. 107, pp. 1–32. Springer, Cham (2015). https://doi.org/10.1007/978-3-319-23315-4_1

9. Liang, C., Mourrain, B., Pavone, J.P.: Subdivision methods for the topology of 2D and 3D implicit curves. In: Juettler, B., Piene, R. (eds.) Geometric Modeling and Algebraic Geometry, pp. 199–214. Springer, Heidelberg (2007). https://doi.org/10.1007/978-3-540-72185-7_11

10. Mantzaflaris, A., Mourrain, B.: A subdivision approach to planar semi-algebraic sets. In: Mourrain, B., Schaefer, S., Xu, G. (eds.) GMP 2010. LNCS, vol. 6130, pp. 104–123. Springer, Heidelberg (2010). https://doi.org/10.1007/978-3-642-13411-1_8
11. Mourrain, B., Pavone, J.P.: Subdivision methods for solving polynomial equations. J. Symb. Comput. **44**(3), 292–306 (2009)
12. Schroeder, W., Martin, K., Lorensen, B.: The Visualization Toolkit, 4th edn. Kitware, Clifton Park (2006)
13. Wintz, J., Kloczko, T., Niclausse, N., Rey, D.: dtk - a metaplatform for scientific software development. ERCIM News **2012**(88) (2012). http://ercim-news.ercim.eu/en88/ri/dtk-a-metaplatform-for-scientific-software-development

Efficient and Secure Delegation
to a Single Malicious Server:
Exponentiation over Non-abelian Groups

Giovanni Di Crescenzo[1]([✉]), Delaram Kahrobaei[2], Matluba Khodjaeva[3],
and Vladimir Shpilrain[4]

[1] Perspecta Labs, Basking Ridge, NJ, USA
gdicrescenzo@perspectalabs.com
[2] City University of New York, New York, NY, USA
DKahrobaei@gc.cuny.edu
[3] John Jay College, City University of New York, New York, NY, USA
mkhodjaeva@jjay.cuny.edu
[4] City University of New York, New York, NY, USA
shpil@groups.sci.ccny.cuny.edu

Abstract. Group exponentiation is an important and expensive operation used in many public-key cryptosystems and, more generally, cryptographic protocols. To expand the applicability of these solutions to computationally weaker devices, it has been advocated that this operation is delegated from a computationally weaker client to a computationally stronger server. Solving this problem in the case of a single, possibly malicious, server, has remained open since a formal model was introduced in [8]. Recently, in [10] we proposed practical and secure solutions applicable to a class of cyclic groups. In this paper, we propose efficient and secure solutions applicable to a large class of multiplicative groups, possibly beyond groups currently subject to quantum cryptanalysis attacks.

1 Introduction

In emerging applications related to Cloud Computing and the Internet of Things, including RFID networks, interest is growing on deploying cryptography solutions onto computationally weaker devices. To achieve that goal, it has been advocated that the most expensive cryptographic operations are delegated from a computationally weaker client to a computationally stronger server. Group exponentiation is an important operation and among the most expensive ones used in many public-key cryptosystems and, more generally, cryptographic protocols. Many studies have already been performed towards various types of delegation of group exponentiation, but almost exclusively in the case of abelian

D. Kahrobaei and V. Shpilrain—Research of Delaram Kahrobaei was partially supported by a PSC-CUNY grant from the CUNY research foundation, as well as the City Tech foundation. Research of Vladimir Shpilrain was partially supported by the NSF grant CNS-1117675. Research of Delaram Kahrobaei and Vladimir Shpilrain was also supported by the ONR (Office of Naval Research) grant N000141210758.

© Springer International Publishing AG, part of Springer Nature 2018
J. H. Davenport et al. (Eds.): ICMS 2018, LNCS 10931, pp. 137–146, 2018.
https://doi.org/10.1007/978-3-319-96418-8_17

groups; specifically, groups related to discrete logarithm or factoring problems (see, e.g., [4,5,8,10] and references therein).

As progresses are being made towards building a large-scale quantum computer, much attention is being devoted in the cryptography community to early quantum computer algorithms such as Shor's [9], capable of solving in quantum polynomial time both the discrete logarithm and the factoring problem. More specifically, the problem at the heart of Shor's algorithms, also known as the hidden subgroup problem, can be solved in quantum polynomial time over any finite abelian group, but currently seems much harder over non-abelian groups. Therefore, the study of cryptographic solutions over non-abelian, or just general, groups is an appealing research direction within quantum-resistant cryptography (see, e.g., [1,6,7] and references therein).

In this paper we consider the delegation of group exponentiation over a large class of general multiplicative groups, not limited to abelian groups and thus going beyond groups currently subject to quantum cryptanalysis attacks.

Our Contributions. We show two interactive protocols allowing a client to delegate exponentiation in a general class of groups to a single, possibly malicious, server, while satisfying natural requirements of correctness (i.e., if client and server follow the protocol, then at the end of the protocol execution, the client's output is the desired exponentiation), security (i.e., if the client follows the protocol, no malicious adversary corrupting the server can convince the client of an incorrect exponentiation, except with small probability), privacy (i.e., if the client follows the protocol, no malicious adversary corrupting the server can obtain some information about the client's input exponent), and efficiency (i.e., the client's runtime is smaller than in a non-delegated computation of the exponentiation). Our first protocol, in Sect. 3.1, consists of a direct parallel repetition of (a slightly simplified version of) a protocol from [3] that achieves security probability 1/2. Our main result, in Sect. 3.2, is a parameterized class of protocols where, for some parameter values, the security probability is reduced more efficiently than by direct parallel repetition. Their privacy and security properties are satisfied even if the adversary corrupting the server is not limited to run in (classical or quantum) polynomial time, and they achieve an efficiency tradeoff, in that they improve the client's runtime during the online protocol phase, while increasing the server's runtime and requiring offline computations returning data to be stored on the client's device. Our theoretical analysis, only considering group exponentiations and multiplications, and neglecting simpler operations such as equality checks and random element generations, suggests that our first (resp., second) protocol reduces the client's online runtime by 1 (resp., 2) orders of magnitude with respect to the textbook exponentiation algorithm, while increasing the server runtime and the protocol communication complexity by 2 (resp., 1) orders of magnitude and the offline client runtime between a constant and 1 order of magnitude. Our software implementation, in Python 3.6, using commodity computing resources and the gmpy2 package, confirms that both our protocols improve the client's online runtime with respect to the exponentiation algorithm available in the same package.

As in all previous work in the area, we consider a model with an offline phase, where a client or another party can precompute fixed-base exponentiations to random exponents, and store them on the client's device to be later used in the online protocol phase. We also consider a model where a client can efficiently run group multiplications, which is partially justified by known application on some RFID devices results (see, e.g., [2]). Our protocols are written so to delegate $F_{G,exp,g}(x) = g^x$ (i.e., variable-exponent, fixed-base exponentiation over multiplicative group G), but can be reformulated so to delegate function $F_{G,exp,k}(x) = x^k$ (i.e., fixed-exponent, variable-base exponentiation).

2   Models and Definitions

In this section we define delegation protocols, and their correctness, security, privacy and efficiency requirements, building on the definitional approach from [3] (also based on [5,8]), and describe group notations and protocol preliminaries.

Participant and Protocol Models. We consider two types of parties: clients and servers, where a client's computational resources are expected to be more limited than a server's ones, and therefore clients are interested in delegating the computation of specific functions to servers. In all our solutions, we consider a single *client*, denoted as C, and a single *server*, denoted as S. We assume that the communication link between each C and S is not subject to confidentiality, integrity, or replay attacks, and note that such attacks can be separately addressed using well-known cryptography techniques. A *client-server protocol for the delegated computation of function F* is an interactive protocol between C and S, where both parties have a description of a function F, C knows an input x, and at the end of a protocol execution, C outputs a value y (intended to be $= F(x)$). The protocol can have two phases: an *offline phase*, including expensive computations not based on input x, such as evaluating F on other inputs, and an *online phase*, where C's computations are based on input x but take less time than what required to compute $F(x)$. We require such protocols to satisfy the following requirements of *correctness, security, privacy* and *efficiency*.

Correctness. Informally speaking, the correctness requirement states that if both parties follow the protocol, at the end of the protocol execution, C's output y is, with high probability, equal to the output of function F on C's input x.

Security. Informally speaking, the security requirement states that if C follows the protocol, a malicious adversary corrupting S and even choosing C's input x can only convince C with a small probability to output, at the end of the protocol, some y' different from value $y = F(x)$ or some failure symbol $\perp$. We will also call this probability as the *security probability*, and denote it as ϵ_s. A desirable value for it will be $2^{-\lambda}$, for some *statistical security parameter* λ, concretely set as, for instance, 128.

Privacy. Informally speaking, the privacy requirement states the following: if C follows the protocol, a malicious adversary corrupting S cannot obtain any information about C's input x from a protocol execution. This is formalized by extending the indistinguishability-based approach typically used in formal definitions for encryption schemes. That is, the adversary can pick two inputs x_0, x_1, then one of these two inputs is chosen at random and used by C in the protocol with the adversary acting as S, and then at the end of the protocol the adversary can only guess which input was used by C with probability $1/2$.

Efficiency. We measure the efficiency of a client-server protocol (C, S) for the delegated computation of function F by the *efficiency metrics* (t_F, t_P, t_C, t_S, cc), meaning that F can be computed (without delegation) using t_F atomic operations, the offline phase requires t_P atomic operations, C requires t_C atomic operations in the online phase, S requires t_S atomic operations, and C and S exchange messages of total length at most cc. In our theoretical analysis, we only consider the most expensive group operations as atomic operations (e.g., group multiplications and/or exponentiation), and neglect lower-order operations (e.g., equality testing, random element generations, additions and subtractions over $\mathbb{Z}_n$-type groups). While we naturally try to minimize all these efficiency metrics, our main goal is to design protocols where $t_C << t_F$, even if possibly resulting in t_S being somewhat larger than t_F and cc being somewhat larger than the length of F's input and output. We note that, according to the textbook 'square-and-multiply' algorithm, t_F is, on average, $= 1.5\sigma$ group multiplications, where σ denotes the length of the binary representation of a group element. Our theoretical target are protocols where t_C is smaller than σ group multiplications.

Group Notations. Let ℓ denote the length of the binary representation of a group's elements. We say that a group is *efficient* if its description is short (i.e., has length polynomial in ℓ), its associated operation $*$ and the inverse operation are efficient (i.e., they can be executed in time polynomial in ℓ). The security parameter σ and the group element length ℓ are typically set as the same value. Let $(G, *)$ be an efficient group, and let g be an element with order q, for some large integer q known to the client, and let $y = g^x$ denote the *exponentiation (in G)* of g to the x-th power; i.e., the value $y \in G$ such that $g * \cdots * g = y$, where the multiplication operation $*$ is applied $x - 1$ times. Also, let $\mathbb{Z}_q = \{0, 1, \ldots, q-1\}$ and let $F_{G,exp,g} : \mathbb{Z}_q \to G$ denote the function that maps every $x \in \mathbb{Z}_q$ to the exponentiation (in G) of g to the x-th power.

Protocol Preliminaries. In all our protocols, inputs common to client and server include a description of the function $F_{G,exp,g}$ to be delegated, a description of group G, a group element g, a computational parameter 1^σ and a security parameter 1^λ. Other inputs to the client include g's order q and exponent $x \in \mathbb{Z}_q$.

3   Delegating Exponentiation in General Groups

In this section we present our protocols for the delegation of exponentiation in a general class of groups to a single (possibly malicious) server. We note that general conversion techniques are known in the cryptography literature to transform a protocol secure against a honest adversary into one secure against a malicious adversary. Typically these techniques are based on zero-knowledge proofs of knowledge of secrets that certify computation correctness. In their most general version, these techniques do not perform well with respect to many efficiency metrics. Even considering their most simplified version, basic proofs of knowledge of exponents in the literature require the verifier to perform group exponentiations, which is precisely what the client is trying to delegate in our protocols. Accordingly, new techniques are needed. Our 1st protocol, in Sect. 3.1, uses a direct parallel repetition of an efficient subprotocol with security probability $1/2$, this latter subprotocol being an improved version of our scheme from (Sect. 5 of) [3]. Our 2nd protocol, in Sect. 3.2, is actually a parameterized class of protocols where, for some values of two parameters c, m, the security probability is reduced more efficiently than by direct parallel repetition.

3.1   Delegating Exponentiation: A Cut-and-choose Approach

We first describe a basic protocol (bC_1, bS_1) with constant security probability (obtained by simplifying the protocol in Sect. 5 of [3]) and then the final protocol (fC_1, fS_1), obtained as a parallel repetition of the basic protocol.

A protocol (bC_1, bS_1) with constant security probability. In an offline phase, bC_1 randomly chooses $u_0, u_1 \in \mathbb{Z}_q$ and computes $v_0 = g^{u_0}$ and $v_1 = g^{u_1}$. In the delegation phase, bC_1 randomly chooses bit $b \in \{0, 1\}$ and computes $z_b = u_b$ and $z_{1-b} = x - u_{1-b} \mod q$, and sends (z_0, z_1) to bS_1. Next, bS_1 computes $w_i = g^{z_i}$, for $i = 0, 1$ and sends (w_0, w_1) to bC_1. Finally, bC_1 checks that $w_b = v_b$; if not, bC_1 returns failure symbol $\perp$; otherwise, bC_1 returns $y = w_{1-b} * v_{1-b}$.

We now show that protocol (bC_1, bS_1) satisfies correctness, privacy, security (with probability $1/2$), and efficiency (with $t_C = 1$ multiplication plus 1 subtraction, $t_S = 2$ exponentiations, and $t_P = 2$ exponentiations).

The *efficiency* properties are verified by protocol inspection. The *correctness* property follows by observing that if bC_1 and bS_1 follow the protocol, bC_1's equality verification is satisfied, and thus C's output y satisfies $y = w_{1-b} * v_{1-b} = g^{z_{1-b}} * g^{u_{1-b}} = g^{x-u_{1-b}} * g^{u_{1-b}} = g^x$, which implies that $y = F_{G,exp,g}(x)$ for each $x \in G$. The *privacy* property follows by observing that the message z_0, z_1 sent by bC_1 does not leak any information about x, since they are randomly and independently distributed in $\mathbb{Z}_q$, as so are chosen u_0 and u_1. To see that the *security* property is satisfied, for any probabilistic polynomial-time adversary corrupting bS_1, consider the values w_0, w_1 returned by the adversary to bC_1. If the adversary honestly computes $w_i = g^{z_i}$ for both $i = 0, 1$, then the probability it fools bC_1 into an incorrect output y is 0. Thus, assume the adversary computes $w_c \neq g^{z_c}$, for some bit $c \in \{0, 1\}$. Then note that bC_1 will find this out and return failure symbol $\perp$ when $b = c$, and, since the message (z_0, z_1) leaks no information about b, the equality $b = c$ holds with probability at least $1/2$. This implies that the probability that the adversary fools bC_1 into an incorrect output y is $\leq 1/2$.

A protocol (fC_1, fS_1) with exponentially small security probability. Protocol (fC_1, fS_1) consists of λ parallel executions of the basic protocol (bC_1, bS_1), with the only additional modification that the output of fC_1 is defined as y if in all

λ parallel executions bC_1 would return the same value y, or as failure symbol $\perp$ otherwise (that is, if bC_1 returns $\perp$ in any one of the parallel executions, or two different values $\neq \perp$ in any two of the parallel executions).

Protocol (fC_1, fS_1) satisfies correctness, privacy, security (with probability $1/2^\lambda$), and efficiency (with $t_C = \lambda$ multiplications plus λ subtractions, $t_S = 2\lambda$ exponentiations, $t_P = 2\lambda$ known-base exponentiations to random exponents, and $cc = O(\lambda\sigma)$). The proof of these properties is a direct extension of the proofs for the properties of (bC_1, bS_1).

We remark that for the typical setting $\lambda = 128$, C only performs 128 group multiplications and 128 subtractions modulo q. This is about 1 order of magnitude smaller than 1.5σ, the average number of group multiplications in the square-and-multiply algorithm, which can be $= 3072$, for the setting $\sigma = 2048$ which has been recommended on some commonly used groups in cryptography.

3.2   Delegating Exponentiation: Improved Probability Reduction

In this subsection we improve the approach in Sect. 3.1 by a computation-efficient (in terms of C's parameters t_P, t_C) reductions of the security probability ϵ_s. Our overall approach towards this goal can be briefly summarized as follows: first, we propose a basic protocol (bC_2, bS_2) with improved constant security probability and then define a final protocol (fC_2, fS_2) that performs a suitable parallel repetition (with a smaller number of repetitions) of this basic protocol.

Informal Discussion. Our main approach consists of reducing the security probability by a more time-efficient approach than the direct parallel repetition approach in Sect. 3.1. While we do not know how to avoid the above parallel repetition, we show that we can reduce the number of repetitions by designing a more efficient protocol with security probability much smaller than $1/2$. As a first simple example of this approach, by starting from protocol (bC_1, bS_1) with security probability $1/2$ from Sect. 3.1, and including 2 random 'decoy' values in $\mathbb{Z}_q$ in the client's message to the server, we obtain a protocol with the following properties: (1) it does *not* increase the client's number of multiplications, (2) it only slightly increases computation by the server; (3) it can be seen to reduce the security probability from $1/2$ to $1/3$. Our protocol generalizes this idea of using random decoy values in $\mathbb{Z}_q$ to a parameterized number m, also representing an upper bound on the number of values that the client sends to the server. This generalization reduces the security probability, even though not as much as we would like. Accordingly, the other idea is that of increasing the number of equality checks, and introducing a second parameter c, representing an upper bound on the number of equality checks that the client wants to execute (and thus, the number of pre-computed exponentiations that the client can afford). Specifically, in the resulting protocol, of the m values in $\mathbb{Z}_q$ sent by the client to the server, one value is used to compute the function output, $c - 1$ values are used to perform equality checks, and $m - c$ values are decoy values. The resulting protocol achieves a security probability which is, very roughly speaking, linear in $1/c$, and thus the number of repetitions to reduce the probability to $2^{-\lambda}$, can be reduced to about $\lambda/\log_2 c$. We actually define a class of protocols that is parameterized by c and m and analyze what values for these parameters give us a more time-efficient reduction of the security probability than what achieved in Sect. 3.1. The two main high-level takeaways on that analysis are: (1) a somewhat large value for m is just as good as a huge value; (2) values of $c \in \{4, \ldots, 9\}$ result in a reduced number of group multiplications from the client.

A protocol (bC_2, bS_2) with constant security probability. We first formally describe the basic protocol (bC_2, bS_2) and then discuss its properties.

Offline instructions:

1. bC_2 randomly chooses distinct $j_1, \ldots, j_m \in \{1, \ldots, m\}$
2. bC_2 randomly chooses $u_i \in \mathbb{Z}_q$, sets $v_i = g^{u_i}$ and $z_{j_i} = u_i$, for $i = 1, \ldots, c$
3. bC_2 randomly and independently chooses $z_{j_{c+1}}, \ldots, z_{j_m} \in \mathbb{Z}_q$

Online instructions:

1. bC_2 sets $z_{j_c} = (x - u_c) \mod q$ and sends $z_1, \ldots, z_m$ to bS_2
2. bS_2 computes $w_j = g^{z_j}$ for $j = 1, \ldots, m$
   bS_2 sends $w_1, \ldots, w_m$ to bC_2
3. if $w_{j_1} \neq v_{j_1}$ or $w_{j_2} \neq v_{j_2}$ or $\ldots$ or $w_{j_{c-1}} \neq v_{j_{c-1}}$ then
       bC_2 **returns:** $\perp$ and the protocol halts
       bC_2 computes $y = w_{j_c} * v_c$ and **returns:** y

We now observe that protocol (bC_2, bS_2) satisfies correctness, privacy, security (with probability $O(1/c)$), and efficiency (with $t_C = 1$ multiplication in G plus 1 subtraction in $\mathbb{Z}_q$, $t_S = m$ exponentiations, and $t_P = m$ exponentiations).

The *efficiency* properties of (bC_2, bS_2) are verified by protocol inspection. The *correctness* properties follows by observing that if bC_2 and bS_2 follow the protocol, none of the inequality verifications in step 3 will be satisfied. Thus, bC_2's output is $\neq \perp$ and is equal to $y = w_{j_c} * v_c = g^{z_{j_c}} * v_c = g^{z_{j_c}} * g^{u_c} = g^{x-u_c} * g^{u_c} = g^x$, which implies that bC_2's output is $= F_{G,exp,g}(x)$ for each $x \in \mathbb{Z}_q$. The *privacy* property follows by observing that the message $z_1, \ldots, z_m$ sent by bC_2 is distributed as m random and independent group values and therefore does not leak any information about x.

To prove the *security* property against a malicious bS_2 we compute an upper bound ϵ_s on the security probability that bS_2 convinces bC_2 to output a y such that $y \neq F_{G,exp,g}(x)$. This is performed by a long case analysis (omitted here for lack of space) for all c and m, and depending on whether the server replies to the client with correct or incorrect answers. Examples of results from this analysis include the following: (1) when $c = 2$, using $m - 2$ decoy elements in $\mathbb{Z}_q$, we have that ϵ_s gets very close to $1/4$ as m grows; (2) when $c = 3$, using $m - 3$ decoy elements in $\mathbb{Z}_q$, we obtain $\epsilon_s < 1/6$ when $m = 100$, and increasing m to 1000 does not reduce ϵ_s significantly. For general c, m, we computed the exact value for ϵ_s for all values of c that guarantee some improved efficiency on the number t_C (of client's group multiplications during the protocol). Specifically, we looked at all values of c such that the obtained ϵ_s is smaller than what could be obtained by a parallel repetition of $\lfloor c/2 \rfloor$ executions of the atomic protocol from Sect. 3.1 with security probability $1/2$. It turns out that only values $c = 4, 5, \ldots, 9$ guarantee some improved efficiency on t_C, with respect to the protocol in Sect. 3.1. The obtained values for ϵ_s when $c = 4, \ldots, 10$ are in Table 1 below. Note that when $c = 4, \ldots, 9$ the obtained value for ϵ_s is strictly smaller than the value $2^{-\lfloor c/2 \rfloor}$ that could be obtained using the protocol from Sect. 3.1. Instead, when $c = 10$, the value $\epsilon_s = 0.03894$ is $> 0.03125 = 2^{-5}$, and the protocol from Sect. 3.1 starts offering a much better efficiency tradeoff.

Table 1. Values of ϵ_s for protocol (bC_2, bS_2), for $c = 4, \ldots, 10$ and $m = 100, 1000$

$c =$	4	5	6	7	8	9	10
$m = 100, \epsilon_s =$	.10763	.08403	.06719	.05875	.05118	.04538	.04080
$m = 1000, \epsilon_s =$	.10568	.08213	.06529	.05686	.04929	.04351	.03894

A protocol (fC_2, fS_2) with exponentially small security probability. Protocol (fC_2, fS_2) consists of $r = \lceil \lambda / \log(1/\epsilon_s) \rceil$ parallel executions of the basic protocol (bC_2, bS_2), with the only additional modification that the output of fC_1 is defined as y if in all λ parallel executions bC_1 would return the same value y, or as failure symbol $\perp$ otherwise (that is, if bC_1 returns $\perp$ in any one of the parallel executions, or two different values $\neq\perp$ in any two of the parallel executions).

Protocol (fC_2, fS_2) satisfies correctness, privacy, security (with probability $1/2^\lambda$), and efficiency (with $t_C = r$ group multiplications, $rc - r$ group equality checks and r subtractions in $\mathbb{Z}_q$; $t_S = mr$ group exponentiations, $t_P = rc$ known-base group exponentiations to random exponents, and $cc = O(mr\sigma)$). The proof of these properties is obtained by extension of the proofs for the properties of (bC_2, bS_2). We remark that for the typical setting $\lambda = 128$, C performs about 30 group multiplications, 30 subtractions in $\mathbb{Z}_q$, and less than 300 group equality checks. The number of group multiplications is about 2 orders of magnitude smaller than 1.5σ, the average number of group multiplications in the square-and-multiply algorithm, which can be $= 3072$, for the setting $\sigma = 2048$, recommended on some commonly used groups in cryptography.

3.3 Implementation and Performance Results

We implemented our protocols in Sects. 3.1 and 3.2 for the multiplicative group $(\mathbb{Z}_p^*, \cdot \mod p)$, for $p = 2q + 1$, and p, q are large primes such that $|p| = 2048$. Our implementation of the offline phase, the client's online program and the server's program was carried out on a macOS High Sierra Version 10.13.4 laptop with 2.7 GHz Intel Core i5 processor with memory 8 GB 1867 MHz DDR3. The protocols were coded in Python 3.6 using the gmpy2 package. Table 2 contains parameters c, m, r, running times t_F, t_P, t_C, t_S and improvement ratio t_F/t_C

for protocol (fC_1, fS_1) from Sect. 3.1 and protocol (fC_2, fS_2) from Sect. 3.2. Here, parameter r represents the number of parallel repetitions of (bC_1, bS_1) and (bC_2, bS_2) needed to get desired security probability $\epsilon_s = 2^{-128}$ in protocols (fC_1, fS_1) and (fC_2, fS_2), respectively. The main takeaway is that in the two protocols, the client's online running time is better than non-delegated computation by half or one order of magnitude, respectively.

Table 2. Performance of protocols (fC_1, fS_1) and (fC_2, fS_2), for $\epsilon_s = 2^{-128}$

t_F	.003838							
	(fC_1, fS_1)	(fC_2, fS_2)						
c	n/a	5		6		7		8
m	n/a	60	100	60	100	60	100	60	100
r	128	36	36	34	33	32	32	30	30
t_P	.953298	.686819	.684862	.769721	.770282	.850836	.862347	.910533	.962034
t_C	.000779	.000393	.000268	.000443	.000270	.000378	.000289	.000367	.000304
t_S	.957278	8.05238	13.2509	7.58752	12.4730	7.15478	12.1795	6.70609	11.7280
$\frac{t_F}{t_C}$	4.92654	9.76534	14.3201	8.66315	14.2140	10.1528	13.2795	10.4572	12.6243

4   Conclusions

We studied the problem of a computationally weak client delegating group exponentiation to a single, possibly malicious, computationally powerful server, as originally left open in [8]. We solved this problem by two protocols that provably satisfy formal correctness, privacy (against adversaries of unlimited power), security (with exponentially small probability) and efficiency requirements, in a general class of multiplicative groups, possibly going beyond groups on which quantum cryptanalysis attacks are currently known. Problems of both theoretical and practical interest include: (a) achieving better efficiency tradeoffs as done in [10] for discrete logarithm groups; and (b) reducing the dependency of the offline computations on the number of delegated computations of F.

References

1. Anshel, I., Atkins, D., Goldfeld, D., Gunnels, P.E.: Post Quantum Group Theoretic Cryptography, November 2016. https://bit.ly/2svnv8z
2. Arbit, A., Livne, Y., Oren, Y., Wool, A.: Implementing public-key cryptography on passive RFID tags is practical. Int. J. Inf. Sec. **14**(1), 85–99 (2015)
3. Cavallo, B., Di Crescenzo, G., Kahrobaei, D., Shpilrain, V.: Efficient and secure delegation of group exponentiation to a single server. In: Mangard, S., Schaumont, P. (eds.) RFIDSec 2015. LNCS, vol. 9440, pp. 156–173. Springer, Cham (2015). https://doi.org/10.1007/978-3-319-24837-0_10

4. Dijk, M., Clarke, D., Gassend, B., Suh, G., Devadas, S.: Speeding up exponentiation using an untrusted computational resource. Des. Codes Crypt. **39**(2), 253–273 (2006)
5. Gennaro, R., Gentry, C., Parno, B.: Non-interactive verifiable computing: outsourcing computation to untrusted workers. In: Rabin, T. (ed.) CRYPTO 2010. LNCS, vol. 6223, pp. 465–482. Springer, Heidelberg (2010). https://doi.org/10.1007/978-3-642-14623-7_25
6. Gryak, J., Kahrobaei, D.: The status of polycyclic group-based cryptography: a survey and open problems. Groups Complexity Cryptology **8**(2), 171–186 (2016)
7. Hart, D., Kim, D.H., Micheli, G., Pascual-Perez, G., Petit, C., Quek, Y.: A practical cryptanalysis of WalnutDSA^{TM}. In: Abdalla, M., Dahab, R. (eds.) PKC 2018. LNCS, vol. 10769, pp. 381–406. Springer, Cham (2018). https://doi.org/10.1007/978-3-319-76578-5_13
8. Hohenberger, S., Lysyanskaya, A.: How to securely outsource cryptographic computations. In: Kilian, J. (ed.) TCC 2005. LNCS, vol. 3378, pp. 264–282. Springer, Heidelberg (2005). https://doi.org/10.1007/978-3-540-30576-7_15
9. Shor, P.W.: Algorithms for quantum computation: discrete logarithms and factoring. In: Proceedings of 35th IEEE Symposium on Foundations of Computer Science (FOCS 1994), pp. 124–134 (1994)
10. Di Crescenzo, G., Khodjaeva, M., Kahrobaei, D., Shpilrain, V.: Practical and secure outsourcing of discrete log group exponentiation to a single malicious server. In: Proceedings of 9th ACM Cloud Computing Security Workshop (CCSW), pp. 17–28 (2017)

NLP-Based Detection of Mathematics Subject Classification

Yihe Dong$^{(\boxtimes)}$

Wolfram Research, Champaign, USA
yihed@wolfram.com

Abstract. We present a classifier for the Mathematics Subject Classification (MSC) system, combining techniques in unsupervised learning such as nearest neighbors, and supervised learning such as neural networks. We will discuss the challenges presented in the classification task, such as the large number of possible classes, many with overlapping scope; and describe the data processing and experimental methodologies employed.

Keywords: Text classification · NLP · Neural networks

1 Introduction

Mathematics Subject Classification (MSC) system consists of five-digit alphanumeric sequences, such as 14D24 or 34L20, used to index a mathematical subject. They are widely used by a variety of journals in mathematics and mathematical physics. For authors, it can be time-consuming to find the right classification amongst thousands of choices, despite the fact that MSC labels naturally follow a hierarchical structure, and are usually presented in a sorted manner. Furthermore, often a piece of research involves several subtopics in addition to the main area of focus, thus having an automatic classifier that finds the non-focal yet related topics can be of great assistance.

In this paper, we present a classifier for 4575 MSC classes, using a combination of techniques such as K-nearest neighbors (KNN) and neural networks (NN). We devise an accurate classifier with the goal of speed and simplicity during both training and prediction. The models are trained and tested using papers obtained through the arXiv Bulk Data Access program hosted on Amazon S3 [4], out of which 160471 papers that contain MSC labels are used as training data, and 1000 as test data. Note that each paper serves as a sample point for multiple classes, as each paper almost always contains multiple MSC labels.

Previously, Řehůřek and Sojka [1] compared and contrasted methods including naive Bayes, KNN, and support vector machines to classify documents amongst 31 two-digit MSC prefix classes that had sufficient samples from two retrodigitization projects. The current work differs in its use of a larger training set from the arXiv, different text preprocessing techniques, and neural networks, as well as its classification amongst 4575 five-digit classes.

© Springer International Publishing AG, part of Springer Nature 2018
J. H. Davenport et al. (Eds.): ICMS 2018, LNCS 10931, pp. 147–155, 2018.
https://doi.org/10.1007/978-3-319-96418-8_18

2   Underlying Theory

2.1   Data Processing and Representation

To classify text, we need to first represent strings of texts in a computable form. We can vectorize texts treating either characters or words as the atomic units. Character-based models are often more powerful, as they are not constrained by a predefined vocabulary, and can better make inference on unseen words in test sets that share spelling similarities with words in the training set. But character-based models require larger training sets and longer training times, since these models need to learn how characters fit together and represent a semantic unit in the human language.

Given the high number of classes to training data ratio, we choose words as the atomic units, this gives the net the prior knowledge to treat each word as a semantic unit. The dependence on any particular form of a word is alleviated by uniformizing vocabulary words to their stems.

Term Selection. An initial list of terms is chosen based on term frequencies in the arXiv corpus, where two thresholds, a high and a low, are used to exclude the most and least frequent terms. The most frequent terms are excluded as they do not sufficiently discriminate amongst different subject classes, since they are ubiquitously used under many different contexts, e.g. "map." The most infrequent terms are excluded as they are not sufficiently frequent to be a general feature of a subject class. A nontrivial subset of these infrequent terms are rare spelling variations of more frequent words or typos. These two thresholds are determined by hand inspection.

In addition, the frequencies of terms in the arXiv corpus are compared with their frequencies in the 450 million-word Corpus of Contemporary American English [8], where terms with a large frequency difference are counted as mathematical terms. For instance, "function," while being a common English word, occurs much more frequently in a mathematical corpus than a generic English corpus. A score threshold dividing selected and unselected terms is determined by hand inspection.

Term selection also includes n-grams for $n \in \{2, 3\}$, for instance "Dedekind domain" and "cyclotomic field." n-grams are gathered based on the frequencies of groups of words occurring together, as well as conditional probabilities of a word w_{i+1} occurring immediately following a given word w_i, to detect n-grams containing the word sequence "$w_i w_{i+1}$." For instance, if 40% of all words following "Dedekind" is "domain," "Dedekind domain" is determined to be a 2-gram. To select 3-gram, the frequencies of all third words following any given two-tuple are used.

Words are normalized by techniques such as stemming and desingularization, e.g. "annihilate," "annihilator," "annihilating" are all normalized to "annihilat." Additionally, a set of key terms are automatically harvested from some hand-selected Wikipedia pages on mathematics, such as the page on list of theorems [3]. A list of the most frequent English words, verified by inspection to exclude any potentially useful mathematical term, is used as stop words to be excluded from the corpus. These processes combined created a vocabulary consisting of 23852 key terms.

Term Vectorization. To represent vocabulary words as fixed-length vectors, there exist two major approaches:

- Assign each word w with an index i_w, where $i_w \in [0, \text{Size}(\text{vocabulary}))$, then the vector representation for the word w is the norm-1 vector with 1 at index i_w, and 0 elsewhere. A variation of this approach is to assign a word weight instead of 1 at the index i_w: this is our approach.
- Embed words into a vector space, e.g. word2vec [5], GloVe [6]. These vectors are produced by unsupervised models trained on texts such as news corpora (word2vec). The resulting vectors are able to capture much semantic meaning of words, e.g. the vector for "dentist" has cosine distance 0.72 to the vector for "doctor," but has distance 0.51 to the vector for "dinosaur."

Challenges. There are a number of challenges underlying the MSC classification task:

- *Small training data size to number of classes ratio.* The training data consist of 160471 MSC-labeled papers amongst 4575 classes. This ratio is small despite the fact that each paper almost always contains multiple classes, thus simultaneously serving as a data point for multiple classes.
- *Imbalanced class representation.* For instance, classes such as 53D (symplectic geometry) appear more often than classes such as 70F (dynamics of a system of particles) by an order of magnitude.
- *Overlapping classes.* Many classes share overlapping content. For example, many documents involving fiber bundles can belong to either 55R, which falls under algebraic topology, or 18F, which falls under category theory.

To address these challenges, we devise a hybrid approach that uses neural networks to predict the three-digit prefixes of the most likely classes, and K-nearest neighbors to predict the five-digit MSC starting with these top prefixes.

2.2   K-Nearest Neighbors (KNN)

Due to the large number of classes relative to the limited set of hand-labeled data, training neural networks directly on all the five-digit classes does not produce a model that captures the characteristics of each subject class, leading to inaccurate predictions. Instead, we use the fact that a test sample is highly likely to share subject classes with the samples in the training data that are the most similar with regard to their terms vector representations.

Therefore, given a paper p to be classified, we take the top k papers in the training set whose vector representations are closest to that of p. All subject classes of the k closest papers are taken, with multiplicity, and weighted inversely proportionally to their rank amongst the closest papers. Then the m classes with the highest scores are taken, where m is the desired number of subject classification recommendations for p.

Dimension Reduction. Due to the high dimension of the feature space (number of key terms, which is 23852), taking the Euclidean distance between the query vector with every training vector at runtime is computationally expensive. Instead, we reduce the dimension of the feature space by projecting it down to the space spanned by the singular vectors corresponding to the largest singular values.

Specifically, let M be the term-document matrix, whose rows correspond to terms, and columns correspond to papers in the arXiv corpus. To reduce the dimension of the row space of M, M can be approximated by a truncated singular value decomposition (SVD) that keeps the n largest singular values:

$$M \to UDV^T.$$

Here M has dimension 23852×17027 for 17027 randomly selected papers, U and V^T are orthogonal matrices of dimensions $23852 \times n$ and $n \times 17027$, respectively, and D is a diagonal matrix of dimension $n \times n$.

By inspecting the absolute values of all the singular values (diagonal entries of D), n is determined to be 150, to retain the singular values above a certain magnitude threshold. This reduces the representation of each document to a 150-dimensional vector, and KNN is performed in this space. This technique is known as latent semantic analysis [2], and has the benefit of consolidating terms that share semantic similarity besides reducing the feature space dimension.

When a query paper p with vector representation $v_p \in \mathbb{R}^{23852}$ needs to be classified, v_p is projected to the reduced space $\mathbb{R}^n$ by $\hat{v}_p = D^{-1}U^T v$, and the closest column vectors in V^T (since its column space represents the documents space) with respect to Euclidean distance are selected as the papers most similar to p.

Note that instead of taking all 160471 papers in the training set as columns of the term-document matrix M, 17027 were randomly selected. In experiments, KNN prediction accuracy with respect to the number of papers represented in M reaches the point of diminishing returns at this point, indicating that the additional data samples do not drastically affect the projection onto the subspace corresponding to the top n singular values in the truncated SVD for our selected n.

Reducing the column space dimension of M, and hence V^T, from 160471 to 17027 significantly reduces the classifier runtime cost of searching for the query vector's nearest representation in the corpus. It also avoids the large bottleneck, albeit during precomputation, of performing SVD on a 23852×160471 matrix.

2.3 Neural Architecture

While KNN can often provide accurate predictions, sometimes the top predictions are only indirectly related to the query paper. This could be due to several reasons: while the projected vectors capture most of the semantic information of a document, some is bound to be lost in the reduction. This is corroborated by the fact that not reducing the dimension at all, or keeping a higher number of singular values, improves the prediction accuracy.

Also, KNN is an instance-based method, and as such, subject to peculiarities of instances. For example, a paper that contains a portion that is relevant to the query paper, but contains a nontrivial amount of less related content, is likely to not be the closest in distance, even though it carries target MSC labels. This applies to any distance metric and normalization method.

Due to this dependence on the particulars of the few papers that are closest to the query paper, KNN predictions do not utilize the full spectrum of features characterizing a given subject class, and so has less interpolation power than approaches such as neural networks. Thus, taking advantage of the fact that MSC's are organized hierarchically with respect to their two- and three-digit prefixes, we train two neural networks to predict the two- and three-digit subject class prefixes, henceforth referred to as the two-digit net and the three-digit net.

The network is based on a modified version of fastText [9], which is an efficient architecture for training on textual data. Specifically, our network consists of a trainable embedding layer, an average pooling layer, and a dense linear layer. The same network architecture is used for both the two- and three-digit nets.

To train, the net takes an input consisting of a variable-length sequence of word indices; to create the target output, the MSC labels of each document are turned into a vector of dimension equal to the number of all two-digit or three-digit MSC classes for the two-digit or three-digit nets, respectively, this vector has entry 1 at every index corresponding to a labeled class, and 0 otherwise.

Layer Details. The **embedding layer** takes an input consisting of a variable-length sequence of word indices (ordering does not matter), and outputs an N-dimensional dense vector, where $N = 128$ for the two-digit network, and $N = 256$ for the three-digit network.

For example, take the tautological statement "a unitary matrix is unitary," and suppose "unitary" and "matrix" have term indices 3 and 5, respectively, then this statement corresponds to the sequence $\{3, 5, 3\}$ as input to the embedding layer. Note the non-key terms "a" and "is" are redacted since they do not belong to the vocabulary, and term multiplicity is preserved, thus giving more frequently-occurring key words more weight. The input sequence includes the indices for all key words in a paper.

The embedding layer initializes the vector representation of each word uniformly randomly, and trains these representations to best approximate the true classification function as the training error is backpropagated.

The **average pooling layer** computes the mean of the n word embeddings, where n is the length of the word list for a paper. While this averaging procedure loses the sequential information for a paper's key words, we found it still effectively retains the characteristics of papers for determining MSC.

Finally, using a **dense layer** with linear activation, where the number of neurons is the number K of possible output classes (68 two-digit and 622 three-digit classes), the N dimensional vector is turned into a vector of dimension equal to K. The most probable classes are those whose indices correspond to the largest entries in the output vector.

The use of the final dense layer avoids a costly softmax computation, which normalizes the probability across all classes, this is especially useful when the number of output classes is high.

The models are trained with the mean squared error loss function, where the loss is minimized with the optimizer algorithm Adam [7].

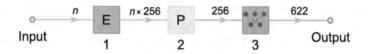

Fig. 1. Overview of the three-digit net architecture. The input to the embedding layer is a variable-length sequence of word indices, each of which represents a processed key word in the text. The vectors are subsequently averaged in a pooling layer, before being converted by a dense layer into a vector of dimension equal to the total number of classes. This architecture was devised with the goal of simplicity and speed.

Experimentation shows that combining two neural networks produces more accurate results than the three-digit net alone. The results are combined by taking the top k_2 predictions of the two-digit net, and then selecting the top k_3 predictions of the three-digit net whose first two digits coincide with the two-digit predictions. k_2 and k_3 are determined based on the number of desired five-digit predictions, in particular, they can be taken to be the same as this desired number.

The recommendations of KNN are then filtered by predictions of the two NN, where only the top KNN results that have the same three-digit prefix as the NN predictions are taken.

3   Results

We use a test set of 1000 papers, whose MSC labels span 1878 two-digit classes, 2313 three-digit classes, and 2747 five-digit classes. (All three-digit classes are also counted towards two-digit classes, similarly for five. Some papers do not specify the full five-digit classes.) When the classifier is allowed to make 8 five-digit predictions, the results contained 122 false negative (is actual label, but missed by classifier) two-digit classes, 220 false negative three-digit classes, and 329 false negative five-digit classes. Note that the three-digit false negatives include all two-digit false negatives, similarly for five.

This gives a recall rate (ratio between true positives and total predicted positives) of 0.88 for five-digit, 0.90 for three-digit, 0.93 for the two-digit classification. Recall that the recall rate measures the extent to which target classes are predicted.

While 8 classes are more than most authors need to specify, this number is to account for the large overlaps amongst classes. Also, it is much easier for a human to quickly decide if a prediction should be discarded based on a one-line description, than to realize a class is missing. The somewhat arbitrariness of this number renders the false positive rate less indicative of the classifier's performance, hence we focus on the recall rate.

Note that due to inherent subjectivity in human-labeled classes, predictions can't reach 100% accuracy on the test set. This is particularly true for this task, since there exist large overlaps amongst classes.

4   Experimentation

4.1   Text Processing

A vector representation based on a list of 23852 vocabulary words is created for each paper in the arXiv corpus that contain MSC classes. Each word in the vocabulary has been stemmed and desingularized, and can be the representative of several words.

Stemming is performed by creating a trie from all words in the corpus, where stems are detected by a roughly uniformly-split dropoff in frequency from a parent node to children nodes. For instance, suppose the trie nodes for "annihilate" and "annihilator" contains frequencies 1541 and 1104, respectively c, and the trie node corresponding to their common prefix "annihilat" contains frequency 2645, the trie-based algorithm then uniformizes "annihilate" and "annihilator" to their stem "annihilat."

Desingularization is performed with a heuristic that considers common pluralization e.g. "ring" → "rings," "proxy" → "proxies," as well as unusual cases such as "matrix" → "matrices."

4.2   Neural Networks

To create a more balanced training set across subject classes, classes that have relatively few representatives in the training set are oversampled, wherein the key terms for papers in these classes are randomly sampled (without replacement), to generate sets of terms, where each set represents an artificially generated paper. Since this procedure does not enrich the information entropy of a subject class, we do not oversample to the extent of equalizing the number of training samples per class, but rather bring the number of samples per class to a minimum percentage of the original average number per class.

The predictions by the three-digit net filtered by the predictions of the two-digit net are more accurate than that of the three-digit net alone. One explanation is that it is easier for the network to get the top two-digit labels correct, since the difference amongst the two-digit classes is more pronounced than the difference amongst three-digit classes, by the virtue of there being 622 three-digit classes but only 68 two-digit ones. In addition, the three digit classes are naturally grouped together by their two-digit prefixes – a pattern that the two-digit net does not need to recognize since it is not applicable, but which the three-digit net needs to learn.

The nets are trained with CPU on a 24-core machine with 64 GB RAM using the Wolfram Language, and can be recreated with any machine learning framework. The two-digit net is trained until the batch loss falls below 10^{-2}, which takes on average 2 h. The three-digit net is trained until the batch loss falls below 2×10^{-3}, which takes on average 4 h 20 min.

Note that a minimum number of features is required within a piece of text for the classifier to make a meaningful prediction, the classifier is wrapped by a conditional that checks if the number of features is above an experimentally-determined threshold.

The classifier will be made available to the public on a standalone web platform.

5 Future Improvements

In lieu of using KNN to predict the five-digit subject classes which share the three-digit prefixes predicted by the NN, one can train gradient-boosted trees for each of the 68 two-digit classes. This may generate better accuracy, since it addresses KNN's dependence on the particulars of a small subset of all data within a class, as it learns based on all individuals in the class.

6 Conclusion

We presented a way to automatically classify a piece of mathematical text amongst 4575 MSC classes. The automatic detection is especially useful in finding classes that are outside the immediate focus of the paper, but are related to its scope. This in turn improves any application that uses classification-based indexing and search.

Acknowledgments. We would like to thank Jeremy Michelson and Michael Trott for continuously lending their ears and ideas throughout this project, and the ICMS reviewer for constructive comments on an earlier draft of this paper.

References

1. Řchůřck, R., Sojka, P.: Automated classification and categorization of mathematical knowledge. In: Autexier, S., Campbell, J., Rubio, J., Sorge, V., Suzuki, M., Wiedijk, F. (eds.) CICM 2008. LNCS (LNAI), vol. 5144, pp. 543–557. Springer, Heidelberg (2008). https://doi.org/10.1007/978-3-540-85110-3_44

2. Furnas, G., Dumais, S., Landauer, T., Harshman, R., Streeter, L., Lochbaum, K.: Information retrieval using a singular value decomposition model of latent semantic structure. In: Proceedings of SIGIR (1998)
3. List of theorems. https://en.wikipedia.org/wiki/List_of_theorems
4. arXiv Bulk Data Access. https://arxiv.org/help/bulk_data_s3
5. Mikolov, T., Chen, K., Corrado, G., Dean, J.: Efficient estimation of word representations in vector space. https://arxiv.org/abs/1301.3781
6. Pennington, J., Socher, R., Manning,. C.D.: GloVe: global vectors for word representation. In: EMNLP (2014)
7. Kingma, D., Lei Ba, J.: Adam: a method for stochastic optimization. In: ICLR (2015)
8. Corpus of Contemporary American English. https://corpus.byu.edu/coca
9. Joulin, A., Grave, E., Bojanowski, P., Mikolov, T.: Bag of tricks for efficient text classification. In: Proceedings of the 15th Conference of the European Chapter of the Association for Computational Linguistics (2017)

NLP and Large-Scale Information Retrieval on Mathematical Texts

Yihe Dong[(✉)]

Wolfram Research, Champaign, USA
yihed@wolfram.com

Abstract. We present a recommender system covering math and math physics papers from the arXiv, to assist researchers to quickly retrieve theorems and discover similar results from this vast corpus. The retrieval aims to discover not just syntactic, but also semantic similarity. We will discuss the challenges encountered and the experimental methodologies used.

Keywords: NLP · Information retrieval

1 Introduction

As the number of academic publications grows rapidly, it becomes increasingly important to categorize and index these bodies of knowledge for fast and accurate information retrieval.

Tools such as Google Scholar [1] and Microsoft Academic [2] have proven to be very useful in finding academic sources relevant to a given query. These retrieval systems focus on entire documents. We present *Mathematical Theorem Search*, henceforth referred to as *Theorem Search*, a framework for analyzing and searching only the most important statements in a paper: theorems, propositions, conjectures, lemmas, and definitions, henceforth broadly referred to as "theorems."

Theorem Search complements existing tools, since it allows the query to specifically target theorems only, and all statements are presented in their entireties on the same page, instead of in fragments as in existing tools.

By providing entire theorem statements on or related to a particular query, such a platform is useful both for quickly gleaning knowledge about a topic, as well as checking if a particular or related statement has been proved before in the search corpus.

The search corpus covers all math and math physics papers on arXiv to date. The corpus is obtained through the arXiv Bulk Data Access program [4]. A total of 2,335,254 theorems are extracted from over 770,000 papers.

2 Semantic Search – Textual Understanding and Representation

We present a multipronged approach for retrieving and ranking relevant results, combining approaches based on words and semantics. The words-based approach

© Springer International Publishing AG, part of Springer Nature 2018
J. H. Davenport et al. (Eds.): ICMS 2018, LNCS 10931, pp. 156–164, 2018.
https://doi.org/10.1007/978-3-319-96418-8_19

retrieves results containing the query words or words related to the query, and the semantics-based approach ranks these results based on how closely their semantic structures align with that of the query.

There are numerous challenges present in this task: theorems and their contextual statements need to be extracted from non-standardized raw TeX files, and then parsed to gather semantic information; the large amount of files and data require a robust preprocessing pipeline, as well as algorithms and data structures for efficient retrieval, while respecting memory constraints.

2.1   Parsing

To effectively retrieve a piece of text, the system needs to understand that text, such as knowing the objects discussed in the text, the relations between them, and whether the text is hypothetical (e.g. "Suppose property p holds"), or assertive (e.g. "X is uniquely determined"). Each piece of text is parsed to understand the objects in the text, and the relations between them.

Objects. A vocabulary is created to represent known objects. An initial list of terms is chosen based on term frequencies in the arXiv corpus, where two thresholds, a high and a low, are used to exclude the most and least frequent terms. These two thresholds are determined by hand inspection.

In addition, the frequencies of terms in the arXiv corpus are compared with their frequencies in the 450 million-word Corpus of Contemporary American English [6], where terms with a large frequency difference are counted as math terms. For instance, "function," while being a common English word, occurs much more frequently in a math corpus than a generic English corpus.

Words are normalized by techniques such as stemming and desingularization, e.g. "annihilate," "annihilator," "annihilating" are normalized to "annihilat." Thus, each word in the vocabulary can be the representative of several words.

N-grams are gathered based on the frequencies of groups of words occuring together, as well as conditional probabilities of a word w_{i+1} occurring immediately following a given word w_i; this detect n-grams containing the word sequence "$w_i w_{i+1}$." Additionally, a set of key terms are automatically harvested from select Wikipedia pages on math. These processes combined created a vocabulary consisting of 23,852 key terms. Each vocabulary word is assigned an index, which will be used throughout.

For parsing, the likely parts of speech (POS) are recorded for the most frequently-occurring vocabulary words. These POS tags are obtained either by consulting with a list of common English words and their possible POS [7], or created by hand.

Relations. To understand the relations between objects, a parse tree is created for each piece of text (Fig. 1). The parser implements a bottom-up dynamic programming algorithm.

First, the text is tokenized. A token usually corresponds to a term (could be an n-gram) that serves as one POS. For instance, "the topological group G is extremely amenable," is tokenized into {{the}, {topological group}, {G}, {is}, {extremely amenable}}, with tags {article, entity, symbol, verb singular, adjectivial qualifier}, respectively.

The possible POS for each term is either retrieved from the vocabulary list from above, or determined based on the Stanford POS tagger [10]. A token can also carry multiple POS tags, in which case ambiguities are resolved by the main parser.

The tokens and associated data are then given to the main parser, which uses a set of hand-created context-free grammar (CFG) rules to combine tokens into increasingly larger grammatical units: words into phrases, phrases into clauses, clauses into sentences.

This grammar-rule-reduction is iterated until all tokens in a statement have reduced to a single head symbol, where the set of tokens can follow multiple paths to different head symbols, since at each stage multiple grammar rules can apply. Each head symbol serves as the root of its associated parse tree. At each step, the least-probable combinations are pruned via beam search, to avoid combinatorial explosion of symbol formations. Some examples of grammar rules are $verbphrase \rightarrow verb\_entity$, $assert \rightarrow entity\_verbphrase$, $pobj \rightarrow poss\_csubj$, where each underscore joins the components being reduced. Note by convention, the rule is written where the RHS combination reduces to the LHS.

Each grammar rule is associated with a probability, which is determined based on a combination of hand curation and the Universal Dependencies framework [5], a system of grammatical annotations. Adjustments are made whenever empirical observations on parses warrant a precedence change. The probabilities not only serve to rank the resulting trees according to likelihood, but also allow the formation of long combinations that rely on an unlikely, but possible, link. The low probability associated with such a link ensures that it does not interfere when more likely rules apply.

Once a winning parse tree is selected, another set of rules are used to categorize the tree into a set of statement types. Some possible types of statements include: "ThereExists ...," "... HasProperty ...."

This set of rules serves to uniformize text amongst a known set of semantic structures. For instance, "the ring R has finite type" and "the ring R is of finite type" are both uniformized to the "HasProperty" relation, specifically {the ring R} ~HasProperty~ {finite type}. Statement type assignment is done by matching (with backtracking) rule components with tree components. Given a query, one weight in the search scoring system depends on whether a result candidate's structure aligns with that of the query.

2.2   Context Vectors

Vector representations of text capturing at least partial object relations can be created based on the parse tree, henceforth referred to as context vectors. Specifically, given a parse tree, the entry at the index corresponding to a qualifier is the index

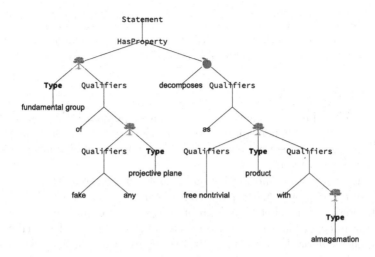

Fig. 1. The parse tree generated for "The fundamental group of any fake projective plane decomposes as a free nontrivial product with amalgamation." A tree symbol represents an entity, and apple symbol a property. This bottom-up parser uses context-free grammar rules to progressively combine token symbols to create subtrees, selecting the highest-scoring root symbol as the winning tree.

of the object it's qualifying. For any given nonzero entry at index i, all previously-trivial terms related to the term corresponding to i will also share this entry value.

During search, a context vector v_q is formed for the query, and a score is created between v_q and candidate context vectors from the corpus, by counting the non-zero coincidences between v_q and each candidate context vector. In other words, this counts the number of qualifier relation coincidences between the query and theorems in the corpus. Since the context vectors are sparse, for efficient memory representation, the nonzero entries for each vector are stored in a hashmap.

2.3   Related Terms

For both terms-based and semantics-based search, knowing which terms are related can be useful in finding results with the same meaning, but different phrasing. Related terms are found by training a neural network to create semantic embeddings of words using the word2vec model architecture [3], where the training data are created by removing TeX expressions and markup from all texts, and canonicalizing the remaining words to their lexemes. Future work may cross-reference these embeddings with the arXiv word embeddings by the KWARC group [8].

For each vocabulary word, its related terms are selected as all words whose vector representations have a cosine distance above an experimentally-determined threshold with the vocabulary word. Related terms are terms that occur under similar contexts, but which are not necessarily synonymous in meaning.

For instance, the terms {"artinian," "noetherian," "dualizing"} are found as related, not because "noetherian" and "artinian" are synonymous, but because both are often used before "ring" and "module."

2.4 TeX Parsing

Context. TeX expressions constitute a substantial portion of theorems, and are integral in conveying their meaning. For example, suppose the statement "G is abelian" appears in a theorem, the standalone theorem is ambiguous if it doesn't also contain what G refers to.

To get the contextual information on symbols referred to in TeX expressions, we parse the entire paper. As the parser parses surrounding texts, it picks up variable definitions such as "Let G be a group," "given a group G," etc. The parser attaches this information to a stateful object as it moves along, and when it encounters an unknown variable in a theorem, it first looks to see if the variable is defined locally anywhere in the theorem, which could happen after a variable's first mention. If not, the parser selects the most recently-encountered definition for a variable. Practically, this is implemented by parsing the theorem twice, first time to collect variable definitions within the theorem, and then resolve unknown variables.

Display. Due to the immense variety of TeX conventions and styles used in the corpus, any platform that renders the search results properly must understand the definitions of custom TeX commands, which are usually defined in the preamble of the TeX document. Therefore, a parser was implemented to parse TeX macros specified by the likes of such as "\newcommand \plusbinomial [3][2]{(#2+#3)^#1}" or "\def\plus[#1]#2{#1+#2}."

Throughout each document, the TeX parser scans all TeX expressions and replaces any occurrence of commands defined by macros with their full definitions.

3   Other Search Methods

3.1   Term-Based Search

The terms-based intersection search takes the intersection of sets of theorems, where each set corresponds to one term in the query, and all theorems in that set contains its term. Each term can be an n-gram for $n \in \{1, 2, 3\}$. Each term's related terms are also used to retrieve theorems. This is useful for prioritizing results that only contain a proper subset of the query terms literally, but which also contain terms related to the complement of that subset, over those that only contain that subset and no related terms. One example of this is, for the query "symmetric matrix", the algorithm prioritizes theorems containing "Hermitian matrix" over those containing just "matrix", because "Hermitian" is related to "symmetric".

For each term, every theorem containing that term is weighted by the score of the term, and each theorem's score is tallied based on all terms it contains.

3.2   Latent Semantic Analysis

If the terms-based and semantics-based searches do not produce sufficiently many results, the theorems closest to the query in terms of reduced term-frequency-inverse-document-frequency (tf-idf) vector representation are returned. This technique is known as latent semantic analysis [9]. The tf-idf statistic measures a word's importance to a document within a corpus, it is proportional to the number of times a term appears in a document, and inversely proportional to the number of times it appears in the corpus.

A vector representation is formed for each theorem in the corpus, where the entry corresponding to a term in the vocabulary is filled with its tf-idf score.

Dimension Reduction. Due to the large number of theorems in the corpus and the high dimension of the feature space (number of key terms, which is 23,852), taking the distance, regardless of metric, between the query vector with every corpus vector in the original dimension at runtime is computationally expensive.

Instead, we reduce the dimension of the feature space by projecting it down to the space spanned by the singular vectors corresponding to the largest singular values. This rank lowering process also consolidates the dimensions associated with terms that have similar meanings, which helps retrieve semantically related results that are not literal matches.

To perform the dimension reduction, let M be a tf-idf term-document matrix, whose rows correspond to terms, and columns correspond to 49256 uniformly randomly selected documents in the arXiv corpus. A subset instead of all papers is used, to ensure computational feasibility of the dimension reduction. To reduce the dimension of the row space of M, M can be approximated by a truncated singular value decomposition (SVD) that keeps the n largest singular values:

$$M \to UDV^T.$$

Here M has dimension 23852×49256, U and V^T are orthogonal matrices of dimensions $23852 \times n$ and $n \times 49256$, respectively, and D is a diagonal matrix of dimension $n \times n$.

n is determined to be 35, by inspecting the absolute values of all the singular values (diagonal entries of D), to retain the eigenvalues above a given magnitude threshold. This low dimension is needed so search results are produced efficiently in terms of speed, since distances are taken between the query vector and all 2,335,254 corpus vectors. This reduces the representation of each document to a 35-dimensional vector. Each vector $v_p \in \mathbb{R}^{23852}$ in the corpus is projected to $\mathbb{R}^{35}$ by the operation $\hat{v}_p = D^{-1}U^T v$.

Given a search query q, a vector representation based on the query terms is created and projected down to $\mathbb{R}^{35}$, in the exact same way as how the corpus vectors are processed. The search results are taken to be the corpus vectors that are the nearest in terms of cosine distance.

4   Results Ranking

The scoring mechanism combines the above methods together into one ranking. A list of scores is consulted in decreasing priority, where each subsequent score is used to tie-break the ranking generated by the previous scores.

The terms-based score is used first to rank theorems that contain any of the query terms, then context vectors and structural similarities as determined by the parse are considered, followed by the count of the query terms that appear in each result candidate. Finally, the latent-semantics-based ranking is used to return additional results when the above methods do not generate sufficiently many.

4.1   Examples

Queries can take any colloquial or technical form. Simple queries consisting entirely of mathematical structures and concepts, such as "fundamental theorem of calculus," yield results including both the statements of the theorem, and theorems that apply it. Similarly, research results can be retrieved on queries such as "reductive group action on complete variety," "Mordell-Weil group of elliptic curve," or "smooth intersection of quadrics." Results also include variations on the query terms: searching for "Lyubeznik number" also brings up the more general Hartshorne-Speiser-Lyubeznik invariant.

Queries containing logical implications yield results based on both the terms contained and the statement logic. For example, searching for "curvature is module map" returns results such as "the curvature R_E is a right module map if ... " amongst the top hits. Querying "group representation is completely reducible" gives conditions on groups for which this is true.

Queries for theorems by particular authors are also supported.

5   Notes on Implementation

5.1   Preprocessing

Most of the data statistics on the corpus are precomputed for better search efficiency at runtime. In particular, the entire corpus is parsed, context vectors and structural metadata are created for each theorem, the indices of all theorems containing a given word are recorded, and the tf-idf vector representations are computed. The preprocessing takes ~5 days on a 24-core CentOS machine with 64 GB RAM.

5.2   Scalability

A total of 2,335,254 theorems are extracted from over 770,000 math and math physics papers. This presents the challenge that not all theorems and their stateful information can fit in the random access memory (RAM) at the same time. While this problem can be alleviated by increasing the amount of RAM, adding memory is not sustainable as the corpus continues to grow.

A MySQL database is created to host the preprocessed data associated with each theorem. An added benefit of the database is that the SQL querying structure naturally supports queries such as "theorems on X by author p," or "theorems on X between years t_1 and t_2."

An earlier iteration of the platform used a FIFO (first-in-first-out) cache instead of a database, where bundles each containing 10,000 theorems and associated data are loaded and ejected from the cache. The cache suffers from the issue of latency whenever a new bundle of theorems needs to be loaded in memory, where latency is caused by data deserialization during loading. While each fetch from a database takes slightly longer than the time the cache takes to fetch a theorem from memory, the database does not suffer the noticeable latency that affects select queries, since it fetches one piece of theorem data at a time.

All together, search results return within a few hundred miliseconds per query. The timing variation depends on the length of the query, and how many search mechanisms are triggered, where queries consisting of simple combinations of phrases contained in the vocabulary list yield results the fastest.

The search platform will be made available to the public on a standalone web platform.

6   Future Improvements

Context-free grammars are limited by their context-freeness, unable to take priorities into account amongst reductions with the same scores. For example, in parsing "A extension of fields over $\mathbb{Q}$," the rule $entp \rightarrow entity\_prep$, where $prep$ denotes a reduced preposition and entity combination (i.e. "of fields" and "over $\mathbb{Q}$"), can be first applied to reduce either "an extension of fields" or "fields over $\mathbb{Q}$," both of which result in the same score. To address this, taking into account parse subtrees of depth 3 and beyond (CFG reductions correspond to depth-2 parse trees) has been shown to significantly improve parsing [11], Future work will explore how using deeper subtrees can produce more semantically accurate parse trees and vector representations of theorems, potentially leading to improved search results.

Recent advancements in machine learning have created vector embeddings of entire sentences [12–14]. These sentence embeddings have shown promising results on tasks such as sentiment analysis and paraphrase detection. Our experiments on adopting sentence embeddings on *Theorem Search* have not yielded better results than the approaches outlined here, largely due to length differences amongst theorems leading to large distances between their embeddings, even for theorems that share strong semantic connections. We hope to improve upon these experiments and use sentence embeddings to improve semantic matching.

Acknowledgments. We would like to thank Jeremy Michelson and Michael Trott for continuously lending their ears and ideas throughout this project, as well as Rob Y. Lewis and the ICMS reviewer for helpful comments on an earlier draft of this paper.

References

1. Google Scholar. https://scholar.google.com
2. Microsoft Academic. https://academic.microsoft.com
3. Mikolov, T., Chen, K., Corrado, G., Dean, J.: Efficient estimation of word representations in vector space. https://arxiv.org/abs/1301.3781
4. arXiv Bulk Data Access. https://arxiv.org/help/bulk_data_s3
5. Universal Dependencies. http://universaldependencies.org/
6. Corpus of Contemporary American English. https://corpus.byu.edu/coca/
7. Word frequency data. https://www.wordfrequency.info/
8. arXMLiv Word Embeddings. https://sigmathling.kwarc.info/resources/arxmliv-embeddings-082017/
9. Furnas, G., Dumais, S., Landauer, T.K., Harshman, R.A., Streeter, L.A., Lochbaum, K.E.: Information retrieval using a singular value decomposition model of latent semantic structure. In: Proceedings of SIGIR (1998)
10. Toutanova, K., Klein, D., Manning, C., Singer, Y.: Feature-rich part-of-speech tagging with a cyclic dependency network. In: Proceedings of HLT-NAACL, pp. 252–259 (2003)
11. Kaliszyk, C., Urban, J., Vyskočil, J.: Automating formalization by statistical and semantic parsing of mathematics. In: Ayala-Rincón, M., Muñoz, C.A. (eds.) ITP 2017. LNCS, vol. 10499, pp. 12–27. Springer, Cham (2017). https://doi.org/10.1007/978-3-319-66107-0_2
12. Conneau, A., Kiela, D., Schwenk, H., Barrault, L., Bordes, A.: Supervised learning of universal sentence representations from natural language inference data. https://arxiv.org/abs/1705.02364
13. Arora, S., Liang, Y., Ma, T.: A simple but tough-to-beat baseline for sentence embeddings. In: ICLR (2017)
14. Pagliardini, M., Gupta, P., Jaggi, M.: Unsupervised learning of sentence embeddings using compositional n-gram features. https://arxiv.org/abs/1703.02507

Machine Learning for Mathematical Software

Matthew England$^{(\boxtimes)}$

Coventry University, Coventry, UK
`Matthew.England@coventry.ac.uk`

Abstract. While there has been some discussion on how Symbolic Computation could be used for AI there is little literature on applications in the other direction. However, recent results for quantifier elimination suggest that, given enough example problems, there is scope for machine learning tools like Support Vector Machines to improve the performance of Computer Algebra Systems. We survey the author's own work and similar applications for other mathematical software.

It may seem that the inherently probabilistic nature of machine learning tools would invalidate the exact results prized by mathematical software. However, algorithms and implementations often come with a range of choices which have no effect on the mathematical correctness of the end result but a great effect on the resources required to find it, and thus here, machine learning can have a significant impact.

Keywords: Machine learning · Mathematical software

1 Introduction

Machine Learning, refers to tools that use statistical techniques to give computer systems the ability to *learn* rules from data; that is, improve their performance on a specific task, without changing their explicit programming. Although many of the core approaches date back decades, machine learning has found great success in recent years, driven by the advances in both computer hardware and the availability of data. There have well publicised successes of machine learning recently such as Google's ALPHAGO being the first to beat a professional human Go player[1]. We are all likely to have interactions with software that at least partially learns on a daily basis, whether through traffic signal control [40] or the extraction and interpretation of our views [39].

Most industries have felt some effect from the advance of these tools, and software engineering itself is no different. Indeed, the idea of using machine learning in the software development process is not a new one [41]. In particular, machine learning is now a common tool in the testing and security analysis of

---

[1] https://research.googleblog.com/2016/01/alphago-mastering-ancient-game-of-go.html.

© Springer International Publishing AG, part of Springer Nature 2018
J. H. Davenport et al. (Eds.): ICMS 2018, LNCS 10931, pp. 165–174, 2018.
https://doi.org/10.1007/978-3-319-96418-8_20

software [21]. Machine learning is at its most attractive when the underlying functional relationship to be modelled is complex or not well understood. It may seem that machine learning is hence not relevant to the sub-field of mathematical software where underlying functional relationships are the key object of study. Further, the inherently probabilistic nature of machine learning tools seems like it would invalidate the exact mathematical results prized by such software.

However, as most developers would acknowledge, mathematical software often comes with a range of choices which, while having no effect on the correctness of the end result, could have a great effect on the resources required to find it. These choices range from the low level (in what order to perform a search that may terminate early) to the high (which of a set of competing exact algorithms to use for this problem instance). In making such choices we may be faced with decisions where the underlying relationships are not fully understood, but are not themselves the key object of study. Thus in practice we will use a, usually fairly crude, man-made heuristic in order to proceed with the implementation.

It is possible that many of these decisions could be improved by allowing learning algorithms to analyse the data. It is even possible that such study could lead to a better understanding of the underlying relationship. For example, a standard step in the use of machine learning is feature selection: identifying a minimal number of features about the data to use in making the decision. The primary reasons for this are to reduce the resources required to train a classifier, and reduce the risk of over-fitting. However, in identifying the most important features the developers of mathematical software may also get insight on new mathematical results, or at least hypotheses to guide future development.

We proceed by surveying the author's own work applying machine learning in one particular area of symbolic computation. We then consider where else in computer algebra and mathematical software more broadly there may be potential applications and existing inspiration.

2   Machine Learning for CAD

The author has been involved in two applications of machine learning [22,23] to improve the performance of a particular algorithm. *Cylindrical Algebraic Decomposition* (CAD) refers to both a mathematical object and the algorithms to produce them, both first introduced by Collins in the 1970s. Here:

- *decomposition* means a partition of $\mathbb{R}^n$ into connected subsets called *cells*;
- *algebraic* is short for *semi-algebraic* and means that each cell may be described by a conjunction of polynomial constraints;
- *cylindrical* refers to the structure of the decomposition: the projections of any two cells, onto a lower coordinate space with respect to the given variable ordering, are either identical or disjoint.

CADs were originally produced as *sign-invariant* for a set of input polynomials[2], meaning each polynomial is to have constant sign on each cell. However, for

---

[2] See for example [2] for a description of the original CAD algorithm.

almost all applications what is truly required is a decomposition *truth-invariant* for logical formulae: where each formula has constant truth value on each cell. A sign-invariant decomposition for the polynomials in the formulae produces truth invariance, but it can be achieved more efficiently [5, 18, 28].

In either case, the invariance properties mean only a finite number of sample points need to be queried to solve problems. In particular, CADs offer a tool to perform Quantifier Elimination (QE). Through QE there are a multitude of applications throughout engineering and the sciences (see for example [33]). Additional application of CAD directly include identification of steady states in biological networks [4], and programming with multi-valued functions [14].

However, CAD is well known for its worst case complexity doubly exponential [15][3]. Hence it is important to optimise how CAD is used, such as the setting of any optional parameters and the presentation of input.

2.1 Deciding Whether to Pre-condition

One choice a user could make is whether to give their problem to CAD directly, or to first precondition it. One common technique for input formulae with multiple equations is the use of a *Gröbner Basis* (GB). A GB is a particular generating set of an ideal with useful properties: although our task is not to study the ideal it turns out the GB can give a simpler representation for CAD to work with.

To be precise: let $E = \{e_1, e_2, \dots\}$ be a set of polynomials; $G = \{g_1, g_2, \dots\}$ be a GB for E; and B be any Boolean combination of constraints. Then

$$\Phi = (e_1 = 0 \wedge e_2 = 0 \wedge \dots) \wedge B \text{ and}$$
$$\Psi = (g_1 = 0 \wedge g_2 = 0 \wedge \dots) \wedge B$$

are equivalent, and a CAD truth-invariant for Ψ can solve problems involving Φ.

This was studied first in 1991 [10] and then again in 2012 [36]. In both cases the conclusion was that usually GB pre-conditioning is beneficial for CAD, but there are some examples where it is greatly detrimental. In [22] we considered using machine learning to decide when to use GB. On a dataset of over a thousand randomly generated problems with multiple equations we found 75% were easier to study after a GB was taken. We trained a *Support Vector Machine* (SVM) classifier [13] with radial basis function (see for example [31]) to make the decision. We used as problem features simple algebraic properties (degrees, density of occurrence of variables etc.) of both the input polynomials and the GB. Only when including those of the GB could the classifier make good decisions: not a problem since for any problem where CAD is tractable GB is trivial. The classifier chooses, not whether to construct the basis, but whether to use it. In [22] we also showed how feature selection experiments (identifying a minimal subset of the features) could improve accuracy (reducing the risk of over-fitting).

---

[3] Doubly exponential usually in the number of variables, although the logical structure can be used to improve this somewhat [5, 18, 19].

2.2 Choosing a Variable Ordering

Another choice a user may have to make for CAD is the variable ordering, used in the definition of cylindricity, and crucial to the computational path of the algorithm. Depending on the application this may be free, constrained or fixed.

For example, for QE one must order variables as they are quantified; but there is no restriction on free variables and adjacent quantifiers of the same type may be swapped. It is well known that this choice can dramatically affect the feasibility of a problem. In fact, there are a class of problems in which one variable ordering gives output of double exponential complexity in the number of variables and another output of a constant size [9]. There are heuristics available to make the choice but each can be misled by certain examples.

In [23] we investigated machine learning for this choice. In this case the choice is not binary but from many different orderings[4], not a typical context for machine learning classification. Instead of the ordering itself, we aimed for machine learning to pick which of three existing heuristics [6,8,16] we should follow. Experiments on over 7000 problems identified substantial subclasses on which each of the three heuristics made the best decision. This time we trained three SVM classifiers, one for each heuristic, and used the relative magnitude of their margin values to choose the one to follow for each problem. We found this machine learned choice did significantly better than any one heuristic overall.

3 Potential Use in Symbolic Computation

3.1 Machine Learning Elsewhere in CAD/QE

It seems [23] was the first publication on the application of machine learning to symbolic computation. The only similar work since is [24] which applied machine learning to decide the order of sub-formulae solving for their QE procedure[5].

There are certainly other decisions to be made when using CAD: such as the designation of equational constraints [5,18,28]; and for some CAD algorithms even the order of polynomials and formulae [17]. Perhaps of most importance is the high level choice of which CAD implementation to use for a problem: most comparison experiments will show problem instances where different solvers prosper. Looking wider still, if the application problem were Quantifier Elimination then there are a multitude of non-CAD approaches, such as virtual substitution [33] or QE by comprehensive GB [20], superior for classes of input. The author will be leading an upcoming EPSRC project (EP/R019622/1) on these topics.

3.2 Machine Learning Elsewhere in Computer Algebra

Computer Algebra Systems (CASs) often have a choice of algorithms to use when solving a problem. Since a single one is rarely the best for the entire problem

---

[4] If the choice is completely free then n variables have $n!$ possible orderings.

[5] The feature set they used for their SVM was seeded from those in [23].

space, CASs usually use *meta-algorithms* to choose, where decisions are often based on some numerical parameters [11]. A prominent example would be how and when to simplify mathematical expressions (see [32] and references within). Could machine learning be more effective? In a presentation at ICMS 2016 it was reported that MAPLE's user level symbolic integration command calls 16 different integration procedures in sequence until one returns an answer. It is likely that the optimal order of calls would vary with problem instance. Ever broader, a generic command like MAPLE's `solve` or MATHEMATICA's `Solve` has to contend with not knowing exactly what the user means by "*solve*", inferring from the input. Machine learning could possibly assist with this, perhaps not just by viewing the input, but also the user's session history.

4   Machine Learning Elsewhere in Mathematical Software

4.1   Satisfiability Checking

There has been some use of machine learning within the satisfiability checking community for their SAT-solvers [3]. These are tools dedicated to the solution of the Boolean SAT problem (given a Boolean formula decide if there is an allocation of values to variables that satisfies it). Despite the SAT problem being NP-Complete, there exist solvers which can process formulae with millions of variables, and they are a common tool in many industries.

There is rarely a single dominant SAT solver; instead, different solvers perform best on different instances. The portfolio solver SATZILLA [38] takes sets of problem instances and solvers, and constructs a portfolio optimizing a given objective function (such as mean runtime, percent of instances solved, or score in a competition). SATZILLA did well in SAT competitions[6].

Machine learning within the actual search algorithms was a prominent part of the MAPLESAT [27] solver. The developers view the question of solver branching as an optimisation problem where the objective is to maximize the learning rate, defined as the propensity for variables to generate learnt clauses. Experiments showed this to correlate well with efficiency, but the cost of an absolute solution could outweigh the savings. Hence the chosen approach was to use machine learning to gain a heuristic solution to the optimisation problem.

Another use of machine learning in SAT is the choice of initial value to variable allocation to begin the search. In [37] the author describes using a logistic regression model to predict the satisfiability of formulae after fixing the values of a certain fraction of the variables and adapting MINISAT to determine the preferable initial values using this and a Monte-Carlo approach. The author reported a high accuracy in the setting of backbone variables (variables that have the same value in all solutions of the formula) on initiation.

---

[6] Although, because problems change little between competitions there is a risk of over-fitting being rewarded: www.msoos.org/2018/01/predicting-clause-usefulness.

4.2   Satisfiability Modulo Theories

SAT-solvers can be applied to problems outside of the Boolean domain. The approach, called Satisfiability Module Theories (SMT), is to iteratively use a SAT solver to find solutions to the Boolean skeleton of a formula and then query whether this is a solution in the domain, learning new logical restrictions when not [Chap. 26][3]. In the domain of non-linear real arithmetic, symbolic algorithms developed for QE are the basis of these theory solvers and so the results and potentials in Sects. 2 and 3.1 all apply. There are likely similar questions of which tool to use for an instance in many of the other domains also.

Machine learning can also be applied to fundamental questions regarding the Boolean encoding. In [30] the authors studied whether it was best to encode atomic subformulas with Boolean variables, or to encode integer variables as bit-vectors, for working in separation logic with uninterpreted functions. They concluded that a hybrid approach was needed after evaluating a wide range of benchmarks and used statistical techniques to decide what to do: an early application of a machine learning approach to SMT-solvers.

4.3   Mathematical Knowledge Management

Perhaps the area of mathematical software with the greatest potential for machine learning applications is Mathematical Knowledge Management (MKM) [12] since many of the tasks are similar to Natural Language Processing (NLP) where machine learning has seen extensive use. For example, [35] describes the automatic identification of a suitable top level from the Mathematics Subject Classification (MSC) system for thousands of articles using an SVM; while [29] describes how NLP techniques were adapted to build a part of speech tagger used for key phrase extraction in the database zbMATH.

4.4   Automated Reasoning

Theorem Provers (TPs) prize correctness to a greater extent than even computer algebra systems. They piece together mathematical results from the most basic rules of logic to give a certificate of correctness. The search space for proofs can be huge so we need techniques to cut it down or guide searches through. So it is perhaps not surprising that Automated Reasoning has been looking at how best to use machine learning for some time.

The work surveyed in Sect. 2 followed [7] which used SVMs and Gaussian processes to select from different search strategies for the E prover (see references within for other studies). Elsewhere, machine learning is used to select the most relevant theorems and definitions to use when proving a new conjecture in the MALAREA system [34]. An overview of such *premise selection* approaches is given in [26] with the first deep learning approach detailed in [1].

These approaches are relevant also for proof assistants. For example, Sledgehammer allows for ISABELLE/HOL to send goals to a variety of automated TPs and SMT solvers. A relevance filter heuristically ranks the thousands of facts available and selects a subset based on syntactic similarity to the goal, with the MaSh option based on machine learning outperforming the standard [25].

5 Summary

There are challenges in applying machine learning to mathematical software:

- Formulating choices in a way suitable for machine learning: e.g. choosing from existing heuristics rather than an ordering directly (Sect. 2.2).
- Obtaining datasets of sufficient size for training: for the work in Sect. 2.1 we had to build random polynomials while for that in Sect. 2.2 we borrowed benchmark sets from another discipline (SMT).
- Making related choices in tandem: for example the best variable ordering for CAD may change after GB preconditioning! How best to deal with this?

However, we have described successful applications in diverse areas and noted some potentials − an ICMS 2018 session should provide further inspiration.

Acknowledgements. Surveyed work in Sect. 2 was supported by the European Union's Horizon 2020 research and innovation programme under grant agreement No H2020-FETOPEN-2015-CSA 712689 (SC^2); and EPSRC grant EP/J003247/1. The author is now supported by EPSRC grant EP/R019622/1.

References

1. Alemi, A., Chollet, F., Een, N., Irving, G., Szegedy, C., Urban, J.: DeepMath - deep sequence models for premise selection. In: Proceedings 30th International Conference on Neural Information Processing Systems (NIPS 2016), pp. 2243–2251. Curran Associates Inc. (2016)
2. Arnon, D., Collins, G., McCallum, S.: Cylindrical algebraic decomposition I: the basic algorithm. SIAM J. Comput. **13**, 865–877 (1984). https://doi.org/10.1137/0213054
3. Biere, A., Heule, M., van Maaren, H., Walsh, T.: Handbook of Satisfiability, vol. 185. Frontiers in Artificial Intelligence and Applications. IOS Press (2009)
4. Bradford, R., Davenport, J., England, M., Errami, H., Gerdt, V., Grigoriev, D., Hoyt, C., Košta, M., Radulescu, O., Sturm, T., Weber, A.: A case study on the parametric occurrence of multiple steady states. In: Proceedings of 2017 ACM International Symposium on Symbolic and Algebraic Computation (ISSAC 2017), pp. 45–52. ACM (2017). https://doi.org/10.1145/3087604.3087622
5. Bradford, R., Davenport, J., England, M., McCallum, S., Wilson, D.: Truth table invariant cylindrical algebraic decomposition. J. Symbolic Comput. **76**, 1–35 (2016). https://doi.org/10.1016/j.jsc.2015.11.002
6. Bradford, R., Davenport, J.H., England, M., Wilson, D.: Optimising problem formulation for cylindrical algebraic decomposition. In: Carette, J., Aspinall, D., Lange, C., Sojka, P., Windsteiger, W. (eds.) CICM 2013. LNCS (LNAI), vol. 7961, pp. 19–34. Springer, Heidelberg (2013). https://doi.org/10.1007/978-3-642-39320-4_2

7. Bridge, J., Holden, S., Paulson, L.: Machine learning for first-order theorem proving. J. Autom. Reasoning **53**(2), 141–172 (2014). https://doi.org/10.1007/s10817-014-9301-5

8. Brown, C.: Companion to the tutorial: cylindrical algebraic decomposition. In: Presented at ISSAC 2004 (2004). http://www.usna.edu/Users/cs/wcbrown/research/ISSAC04/handout.pdf

9. Brown, C., Davenport, J.: The complexity of quantifier elimination and cylindrical algebraic decomposition. In: Proceedings of 2007 International Symposium on Symbolic and Algebraic Computation (ISSAC 2007), pp. 54–60. ACM (2007). https://doi.org/10.1145/1277548.1277557

10. Buchberger, B., Hong, H.: Speeding up quantifier elimination by Gröbner bases. Technical report, 91–06. RISC, Johannes Kepler University (1991)

11. Carette, J.: Understanding expression simplification. In: Proceedings of the 2004 International Symposium on Symbolic and Algebraic Computation (ISSAC 2004), pp. 72–79. ACM (2004). https://doi.org/10.1145/1005285.1005298

12. Carette, J., Farmer, W.M.: A review of mathematical knowledge management. In: Carette, J., Dixon, L., Coen, C.S., Watt, S.M. (eds.) CICM 2009. LNCS (LNAI), vol. 5625, pp. 233–246. Springer, Heidelberg (2009). https://doi.org/10.1007/978-3-642-02614-0_21

13. Cortes, C., Vapnik, V.: Support-vector networks. Mach. Learn. **20**(3), 273–297 (1995). https://doi.org/10.1023/A:1022627411411

14. Davenport, J., Bradford, R., England, M., Wilson, D.: Program verification in the presence of complex numbers, functions with branch cuts etc. In: 14th International Symposium on Symbolic and Numeric Algorithms for Scientific Computing, SYNASC 2012, pp. 83–88. IEEE (2012). https://doi.org/10.1109/SYNASC.2012.68

15. Davenport, J., Heintz, J.: Real quantifier elimination is doubly exponential. J. Symbolic Comput. **5**(1–2), 29–35 (1988). https://doi.org/10.1016/S0747-7171(88)80004-X

16. Dolzmann, A., Seidl, A., Sturm, T.: Efficient projection orders for CAD. In: Proceedings of 2004 International Symposium on Symbolic and Algebraic Computation (ISSAC 2004), pp. 111–118. ACM (2004). https://doi.org/10.1145/1005285.1005303

17. England, M., et al.: Problem Formulation for truth-table invariant cylindrical algebraic decomposition by incremental triangular decomposition. In: Watt, S.M., Davenport, J.H., Sexton, A.P., Sojka, P., Urban, J. (eds.) CICM 2014. LNCS (LNAI), vol. 8543, pp. 45–60. Springer, Cham (2014). https://doi.org/10.1007/978-3-319-08434-3_5

18. England, M., Bradford, R., Davenport, J.: Improving the use of equational constraints in cylindrical algebraic decomposition. In: Proceedings of 2015 International Symposium on Symbolic and Algebraic Computation (ISSAC 2015), pp. 165–172. ACM (2015). https://doi.org/10.1145/2755996.2756678

19. England, M., Davenport, J.H.: The complexity of cylindrical algebraic decomposition with respect to polynomial degree. In: Gerdt, V.P., Koepf, W., Seiler, W.M., Vorozhtsov, E.V. (eds.) CASC 2016. LNCS, vol. 9890, pp. 172–192. Springer, Cham (2016). https://doi.org/10.1007/978-3-319-45641-6_12

20. Fukasaku, R., Iwane, H., Sato, Y.: Real quantifier elimination by computation of comprehensive Gröbner systems. In: Proceedings of 2015 International Symposium on Symbolic and Algebraic Computation (ISSAC 2015), pp. 173–180. ACM (2015). https://doi.org/10.1145/2755996.2756646

21. Ghaffarian, S., Shahriari, H.: Software vulnerability analysis and discovery using machine-learning and data-mining techniques: a survey. ACM Comput. Surv. **50**(4) (2017). 36 pages, Article no. 56, https://doi.org/10.1145/3092566

22. Huang, Z., England, M., Davenport, J., Paulson, L.: Using machine learning to decide when to precondition cylindrical algebraic decomposition with Groebner bases. In: 18th International Symposium on Symbolic and Numeric Algorithms for Scientific Computing (SYNASC 2016), pp. 45–52. IEEE (2016). https://doi.org/10.1109/SYNASC.2016.020

23. Huang, Z., et al.: Applying machine learning to the problem of choosing a heuristic to select the variable ordering for cylindrical algebraic decomposition. In: Watt, S.M., Davenport, J.H., Sexton, A.P., Sojka, P., Urban, J. (eds.) CICM 2014. LNCS (LNAI), vol. 8543, pp. 92–107. Springer, Cham (2014). https://doi.org/10.1007/978-3-319-08434-3_8

24. Kobayashi, M., Iwane, H., Matsuzaki, T., Anai, H.: Efficient subformula orders for real quantifier elimination of non-prenex formulas. In: Kotsireas, I.S., Rump, S.M., Yap, C.K. (eds.) MACIS 2015. LNCS, vol. 9582, pp. 236–251. Springer, Cham (2016). https://doi.org/10.1007/978-3-319-32859-1_21

25. Kühlwein, D., Blanchette, J.C., Kaliszyk, C., Urban, J.: MaSh: machine learning for Sledgehammer. In: Blazy, S., Paulin-Mohring, C., Pichardie, D. (eds.) ITP 2013. LNCS, vol. 7998, pp. 35–50. Springer, Heidelberg (2013). https://doi.org/10.1007/978-3-642-39634-2_6

26. Kühlwein, D., van Laarhoven, T., Tsivtsivadze, E., Urban, J., Heskes, T.: Overview and evaluation of premise selection techniques for large theory mathematics. In: Gramlich, B., Miller, D., Sattler, U. (eds.) IJCAR 2012. LNCS (LNAI), vol. 7364, pp. 378–392. Springer, Heidelberg (2012). https://doi.org/10.1007/978-3-642-31365-3_30

27. Liang, J.H., V.K., H.G., Poupart, P., Czarnecki, K., Ganesh, V.: An empirical study of branching heuristics through the lens of global learning rate. In: Gaspers, S., Walsh, T. (eds.) SAT 2017. LNCS, vol. 10491, pp. 119–135. Springer, Cham (2017). https://doi.org/10.1007/978-3-319-66263-3_8

28. McCallum, S.: On projection in CAD-based quantifier elimination with equational constraint. In: Proceedings of 1999 International Symposium on Symbolic and Algebraic Computation (ISSAC 1999), pp. 145–149. ACM (1999). https://doi.org/10.1145/309831.309892

29. Schöneberg, U., Sperber, W.: POS tagging and its applications for mathematics. In: Watt, S.M., Davenport, J.H., Sexton, A.P., Sojka, P., Urban, J. (eds.) CICM 2014. LNCS (LNAI), vol. 8543, pp. 213–223. Springer, Cham (2014). https://doi.org/10.1007/978-3-319-08434-3_16

30. Seshia, S., Lahiri, S., Bryant, R.: A hybrid SAT-based decision procedure for separation logic with uninterpreted functions. In: Proceedings of 2003 Design Automation Conference, pp. 425–430 (2003). https://doi.org/10.1145/775832.775945

31. Shawe-Taylor, J., Cristianini, N.: Kernel methods for pattern analysis. CUP (2004)

32. Stoutemyer, D.: Ten commandments for good default expression simplification. J. Symbolic Comput. **46**(7), 859–887 (2011). https://doi.org/10.1016/j.jsc.2010.08.017

33. Sturm, T.: A survey of some methods for real quantifier elimination, decision, and satisfiability and their applications. Math. Comp. Sci. **11**(3), 483–502 (2017). https://doi.org/10.1007/s11786-017-0319-z

34. Urban, J.: MaLARea: a metasystem for automated reasoning in large theories. In: Empirically Successful Automated Reasoning in Large Theories (ESARLT 2007), CEUR Workshop Proceedings, vol. 257, 14 pages. CEUR-WS (2007)

35. Řehůřek, R., Sojka, P.: Automated classification and categorization of mathematical knowledge. In: Autexier, S., et al. (eds.) CICM 2008. LNCS (LNAI), vol. 5144, pp. 543–557. Springer, Heidelberg (2008). https://doi.org/10.1007/978-3-540-85110-3_44

36. Wilson, D.J., Bradford, R.J., Davenport, J.H.: Speeding up cylindrical algebraic decomposition by Gröbner Bases. In: Jeuring, J., et al. (eds.) CICM 2012. LNCS (LNAI), vol. 7362, pp. 280–294. Springer, Heidelberg (2012). https://doi.org/10.1007/978-3-642-31374-5_19

37. Wu, H.: Improving SAT-solving with machine learning. In: Proceedings of the 2017 ACM SIGCSE Technical Symposium Computer Science Education, pp. 787–788. ACM (2017). https://doi.org/10.1145/3017680.3022464

38. Xu, L., Hutter, F., Hoos, H., Leyton-Brown, K.: SATzilla: portfolio-based algorithm selection for SAT. J. Artif. Intell. Res. **32**, 565–606 (2008)

39. Yadollahi, A., Shahraki, A., Zaiane, O.: Current state of text sentiment analysis from opinion to emotion mining. ACM Comput. Surv. **50**(2) (2017). 33 pages, Article no. 25, https://doi.org/10.1145/3057270

40. Yau, K.L., Qadir, J., Khoo, H., Ling, M., Komisarczuk, P.: A survey on reinforcement learning models and algorithms for traffic signal control. ACM Comput. Surv. **50**(3) (2017). 38 pages, Article no. 34, https://doi.org/10.1145/3068287

41. Zhang, D., Tsai, J.: Machine learning and software engineering. Software Qual. J. **11**(2), 87–119 (2003). https://doi.org/10.1109/TAI.2002.1180784

A New Style of Mathematical Proof

William M. Farmer$^{(\boxtimes)}$

McMaster University, Hamilton, Canada
wmfarmer@mcmaster.ca
http://imps.mcmaster.ca/wmfarmer/

Abstract. Mathematical proofs will play a crucial role in building a *universal digital mathematics library (UDML)*. Traditional and formal style proofs do not adequately fulfill all the purposes that mathematical proofs have. We propose a new style of proof that fulfills seven purposes of mathematical proofs. We believe this style of proof is needed to build a highly interconnected UDML.

Keywords: Mathematical proof · Traditional proof style
Formal proof style · Universal digital mathematics library
Little theories method · Theory graphs · Flexiformalization
Cross checks

1 Introduction

Over the course of the next few decades, mathematical software systems will revolutionize how mathematical knowledge is expressed, organized, and applied. The end product of this revolution will be a *universal digital mathematics library (UDML)* containing vast amounts of highly interconnected mathematical knowledge.

We believe that the mathematical knowledge in a UDML should be represented in accordance with the *little theories method* [2] as a *theory graph* [4] consisting of axiomatic theories as nodes and theory morphisms as directed edges. The theories—which may have different underlying logics—serve as abstract mathematical models. The morphisms—which are meaning-preserving mappings from the formulas of one theory to the formulas of another—serve as information conduits that enable theory components such as definitions and theorems to be transported across the graph [1]. A theory graph enables mathematical knowledge to be formalized in the most convenient underlying logic at the most convenient level of abstraction using the most convenient vocabulary and then applied in many different contexts. In addition, the morphisms and other connections in a theory graph provide an infrastructure for finding relevant concepts and facts in the theory graph, e.g., all the definitions that are equivalent to a given definition.

This research was supported by NSERC.

© Springer International Publishing AG, part of Springer Nature 2018
J. H. Davenport et al. (Eds.): ICMS 2018, LNCS 10931, pp. 175–181, 2018.
https://doi.org/10.1007/978-3-319-96418-8_21

As one would expect, mathematical proofs will have a crucial role to play in the building of a UDML. They will serve as threads that tie the knowledge in a UDML together. We will argue that both the traditional proofs that appear in mathematical books and articles and the formal proofs developed using proof assistants are not adequate for the job and that a new style of proof is needed.

2   Styles of Mathematical Proof

A *proof* is an argument intended to show that a mathematical statement is a logical consequence of a set of premises. There are many styles of proof. Some proofs *describe* a deduction of the statement from the premises, while other proofs *prescribe* the steps needed to produce the deduction. Many proofs are presented in a *two-column format* where each line in the left column is an intermediate result in a deduction and the corresponding line in the right column explains why the result is justified. Some proofs contain *computations* (e.g., numeric or algebraic simplifications) or *constructions* (e.g., via straightedge and compass). *Geometry proofs* are deductions guided by a geometric drawing. *Visual proofs* are presented by a series of diagrams or an animation.

The proofs presented in mathematical books and articles usually exhibit a particular style that we call the *traditional proof style*. Proofs of this style are arguments written in a stylized form of natural language with a heavy use of special symbols. In traditional proofs the terminology and notation may be ambiguous, assumptions may be unstated, and the argument may contain logical gaps. However, the reader is expected to be able to resolve the ambiguities, identify the unstated assumptions, and fill in the gaps in the argument. The writer—whose purpose is to serve some particular community of readers—has the freedom to express the argument in whatever manner is deemed most effective. This includes exhibiting other styles of proof within the traditional style.

The *formal proof style* is to present a proof as a derivation in a proof system for a formal logic. Using software systems, formal proofs can be interactively developed and mechanically checked. This style of proof is highly constrained by the logic, proof system, and the fact that every detail must be verified. On the other hand, there is a very high level of assurance that the statement proved is indeed a theorem of the proof system. Although the traditional proof style dominates mathematics, the formal proof style is beginning to make some modest inroads in mathematical practice.

3   Purposes of Mathematical Proof

Mathematical proofs serve (at least) seven purposes. For each of the seven, we describe what the purpose is and compare how well traditional and formal proofs fulfill the purpose.

Purpose 1: Communication

The main purpose of a proof given in a textbook or scientific article is to *communicate* to the reader why a mathematical statement follows from a set of premises. Proofs constructed for communication are used to convey insight and to build intuition. The highly flexible style of traditional proofs is usually a much better vehicle for communication than the highly constrained style of formal proofs. This is especially true when the writer is more concerned about high-level ideas than low-level details (that often can be mechanically checked by computation). However, formal proofs can be much more effective at presenting intricate syntactic manipulations than traditional proofs.

Purpose 2: Certification

Another important purpose of a proof is to *certify* that a mathematical statement follows from a set of premises. Such a proof serves as a certificate that can be independently checked. Since a traditional proof is written for a particular audience, it may not be easily checked by someone outside of this audience. Moreover, a traditional proof may contain mistakes that are not easily noticed by a reader, even a reader in the intended audience. In contrast, a formal proof can be mechanically checked by software alone. A formal proof thus offers the highest level of certification.

Purpose 3: Discovery

A proof is often formulated to be a provisional argument that a mathematician can use to *discover* new theorems. This idea is brilliantly expressed in *Proofs and Refutations* by Imre Lakatos [7]. See also Yehuda Rav, "Why Do We Prove Theorems?" [8]. Traditional proofs are well suited for expressing provisional arguments that can be analyzed by humans. Formal proofs are too rigid to express provisional arguments and thus are poorly suited for this task. On the other hand, machines can be used to discover various kinds of structure embodied in a formal proof, but it is much more difficult to analyze traditional proofs in this way.

Purpose 4: Learning

The most effective way to *learn* mathematics is to read and write proofs. Traditional proofs are today generally much easier to read and write than formal proofs. However, a reader of a traditional proof may have to work harder on resolving ambiguities, identifying unstated assumptions, and filling in the gaps in the argument, and a writer may have to work harder on verifying that each step of the argument is valid. With effective software support, reading and writing formal proofs could become almost as easy as reading and writing traditional proofs.

Purpose 5: Universality

A proof is *universal* if it is expressed without any superfluous ideas and can thus be applied in any context in which the conditions of the proof hold. Traditional proofs can be expressed in a universal manner, but the underlying mathematical foundation is usually implicit. Traditional proofs are thus untethered; they do not have a precise mathematical home. Formal proofs have a precise mathematical home, but the home is usually not connected to many other contexts in which the proof can be applied. Hence both traditional and formal proofs fall short in achieving universality.

Purpose 6: Coherency

A theorem is *coherent* with a body of mathematical knowledge if it properly fits into the body without any contradictions or unexpected relationships. A proof by itself does not establish that the theorem it proves is coherent. Most mathematicians are reluctant to accept a theorem on only the basis of its proof. There is always the possibility of error, especially if the proof is not machine checked. Georg Kreisel has noted in several of his papers, e.g., in [5, p. 126] and [6, p. 145], that a better way to avoid error than carefully checking a proof is to use *cross checks* to compare the result with known facts. For example, the proof can be checked against similarly structured proofs and the theorem can be compared with consequences of the theorem or related versions of the theorem that have been independently proven. Although cross checks are very important, they are rarely written down and are not considered as part of either a traditional or a formal proof.

Purpose 7: Beauty

Mathematics is a utilitarian art form like architecture or industrial design. The desire to create *beauty* (what mathematicians call *elegance*) is one of the strongest driving forces in mathematics. Mathematicians seek to develop proofs that are beautiful as well as correct. Indeed some mathematicians will not accept a theorem until an elegant proof of the theorem has been found. It is safe to say that most mathematicians find it easier to write beautiful proofs in the highly flexible traditional proof style than in the highly constrained formal proof style.

Summary

Table 1 summarizes the differences between traditional and formal proofs. As can be seen, neither traditional proofs nor formal proofs fulfill all the purposes that mathematical proofs have. Furthermore, both styles lack the capacity to fully achieve universality and coherency.

Table 1. Traditional vs. formal proofs

	Traditional proofs	Formal proofs
Communication	●	◖
Certification	◖	●
Discovery (human)	●	○
Discovery (machine)	○	●
Learning (reading)	◗	◖
Learning (writing)	◗	◖
Universality	◖	◖
Coherency	○	○
Beauty	●	○

●: high; ◗: medium high; ◖: medium low; ○: low.

4 A Proposed New Style of Proof

Since traditional and formal proofs do not adequately achieve universality and coherency, they are not adequate for building a highly interconnected UDML. We therefore propose a new style of proof that is better suited for threading together the concepts and facts in a UDML. This new proof style has four components:

1. A *home theory* HT consisting of a formal logic Log, a language Lang in Log, and a set Axms of formulas in Lang that serve as the axioms of the theory.
2. A *theorem* Thm that is a formula in Lang purported to be a logical consequence of Axms.
3. An *argument* that shows Thm is a logical consequence of Axms.
4. A set CC of *cross checks* that compare the argument with similar arguments and the theorem with related theorems.

The home theory is a node in a UDML and a formal context for the proof. It is connected via meaning-preserving morphisms to other theories in the UDML. Ideally, the home theory is at the optimal level of abstraction for the proof and contains only the concepts and assumptions needed to express the proof's argument and theorem.

The theorem is a formal statement of what the proof's argument shows. It can be transported via appropriate morphisms to other theories in which the conditions of the proof hold. The home theory HT and the theorem Thm together thus serve as a specification of the set of theories T and formulas A in the UDML's theory graph such that T is an instance of HT, A is an instance of Thm, and A is a theorem of T. In this way, the proof fulfills the purpose of universality.

The argument has both a traditional component for communication, human-oriented discovery, learning, and beauty and a formal component for certification, machine-oriented discovery, and learning. The two components are tightly integrated so that, for example, a reader of the traditional component can switch, if desired, to the formal component when a gap in the argument is reached. It is not necessary that the formal component is a complete formal proof of the theorem. The formal component can even be totally absent. Thus the proof is *flexiformal* [3].

The set of cross checks should be carefully chosen to show that the theorem is coherent with the web of previously established facts in the UDML. There are various kinds of cross checks that can be in CC. One kind is a similar proof of a similar theorem. A second kind is a logical consequence of Thm in HT that has been proved independently of Thm. For example, the logical consequence could be a special case of Thm or a corollary of Thm. A third kind is an instance of Thm in an instance of HT that has been proved independently of Thm. For example, the instance of Thm could be an expression of Thm in a more concrete setting than HT or the dual of Thm in HT under some notion of duality. With the set CC the proof thus fulfills the purpose of coherency.

Of course, it is possible that a cross check fails. This could indicate that a mistake has been made or that something is not adequately understood. Thus failed cross checks are valuable because they can lead to finding hidden mistakes and making new discoveries.

In summary, the new style of proof we propose is a mixture of the traditional and formal proof styles in which the context of the proof and the statement proved are formal, the argument of the proof is expressed in a traditional style, and parts of the argument may be integrated with formal derivations. The home theory of the proof is a node in a theory graph of a UDML that is an optimal expression of the context of the proof. And the cross checks of the proof connect the proof and the theorem to similar proofs and related theorems in the theory graph.

5   Conclusion

We have proposed a new style of proof that contains elements of the traditional and formal styles of proof. It fulfills the seven purposes of mathematical proofs including universality and coherency. We believe this proof style is the thread that is needed to interconnect the concepts and facts in a UDML. We also believe its use will promote the formalization of mathematical knowledge while preserving the benefits of both traditional and formal proofs.

Acknowledgments. The author would like to thank the referees for their comments. This research was supported by NSERC.

References

1. Barwise, J., Seligman, J.: Information Flow: The Logic of Distributed Systems. Tracts in Computer Science, vol. 44. Cambridge University Press, Cambridge (1997)
2. Farmer, W.M., Guttman, J.D., Javier Thayer, F.: Little theories. In: Kapur, D. (ed.) CADE 1992. LNCS, vol. 607, pp. 567–581. Springer, Heidelberg (1992). https://doi.org/10.1007/3-540-55602-8_192

3. Kohlhase, M.: The flexiformalist manifesto. In: Voronkov, A., Negru, V., Ida, T., Jebelean, T., Petcu, D., Watt, S.M., Zaharie, D. (eds.) 14th International Workshop on Symbolic and Numeric Algorithms for Scientific Computing (SYNASC 2012), pp. 30–36. IEEE Press (2013)

4. Kohlhase, M.: Mathematical knowledge management: transcending the one-brain-barrier with theory graphs. Eur. Math. Soc. (EMS) Newsl. **92**, 22–27 (2014)

5. Kreisel, G.: Some uses of proof theory for finding computer programs. In: Colloque international de logique: Clermont-Ferrand 18–25 juillet 1975, Colloques internationaux du Centre national de la recherche scientifique, vol. 249, pp. 123–133. Centre national de la recherche scientifique (1977)

6. Kreisel, G.: Mathematical logic: tool and object lesson for science. Synthese **62**, 139–151 (1985)

7. Lakatos, I.: Proofs and Refutations. Cambridge University Press, Cambridge (1976)

8. Rav, Y.: Why do we prove theorems. Philosophia Mathematica **7**, 5–41 (1999)

Neural Ideals in SageMath

Ethan Petersen[1], Nora Youngs[2], Ryan Kruse[3], Dane Miyata[4],
Rebecca Garcia[5], and Luis David García Puente[5(✉)]

[1] Department of Mathematics, Rose-Hulman Institute of Technology,
Terre Haute, IN, USA
`peterseo@rose-hulman.edu`
[2] Department of Mathematics and Statistics, Colby College,
Waterville, ME, USA
`nora.youngs@colby.edu`
[3] Mathematics Department, Central College, Pella, IA, USA
`kruser1@central.edu`
[4] Department of Mathematics, University of Oregon, Eugene, OR 97403-1222, USA
`dmiyata@willamette.edu`
[5] Department of Mathematics and Statistics, Sam Houston State University,
Huntsville, TX, USA
`{rgarcia,lgarcia}@shsu.edu`

Abstract. A major area in neuroscience research is the study of how the brain processes spatial information. Neurons in the brain represent external stimuli via neural codes. These codes often arise from stereotyped stimulus-response maps, associating to each neuron a convex receptive field. An important problem consists in determining what stimulus space features can be extracted directly from a neural code. The neural ideal is an algebraic object that encodes the full combinatorial data of a neural code. This ideal can be expressed in a canonical form that directly translates to a minimal description of the receptive field structure intrinsic to the code. Here, we describe a SageMath package that contains several algorithms related to the canonical form of a neural ideal.

Keywords: Canonical form · Neural codes · Neural ideal · Neural ring

1 Introduction

Due to many recent technological advances in neuroscience, the ability to collect neural data has increased dramatically. With this comes a need for new methods to process and understand this data. One major question faced by researchers is to determine how the brain encodes spatial features of its environment through patterns of neural activity, as with place cell codes [2]. Curto et al. [1] phrase this question as, "What can be inferred about the underlying stimulus space from neural activity alone?"

To answer this question, Curto et al. [1] introduced the *neural ideal*—an algebraic object encoding the full combinatorial data of a neural code. The neural

J. H. Davenport et al. (Eds.): ICMS 2018, LNCS 10931, pp. 182–190, 2018.
https://doi.org/10.1007/978-3-319-96418-8_22

ideal can be expressed in a *canonical form* that directly translates to a minimal description of the receptive field structure intrinsic to the code.

In this article we describe a SageMath [3] package containing several algorithms to compute the neural ideal, and its canonical form. We also leverage the power of specialized algorithms in SageMath to compute Gröbner bases and other algebraic objects related to the neural ideal. Our main contribution is a new algorithm to compute canonical forms. This algorithm is significantly more efficient than the original method outlined by Curto et al. [1].

The outline of this paper is as follows. Section 2 introduces the algebraic geometry of neural codes. Section 3 describes the new iterative algorithm to compute canonical forms. Section 4 gives a tutorial of our SageMath package. Section 5 proves the correctness of the algorithm introduced in Sect. 3.

2 Background

A *neural code* $\mathcal{C} \subseteq \{0,1\}^n$ is a set of binary strings that represent neural activity. A '1' represents a firing neuron, while a '0' represents an idle neuron. Given $v \in \{0,1\}^n$, denote by ρ_v the characteristic function for v:

$$\rho_v = \prod_{i=1}^{n}(1 - v_i - x_i) = \prod_{\{i|v_i=1\}} x_i \prod_{\{j|v_j=0\}} (1 - x_j).$$

Note that $\rho_v(v) = 1$ and $\rho_v(x) = 0$ for any $x \neq v \in \{0,1\}^n$. Given a neural code $\mathcal{C} \subseteq \{0,1\}^n$, the *neural ideal* $J_{\mathcal{C}}$ is the ideal in $\mathbb{F}_2[x_1, \ldots, x_n]$ generated by the polynomials ρ_v with $v \notin \mathcal{C}$, that is,

$$J_{\mathcal{C}} = \langle \rho_v \mid v \notin \mathcal{C} \rangle.$$

Many systems of neurons react to stimuli which have a natural geographic association. For example, place cells in rats are associated to place fields or regions of the rat's 2-dimensional environment [2]. In such a geographic setup, we would assume that if two neurons are observed to fire together, then the sets of stimuli for these neurons must overlap. The idea of a *realization for a code* formalizes this notion. Suppose $\mathcal{U} = \{U_1, \ldots, U_n\}$ is a collection of open subsets of a set $X \subset \mathbb{R}^n$. Here, X represents the space of possible stimuli, and U_i is the receptive field of the ith neuron, i.e., the set of stimuli which will cause that neuron to fire. We say that $\mathcal{U}$ is a *realization* for a code $\mathcal{C}$, or that $\mathcal{C} = \mathcal{C}(\mathcal{U})$, if

$$\mathcal{C} = \{v \in \{0,1\}^n \mid (\bigcap_{v_i=1} U_i) \backslash \bigcup_{v_j=0} U_j \neq \emptyset\}.$$

Given any code $\mathcal{C}$, a realization of $\mathcal{C}$ in $\mathbb{R}^1$ always exists. However, such realization will not generally be representative of the space of stimuli in any geometric sense unless we place additional restrictions on the sets in $\mathcal{U}$, such as convexity or connectedness. The ideal $J_{\mathcal{C}}$ for the code $\mathcal{C}$ completely determines the interaction of the U_i in any realization $\mathcal{U}$ of $\mathcal{C}$. This information is more simply described by the *canonical form* of the neural ideal $J_{\mathcal{C}}$.

A polynomial $f \in \mathbb{F}_2[x_1, \ldots, x_n]$ is a *pseudo-monomial* if f has the form

$$f = \prod_{i \in \sigma} x_i \prod_{j \in \tau} (1 - x_j),$$

where $\sigma \cap \tau = \emptyset$. An ideal $J \subseteq \mathbb{F}_2[x_1, \ldots, x_n]$ is a *pseudo-monomial ideal* if J can be generated by a set of pseudo-monomials. Note that the neural ideal $J_{\mathcal{C}}$ is a pseudo-monomial ideal. Let $J \subseteq \mathbb{F}_2[x_1, \ldots, x_n]$ be an ideal, and $f \in J$ a pseudo-monomial. Then f is a *minimal pseudo-monomial* of J if there is no other pseudo-monomial $g \in J$ with $\deg(g) < \deg(f)$ such that $f = gh$ for some $h \in \mathbb{F}_2[x_1, \ldots, x_n]$. The *canonical form* of a pseudo-monomial ideal J, denoted $\mathrm{CF}(J)$, is the set of *all* minimal pseudo-monomials of J.

Given a pseudo-monomial ideal J, the canonical form $\mathrm{CF}(J)$ is unique and $J = \langle \mathrm{CF}(J) \rangle$. Even though $\mathrm{CF}(J)$ consists of minimal pseudo-monomials, it is not necessarily the case that $\mathrm{CF}(J)$ is a minimal generating set for J.

The following theorem describes the set of relations on any realization $\mathcal{U}$ of $\mathcal{C}$ provided by $\mathrm{CF}(J)$. Given $\sigma \subseteq \{1, \ldots, n\}$, let $U_\sigma := \bigcap_{i \in \sigma} U_i$, with $U_\emptyset = X$.

Theorem 1 ([1, **Theorem 4.3**])**.** *Let $\mathcal{C} \subseteq \{0,1\}^n$ be a neural code, and let $\mathcal{U} = \{U_1, \ldots, U_n\}$ be any collection of open sets in a nonempty stimulus space X such that $\mathcal{C} = \mathcal{C}(\mathcal{U})$. The canonical form of $J_{\mathcal{C}}$ is the union of the following sets:*

$$\{x_\sigma \mid \sigma \text{ is minimal w.r.t. } U_\sigma = \emptyset\},$$

$$\left\{ x_\sigma \prod_{i \in \tau}(1 - x_i) \;\middle|\; \sigma, \tau \neq \emptyset, \; \sigma \cap \tau = \emptyset, \; U_\sigma \neq \emptyset, \; \bigcup_{i \in \tau} U_i \neq X, \right.$$

$$\left. \text{and } \sigma, \tau \text{ are minimal w.r.t. } U_\sigma \subseteq \bigcup_{i \in \tau} U_i \right\},$$

$$\left\{ \prod_{i \in \tau}(1 - x_i) \;\middle|\; \tau \text{ is minimal w.r.t. } X \subseteq \bigcup_{i \in \tau} U_i \right\}.$$

We call the above three (disjoint) sets of relations comprising $\mathrm{CF}(J_{\mathcal{C}})$ the minimal Type 1, Type 2 and Type 3 relations, respectively. Since the canonical form is unique, by Theorem 1, any receptive field representation of the code $\mathcal{C} = \mathcal{C}(U)$ satisfies the following relationships:

Type 1: $x_\sigma \in \mathrm{CF}(J_{\mathcal{C}})$ implies $U_\sigma = \emptyset$, and $U_\gamma \neq \emptyset$, for all $\gamma \subsetneq \sigma$.
Type 2: $x_\sigma \prod_{i \in \tau}(1 - x_i) \in \mathrm{CF}(J_{\mathcal{C}})$ implies $U_\sigma \subseteq \bigcup_{i \in \tau} U_i$, with σ and τ minimal.
Type 3: $\prod_{i \in \tau}(1 - x_i) \in \mathrm{CF}(J_{\mathcal{C}})$ implies $X \subseteq \bigcup_{i \in \tau} U_i$, and τ is minimal.

The minimality above refers to the fact that the inclusions fail for any proper subset of σ and τ.

3   The Iterative Algorithm

Curto et al. [1] detailed an algorithm to obtain the canonical form via the primary decomposition of the neural ideal. Here we present an iterative algorithm that requires only simple polynomial arithmetic. This algorithm begins with the canonical form for a code consisting of a single codeword, and iterates by adding the remaining codewords one by one and adjusting the canonical form accordingly. Algorithm 1 describes the process for adding in a new codeword.

Algorithm 1. Iterative step to update CF after adding code word c

---

Input   : $\mathrm{CF}(J_{\mathcal{C}}) = \{f_1, \ldots, f_k\}$, where $\mathcal{C} \subseteq \{0,1\}^n$ is a code on n neurons, and a codeword $c \in \{0,1\}^n$

Output: $\mathrm{CF}(J_{\mathcal{C} \cup \{c\}})$

begin

    $L \longleftarrow \{\}, M \longleftarrow \{\}, N \longleftarrow \{\}$

    for $x \longleftarrow 1$ **to** k **do**

        if $f_i(c) = 0$ **then**

            $L \longleftarrow L \cup \{f_i\}$

        else

            $M \longleftarrow M \cup \{f_i\}$

        end

    end

    for $f \in M$ **do**

        for $j \longleftarrow 1$ **to** n **do**

            if $(x_j - c_j)f$ *is not a multiple of an element of* L **and** $(x_j - c_j - 1) \nmid f$ **then**

                $N \longleftarrow N \cup \{(x_j - c_j)f\}$

            end

        end

    end

    return $L \cup N = \mathrm{CF}(J_{\mathcal{C} \cup \{c\}})$

end

Each $f \in \mathrm{CF}(J_{\mathcal{C}})$ for which $f(c) = 0$ is included in $\mathrm{CF}(J_{\mathcal{C} \cup \{c\}})$. For each $f \in \mathrm{CF}(J_{\mathcal{C}})$ with $f(c) = 1$, we consider the product of f with each linear term $(x_j - c_j)$, since $(x_j - c_j)f$ evaluated at c is 0. We add each of those products to $\mathrm{CF}(J_{\mathcal{C} \cup \{c\}})$ unless it is a multiple of a pseudo-monomial already in the canonical form, or a multiple of a Boolean polynomial $x_j^2 - x_j$. Certainly, any polynomial f returned by Algorithm 1 satisfies $f(v) = 0$ for all $v \in \mathcal{C} \cup \{c\}$. A proof that this algorithm outputs exactly $\mathrm{CF}(J_{\mathcal{C} \cup \{c\}})$ is found in Sect. 5.

Algorithm 1 has been implemented in Matlab [4] and also in our SageMath package. Table 1 displays some runtime statistics regarding the SageMath implementation. These runtime statistics summarize the running times on 100 randomly generated sets of codewords for each dimension $n = 4, \ldots, 10$. These computations were performed on SageMath 7.2 running on a Macbook Pro with a 2.8 GHz Intel Core i7 processor and 16 GB of memory. We performed a similar test for our implementation of the original canonical form algorithm [1] and also on the Matlab implementation of our iterative method. However, even in dimension 5 the performance of the original algorithm is subpar. In our tests, we found several codes for which the original algorithm took hundreds or even thousands of seconds to finish. For example, the iterative algorithm takes 0.01 s to compute the canonical form of the code below, but the original method takes 1 h and 8 min to perform the same computation.

10000, 10001, 01011, 01010, 10010, 01110, 01101, 01100, 11111,

11010, 11011, 01000, 01001, 00111, 00110, 00001, 00010, 00011, 00101.

We found several codes in dimension 5 for which the original algorithm halts due to lack of memory. We also found examples in dimension 6 for which our Matlab implementation took thousands of seconds to finish.

Table 1. Runtime statistics (in seconds) for the iterative CF algorithm in SageMath.

Dimension	min	max	mean	median	std
4	0.000077	0.0034	0.0016	0.0018	0.00076
5	0.000087	0.014	0.0076	0.0082	0.0034
6	0.00012	0.108	0.049	0.051	0.024
7	0.00012	0.621	0.298	0.323	0.135
8	0.000097	4.011	1.964	2.276	1.036
9	0.698	39.28	24.86	27.38	9.976
10	0.229	350.5	237.45	271.3	87.1

4   SageMath Tutorial

The latest stable version of the `NeuralIdeals` package can be downloaded from https://github.com/e6-1/NeuralIdeals. We will assume that SageMath is properly installed on the system and that the files `iterative_canonical.spyx`, `neuralcode.py` and `examples.py` are downloaded in the folder `NeuralIdeals`. `NeuralIdeals` can also be installed in **CoCalc** (https://cocalc.com/) following similar commands. The package is loaded by:

```
sage: load("NeuralIdeals/iterative_canonical.spyx")
sage: load("NeuralIdeals/neuralcode.py")
sage: load("NeuralIdeals/examples.py")
```

The first file contains the iterative algorithm in Cython, so loading it requires a C compiler. The file `neuralcode.py` contains all other functions in the package. The file `examples.py` has some additional examples that can be loaded with `sage: neuralcodes()`.

Now, we exemplify the main commands of `NeuralIdeals`. First, we define a neural code:

```
sage: neuralCode = NeuralCode(['001','010','110'])
```

We can compute the neural ideal:

```
sage: neuralIdeal = neuralCode.neural_ideal()
sage: neuralIdeal
Ideal (x0*x1*x2 + x0*x1 + x0*x2 + x0 + x1*x2 + x1 + x2 + 1,
x0*x1*x2 + x1*x2, x0*x1*x2 + x0*x1 + x0*x2 + x0,
x0*x1*x2 + x0*x2, x0*x1*x2) of Multivariate Polynomial Ring
in x0, x1, x2 over Finite Field of size 2
```

We can compute the primary decomposition using a custom algorithm:

```
sage: pm_primary_decomposition(neuralIdeal)
[Ideal (x2 + 1, x1, x0), Ideal (x2, x1 + 1)]
```

We can compute the canonical form of the neural ideal.

```
sage: canonicalForm = neuralCode.canonical()
sage: canonicalForm
Ideal (x1*x2, x1*x2 + x1 + x2 + 1, x0*x1 + x0, x0*x2)
```

The method `canonical()` uses the iterative algorithm by default. In order to use the procedure described in [1], we need the following flag

```
sage: neuralCode.canonical(algorithm="original")
Ideal (x1*x2, x0*x1 + x0, x1*x2 + x1 + x2 + 1, x0*x2)
```

This procedure uses by default the specialized primary decomposition of pseudo-monomial ideals outlined in [1]. One can make this explicit with the flag `canonical(algorithm="original", decomposition_algorithm="pm")`. The method `canonical` can also use the general primary decomposition methods implemented in SageMath, namely, Shimoyama-Yokoyama and Gianni-Trager-Zacharias with the flags `sy` and `gtz`, respectively. We also compared these primary decomposition algorithms. In our tests, `gtz` generally outperforms `pm`. However, `gtz` is meant to be used in characteristic 0 or in large positive characteristic. But in small characteristic it may not terminate.

The method `canonical()` returns an ideal whose generators are not factored, and hence not easy to interpret in our context. The following command outputs these polynomials in factored form:

```
sage: neuralCode.factored_canonical()
[x2 * x1, (x1 + 1) * x0, (x2 + 1) * (x1 + 1), x2 * x0]
```

From this output we can easily read off the RF structure of the neural code. But this can be obtained directly with the following command

```
sage: neuralCode.canonical_RF_structure()
Intersection of U_['2', '1'] is empty
X = Union of U_['2', '1']
Intersection of U_['0'] is a subset of Union of U_['1']
Intersection of U_['2', '0'] is empty
```

We can also compute Gröbner bases and the Gröbner fan of a neural ideal. We could compute the neural ideal and use the built in `groebner_basis()` method, but that approach does not impose the Boolean relations ($x^2 + x = 0$). Our method uses the specialized and very efficient Gröbner basis algorithm for ideals in Boolean rings; thus taking into account the Boolean relations.

```
sage: neuralCode.groebner_basis()
Ideal (x0*x2, x1 + x2 + 1)
sage: neuralIdeal.groebner_basis()
[x0*x2, x1 + x2 + 1, x2^2 + x2]
```

The current stable version of `NeuralIdeals` dates from 2016. However, this package continues to be actively developed. Currently we are implementing methods to test convexity and also to draw realizations in $\mathbb{R}, S^1$ and $\mathbb{R}^2$.

5   Correctness Proof for Algorithm 1

Here, we show that the process described in Algorithm 1 gives $\mathrm{CF}(J_{\mathcal{C} \cup \{c\}})$ from $\mathrm{CF}(J_{\mathcal{C}})$ and c. Throughout, we use the following conventions and terminology: $\mathcal{C}$ and $\mathcal{D}$ are neural codes on n neurons; so, $\mathcal{C}, \mathcal{D} \subseteq \{0,1\}^n$. A polynomial is *multilinear* if it is linear in each of its variables. Note that there is a unique multilinear representative of every equivalence class of $\mathbb{F}_2[x_1, \ldots, x_n]/\langle x_i(1-x_i)\rangle$. For $h \in \mathbb{F}_2[x_1, \ldots, x_n]$, let h_R denote the unique multilinear representative of the equivalence class of h in $\mathbb{F}_2[x_1, \ldots, x_n]/\langle x_i(1-x_i)\rangle$.

Assuming $\mathrm{CF}(J_{\mathcal{C}}) = \{f_1, \ldots, f_r\}$ and $\mathrm{CF}(J_{\mathcal{D}}) = \{g_1, \ldots, g_s\}$, we define the set of *reduced products*

$$P(\mathcal{C}, \mathcal{D}) \stackrel{\text{def}}{=} \{(f_i g_j)_R \mid i \in [r], j \in [s]\}.$$

Note that since pseudo-monomials are multilinear, for each pair i, j we have either $(f_i g_j)_R = 0$ or $(f_i g_j)_R$ is a multiple of both f_i and g_j. We define the *minimal reduced products* as

$$\mathrm{MP}(\mathcal{C}, \mathcal{D}) \stackrel{\text{def}}{=} \{h \in P(\mathcal{C}, \mathcal{D}) \mid h \neq 0 \text{ and } h \neq fg \text{ for any } f \in P(\mathcal{C}, \mathcal{D}), \deg g \geq 1\}.$$

Lemma 1. *If $\mathcal{C}, \mathcal{D} \subseteq \{0,1\}^n$, then $\mathrm{CF}(J_{\mathcal{C} \cup \mathcal{D}}) = MP(\mathcal{C}, \mathcal{D})$.*

Proof. First, we show $\mathrm{MP}(\mathcal{C}, \mathcal{D}) \subseteq J_{\mathcal{C} \cup \mathcal{D}}$. For any $h \in \mathrm{MP}(\mathcal{C}, \mathcal{D})$, there is some $f_i \in \mathrm{CF}(J_{\mathcal{C}})$ and $g_j \in \mathrm{CF}(J_{\mathcal{D}})$ so $h = (f_i g_j)_R$. In particular, $h \in J_{\mathcal{C}}$ as h is a multiple of f_i, and $h \in J_{\mathcal{D}}$ as it is a multiple of g_j. Thus $h(c) = 0$ for all $c \in \mathcal{C} \cup \mathcal{D}$, so $h \in J_{\mathcal{C} \cup \mathcal{D}}$.

Suppose $h \in \mathrm{CF}(J_{\mathcal{C} \cup \mathcal{D}})$. Then as $J_{\mathcal{C} \cup \mathcal{D}} \subseteq J_{\mathcal{C}}$, there is some $f_i \in \mathrm{CF}(J_{\mathcal{C}})$ so that $h = h_1 f_i$, and likewise there is some $g_j \in \mathrm{CF}(J_{\mathcal{D}})$ so $h = h_2 g_j$ where h_1, h_2 are pseudo-monomials. Thus h is a multiple of $(f_i g_j)_R$ and hence is a multiple of some element of $\mathrm{MP}(\mathcal{C}, \mathcal{D})$. But as every element of $\mathrm{MP}(\mathcal{C}, \mathcal{D})$ is an element of $J_{\mathcal{C} \cup \mathcal{D}}$, and $h \in \mathrm{CF}(J_{\mathcal{C} \cup \mathcal{D}})$, this means h itself must actually be in $\mathrm{MP}(\mathcal{C}, \mathcal{D})$. Thus, $\mathrm{CF}(J_{\mathcal{C} \cup \mathcal{D}}) \subseteq \mathrm{MP}(\mathcal{C}, \mathcal{D})$. For the reverse containment, suppose $h \in \mathrm{MP}(\mathcal{C}, \mathcal{D})$; by the above, $h \in J_{\mathcal{C} \cup \mathcal{D}}$. It is thus the multiple of some $f \in \mathrm{CF}(J_{\mathcal{C} \cup \mathcal{D}})$. But we have shown that $f \in \mathrm{MP}(\mathcal{C}, \mathcal{D})$, which contains no multiples. So $h = f$ is in $\mathrm{CF}(J_{\mathcal{C} \cup \mathcal{D}})$.

*Proof (**Correctness Proof for Algorithm 1**).* Note that if $c \in \mathcal{C}$, then $L = \mathrm{CF}(J_{\mathcal{C}})$, so the algorithm ends immediately and outputs $\mathrm{CF}(J_{\mathcal{C}})$; we will generally assume $c \notin \mathcal{C}$.

To show that Algorithm 1 produces the correct canonical form, we apply Lemma 1; it suffices to show that the set $L \cup N$ equals $\mathrm{MP}(\mathcal{C}, \{c\})$. We must consider all the products in $P(\mathcal{C}, \{c\})$, and then remove any redundancies. Note that $\mathrm{CF}(J_{\{c\}}) = \{x_i - c_i \mid i \in [n]\}$.

We will look at L and M separately. Let $g \in L$. Since $g(c) = 0$, we know $(g \cdot (x_i - c_i))_R = g$ for at least one i. So $g \in \mathrm{MP}(\mathcal{C}, \{c\})$. Any other product $(g \cdot (x_j - c_j))_R$ will either be 0, g, or a multiple of g, and hence will not appear in $\mathrm{MP}(\mathcal{C}, \{c\})$. Thus, all products of linear terms with elements of L are considered, and all multiples or zeros are removed. It is impossible for elements of L to be multiples of one another, as $L \subseteq \mathrm{CF}(J_{\mathcal{C}})$.

Now consider all products of elements of M with the linear elements of $\mathrm{CF}(J_{\{c\}})$. We discard them if their reduction is 0, or if they are a multiple of something in L. If neither holds, we add them to N. So it remains to show that no element of N can be a multiple of any other element in N, and no element of N can be a multiple of anything in L, and thus that we have removed all possible multiples. First, no element of N may be a multiple of an element of L, since if $g \in L$, $f \cdot (x_i - c_i) \in N$, and $f \cdot (x_i - c_i) \cdot p = g$ for some pseudo-monomial p, then $f \mid g$. But this is impossible as f, g are both in $\mathrm{CF}(J_{\mathcal{C}})$. Now, suppose $f \cdot (x_i - c_i) = h \cdot g \cdot (x_j - c_j)$ for $f, g \in \mathrm{CF}(J_{\mathcal{C}})$ and $f \cdot (x_i - c_i), g \cdot (x_j - c_j) \in N$, and h a pseudo-monomial. Then as $f \nmid g$ and $g \nmid f$, we have $i \neq j$, and so $(x_j - c_j) \mid f$. But this means $f \cdot (x_j - c_j) = f$ and therefore $f \in L$, which is a contradiction. So no elements of N may be multiples of one another.

References

1. Curto, C., Itskov, V., Veliz-Cuba, A., Youngs, N.: The neural ring: an algebraic tool for analyzing the intrinsic structure of neural codes. Bull. Math. Biol. **75**(9), 1571–1611 (2013). https://doi.org/10.1007/s11538-013-9860-3
2. O'Keefe, J., Dostrovsky, J.: The hippocampus as a spatial map. Preliminary evidence from unit activity in the freely-moving rat. Brain Res. **34**(1), 171–175 (1971). http://www.sciencedirect.com/science/article/pii/0006899371903581

3. Stein, W., et al.: Sage Mathematics Software (Version 7.2.0). The Sage Developers (2016). http://www.sagemath.org
4. Youngs, N.: Neural ideal: a Matlab package for computing canonical forms (2015). http://github.com/nebneuron/neural-ideal

Universal Gröbner Basis for Parametric Polynomial Ideals

Amir Hashemi[1], Mahdi Dehghani Darmian[2(⊠)], and Marzieh Barkhordar[3]

[1] Department of Mathematical Sciences, Isfahan University of Technology,
84156-83111 Isfahan, Iran
[2] School of Mathematics, Institute for Research in Fundamental Sciences (IPM),
19395-5746 Tehran, Iran
m.dehghanidarmian@ipm.ir
[3] Department of Computer Engineering, Sharif University of Technology,
11365-8639 Tehran, Iran

Abstract. In this paper, we introduce the concept of universal Gröbner basis for a parametric polynomial ideal. In this direction, we present a new algorithm, called UGS, which takes as input a finite set of parametric polynomials and outputs a universal Gröbner system for the ideal generated by input polynomials, by decomposing the space of parameters into a finite set of parametric cells and for each cell associating a finite set of parametric polynomials which is a universal Gröbner basis for the ideal corresponding to that cell. Indeed, for each values of parameters satisfying a condition set, the corresponding polynomial set forms a universal Gröbner basis for the ideal. Our method relies on the parametric variant of the Gröbner basis conversion and also on the PGBMain algorithm due to Kapur et al. to compute parametric Gröbner bases. The proposed UGS algorithm has been implemented in MAPLE-SAGE and its performance is investigated through an example.

Keywords: Parametric polynomials · Gröbner bases
Gröbner systems · Gröbner fan · Universal Gröbner bases
Universal Gröbner systems

1 Introduction

Gröbner bases are a powerful computational tool in computer algebra with many interesting applications in Mathematics, science, and engineering. These bases and the first algorithm to compute them were introduced by Buchberger in his PhD thesis [3]. On the other hand, a *universal Gröbner basis* (UGB) introduced by Schwartz [14] is a finite basis for a polynomial ideal which remains a Gröbner basis with respect to any monomial order. However, his results are not constructive. Later, Weispfenning [18] presented an approach to construct UGBs. We refer to [16] for more details on UGBs and their applications.

In this paper, we are interested in computing a universal Gröbner system (UGS) for a given parametric polynomial ideal, i.e. we want to decompose the

© Springer International Publishing AG, part of Springer Nature 2018
J. H. Davenport et al. (Eds.): ICMS 2018, LNCS 10931, pp. 191–199, 2018.
https://doi.org/10.1007/978-3-319-96418-8_23

space of parameters into a finite set of cells and for each cell we want to give a finite set of parametric polynomials which forms a UGB for the ideal corresponding to that cell. Indeed, UGS is a natural generalization of the concept of *Gröbner system* (GS). The concept of GS for a parametric ideal was introduced by Weispfenning in [19]. Montes in [12] proposed an efficient algorithm based on Buchberger's algorithm to compute a GS for a parametric ideal (see also [5]). Kapur et al. [11] gave a new algorithm for computing GS by combining Weispfenning's algorithm with the Suzuki and Sato algorithm [17].

In this paper, after reviewing some basic notations and definitions related to Gröbner bases, GSs and UGSs (see Sect. 2) we describe an algorithm to compute a UGS for a parametric ideal (see Sect. 3). This algorithm has been implemented in MAPLE-SAGE and it is illustrated through an example (see Sect. 4).

2   Preliminaries

In this section, we review the required concepts such as Gröbner basis, GS and Gröbner fan for a polynomial ideal. Let $\mathcal{R} = \mathbb{K}[\mathbf{x}]$ be the polynomial ring over a field $\mathbb{K}$ in $\mathbf{x} = x_1, \ldots, x_n$. Let $\mathcal{I} = \langle f_1, \ldots, f_k \rangle$ be the ideal of $\mathcal{R}$ generated by the f_i's. Also let $f \in \mathcal{R}$ and $\prec$ be a monomial order on $\mathcal{R}$. The leading monomial of f is the greatest monomial (with respect to $\prec$) appearing in f, where we denote it by $\mathrm{LM}(f)$. The leading monomial ideal of $\mathcal{I}$ is defined to be $\mathrm{LM}(\mathcal{I}) = \langle \mathrm{LM}(f) | f \in \mathcal{I} \rangle$. A finite subset $\{g_1, \ldots, g_t\} \subset \mathcal{I}$ is called a Gröbner basis for $\mathcal{I}$ w.r.t. $\prec$ if $\mathrm{LM}(\mathcal{I}) = \langle \mathrm{LM}(g_1), \ldots, \mathrm{LM}(g_t) \rangle$. The construction of a Gröbner basis depends on the choice of a monomial order. However, a Gröbner basis is universal if it has the Gröbner property w.r.t. each monomial order.

Definition 1. *A* universal Gröbner basis *(UGB) of* $\mathcal{I}$ *is a finite basis which is a Gröbner basis for* $\mathcal{I}$ *with respect to all monomial orders.*

Schwartz in [14, Corollary 31] showed that any polynomial ideal possesses a finite UGB. Then, Weispfenning in [18] studied the construction of these bases, together with upper complexity bounds. Bobson et. al. in [2, Theorem 4.2] proved that there is a polynomial time algorithm for computing a UGB for a zero-dimensional polynomial ideal, i.e. any ideal having a finite set of common zeros. Computation of UGBs is based on the computation of Gröbner fans. The Gröbner fan of a polynomial ideal was introduced by Mora and Robbiano [13]. For recalling the concept of Gröbner fan, it is required to mention some preliminaries on convex geometry but due to the lack of space we give a short review.

Definition 2. *Let* $G = \{\underline{x^{\alpha_i}} + \sum_j c_{ij} x^{\beta_{ij}}\}_{i=1}^{\ell}$ *be the reduced Gröbner basis of* $\mathcal{I}$ *w.r.t. a monomial order* $\prec$ *(where the leading terms w.r.t* $\prec$ *are underlined and in this case* G *is called a marked basis). The Gröbner cone of* $\mathcal{I}$ *associated to* $\prec$ *is* $C_{\prec}(\mathcal{I}) = \{w \in \mathbb{R}_+^n \mid w.(\alpha_i - \beta_{ij}) \geq 0, \text{for each } i \text{ and } j\}.$

According to [13, Property 4.1] and [16, Theorem 1.2], there exist only finitely many different reduced Gröbner bases for a given ideal $\mathcal{I}$. So, the *Gröbner fan* of $\mathcal{I}$ is a fan consisting of the cones corresponding to different monomial orders for $\mathcal{I}$. In other words, the set of all distinct Gröbner cones of an ideal $\mathcal{I}$ is called the Gröbner fan of $\mathcal{I}$. It is shown that the Gröbner fan of an ideal is a fan (see [7,16]). An implementation of the Gröbner fan is available via the package Gfan [9], based on the article of Fukuda et al. [7] which is included in some computer algebra systems such as Singular [4] and SAGE [15]. Computing a Gröbner fan is equivalent to computing all reduced Gröbner bases of the ideal. Thus, one can define and compute a UGB based on Gröbner fans. Below, we will explain how one can apply the GFAN algorithm from [7] to compute UGBs. In the following algorithm, there is a global variable U which is initially the empty set and at each iteration of the algorithm, a new reduced Gröbner basis is added into this set and at the end it is a UGB of the input ideal.

Algorithm 1. UGB

Require: G; the reduced Gröbner basis of $\mathcal{I}$ w.r.t. a monomial order $\prec$
Ensure: a UGB of $\mathcal{I}$
  $Facets :=$ the list of all facets of $C_\prec(G)$
  for F in $Facets$ **do**
    if F is a valid facet **then**
      $G' :=$ CONVERT(G, F)
      Add G' to U
      UGB(G')
    end if
  end for
  Return(U)

We refer to the PhD thesis of Jensen [10] for the proof of termination and correctness of this algorithm. He associated an undirected graph Γ to the Gröbner fan of a polynomial ideal with two maximal cones being connected if they share a common facet. Then he proved that the reverse search technique [1] can be used for traversing the graph Γ (reverse search technique is a strategy which is applied for enumerating vertices of a polytope). In fact, he showed that the graph Γ associated to the Gröbner fan of a polynomial ideal may be oriented easily without cycles and with a unique sink. His idea is to define a spanning tree of the graph which can be easily traversed by the reverse search technique. In the above algorithm, a facet F is valid iff

- the associated edge to facet F is incoming,
- the corresponding cone to the new Gröbner basis obtained by CONVERT(G, F) has not been visited already.

Furthermore, CONVERT is the following local Gröbner basis change procedure which is one iteration of the generic Gröbner walk algorithm [6].

Algorithm 2. CONVERT

Require: G; a reduced Gröbner basis w.r.t. a monomial order $\prec$ and F; a facet of $C_\prec(G)$
Ensure: G'; a reduced Gröbner basis different from G whose Gröbner cone also has F as a facet
  $\mathrm{LP}_v(G) := \{\mathrm{LP}_v(g) \mid g \in G\}$ where v is an interior point of F
  $H :=$ marked reduced Gröbner basis of $\mathrm{LP}_v(G)$ w.r.t. $\prec_{-\alpha}$ where α is the normal vector of F
  $H' := \{f - \mathrm{NormalForm}(f, G) \mid f \in H\}$
  $G' := \mathrm{AutoReduce}(H')$ to transform H' into a reduced Gröbner basis
  Return(G')

Here the *leading part* of g w.r.t. v denoted by $\mathrm{LP}_v(g)$ is the sum of all terms $a_i\mathbf{x}^{\alpha_i}$ in g such that $v.\alpha_i$ is maximal.

Now consider $\mathcal{S} = \mathbb{K}[\mathbf{a}, \mathbf{x}]$ where $\mathbf{a} = a_1, \ldots, a_m$ is a sequence of parameters. Let $\prec_{\mathbf{x}}$ be a monomial order on the variables and $\prec_{\mathbf{a}}$ be a monomial order on the parameters. The lexicographic combination of $\prec_{\mathbf{x}}$ and $\prec_{\mathbf{a}}$ gives rise to an ordering on $\mathcal{S}$, denoted by $\prec_{\mathbf{x},\mathbf{a}}$ which is defined as follows: For all $\alpha, \beta \in \mathbb{N}^n$ and $\gamma, \delta \in \mathbb{N}^m$, $\mathbf{x}^\alpha \mathbf{a}^\gamma \prec_{\mathbf{x},\mathbf{a}} \mathbf{x}^\beta \mathbf{a}^\delta \iff \mathbf{x}^\alpha \prec_{\mathbf{x}} \mathbf{x}^\beta$ or $(\mathbf{x}^\alpha = \mathbf{x}^\beta$ and $\mathbf{a}^\gamma \prec_{\mathbf{a}} \mathbf{a}^\delta)$. Let us consider $\sigma : \mathbb{K}[\mathbf{a}] \to \overline{\mathbb{K}}$ as a specialization of parameters where $\overline{\mathbb{K}}$ is the algebraic closure of $\mathbb{K}$. This morphism can be considered as a substitution of parameters in $f \in \mathbb{K}[\mathbf{a}]$ with elements of $\overline{\mathbb{K}}^m$. Also, for a finite set $F \subset \mathcal{R}$, we call $\mathbb{V}(F)$ the variety of F which is the set of common zeros of F. Now, we recall the definition of a Gröbner system for a parametric polynomial ideal.

Definition 3. *Let $F \subset \mathcal{S}$ and $\mathcal{G} = \{(N_i, W_i, G_i)\}_{i=1}^\ell$ be a finite set of triples where $N_i, W_i \subset \mathbb{K}[\mathbf{a}]$ and $G_i \subset \mathcal{S}$ are finite for $i = 1, \ldots, \ell$. The set $\mathcal{G}$ is called a Gröbner system (GS) of $\langle F \rangle$ w.r.t. $\prec_{\mathbf{x},\mathbf{a}}$ on $V \subseteq \overline{\mathbb{K}}^m$ if for any i we have*

- *For any specialization $\sigma : \mathbb{K}[\mathbf{a}] \to \overline{\mathbb{K}}$ satisfying (N_i, W_i) the set $\sigma(G_i) \subset \overline{\mathbb{K}}[\mathbf{x}]$ is a Gröbner basis of $\langle \sigma(F) \rangle$ w.r.t. $\prec_{\mathbf{x}}$. (We say that σ satisfies (N_i, W_i) if $\sigma(p) = 0$ for all $p \in N_i$ and $\sigma(q) \neq 0$ for some $q \in W_i$)*
- $V \subseteq \bigcup_{i=1}^\ell \mathbb{V}(N_i) \setminus \mathbb{V}(W_i)$

Each (N_i, W_i, G_i) is called a branch of the Gröbner system $\mathcal{G}$ and we can consider (N_i, W_i) as a condition set where N_i is the null condition set and W_i the non-null condition set. Furthermore, $\mathcal{G}$ is called a Gröbner system of F if $V = \overline{\mathbb{K}}^m$.

Weispfenning [19, Theorem 2.7] showed that any parametric polynomial ideal has a GS and described an algorithm to compute it. We use the efficient PGB-MAIN algorithm due to Kapur et al. [11] for computing GSs.

Example 1. Let $F = \{(1 - c)y - ax^2, x + by^2\} \subset \mathbb{K}[a, b, c, x, y]$ where a, b, c are parameters and x, y are variables. Using our implementation of PGBMAIN algorithm in MAPLE, we obtain the following GS for $\langle F \rangle$ w.r.t. the product ordering $y \prec_{lex} x$ and $c \prec_{lex} b \prec_{lex} a$

$$\begin{cases} ([\,], & [ab^2], & [ab^2y^4 - y + cy, x + by^2]) \\ ([ab^2], & [c - 1], & [cy - y, x + by^2]) \\ ([c - 1, ab^2], [\,], & [x + by^2]). \end{cases}$$

For instance, if $a = 2$, $b = 0$ and $c = 3$ then the second branch corresponds to these values of parameters. Therefore, $\{x, y\}$ will be a Gröbner basis for the ideal $\langle F \rangle \mid_{a=2,b=0,c=3} = \langle -2y - 2x^2, x \rangle$.

Similarly the notion of a universal Gröbner system (UGS) may be defined for a parametric ideal.

3   Computation of Universal Gröbner Systems

In this section, we present an algorithm for computing a Gröbner fan of a parametric ideal. For this, we first introduce PCONVERT which is a parametric variant of the local change algorithm based on the article [8].

Algorithm 3. PCONVERT

Require: (N, W, G); a branch of the Gröbner system $\mathcal{G}$ w.r.t. a monomial order $\prec$ and F; a facet of $C_\prec(G)$
Ensure: $\{(N_i, W_i, G_i)\}_{i=1}^{\ell}$; some branches different from $\mathcal{G}$ whose Gröbner cone also has F as a facet.
$\quad \mathcal{G}' := \{ \}$
$\quad \mathrm{LP}_v(G) := \{\mathrm{LP}_v(g) \mid g \in G\}$ where v is an interior point of F
$\quad H := \{(N_i, W_i, G_i)\}_{i=1}^{k}$ a Gröbner system of the parametric ideal $\mathrm{LP}_v(G)$ according to the condition set (N, W) w.r.t. $\prec_{-\alpha}$ where α is the normal vector of F
$\quad$**for** i from 1 to k **do**
$\quad\quad H_i := \{f - \mathrm{NormalForm}(f, G) \mid f \in G_i\}$
$\quad\quad G_i := \mathrm{AutoReduce}(H_i)$
$\quad\quad \mathcal{G}' := \mathcal{G}' \cup \{(N_i, W_i, G_i)\}$
$\quad$**end for**
$\quad$**Return** $(\mathcal{G}')$

The following algorithm takes as input a finite set of parametric polynomials and outputs a UGS for the ideal generated by the input polynomials. For this, we decompose the parameter space into a finite set of parametric cells and for each cell, we give the corresponding UGB. In this algorithm, we use a global variable Sys which is initially the empty set and at each iteration of the algorithm, some new branches are added to the set. Also, we initialize U as the empty set. If the universal Gröbner basis of a parametric cell is completed, it will be added into U and at the end U is a UGS for the input ideal.

Algorithm 4. UGS

Require: $\{f_1, \ldots, f_k\}$; set of parametric polynomials, $\prec_{\mathbf{x}}, \prec_{\mathbf{a}}$; monomial orders
Ensure: a universal Gröbner system for $\mathcal{I} = \langle f_1, \ldots, f_k \rangle$
$\quad \mathcal{G} := \{(N_i, W_i, G_i)\}_{i=1}^{\ell}$ a Gröbner system of $\mathcal{I}$ w.r.t. $\prec_{\mathbf{x}}, \prec_{\mathbf{a}}$
$\quad \mathrm{Sys} := \mathrm{Sys} \cup \{[(N_i, W_i, G_i)]\}_{i=1}^{\ell}$
$\quad$**while** Sys is not empty **do**
$\quad\quad H :=$ the last element of Sys and remove it from Sys
$\quad\quad$**if** All members of H are marked branches **then**
$\quad\quad\quad$Add H to U
$\quad\quad$**else**
$\quad\quad\quad R := \{[(N_i, W_i, G_i)]\}_{i=1}^{|H|}$
$\quad\quad\quad$**for** i from 1 to $|H|$ **do**
$\quad\quad\quad\quad$**if** (N_i, W_i, G_i) is a unmarked branch **then**
$\quad\quad\quad\quad\quad Facets :=$ the list of all facets of the Gröbner cone corresponding to G_i
$\quad\quad\quad\quad\quad$**for** F in $Facets$ **do**
$\quad\quad\quad\quad\quad\quad$**if** F is a valid facet **then**
$\quad\quad\quad\quad\quad\quad\quad \mathcal{G}' := \mathrm{PCONVERT}((N_i, W_i, G_i), F)$
$\quad\quad\quad\quad\quad\quad\quad R := \{R_i \cup [(N_j, W_j, G_j)]_{j=1}^{|\mathcal{G}'|}\}_{i=1}^{|R|}$
$\quad\quad\quad\quad\quad\quad\quad$Remove incompatible members of R
$\quad\quad\quad\quad\quad\quad$**end if**
$\quad\quad\quad\quad\quad$**end for**
$\quad\quad\quad\quad$**end if**
$\quad\quad\quad$**end for**
$\quad\quad\quad \mathrm{Sys} := \mathrm{Sys} \cup R$
$\quad\quad$**end if**
$\quad$**end while**
$\quad$**Return**(U)

It should be noted that for checking the validity of F, we first compute the set of Gröbner cones for each member of R. Then, two points mentioned in the previous section are considered for each set separately. In the above algorithm, a marked branch is a visited branch, if it has been underlined in an iteration of the algorithm. Also, a member of R will be incompatible if at least one of its branches is incompatible (A branch (N, W, G) is incompatible if $\mathbb{V}(N) \setminus \mathbb{V}(W) = \emptyset$).

Theorem 1. UGS *algorithm terminates after finitely many steps and is correct.*

Proof. At the beginning of the algorithm, we compute a Gröbner system $\mathcal{G}$ of the ideal generated by the input polynomials w.r.t. $\prec_{\mathbf{x}}, \prec_{\mathbf{a}}$. According to the PGBMAIN algorithm, $\mathcal{G}$ is a set of branches with decomposition of the space of parameters into a finite set of parametric cells. Each of these branches is a candidate for a UGB w.r.t. its specific parametric cell. We define a path as a set of branches with compatible null and non-null conditions which is converted into a UGB after execution of some iteration of the algorithm. So, we consider each branch of $\mathcal{G}$ as a separate path and run the GFAN algorithm [10] for them separately. In each iteration of the algorithm, when we consider a branch (N, W, G), this branch is converted, w.r.t. some of its valid facets, into some new branches, say $\{(N_i', W_i', G_i')\}_{i=1}^k$ based on [8], where condition (N, W) is decomposed into some sub conditions $(N_i', W_i')_{i=1}^k$. These new branches create new paths in the algorithm. For example, according to the following picture, suppose there is a path with one branch A and A has two valid facets F_1 and F_2. Also, suppose the branch A with facets F_1 and F_2 is converted into two branches sets $\{B, C\}$ and $\{D, E\}$ respectively. Now, four paths $\{A, B, D\}$, $\{A, B, E\}$, $\{A, C, D\}$ and $\{A, C, E\}$ are added in the algorithm and the GFAN algorithm must be considered for them separately.

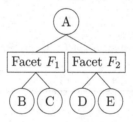

In fact, it is possible that a path in an iteration of the algorithm is separated to some new paths. At the end of the algorithm, each completed path is a UGB w.r.t. a specific parametric cell. It is necessary to mention two points. Firstly, each branch of a path is considered just once in the algorithm and if a branch has been considered before, it will be not considered any more. Secondly, if there is at least one incompatible branch in a path, the path will be removed from the algorithm. As a result, since the algorithm UGS is a combination of the GFAN algorithm from [10] and the PGGW algorithm from [8] (an algorithm for converting a GS of an arbitrary parametric polynomial ideal into a GS of the ideal w.r.t. another monomial order), the termination and correctness of the new algorithm is warranted by the termination and correctness of GFAN and PGGW, which ends the proof. □

4   A Simple Example

We have implemented the UGS algorithm in MAPLE (by applying some auxiliary functions from SAGE) and the code of the implementation of our algorithms is available at http://amirhashemi.iut.ac.ir/softwares. To illustrate the execution of the UGS algorithm, let $\mathcal{I} = \{ax + byx, cuy + z - 2u\} \subset \mathbb{K}[\mathbf{a}, \mathbf{x}] = \mathbb{K}[a, b, c, x, y, z, u]$ where a, b, c are parameters and x, y, z, u are variables. First, we compute a GS for $\mathcal{I}$ w.r.t. the lexicographic combination of $\prec_\mathbf{x}$: $u \prec_{lex} z \prec_{lex} y \prec_{lex} x$ and $\prec_\mathbf{a}$: $c \prec_{lex} b \prec_{lex} a$ as follows.

$$\mathcal{G} = \begin{cases} (N_1, W_1, G_1) = ([b], & [a, c], & [\underline{cuy} - 2u + z, \underline{x}]) \\ (N_2, W_2, G_2) = ([b, a], & [c], & [\underline{cuy} - 2u + z]) \\ (N_3, W_3, G_3) = ([c, b, a], [\ ], & [\underline{z} - 2u]) \\ (N_4, W_4, G_4) = ([c, b], & [a], & [\underline{z} - 2u, \underline{x}]) \\ (N_5, W_5, G_5) = ([c], & [b], & [\underline{z} - 2u, \underline{bxy} + ax]) \\ (N_6, W_6, G_6) = ([\ ], & [b, c], & [\underline{cuy} - 2u + z, \underline{bxz} - acux - 2bux, \underline{bxy} + ax]) \end{cases}$$

For this, we called PGBMain($[\ ], [1], \mathcal{I}, \prec_\mathbf{a}, \prec_\mathbf{x}$) from our code. So, we obtain

$$\text{Sys} = \{[(N_1, W_1, G_1)], [(N_2, W_2, G_2)], [(N_3, W_3, G_3)], [(N_4, W_4, G_4)],$$
$$[(N_5, W_5, G_5)], [(N_6, W_6, G_6)]\}.$$

By selecting the last member of Sys, we have $H = [(N_6, W_6, G_6)]$. Since the branch (N_6, W_6, G_6) has not been considered yet, it is underlined and added to R. Using the SAGE function GroebnerCone($G_6, \{\mathbf{x}\}$) two facets F_1 and F_2 of this branch with normal vectors $\alpha_1 = [0, 0, 1, -1]$ and $\alpha_2 = [0, 1, -1, 1]$ are computed. F_1 is valid (the test of validity is done by checkValid($R, \alpha_1, v, \{\mathbf{x}\}$) available in our code where v is an interior point of F_1). Therefore,

$$\mathcal{G}' = \begin{cases} (N_1', W_1', G_1') = ([\ ], [b, c, ac + 2b], [xu(ac + 2b) - bzx, \underline{bxy} + ax, \underline{cuy} - 2u + z]) \\ (N_2', W_2', G_2') = ([ac + 2b], [b, c], [\underline{bxy} + ax, \underline{cuy} - 2u + z, \underline{bzx}]). \end{cases}$$

This conversion is done by the function flip($G_6, v, \alpha_1, \{\mathbf{a}\}, \{\mathbf{x}\}$) implemented in MAPLE. The second facet also is valid. So, $\mathcal{G}'' = \{(N_1'', W_1'', G_1'') = ([\ ], [b, c], [\underline{z} + cuy - 2u, \underline{bxy} + ax])\}$. Now, the set R is updated as follows

$$R = \begin{cases} [(N_6, W_6, G_6), (N_1', W_1', G_1'), (N_1'', W_1'', G_1'')] \\ [\underline{(N_6, W_6, G_6)}, (N_2', W_2', G_2'), (N_1'', W_1'', G_1'')]. \end{cases}$$

Since all members of R are compatible, they are added to Sys. In the next iteration of the algorithm, we have

$$H = [\underline{(N_6, W_6, G_6)}, (N_2', W_2', G_2'), (N_1'', W_1'', G_1'')].$$

The branch (N_6, W_6, G_6) is marked, but two other unmarked branches must be considered. Each of these branches has an invalid facet. Therefore, the algorithm does not make any new branch and in the next iteration, H is added to U as

$$([ac + 2b], [b, c], \{[cuy - 2u + z, -acux - 2bux + bxz, bxy + ax],$$
$$[bxy + ax, cuy - 2u + z, zbx], [cuy - 2u + z, bxy + ax]\})$$

If we consider all the remaining branches of Sys, we obtain the following universal Gröbner system for $\mathcal{I}$:

- $([c], [b], \{[z - 2u, bxy + ax], [bxy + ax, 2u - z]\})$
- $([c, b], [a], \{[z - 2u, x], [x, 2u - z]\})$
- $([a, b, c], [\ ], \{[z - 2u], [2u - z]\})$
- $([a, b], [c], \{[cuy - 2u + z], [cuy - 2u + z]\})$
- $([b], [a, c], \{[cuy - 2u + z, x], [x, cuy - 2u + z]\})$
- $([ac + 2b], [b, c], \{[cuy - 2u + z, -acux - 2bux + bxz, bxy + ax], [bxy + ax, cuy - 2u + z, zbx], [cuy - 2u + z, bxy + ax]\})$
- $([\ ], [ac + 2b, b, c], \{[cuy - 2u + z, -acux - 2bux + bxz, bxy + ax], [-zbx + xu(ac + 2b), bxy + ax, cuy - 2u + z], [cuy - 2u + z, bxy + ax]\})$.

References

1. Avis, D., Fukuda, K.: Reverse search for enumeration. Discrete Appl. Math. **65**(1–3), 21–46 (1996)
2. Babson, E., Onn, S., Thomas, R.: The Hilbert zonotope and a polynomial time algorithm for universal Gröbner bases. Adv. Appl. Math. **30**, 529–544 (2003)
3. Buchberger, B.: Bruno Buchberger's Ph.D thesis 1965: an algorithm for finding the basis elements of the residue class ring of a zero dimensional polynomial ideal. J. Symb. Comput. **41**(3–4), 475–511 (2006). Translation from the German
4. Decker, W., Greuel, G.-M., Pfister, G., Schönemann, H.: Singular 4-1-0 – A computer algebra system for polynomial computations (2016). http://www.singular.uni-kl.de
5. Dehghani Darmian, M., Hashemi, A., Montes, A.: Erratum to "A new algorithm for discussing Gröbner bases with parameters" (J. Symbolic Comput. **33**(1–2), 183–208 (2002)). J. Symb. Comput. **46**(10), 1187–1188 (2011)
6. Fukuda, K., Jensen, A., Lauritzen, N., Thomas, R.: The generic Gröbner walk. J. Symb. Comput. **42**(3), 298–312 (2007)
7. Fukuda, K., Jensen, A.N., Thomas, R.R.: Computing Gröbner fans. Math. Comput. **76**(260), 2189–2212 (2007)
8. Hashemi, A., Dehghani Darmian, M., Barkhordar, M.: Gröbner systems conversion. Math. Comput. Sci. **11**(1), 61–77 (2017)
9. Jensen, A.N.: Gfan, a software system for Gröbner fans and tropical varieties. http://home.imf.au.dk/jensen/software/gfan/gfan.html
10. Jensen, A.N.: Algorithmic aspects of Gröbner fans and tropical varieties. Technical report (2007)
11. Kapur, D., Sun, Y., Wang, D.: A new algorithm for computing comprehensive Gröbner systems. In: Proceedings of ISSAC 2010. pp. 29–36. ACM Press (2010)
12. Montes, A.: A new algorithm for discussing Gröbner bases with parameters. J. Symb. Comput. **33**(2), 183–208 (2002)
13. Mora, T., Robbiano, L.: The Gröbner fan of an ideal. J. Symb. Comput. **6**(2–3), 183–208 (1988)
14. Schwartz, N.: Stability of Gröbner bases. J. Pure Appl. Algebra **53**(1–2), 171–186 (1988)
15. Stein, W.: Sage: An open source mathematical software (version 7.0). The sage group (2016). http://www.sagemath.org

16. Sturmfels, B.: Gröbner Bases and Convex Polytopes. AMS, American Mathematical Society, Providece (1996)
17. Suzuki, A., Sato, Y.: A simple algorithm to compute comprehensive Gröbner bases using Gröbner bases. In: Proceedings of ISSAC 2006, pp. 326–331. ACM Press (2006)
18. Weispfenning, V.: Constructing universal Gröbner bases. In: Huguet, L., Poli, A. (eds.) AAECC 1987. LNCS, vol. 356, pp. 408–417. Springer, Heidelberg (1989). https://doi.org/10.1007/3-540-51082-6_96
19. Weispfenning, V.: Comprehensive Gröbner bases. J. Symb. Comput. **14**(1), 1–29 (1992)

Certifying Reality of Projections

Jonathan D. Hauenstein[1]($\boxtimes$), Avinash Kulkarni[2], Emre C. Sertöz[3],
and Samantha N. Sherman[1]

[1] Department of Applied and Computational Mathematics and Statistics,
University of Notre Dame, Notre Dame, USA
{hauenstein,ssherma1}@nd.edu
[2] Department of Mathematics, Simon Fraser University, Burnaby, Canada
akulkarn@sfu.ca
[3] Max Planck Institute for Mathematics in the Sciences, Leipzig, Germany
emresertoz@gmail.com
http://www.nd.edu/~jhauenst, http://www.emresertoz.com

Abstract. Computational tools in numerical algebraic geometry can
be used to numerically approximate solutions to a system of polynomial equations. If the system is well-constrained (i.e., square), Newton's
method is locally quadratically convergent near each nonsingular solution. In such cases, Smale's alpha theory can be used to certify that
a given point is in the quadratic convergence basin of some solution.
This was extended to certifiably determine the reality of the corresponding solution when the polynomial system is real. Using the theory of
Newton-invariant sets, we certifiably decide the reality of projections of
solutions. We apply this method to certifiably count the number of real
and totally real tritangent planes for instances of curves of genus 4.

Keywords: Certification · Alpha theory · Newton's method
Real solutions · Numerical algebraic geometry

1 Introduction

For a well-constrained system of polynomial equations f, numerical algebraic
geometric tools (see, e.g., [2,12]) can be used to compute numerical approximations of solutions of $f = 0$. These approximations can be certified to lie in
a quadratic convergence basin of Newton's method applied to f using Smale's
α-theory (see, e.g., [3, Chap. 8]). When the system f is real, α-theory can be
used to certifiably determine if the true solution corresponding to an approximate solution is real [6]. That is, one can certifiably decide whether or not every
coordinate of a solution is real from a sufficiently accurate approximation. It is

JDH was supported by Sloan Fellowship BR2014-110 TR14 and NSF ACI-1460032.
SNS was supported by Schmitt Leadership Fellowship in Science and Engineering.
AK was partially supported by the Max Planck Institute for Mathematics in the
Sciences.

J. H. Davenport et al. (Eds.): ICMS 2018, LNCS 10931, pp. 200–208, 2018.
https://doi.org/10.1007/978-3-319-96418-8_24

often desirable in computational algebraic geometry to instead decide the reality of a projection of a solution of a real polynomial system. In this manuscript, we develop an approach for this situation using Newton-invariant sets [4].

The paper is organized as follows. Section 2 provides a summary of Smale's α-theory and Newton-invariant sets. Section 3 provides our main results regarding certification of reality of projections. Section 4 applies the method to certifying real and totally real tritangents of various genus 4 curves.

2 Smale's Alpha Theory and Newton-Invariant Sets

Our certification procedure is based on the ability to certify quadratic convergence of Newton's method via Smale's α-theory (see, e.g., [3, Chap. 8]) and Newton-invariant sets [4]. This section summarizes these two items following [4].

Assume that $f \colon \mathbb{C}^n \to \mathbb{C}^n$ is an analytic map and consider the Newton iteration map $N_f \colon \mathbb{C}^n \to \mathbb{C}^n$ defined by

$$N_f(x) := \begin{cases} x - Df(x)^{-1} f(x) & \text{if } Df(x) \text{ is invertible,} \\ x & \text{otherwise,} \end{cases}$$

where $Df(x)$ is the Jacobian matrix of f at x. The map N_f is globally defined with fixed points $\{x \in \mathbb{C}^n \mid f(x) = 0 \text{ or } \operatorname{rank} Df(x) < n\}$. Hence, if $Df(x)$ is invertible and $N_f(x) = x$, then $f(x) = 0$.

One aims to find solutions of $f = 0$ by iterating N_f to locate fixed points. To that end, for each $k \geq 1$, define $N_f^k(x) := \underbrace{N_f \circ \cdots \circ N_f}_{k \text{ times}}(x)$.

Definition 1. *A point $x \in \mathbb{C}^n$ is an* approximate solution *of $f = 0$ if there exists $\xi \in \mathbb{C}^n$ such that $f(\xi) = 0$ and $\|N_f^k(x) - \xi\| \leq \left(\frac{1}{2}\right)^{2^k - 1} \|x - \xi\|$ for each $k \geq 1$ where $\|\cdot\|$ is the Euclidean norm on $\mathbb{C}^n$. The point ξ is the associated* solution *to x and the sequence $\{N_f^k(x)\}_{k \geq 0}$ converges quadratically to ξ.*

Smale's α-theory provides sufficient conditions for x to be an approximate solution of $f = 0$ via data computable from f and x. We will use approximate solutions to determine characteristics of the corresponding associated solutions using Newton-invariant sets.

Definition 2. *A set $V \subset \mathbb{C}^n$ is called* Newton invariant *with respect to f if $N_f(v) \in V$ for every $v \in V$ and $\lim_{k \to \infty} N_f^k(v) \in V$ for every $v \in V$ such that this limit exists.*

For example, the set $V = \mathbb{R}^n$ is Newton invariant with respect to a real map f. The algorithm presented in Sect. 3 considers both the set of real numbers as well as other Newton-invariant sets to perform certification together with the following theorem derived from [3, Chap. 8] and [4].

Theorem 1. *Let $f\colon \mathbb{C}^n \to \mathbb{C}^n$ be analytic, let $V \subset \mathbb{C}^n$ be Newton invariant with respect to f, let $x, y \in \mathbb{C}^n$ such that $Df(x)$ and $Df(y)$ are invertible, and let*

$$\alpha(f,x) := \beta(f,x) \cdot \gamma(f,x), \qquad\qquad \beta(f,x) := \|x - N_f(x)\| = \|Df(x)^{-1}f(x)\|,$$

$$\gamma(f,x) := \sup\nolimits_{k \geq 2} \left\| \frac{Df(x)^{-1}D^k f(x)}{k!} \right\|^{\frac{1}{k-1}}, \qquad \delta_V(x) := \inf\nolimits_{v \in V} \|x - v\|$$

where the norms are the corresponding vector and operator Euclidean norms.

1. *If $4 \cdot \alpha(f,x) < 13 - 3\sqrt{17}$, then x is an approximate solution of $f = 0$.*
2. *If $100 \cdot \alpha(f,x) < 3$ and $20 \cdot \|x - y\| \cdot \gamma(f,x) < 1$, then x and y are approximate solutions of $f = 0$ with the same associated solution.*
3. *Suppose that x is an approximate solution of $f = 0$ with associated solution ξ.*
   (a) *$N_f(x)$ is also an approximate solution with associated solution ξ and*

   $$\|x - \xi\| \leq 2\beta(f,x) = 2\|x - N_f(x)\| = 2\|Df(x)^{-1}f(x)\|.$$

   (b) *If $\delta_V(x) > 2\beta(f,x)$, then $\xi \notin V$.*
   (c) *If $100 \cdot \alpha(f,x) < 3$ and $20 \cdot \delta_V(x) \cdot \gamma(f,x) < 1$, then $\xi \in V$.*

The value $\beta(f,x)$ is the Newton residual. When f is a polynomial system, $\gamma(f,x)$ is a maximum over finitely many terms and thus can be easily bounded above [11]. A similar bound for polynomial-exponential systems can be found in [5]. The value $\delta_V(x)$ is the distance between x and V. The special case of $V = \mathbb{R}^n$ was first considered in [6].

The following procedure from [4], which is based on Theorem 1, certifiably decides if the associated solution of a given approximate solution lies in a given Newton-invariant set V.

Procedure $b = \mathbf{Certify}(f, x, \delta_V)$
Input A well-constrained analytic system $f\colon \mathbb{C}^n \to \mathbb{C}^n$ such that $\gamma(f, \cdot)$ can be computed (or bounded) algorithmically, a point $x \in \mathbb{C}^n$ which is an approximate solution of $f = 0$ with associated solution ξ such that $Df(\xi)^{-1}$ exists, and distance function δ_V for some Newton-invariant set V that can be computed algorithmically.
Output A boolean b which is **true** if $\xi \in V$ and **false** if $\xi \notin V$.
Begin
  1. Compute $\beta := \beta(f,x)$, $\gamma := \gamma(f,x)$, $\alpha := \beta \cdot \gamma$, and $\delta := \delta_V(x)$.
  2. If $\delta > 2\beta$, **Return false**.
  3. If $100 \cdot \alpha < 3$ and $20 \cdot \delta \cdot \gamma < 1$, **Return true**.
  4. Update $x := N_f(x)$ and go to Step 1.

3   Certification of Reality

The systems under consideration are well-constrained polynomial systems

$$f(a, b_1, \ldots, b_k, c_1, \ldots, c_\ell, d_1, \ldots, d_\ell) = \begin{bmatrix} g(a) \\ p(a, b_i) \text{ for } i = 1, \ldots, k \\ p(a, c_i) \text{ for } i = 1, \ldots, \ell \\ p(a, d_i) \text{ for } i = 1, \ldots, \ell \end{bmatrix} \tag{1}$$

with variables $a \in \mathbb{C}^m$ and $b_r, c_s, d_t \in \mathbb{C}^q$, and polynomial systems $g \colon \mathbb{C}^m \to \mathbb{C}^u$ and $p \colon \mathbb{C}^{m+q} \to \mathbb{C}^w$ which have real coefficients such that

$$u \le m \quad \text{and} \quad m + (k + 2\ell)q = u + (k + 2\ell)w. \tag{2}$$

The first condition in (2) yields that a is not over-constrained by g while the second condition provides that the whole system is well-constrained.

Example 1. To illustrate the setup, we consider an example with $m = 3$, $k = 0$, $\ell = 1$, $q = 1$, $u = 1$, and $w = 2$ so that (2) holds, resulting in a well-constrained system of 5 polynomials in 5 variables. Namely, we consider

$$f(a, c, d) = \begin{bmatrix} g(a) \\ \hline p(a, c) \\ \hline p(a, d) \end{bmatrix} = \begin{bmatrix} a_1^2 + a_2^2 + a_3^2 - 1 \\ \hline a_1 + (1 - c^2)(a_2 c + a_3 c^2) \\ a_1(3c^2 - 1) + a_2(2c^5 - 4c^3 + 2c - 1) \\ \hline a_1 + (1 - d^2)(a_2 d + a_3 d^2) \\ a_1(3d^2 - 1) + a_2(2d^5 - 4d^3 + 2d - 1) \end{bmatrix}.$$

Since the polynomial system f in (1) has real coefficients, we can use Theorem 1 with $V = \mathbb{R}^n$ where $n = m + (k + 2\ell)q = u + (k + 2\ell)w$ to certifiably determine if all coordinates of the associated solution are simultaneously real.

Example 2. Let f be the polynomial system with real coefficients considered in Example 1 with Newton-invariant set $V = \mathbb{R}^5$. For the points P_1 and P_2, respectively:

$$\left(\frac{1543}{8003} + \frac{\sqrt{-1}}{530485174}, \frac{-34488}{50521} - \frac{\sqrt{-1}}{190996265}, \frac{32768}{46489} - \frac{\sqrt{-1}}{310964547}, \frac{6713}{18120} + \frac{4777\sqrt{-1}}{19088}, \frac{6713}{18120} - \frac{4538\sqrt{-1}}{18133} \right),$$

$$\left(\frac{18245}{111912} - \frac{\sqrt{-1}}{772703930}, \frac{15244}{38793} - \frac{\sqrt{-1}}{307556791}, \frac{27099}{29944} - \frac{\sqrt{-1}}{155308656}, \frac{-44817}{40271} - \frac{\sqrt{-1}}{372454657}, \frac{8603}{8149} + \frac{\sqrt{-1}}{608134511} \right),$$

alphaCertified [6] computed the following information:

j	upper bound of $\alpha(f, P_j)$	$\beta(f, P_j)$	upper bound of $\gamma(f, P_j)$	$\delta_{\mathbb{R}^5}(P_j)$
1	$1.32 \cdot 10^{-5}$	$2.05 \cdot 10^{-8}$	$6.40 \cdot 10^2$	0.35
2	$2.38 \cdot 10^{-4}$	$1.47 \cdot 10^{-8}$	$1.63 \cdot 10^2$	$7.98 \cdot 10^{-9}$

Item 1 of Theorem 1 yields that both points P_1 and P_2 are approximate solutions of $f = 0$. Suppose that ξ_1 and ξ_2, respectively, are the corresponding associated solutions. Items 3b and 3c, respectively, provide that $\xi_1 \notin \mathbb{R}^5$ and $\xi_2 \in \mathbb{R}^5$.

Rather than consider all coordinates simultaneously, the following shows that we can certifiably decide the reality of some of the coordinates.

Theorem 2. *For f as in (1), the set*

$$V = \{(a, b_1, \ldots, b_k, c_1, \ldots, c_\ell, \mathrm{conj}(c_1), \ldots, \mathrm{conj}(c_\ell)) \in \mathbb{R}^m \times (\mathbb{R}^q)^k \times (\mathbb{C}^q)^{2\ell}\} \quad (3)$$

is Newton invariant with respect to f where $\mathrm{conj}()$ denotes complex conjugate.

Proof. Suppose that $v = (a, b_1, \ldots, b_k, c_1, \ldots, c_\ell, d_1, \ldots, d_\ell) \in V$ such that the Jacobian matrix $Df(v)$ is invertible. Let $\Delta v = Df(v)^{-1} f(v)$ and write

$$\Delta v = \begin{bmatrix} \Delta a^T & \Delta b_1^T & \cdots & \Delta b_k^T & \Delta c_1^T & \cdots & \Delta c_\ell^T & \Delta d_1^T & \cdots & \Delta d_\ell^T \end{bmatrix}^T.$$

Since f has real coefficients, we know that

$$\mathrm{conj}(\Delta v) = \mathrm{conj}(Df(v)^{-1} f(v)) = Df(\mathrm{conj}(v))^{-1} f(\mathrm{conj}(v)).$$

Since $v \in V$, $\mathrm{conj}(v) = (a, b_1, \ldots, b_k, d_1, \ldots, d_\ell, c_1, \ldots, c_\ell) \in V$. Based on the structure of f, it immediately follows that

$$\mathrm{conj}(\Delta v) = Df(\mathrm{conj}(v))^{-1} f(\mathrm{conj}(v)) = \begin{bmatrix} \Delta a^T & \Delta b_1^T & \cdots & \Delta b_k^T & \Delta d_1^T & \cdots & \Delta d_\ell^T & \Delta c_1^T & \cdots & \Delta c_\ell^T \end{bmatrix}^T.$$

Hence, $\mathrm{conj}(\Delta a) = \Delta a$, $\mathrm{conj}(\Delta b_i) = \Delta b_i$, and $\mathrm{conj}(\Delta c_j) = \Delta d_j$. Thus, it immediately follows that $N_f(v) = v - \Delta v \in V$.

The remaining condition in Definition 2 follows from the fact that V is closed. $\qquad \square$

All that remains to utilize **Certify** is to provide a formula for δ_V.

Proposition 1. *For any $x = (a, b_1, \ldots, b_k, c_1, \ldots, c_\ell, d_1, \ldots, d_\ell) \in \mathbb{C}^{m+(k+2\ell)q}$ and V as in (3),*

$$\delta_V(x) = \frac{1}{2} \left\| \begin{array}{l} (a - \mathrm{conj}(a), b_1 - \mathrm{conj}(b_1), \ldots, b_k - \mathrm{conj}(b_k), \\ c_1 - \mathrm{conj}(d_1), \ldots, c_\ell - \mathrm{conj}(d_\ell), d_1 - \mathrm{conj}(c_1), \ldots, d_\ell - \mathrm{conj}(c_\ell)) \end{array} \right\|. \quad (4)$$

Proof. The projection of $x = (a, b_1, \ldots, b_k, c_1, \ldots, c_\ell, d_1, \ldots, d_\ell)$ onto V is

$$v = \tfrac{1}{2}(a + \mathrm{conj}(a), b_1 + \mathrm{conj}(b_1), \ldots, b_k + \mathrm{conj}(b_k),$$
$$c_1 + \mathrm{conj}(d_1), \ldots, c_\ell + \mathrm{conj}(d_\ell), d_1 + \mathrm{conj}(c_1), \ldots, d_\ell + \mathrm{conj}(c_\ell)).$$

Thus, $\delta_V(x) = \|x - v\|$ which simplifies to (4). $\qquad \square$

Example 3. For the polynomial system f considered in Example 1, Theorem 2 provides that $V = \{(a, c_1, \mathrm{conj}(c_1)) \in \mathbb{R}^3 \times \mathbb{C} \times \mathbb{C}\}$ is Newton invariant with respect to f. Let ξ_1 be the associated solution of the first point P_1 from Example 1. Since $\delta_V(P_1) = 8.88 \cdot 10^{-9}$, we know $\xi_1 \in V$ using the data from Example 2 together with Item 3c of Theorem 1, i.e., the first three coordinates of ξ_1 are real and the last two coordinates are complex conjugates of each other. Hence, $\xi_1 \in V \setminus \mathbb{R}^5$.

4   Tritangents

We conclude by applying this new certification method to a problem from real algebraic geometry considered in [8,9]. A *smooth space sextic* is a nonsingular algebraic curve $C \subset \mathbb{P}^3$ which is the intersection of a quadric surface Q and cubic surface Γ. The curve C is a curve of degree 6 and genus 4, and every hyperplane of $\mathbb{P}^3$ intersects C in exactly 6 points (counting multiplicities). The problem considered in [8,9] concerns counting the number of hyperplanes which are tangent to C at all points of intersection.

Definition 3. *A plane $H \subset \mathbb{P}^3$ is a* tritangent *plane for C if every point in $C \cap H$ has even intersection multiplicity.*

In the generic case, each tritangent plane intersects C in 3 points, each with multiplicity 2, and there are a total of 120 complex tritangent planes. For simplicity, we henceforth restrict our attention to the generic case. Each of the 120 tritangent planes can be categorized as either totally real, real, or nonreal.

Definition 4. *A tritangent plane H is* real *if it can be expressed as the solution set of a linear equation with real coefficients and* nonreal *otherwise. A real tritangent plane is* totally real *if each point in $C \cap H$ is real.*

Example 4. The smooth space sextic curve $C \subset \mathbb{P}^3$ equal to

$$\{[x_0, x_1, x_2, x_3] \in \mathbb{P}^3 \mid x_0^2 + x_0 x_3 = x_1 x_2, \ x_0 x_2 (x_0 + x_1 + x_3) = x_3 (x_1^2 - x_2^2 + x_3^2)\}$$

has 16 real tritangents, 7 of which are totally real, and 104 nonreal tritangents.

4.1   Counting Real and Totally Real Tritangents

Gross and Harris [7] prove that the number of real tritangents of a genus 4 curve is either 0, 8, 16, 24, 32, 64 or 120. This number depends only on the topological properties of the real part of the curve, as summarized in Table 1.

Example 5. Since the curve C in Example 4 has 16 real tritangents, it follows from [7] that the real part of C consists of two connected components.

In contrast, *totally* real tritangents reflect the *extrinsic* geometry of the real part of the curve. Indeed, Kummer [9] recently obtained bounds on the number of totally real tritangents for each real topological type. We will use our certification procedure to prove results that help close the gaps between the theoretical bounds and instances which have actually been realized.

To that end, we formulate a well-constrained parameterized polynomial system of the form (1) as follows. For a generic smooth space sextic $C = Q \cap \Gamma \subset \mathbb{P}^3$, let q and c be quadric and cubic polynomials that define Q and Γ, respectively. By assuming the coordinates are in general position, we solve in affine space by setting the first coordinate equal to 1. In particular, we are seeking $a \in \mathbb{C}^3$, $x_1, x_2, x_3 \in \mathbb{C}^3$, and $\lambda_1, \lambda_2, \lambda_3 \in \mathbb{C}^2$ such that

$$f(h, x_1, \lambda_1, x_2, \lambda_2, x_3, \lambda_3) = \begin{bmatrix} p(h, x_1, \lambda_1) \\ p(h, x_2, \lambda_2) \\ p(h, x_3, \lambda_3) \end{bmatrix} = 0 \text{ with } p(h, x_i, \lambda_i) = \begin{bmatrix} H(X_i) \\ q(X_i) \\ c(X_i) \\ \begin{bmatrix} \nabla_x H(X_i) \\ \nabla_x q(X_i) \\ \nabla_x c(X_i) \end{bmatrix} \Lambda_i \end{bmatrix} \tag{5}$$

where $H = [1, h] \in \widehat{\mathbb{P}^3}$, $X_i = [1, x_i] \in \mathbb{P}^3$, $\Lambda_i = [1, \lambda] \in \mathbb{P}^2$, and $\nabla_x \zeta([1, x]) \in \mathbb{C}^3$ is the gradient of ζ with respect to x. In particular, f is a system of 18 polynomials in 18 variables with the first 3 polynomials in p enforcing that $X_i \in C \cap H$ and the last 3 polynomials providing that H is tangent to C at X_i. The values of k and ℓ from (1) are dependent on the number of real points in $C \cap H$. A real tritangent H will either have three or one real points in $C \cap H$ corresponding, respectively, to totally real tritangents ($k = 3$ and $\ell = 0$) and real tritangents that are not totally real ($k = \ell = 1$).

Remark 1. For generic quadric q and cubic c, the condition $f = 0$ in (5) has $120 \cdot 3! = 720$ isolated solutions where the factor $3! = 6$ corresponds to trivial reorderings. By selecting one point in each orbit, (5) can be used as a parameter homotopy [10], where the parameters are the coefficients of q and c, to compute tritangents for generic smooth space sextic curves.

4.2   Computational Results

In the following, we utilize Bertini [1] to numerically approximate the tritangents via a parameter homotopy following Remark 1. After heuristically classifying the tritangents as either totally real, real, or nonreal, we use the results from Sect. 3 applied to f in (5) to certify the results using alphaCertified [6]. More computational details for applying our approach to the examples that follow can be found at https://doi.org/10.7274/R0DB7ZW2. The reported timings are based on using either one (in serial) or all 64 (in parallel) cores of a 2.4 GHz AMD Opteron Processor 6378 with 128 GB RAM.

Example 6. For $i = 1, 2$, let $C_i \subset \mathbb{P}^3$ be defined by $q_i = c_i = 0$ where

$q_1(x) = q_2(x) = x_0 x_3 - x_1 x_2$
$c_1(x) = (25x_0^3 - 24x_0^2 x_1 - 89x_0^2 x_2 - 55x_0^2 x_3 - 14x_1^3 - 31x_1^2 x_2 + 86x_1 x_2 x_3 + 74x_2^2 x_3 - 45x_2 x_3^2 - 62x_3^3)/100$
$c_2(x) = (89x_0^3 - 41x_0^2 x_1 - 87x_0 x_1^2 - 26x_0 x_2^2 - 25x_1^2 x_2 + 42x_1^2 x_3 + 56x_1 x_2^2 + 87x_2^3 - 67x_2 x_3^2 - 42x_3^3)/100.$

We first use a parameter homotopy in Bertini following Remark 1 to numerically approximate the solutions of $f = 0$ in (5). Each of these instances took approximately 45 s in serial and 1.5 s in parallel to compute all numerical solutions to roughly 50 correct digits. Converting to rational numbers and applying alphaCertified to each instance shows that all numerical approximations computed by Bertini are approximate solutions in roughly 33 minutes using rational arithmetic with serial processing.

First, we certify that we have indeed computed 120 distinct tritangents up to the action of reordering. This is accomplished by comparing the pairwise distances between the h coordinates corresponding to the tritangent hyperplane with the known error bound 2β from Item 3a of Theorem 1. In both of our examples, $2\beta < 10^{-54}$ while the pairwise distances were larger than 10^{-2} showing that 120 distinct tritangents were computed as expected.

Second, we compare the size of the imaginary parts of the h coordinates with the error bound 2β to certifiably determine which are nonreal tritangets. For both cases, this proves that there are 104 nonreal tritangents leaving 16 tritangents requiring further investigation.

Third, we apply **Certify** with $V = \mathbb{R}^{18}$ to certifiably determine the number of totally real tritangents. This proves that C_1 and C_2 have exactly 0 and 16 totally real tritangents, respectively.

The only remaining item is to show that the 16 tritangents for C_1 are real which follows from our new results in Sect. 3. We reorder the intersection points so that the first one has the smallest imaginary part and apply **Certify** with V as in (3) where $k = 1$ and $\ell = 1$, i.e., one real intersection point and a pair of complex conjugate intersection points.

In summary, these computations prove that both C_1 and C_2 have 16 real tritangents, where none and all of these 16 are totally real, respectively.

Example 6 provides two new instances of results that had not been realized in [8]. Combining these two examples together with results from [8,9] shows that any number between 0 and 16 totally real tritangents can be realized for a smooth sextic curve which has 16 real tritangents. In Table 1 we summarize the theoretical bounds from [9] for the number of totally real tritangents, together with the values that are realized in [8] and our computations (including the computations we describe below). In particular, the bold numbers show new results we obtained using our certification approach. Only 4 open cases remain to be realized or shown to be impossible: 120 real tritangents with between 80 and 83 totally real tritangents.

Table 1. Summary of results for tritangents of genus 4 curves with **bold** numbers showing the new results obtained using our certification approach.

# real tritangents [7]	# connected real components	Dividing type?	Range of # totally real [9]	Realized # totally real ([8] & our results)
0	0	No	$[0,0]$	$[0,0]$
8	1	No	$[0,8]$	$[0,8]$
16	2	No	$[0,16]$	$[\mathbf{0},\mathbf{16}]$
24	3	Yes	$[0,24]$	$[0,\mathbf{24}]$
32	3	No	$[8,32]$	$[\mathbf{8},32]$
64	4	No	$[32,64]$	$[\mathbf{32},\mathbf{64}]$
120	5	Yes	$[80,120]$	$[84,120]$

Typically, our computations to generate these results started with the Cayley cubic $c = -x_0^2 x_2 + x_0^2 x_3 + x_1^2 x_2 + x_1^2 x_3 + x_2^2 x_3 - x_3^3$ and selected quadrics q which intersected various real components of the Cayley cubic surface Γ defined by c. We then randomly perturbed all of the coefficients of q and c to locally explore the surrounding area of the parameter space of the selected instance. As in Example 6, Bertini was used to compute numerical approximations of the solutions with certification provided by alphaCertified.

References

1. Bates, D.J., Hauenstein, J.D., Sommese, A.J., Wampler, C.W.: Bertini: software for numerical algebraic geometry. bertini.nd.edu
2. Bates, D.J., Hauenstein, J.D., Sommese, A.J., Wampler, C.W.: Numerically Solving Polynomial Systems with Bertini. Software, Environments, and Tools, vol. 25. Society for Industrial and Applied Mathematics (SIAM), Philadelphia (2013)
3. Blum, L., Cucker, F., Shub, M., Smale, S.: Complexity and Real Computation. Springer, New York (1998). https://doi.org/10.1007/978-1-4612-0701-6
4. Hauenstein, J.D.: Certification using Newton-invariant subspaces. In: Blömer, J., Kotsireas, I.S., Kutsia, T., Simos, D.E. (eds.) MACIS 2017. LNCS, vol. 10693, pp. 34–50. Springer, Cham (2017). https://doi.org/10.1007/978-3-319-72453-9_3
5. Hauenstein, J.D., Levandovskyy, V.: Certifying solutions to square systems of polynomial-exponential equations. J. Symb. Comput. **79**(3), 575–593 (2017)
6. Hauenstein, J.D., Sottile, F.: Algorithm 921: alphaCertified: certifying solutions to polynomial systems. ACM ToMS **38**(4), 28 (2012)
7. Gross, B.H., Harris, J.: Real algebraic curves. Ann. Sci. École Norm. Sup. (4) **14**(2), 157–182 (1981)
8. Kulkarni, A., Ren, Y., Sayyary Namin, M., Sturmfels, B.: Real space sextics and their tritangents. arXiv:1712.06274 (2017)
9. Kummer, M.: Totally real theta characteristics. arXiv:1802.05297 (2018)
10. Morgan, A.P., Sommese, A.J.: Coefficient-parameter polynomial continuation. Appl. Math. Comput. **29**, 123–160 (1989)
11. Shub, M., Smale, S.: Complexity of Bézout's theorem. I. Geometric aspects. J. Amer. Math. Soc. **6**(2), 459–501 (1993)
12. Sommese, A.J., Wampler II, C.W.: The Numerical Solution of Systems of Polynomials Arising in Engineering and Science. World Scientific Publishing, Hackensack (2005)

3BA: A Border Bases Solver
with a SAT Extension

Jan Horáček$^{(\boxtimes)}$ and Martin Kreuzer

Faculty of Informatics and Mathematics,
University of Passau, 94030 Passau, Germany
{Jan.Horacek,Martin.Kreuzer}@uni-passau.de,
http://www.symbcomp.fim.uni-passau.de/

Abstract. Many search problems over Boolean variables can be formulated in terms of satisfiability of a set of clauses or solving a system of Boolean polynomials. On one hand, there exists a great variety of software coming from different areas such as commutative algebra, SAT or SMT, that can be used to tackle these instances. On the other hand, their approaches to inferring new constraints vary and seem to be complementary to each other. For instance, compare the handling of XOR constraints in SAT solvers to that in computer algebra systems. We present a C++ implementation of a platform that combines the power of the Boolean Border Basis Algorithm (BBBA) with a CDCL SAT solver in a portfolio-based fashion. Instead of building a complete fusion or a theory solver for a particular problem, both solvers work independently and interact through a communication interface. Hence a greater degree of flexibility is achieved. The SAT solver antom, which is currently used in the integration, can be easily replaced by any other CDCL solver. Altogether, this is the first open-source implementation of the BBBA and its combination with a SAT solver.

Keywords: Boolean Border Basis Algorithm · Boolean polynomial
Cryptographic attack · SAT solving

1 Introduction

Solving a Boolean system is a well-established problem in commutative algebra that has interesting connections to many different areas such as cryptography or hardware verification. The Gröbner Basis Algorithm (GBA) is a standard method to tackle this problem. There exist well-tuned libraries that implement the (Boolean) GBA, e.g., PolyBoRi [4], or FGb [5] with its linear algebra package GBLA [3].

On the other hand, the Boolean Border Basis Algorithm (BBBA) is at an early development stage. An implementation of the general border basis algorithm can be found in [14]. Nevertheless, we are not aware of any specialized implementation other than 3BA that provides the BBBA.

© Springer International Publishing AG, part of Springer Nature 2018
J. H. Davenport et al. (Eds.): ICMS 2018, LNCS 10931, pp. 209–217, 2018.
https://doi.org/10.1007/978-3-319-96418-8_25

In this paper we present 3BA, which is a C++ implementation of the BBBA and the integration of the BBBA with a SAT solver. 3BA is based on the theory and the implementations developed successively in [7,9,10,12]. In comparison to previous versions, we have improved the implementation in the following ways. The system of numbering the terms allows us to develop efficient data structures representing Boolean systems such that interreducing the leading terms is very efficient. For this purpose, we introduce the definition of the map of reducers that cache the "pivot polynomials" during the elimination. We have enhanced the SAT support in 3BA to different techniques of clause and polynomial filtration such that the communication between the solvers is now more robust. Because of a new synchronization mechanism, the communication can be run for multiple instances without shutting down the solving processes. The integration can now be executed by running only one bash script, which makes the procedure more user-friendly. Moreover, we extended the experiments to new benchmarks coming from SAT competitions. On top of that, the entire code has been cleaned substantially and documented carefully.

We tried rewriting 3BA using different libraries such as PolyBoRi [4], or LELA [11] respectively, but surprisingly, we did not observe any speed-up, and hence we stayed with a version without any external libraries. That is why 3BA uses only the standard C++ libraries std and boost, which make 3BA very easily portable.

3BA is publicly available upon request via email to the authors or under the following link:

http://www.iti.uni-stuttgart.de/abteilungen/hardware-orientierte-informatik/projekte/
algebraische-fehlerangriffe/source-code.html

The article is structured as follows. In Sect. 2 the functionality of 3BA is described. The main C++ classes of 3BA with underlying theory are explained in Sect. 3. Finally, in Sect. 4 we present new experiments based on some SAT competition benchmarks.

Unless explicitly stated otherwise, we use the basic definitions and results in [7,9].

2   Functionality

In this section we describe the high-level functionality of 3BA.

1. **Basic structures.** 3BA classes realize various mathematical data structures and operations that can be used independently in other projects. All of these structures are provided in separated source files. We mention here only the most import ones together with their operations.
   (a) *Squarefree order ideals.* Computing order ideals minus a monomial ideal, an order ideal membership test, determining a squarefree border, and enlarging order ideals by a set of squarefree terms, are a few functions available in 3BA.

(b) *Boolean polynomials.* The following functions are provided: addition of Boolean polynomials, multiplying a Boolean polynomial by an indeterminate, accessing the indeterminates that occur in its support, getting the degree or the leading term of a Boolean polynomial.

(c) *Sparse matrices over $\mathbb{F}_2$.* These matrices are interpreted as coefficient matrices of a Boolean system w.r.t. the given term ordering (e.g., `lex`, `deglex`). Besides standard methods such as computing the row echelon form, a function for converting a given row to a Boolean polynomial (and vice versa) is provided.

2. **BBBA.** Computation of Boolean border bases of the ideal I generated by a set of Boolean polynomials.

(a) *Standard version.* This variant corresponds to the algorithm described in [9].

(b) *Substitution version.* The previous algorithm is extended by substituting linear polynomials x_i, $x_i + x_j$, $x_i + x_j + 1$, $x_i + x_j$ into the derived polynomials, whenever these special polynomials are found. E.g., if $x_i + x_j + 1$ with its leading term x_i is found in the ideal, we rewrite all polynomials containing x_i found so far by $x_i \mapsto x_j + 1$. Thus we reduce the number of indeterminates in the system by one.

(c) *Signature version.* This algorithm was presented in [10]. It stores a history of creation of derived polynomials, i.e., the signature (i, t) with $i \in \mathbb{N}$ and a squarefree term t associated to a polynomial g indicates that g is some linear reduction of the i-th input polynomial multiplied by the term t. Using this mechanism, many reductions to zero can be predicted and skipped.

3. **Conversion methods.** Different conversion methods from ANF (algebraic normal form) to CNF (conjunctive normal form) and vice versa are available in 3BA. Recall that a logical formula in CNF corresponds to a set of clauses, and a formula in ANF corresponds to a squarefree polynomial over the binary field. Each $\mathbb{F}_2$-rational zero of an ANF then corresponds to a model of a converted CNF and vice versa. For details on the conversions and the definitions of CNF and ANF, we refer the interested reader to [8].

4. **BBBA + SAT.** Using the above conversions, 3BA provides an integration of BBBA with the SAT solver `antom`.

(a) *BBBA supports SAT.* In this case, the BBBA is called by a SAT solver to enhance the conflict analysis, and henceforth, to produce better conflict clauses.

(b) *SAT supports BBBA.* The SAT solver runs in parallel on the converted system of Boolean polynomials. Whenever a short clause (i.e., a clause that contains only a few literals) is generated, it is converted to ANF and added to the ANF database of the BBBA.

Note that Functionality 2.(b) outperforms the standard version 2.(a) if a linear polynomial is derived during the computation. Sometimes one substitution leads to further substitutions, and hence to reducing the complexity of solving the system even more. Similarly, Functionality 2.(c) is strictly better than 2.(a) because many useless reductions in 2.(a) are skipped.

3   Theory and Implementation

The set of **squarefree terms** in the indeterminates $x_1, \ldots, x_n$ is denoted by $\mathbb{S}^n$. A squarefree term $t = x_{i_1} \cdots x_{i_k} \in \mathbb{S}^n$ with $1 \leq i_1 < \cdots < i_k \leq n$ is implemented in the class `SparseTerm` as the sorted **array** of uint $\tilde{t} = (i_1, \ldots, i_k)$. For more details, see [9, Remark 7.1]. The arrays are allocated length $d \in \mathbb{N}$. The number d corresponds to the maximal degree occurring in the run of the algorithm, and it has to be determined before the run of the algorithm. Typically, d is set according to the amount of RAM memory available on a computer. To give a sense of scale, the maximal degree does not exceed 7 for the examples of quadratic systems in [6]. Thus we use only terms in $\mathbb{S}^n$ with degree at most d. This set is denoted by $\mathbb{S}^n_{\leq d}$.

The squarefree terms are ordered by a **degree compatible term ordering** σ because the BBBA is a degree-by-degree algorithm. Moreover, 3BA can rearrange the indeterminates according to how frequently they appear in the input, where x_1 is the most frequent. The rearrangements can speed up the BBBA in some cases. The terms used in the algorithm are "hashed" to numbers such that these numbers reflect the term ordering σ.

Definition 1. *The unique bijective map* $\psi_{\sigma,d} : \mathbb{S}^n_{\leq d} \to \{1, \ldots, \#\mathbb{S}^n_{\leq d}\} \subseteq \mathbb{N}$ *with the property* $\psi_\sigma(t) \leq \psi_\sigma(t')$ *if and only if* $t \leq_\sigma t'$ *for* $t, t' \in \mathbb{S}^n_{\leq d}$ *is called the* **numbering of terms** *in* $\mathbb{S}^n_{\leq d}$ *induced by* σ.

Throughout this section, we use the following numbering of terms in the examples.

Example 1. Using $\sigma = \mathtt{deglex}$, $d = 2$ and $n = 2$, the map $\psi_{\sigma,d}$ defined by $1 \mapsto 1$, $x_2 \mapsto 2$, $x_1 \mapsto 3$, $x_1 x_2 \mapsto 4$ is a numbering of terms in $\mathbb{S}^n_{\leq d}$ induced by σ.

An **order ideal** is a divisor-closed set of terms. We refer the reader to [9, Sect. 7] and [7, Sect. 2] for details, for the data structures which have been used, and how to perform the operations effectively.

Let $\mathbb{F}_2 = \mathbb{Z}/2\mathbb{Z}$ be the binary field and $\mathbb{F}_2[x_1, \ldots, x_n]$ a polynomial ring over $\mathbb{F}_2$. The ring $\mathbb{B}_n = \mathbb{F}_2[x_1, \ldots, x_n]/F$ with $F = \langle x_1^2 + x_1, \ldots, x_n^2 + x_n \rangle$ is called the **ring of Boolean polynomials** in the indeterminates $x_1, \ldots, x_n$.

A Boolean polynomial $g = t_1 + \cdots + t_k$, where $t_1 <_\sigma t_2 <_\sigma \cdots <_\sigma t_k$ are terms in $\mathbb{S}^n_{\leq d}$, is represented in the class `SparseRow` as a `vector<int128_t>` via $\tilde{g} = (\psi_{\sigma,d}(t_1), \ldots, \psi_{\sigma,d}(t_k)) \in \mathbb{N}^k$, where $\psi_{\sigma,d}$ is the numbering of terms induced by σ. Note that we can choose a different map instead of $\psi_{\sigma,d}$, but then accessing the leading term would have linear complexity instead of the constant one.

Example 2. Using $\psi_{\sigma,d}$ from Example 1, we represent $g = 1 + x_2 + x_1 x_2 \in \mathbb{B}_2$ as the vector $\tilde{g} = (1, 2, 4)$.

Multiplication of g by an indeterminate x_i in $\mathbb{B}_n$ is done by translating $\tilde{g}$ back to terms in the support of g via $\psi_{\sigma,d}^{-1}$ and handled there. The result is then converted back using $\psi_{\sigma,d}$. These conversions do not slow down the overall performance according to our profiling. Moreover, the conversions are cached. The BBBA spends more than 95% in addition of two Boolean polynomials, and this is very fast in this representation, namely only a symmetric difference of two sorted vectors.

Example 3. Using $\psi_{\sigma,d}$ from Example 1, we represent $f = 1 + x_2 + x_1 x_2 \in \mathbb{B}_2$ as the vector $\tilde{f} = (1, 2, 4)$ and $g = 1 + x_1 + x_1 x_2 \in \mathbb{B}_2$ as $\tilde{g} = (1, 3, 4)$. The sum $f + g = x_2 + x_1$ in $\mathbb{B}_2$ corresponds to $(2, 3)$.

A system of Boolean polynomials V is implemented in the class `SparseMatrix2` as a `vector<SparseRow>`$\tilde{V}$, where the inner collections represent individual Boolean polynomials. At this point, we may remark that the numbering of terms from Definition 1 can be alternatively defined only for the terms appearing in the system, and hence the representation would consist of smaller numbers. However, the representation of the entire system would have to be rewritten every time when a new term is introduced.

We say that a set of Boolean polynomials $G \subseteq \mathbb{B}_n$ is LT_σ-**interreduced** if $\mathrm{LT}_\sigma(g) \neq \mathrm{LT}_\sigma(g')$ for all $g, g' \in G$ with $g \neq g'$. For more details, see [7, Sect. 3]. The reducers are cached on-the-fly such that the polynomial that has the same leading term as a given polynomial is easily accessible. This idea is captured in the next definition.

Definition 2. *Let $V \subseteq \mathbb{B}_n$ be an LT_σ-interreduced set of Boolean polynomials. A map $\varrho : \mathbb{S}_n \to V \cup \{\square\}$ such that*

$$\varrho(t) = \begin{cases} v \in V & \text{if there exists } v \in V \text{ with } \mathrm{LT}_\sigma(v) = t, \\ \square & \text{otherwise.} \end{cases}$$

holds for $t \in \mathbb{S}_n$ is called a **map of reducers** *of V.*

To illustrate how a system is LT_σ-interreduced and how Definition 2 is used, we present the following example.

Example 4. With the setting of Example 1, we want to LT_σ-interreduce $W = \{f, g, h\}$ with $f = 1 + x_2 + x_1 x_2$, $g = 1 + x_1 + x_1 x_2$ and $h = 1 + x_1$ in $\mathbb{B}_2$. We initialize $\tilde{V} = \emptyset$ and $\varrho = \left(1 : \square \mid x_1 : \square \mid x_2 : \square \mid x_1 x_2 : \square\right)$. Starting with f, we define $\tilde{V} = \left((1, 2, 4)\right)$ and $\varrho = \left(1 : \square \mid x_1 : \square \mid x_2 : \square \mid x_1 x_2 : f\right)$. We continue with g as in Example 3 and get $\tilde{V} = \left((1, 2, 4), (2, 3)\right)$ and $\varrho = \left(1 : \square \mid x_1 : x_2 + x_1 \mid x_2 : \square \mid x_1 x_2 : f\right)$. Finally, we reduce h, and we get $\tilde{V} = \left((1, 2, 4), (2, 3), (1, 2)\right)$ and $\varrho = \left(1 : \square \mid x_1 : x_2 + x_1 \mid x_2 : 1 + x_2 \mid x_1 x_2 : f\right)$. Thus the result is $1 + x_2 + x_1 x_2$, $x_2 + x_1$ and $1 + x_2$.

The conversion methods between CNF and ANF can be found in the class `BooleanPoly2` that implements a Boolean polynomial as the `vector<SparseTerm>` of its terms. This representation is more suitable here. On the other hand, addition of two Boolean polynomials is here far slower than the representation based on $\psi_{\sigma,d}$. Thus we do not use `BooleanPoly2` as the main representation of Boolean polynomials in the BBBA. The theory behind the conversions is described in [8].

The SAT extension to the BBBA is implemented in the class BBA_SAT2 based on [7, Sect. 5]. The most crucial part is how to filter the incoming and outgoing data. The original approach has been improved and scaled by four parameters $a, b, c, d \in \mathbb{N}$ as follows. In principle, the BBBA waits until it receives a set of clauses of cardinality a from the SAT solver. These clauses can be sorted (e.g., w.r.t. to the size, or in the lexicographical order, etc.) after that. Then b is set to be the maximal number of clauses after the filtration. The clause filtration (cf. [8, Algorithm 2]) outputs a set of clauses C of cardinality $\leq b$ such that C shares as many logical variables as possible. After converting C to the ANF and running the BBBA, we use a polynomial filtration to select the polynomials. Again, we may sort polynomials in the database (e.g., according to the degree and then the size of the support, etc.). We consider only the polynomials of degree $\leq c$ and convert them to the CNF on an as-they-come basis. We send only at most d clauses back to the SAT solver, and then we start again from the beginning.

In the following toy example we illustrate the synergy of the algebraic solver with a resolution-based SAT solver.

Example 5. Let $f = 1 + x_2 + x_1 x_2$ and $g = 1 + x_1 + x_1 x_2$ both from $\mathbb{B}_2$. It is easy to verify that there exists no common $\mathbb{F}_2$-rational zero of $\{f, g\}$. Let us compute $f + g = x_2 + x_1$ in order to get rid of the leading term. We convert f (resp. $f + g$) into CNF via the sparse method described in [8, Example 1] and get $\{\{X_1, X_2\}, \{\bar{X}_1, X_2\}, \{\bar{X}_1, \bar{X}_2\}\}$ (resp. $\{\{X_1, \bar{X}_2\}, \{\bar{X}_1, X_2\}\}$). Resolution then yields $\{X_1\}$ and $\{\bar{X}_1\}$, and hence certifies inconsistency. Note that the BBBA can find the element 1 as follows. Firstly, multiply $f + g$ by x_1 and get $x_1 + x_1 x_2$. Secondly, interreduce the latter polynomial with g and get 1.

4   Application and Performance

3BA can be used for two types of inputs: Boolean systems or CNF formulae. In the first case, 3BA actually computes extensions of Boolean Gröbner bases because the order ideals are restricted to a special form (cf. [9, Example 2.3]). Thus 3BA can be used to tackle various problems ranging from hardware verification to algebraic attacks. The BBBA uses a different approach than the GBA (see [9, Sect. 9]) to create low-degree polynomials. To show a comparison between those methods, we refer to the experiments in [9, Sect. 8].

In the second case, 3BA can be used to find a model of an arbitrary CNF formula in the DIMACS format (i.e., the standard format for SAT solvers). This option seems to be effective when the input contains a rich algebraic structure, e.g., many XOR constraints (see [7, Table 2]). Note that such benchmarks are quite common in cryptanalysis. Most SAT solvers do not use the rich XOR structure hidden in modern cryptosystems (e.g., in ARX ciphers or in permutation-substitution networks, etc.) at all. Thus an integration with the BBBA, which naturally works with addition modulo 2, may come in handy. The initial experiments with the integration are provided in [7, Table 2].

The most promising application of **3BA** seems to be Functionality 4.(a) presented in Sect. 2. That is why we focus on this function. Before providing some more experiments, we discuss the impact of a new clause coming from the algebraic reasoning in **3BA** on the SAT solver **antom**. There are three basic scenarios:

(i) The new clause is not satisfied by the partial model constructed by the SAT solver, and hence the SAT solver is forced to backtrack.
(ii) The new clause eliminates a large portion of assignments, and hence these assignments do not have to be considered.

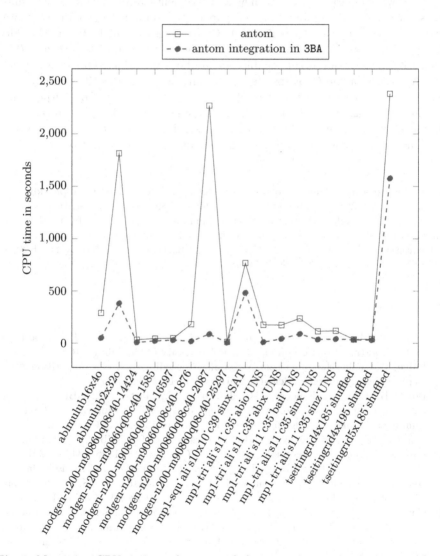

Fig. 1. Measuring CPU timings of **antom** and the **antom** integration in **3BA** tested on various CNF benchmarks.

(iii) The new clause effects the statistics carried out by the SAT solver. That is why a different decision is chosen in the next level of branching by the heuristics.

On one hand, scenarios (i) and (ii) have a positive effect on solving. On the other hand, scenario (iii) may affect the integration in both positive and negative ways, depending on the state of the SAT solver and the example under consideration.

We have extended our experiments to the SAT competition benchmarks [1,2], together with incorporating the new improvements from Sect. 3. The results are shown in Fig. 1. Timings in this paper were obtained on a Linux server having a 2.60 GHz Intel Core i7-5600U CPU and a total of 16 GB RAM. The timeout limit was set to 2500 seconds. We choose the parameters $(a, b, c, d) = (40, 10, 3, 20)$ with the sorting enabled before both filtrations. On the x-axis we put the names of the CNF instances, and on the y-axis the CPU timings are displayed. We focused on the examples where the 3BA integration outperforms the vanilla antom. On one hand, there exist cases where the integration is slower. On the other hand, we found 8 instances, for which the integration finished before the timeout, whereas the vanilla antom did not. (These cases are not displayed in the figure.)

Acknowledgments. The authors are grateful to Tobias Schubert for providing us the SAT solver antom [13] and Jan Burchard for making the communication between antom and 3BA possible. John Abbott and Anna Bigatti gave us useful feedback on our C++ implementation. This work was financially supported by the DFG project "Algebraische Fehlerangriffe" [KR 1907/6-2] and partially supported by the H2020-FETOPEN-2015-CSA project SC^2 (712689).

References

1. Balyo, T., Heule, M. J., Järvisalo, M., et al.: Proceedings of SAT Competition 2016. Department of Computer Science, University of Helsinki (2016)
2. Balyo, T., Heule, M.J., Järvisalo, M., et al.: Proceedings of SAT Competition 2017. Department of Computer Science, University of Helsinki (2017)
3. Boyer, B., Eder, C., Faugère, J.-C., Lachartre, S., Martani, F.: GBLA: Gröbner basis linear algebra package. In: Proceedings of the ACM on International Symposium on Symbolic and Algebraic Computation, pp. 135–142. ACM (2016)
4. Brickenstein, M., Dreyer, A.: PolyBoRi: a framework for Gröbner basis computations with Boolean polynomials. J. Symbolic Comput. **44**, 1326–1345 (2009)
5. Faugère, J.-C.: FGb: a library for computing Gröbner bases. In: Fukuda, K., Hoeven, J., Joswig, M., Takayama, N. (eds.) ICMS 2010. LNCS, vol. 6327, pp. 84–87. Springer, Heidelberg (2010). https://doi.org/10.1007/978-3-642-15582-6_17
6. Gay, M., Burchard, J., Horáček, J., Messeng Ekossono, A.S., Schubert, T., Becker, B., Kreuzer, M., Polian, I.: Small scale AES toolbox: algebraic and propositional formulas, circuit-implementations and fault equations. In: Conference on Trustworthy Manufacturing and Utilization of Secure Devices (TRUDEVICE 2016), Barcelona (2016)

7. Horáček, J., Burchard, J., Becker, B., Kreuzer, M.: Integrating algebraic and SAT solvers. In: Blömer, J., Kotsireas, I.S., Kutsia, T., Simos, D.E. (eds.) MACIS 2017. LNCS, vol. 10693, pp. 147–162. Springer, Cham (2017). https://doi.org/10.1007/978-3-319-72453-9_11

8. Horáček, J., Kreuzer, M.: On conversions from CNF to ANF. In: 2th International Workshop on Satisfiability Checking and Symbolic Computation, SC-square, Kaiserslautern (2017)

9. Horáček, J., Kreuzer, M., Messeng Ekossono, A.S.: Computing Boolean border bases. In: 18th International Symposium on Symbolic and Numeric Algorithms for Scientific Computing, SYNASC, Timisoara, pp. 465–472. IEEE (2016)

10. Horáček, J., Kreuzer, M., Messeng Ekossono, A.S.: A signature based border basis algorithm. In: Conference on Algebraic Informatics, CAI, Kalamata (2017)

11. Hovinen, B., Martani, F.: LELA: library for exact linear algebra (2011). https://github.com/Singular/LELA

12. Kehrein, A., Kreuzer, M.: Computing border bases. J. Pure Appl. Algebra **205**(2), 279–295 (2006)

13. Schubert, T., Reimer, S.: antom (2016). https://projects.informatik.uni-freiburg.de/projects/antom

14. The ApCoCoA Team. ApCoCoA: Applied Computations in Commutative Algebra. http://apcocoa.uni-passau.de

The Hidden Subgroup Problem and Post-quantum Group-Based Cryptography

Kelsey Horan[1]([✉]) and Delaram Kahrobaei[2,3]

[1] The Graduate Center, CUNY, New York, USA
khoran@gradcenter.cuny.edu
[2] The Graduate Center and NYCCT, CUNY, Brooklyn, USA
[3] New York University, New York, USA
dk2572@nyu.edu

Abstract. In this paper we discuss the Hidden Subgroup Problem (HSP) in relation to post-quantum cryptography. We review the relationship between HSP and other computational problems, discuss an optimal solution method, and review results about the quantum complexity of HSP. We also overview some platforms for group-based cryptosystems. Notably, efficient algorithms for solving HSP in the proposed infinite group platforms are not yet known.

Keywords: Hidden Subgroup Problem · Quantum computation
Post-quantum cryptography · Group-based cryptography

1 Introduction

In August 2015 the National Security Agency (NSA) announced plans to upgrade security standards; the goal is to replace all deployed cryptographic protocols with quantum secure protocols. This transition requires a new security standard to be accepted by the National Institute of Standards and Technology (NIST). Proposals for quantum secure cryptosystems and protocols have been submitted to the standardization process. There are six main primitives currently proposed to be quantum-safe: (1) lattice-based (2) code-based (3) isogeny-based (4) multivariate-based (5) hash-based, and (6) *group-based* cryptographic schemes.

One goal of cryptography, as it relates to complexity theory, is to analyze the complexity assumptions used as the basis for various cryptographic protocols and schemes. A central question is determining how to generate intractable

Research of Delaram Kahrobaei was partially supported by a PSC-CUNY grant from the CUNY research foundation. Research of Delaram Kahrobaei was also supported by the ONR (Office of Naval Research) grant N000141512164. We thank E. Kashefi and L. Perret for discussions and hospitality in summer 2017 at University of Sorbonne.

J. H. Davenport et al. (Eds.): ICMS 2018, LNCS 10931, pp. 218–226, 2018.
https://doi.org/10.1007/978-3-319-96418-8_26

instances of these hard problems, upon which to implement an actual cryptographic scheme. The candidates for these instances must be platforms in which the relevant hardness assumption is reasonable. Determining if these group-based cryptographic schemes are quantum-safe begins with determining the groups in which these hardness assumptions are invalid in the quantum setting.

In what follows we address the quantum complexity of the *Hidden Subgroup Problem (HSP)* to determine the groups in which the hardness assumption still stands. *The Hidden Subgroup Problem (HSP) asks the following: given a description of a group G and a function $f : G \to X$ for some finite set X is guaranteed to be strictly H-periodic, i.e. constant and distinct on left (resp. right) cosets of a subgroup $H \leq G$, find a generating set for H.* It is important to note that Simon's problem of computing a XOR-mask, Shor's algorithm for factoring and finding the discrete log, Boneh's algorithm for finding a hidden linear function, and Kitaev's algorithm for the abelian stabilizer problem are all special cases of HSP. Therefore, the HSP is directly related to problems such as breaking one-time pad, discrete logarithm problem, graph isomorphism problem (which is now known to be in quali-polynomial), lattice-based problems, and the problem for factoring for RSA.

The classical complexity of HSP is known [1]: *Suppose that G has a set $\mathcal{H}$ of N subgroups, such that $H_1 \cap H_2 \cap \ldots \cap H_{\mathcal{H}} = e_G$. Then a classical computer must make $\Omega(\sqrt{N})$ queries to solve the HSP.* The classical cases in which HSP is easy are the cases in which G has only a polynomial number of subgroups, allowing brute-force for the function f on all subgroups.

We provide a survey of results regarding the complexity of quantum algorithms for solving HSP in various group platforms. We also provide information on the relationship between HSP and other computational problems. These results provide insight into potential platforms for quantum safe cryptography, when the underlying hard problem is reducible to HSP.

2   Group-Based Cryptography

Group-based cryptography could be shown to be post-quantum if the underlying security problem is NP-complete or intractable; firstly, we need to analyze the problem's equivalence to HSP, then analyze the applicability of Grover's search algorithm. Cryptanalysis based on a reduction to solving HSP creates some obstacles, as the groups under consideration below are mostly infinite and do not have an efficient algorithm for HSP. In the following cryptosystems a connection to HSP can assist in the analysis of security.

For example in [2] a practical cryptanalysis of WalnutDSA was proposed, a post-quantum cryptosystem using the conjugacy search problem (CSP) over braid groups that was submitted to the NIST competition in 2017 [3]. It has been argued since the braid group does not contain any non-trivial finite subgroups, there does not seem to be any viable way to connect CSP with HSP. It has been shown there is no reduction between CSP and HSP, [4,5]. As for analysis via Grover's algorithm [6], it has been mentioned that a majority of the time for signature verification in WalnutDSA is repeated E-Multiplications.

Aside from WalnutDSA, there are alternative group-theoretic problems and classes of groups which have been proposed for post-quantum cryptography. For example, the first proto-cryptosystem based on groups was proposed by Wagner-Magyarik in [7] on the assumption that the word choice problem was hard. Later on Flores-Kahrobaei-Koberda proposed right-angled Artin groups for various other cryptographic protocols [8,9]. Eick and Kahrobaei proposed Polycyclic groups for cryptography, based on the Conjugacy Search Problem [10]. Later, Gryak-Kahrobaei proposed other group-theoretic problems for consideration when using polycyclic groups in their survey [11]. Kahrobaei-Koupparis [12] proposed a post-quantum digital signature using polycyclic groups. Kahrobaei-Khan proposed a public-key cryptosystem using polycyclic groups [13]. Habeeb-Kahrobaei-Koupparis-Shpilrain proposed the use of a semigroup of matrices with a semidirect product structure [14].

Thompson groups have been considered by Shpilrain-Ushakov based on the Decomposition Search Problem [15]. Hyperbolic groups have been proposed by Chatterji-Kahrobaei-Lu using properties of subgroup distortion and the Geodesic Length Problem [16]. Free metabelian groups have been proposed based on the Subgroup Membership Search Problem by Shpilrain-Zapata [17]. Kahrobaei-Shpilirain proposed Free nilpotent p-groups for a semidirect product public key cryptosystem [18]. Linear groups were proposed by Baumslag-Fine-Xu [19]. Grigorchuk groups, have been proposed by [20]. Groups of matrices were proposed by Grigoriev-Ponomarenko for a Homomorphic Encryption scheme [21].

3   Relation of HSP to Other Computational Problems

Many computational problems are special cases of the HSP; in many cases an efficient algorithm for HSP over a certain group implies an efficient algorithm for some other computational problem. It is important to note that one method of determining an efficient quantum solution to a hard problem consists of reducing the problem to an instance of HSP over a group with a known efficient solution. This process consists of determining the appropriate group G, the subgroup H and the strongly H-periodic function f. For example, Simon's problem can be viewed as an instantiation of HSP over $G = \mathbb{Z}_2^n$ with a subgroup H of order 2. Duetsch's algorithm solves a variant of HSP where H is either $\{0\}$ (f is balanced) or $\mathbb{Z}_2$ (f is constant). Shor's algorithm solves period finding as a special case of HSP, allowing for an efficient quantum algorithm for factoring and discrete log.

The graph automorphism (resp. isomorphism) problem can also be framed as an instance of HSP. To solve graph automorphism we consider HSP in the symmetric group on n letters, $G = S_n$, any function f which hides the trivial subgroup is an automorphism. Analogously the graph isomorphism problem is an instance of HSP over the wreath product $G = S_n \wr S_2$ [22]. Also, solutions to HSP can solve the abelian stabilizer problem; when G is acting on a finite set X and where $\mathrm{St}_G(x)$ is the stabilizer of x we have that $f_x : G \to X$ can be defined such that $g \mapsto g(x)$ is strongly $\mathrm{St}_G(x)$-periodic.

A solution to a particular instance of HSP is a solution to the problem of hidden linear functions [23]; if g is a permutation of $\mathbb{Z}_N$ and $h : \mathbb{Z} \times \mathbb{Z} \to \mathbb{Z}_N$ is such that $x, y \mapsto x + ay \mod N$, we have $f = g \circ h$ hiding $\langle (-a, 1) \rangle$. Additionally, self-shift-equivalent polynomials can be framed as an instance of HSP, in this case Grigoriev shows how to compute the hidden subgroup [24].

Regev showed that an efficient solution to the dihedral HSP implies a quantum solution to lattice problems [25]. Specifically, the $g(n)$-Unique Shortest Vector Problem (USVP) is NP-hard for $g(n) = O(1)$, and has a polynomial time classical solution when $g(n)$ is large. A solution to the dihedral HSP based on the standard-method (found in Sect. 4) can be used to solve poly(n)-USVP. HSP over the symmetric and dihedral groups are highly motivated open questions in post-quantum group-based cryptography.

Another related computational problem is the Hidden Shift Problem, which has been proposed as a basis for post-quantum cryptography in symmetric cryptosystems that are quantum-CPA secure [26]. Other than the use of a generalization of Simon's algorithm, and Kuperberg's algorithm discussed above, very little is known about the Hidden Shift Problem. Clearly, this problem is closely related to HSP as some solutions coincide. It is important to note that constructions based on the Hidden Shift problem have also remained quantum secure.

4   Solution Methods

The standard method of solving HSP over G performs the following steps. First, the algorithm queries the H-periodic function f in superposition and discards the register which holds the output. This leaves the first register entangled in a hidden subgroup state, a superposition of coset representatives for some left traversal $K \subset G$. Following this, the state can be sampled using post-processing techniques to determine H. In the following we have $|gH\rangle = |H|^{-1/2} \sum_{h \in H} |gh\rangle$ as the coset state. This approach reduces the problem to a problem of quantum mechanics: how to distinguish the members of an ensemble of quantum hidden subgroup states.

$$|G|^{1/2} \sum_{g \in G} |g, 0\rangle \mapsto |G|^{1/2} \sum_{g \in G} |g, f(g)\rangle \mapsto \rho_H = |H||G|^{-1} \sum_{g \in K} |gH\rangle\langle gH|$$

How do we measure the state? The problem of distinguishing these quantum states has some proposed solutions. Most namely, the often optimal solution entitled Pretty Good Measurement (PGM) can be used. An obstacle to performing PGM is the lack of an efficient QTF/CFT in the underlying group. For these instances we know of no efficient quantum algorithm for solving HSP.

5   Results

Finite Abelian and Finite Near-Abelian. The infamous quantum algorithms of Simon and Shor provide quantum solutions to HSP in the abelian cases where $G = (\mathbb{Z}/2)^n$ and $G = \mathbb{Z}$ respectively. Shor's algorithm extends to the general abelian case as well, providing a polynomial time quantum algorithm with bounded error [27–29]. The probability of success can be improved to 1 when G is *abelian of smooth order*, i.e. if all prime factors of $|G|$ are at most $(\log |G|)^c$ [30].

In the case that G is nearly abelian, i.e. if the value $\kappa(G) = \{\cap_{H \leq G} N(H)\}$ (where $N(H)$ is the normalizer for H) is sufficiently large, then there are established computational bounds on HSP. The size of this intersection relative to the group is a measure of the abelianness of G. If $[G : \kappa(G)] = 2^{O(\log^{\frac{1}{2}} n)}$ then Grigni et al. have an efficient algorithm for solving HSP [31]. These results were improved upon by Gavinsky [32] gave results to show that an efficient algorithm exists when $[G : \kappa(G)] = \text{poly}(\log |G|)$.

There is a polynomial time quantum algorithm which solves HSP when H is a hidden normal subgroup of a solvable group or permutation group, also finding hidden subgroups of groups with small commutator subgroup and of groups admitting an elementary abelian normal 2-subgroup of small index or with cyclic factor group [33]. Subexponential algorithms for HSP in any solvable group have been given by Friedl et al. [34].

When G is a known finite abelian group with a subgroup $H \leq G$, given black-box access to the H-hiding function f, a quantum computer can uniquely and completely determine f in $\text{poly}\log(|G|)$ time and query complexity. When G is "nearly" abelian, or built from abelian parts, one can leverage this fact to obtain an efficient algorithm for HSP.

Finite, Non-Abelian. The finite non-abelian case of HSP is much more elusive. Shor's algorithm extends to any group G when H is normal if the quantum fourier transform (QFT) can be efficiently computed over the group [35]. The algorithm also extends to when H has few conjugates, this time requiring the quantum character transform (QCT) over the group algebra $\mathbb{C}[G]$ [31]. This variation is not applicable when H has many conjugates, as in some of the following cases. Alternatively, when H is normal in G, a black-box group, generators for H can be found in time polynomial in the input size $+ v(G)$ [33] without requiring an efficient QFT over G. Additionally, the quantum computation of the discrete log in semi-groups [36] is an instance of HSP.

When G is the (discrete) Heisenberg group, $H_p = \langle (a,b,c)|a,b,c \in \mathbb{Z}_p \rangle$ with group law $(a_1, b_1, c_1)(a_2, b_2, c_2) = (a_1 + a_2, b_1 + b_2, c_1 + c_2 + a_1 b_2)$, it is sufficient to be able to distinguish cyclic subgroups of order p: $H_{a,b} = \langle (a,b,1) \rangle = \{(a,b,1)^x | x \in \mathbb{Z}_p\}$. Thus, finding an arbitrary H reduces to determining two parameters a, b, given the coset state produced by the standard method with the f which hides $H_{a,b}$. This can be efficiently computed with an overall success probability close to $\frac{1}{2}$.

In a more difficult case we consider instances of HSP in the dihedral group of order $2N$, D_N, where the function f hides H of order 2. In this case H is a hidden reflection and has many conjugates in G. Therefore the QCT based solution is not applicable. Kuperberg stated that *finding an arbitrary hidden subgroup H of D_N reduces to finding the slope of a hidden reflection* and provides a quantum algorithm with both time and query complexity of $2^{O(\sqrt{\log N})}$, applicable to D_N for all values of N but achieving an even tighter complexity bound for specific smooth values of N [37]. Kuperberg's algorithm also provides a solution to the hidden shift problem in an arbitrary finitely generated abelian group G. Regev improved upon the bounds of Kuperberg's original algorithm providing a polynomial space variation to the original superpolynomial space algorithm, which still achieves subexponential complexity [38].

When G is a type of wreath product $W_n = \mathbb{Z}_2^n \wr \mathbb{Z}_2$, Roetteler et al. [39] provide a positive result for finding an efficient solution to the non-abelian HSP within W_n. This result is due to the existence of an efficient non-abelian QFT in W_n. Wreath product groups are in turn a subset of semi-direct product groups. When G is (one of some groups that are) a semidirect product of abelian groups, alternative efficient algorithms have been proposed. The polycyclic HSP has been addressed for $\mathbb{Z}_{p^k} \rtimes \mathbb{Z}_2$ for fixed prime power p^k [40], $\mathbb{Z}_q \ltimes \mathbb{Z}_p$ with $q|(p-1)$ and $q = \frac{p}{\mathrm{polylog}(p)}$, certain affine groups [41], $\mathbb{Z}_{p^r}^m \rtimes \mathbb{Z}_p$ [42], with $p \in \mathbb{P}$, $\mathbb{Z}_{p^r} \rtimes \mathbb{Z}_{q^s}$ where $p^r/q = \mathrm{poly}(\log p^2$ where $p, q \in \mathbb{P}$ and $r, s \in \mathbb{N}$ [43], and $\mathbb{Z}_N \rtimes \mathbb{Z}q^s$ where N has a special prime factorization [44].

In general, when G is a group of finite order HSP has quantum query complexity of $\mathrm{poly}(\log |G|)$, as shown by Ettinger, Høyer and Knill [45]; for any group G, $O(\log |G|)$ queries provides sufficient statistical information to solve HSP. This result provides no guarantees on computational complexity. The new problem is determining how to implement queries efficiently, as well as how to control the amount of postprocessing required by the algorithm. In the case of the dihedral group, an algorithm with the lower bound on query complexity has been constructed, but the postprocessing required is exponential. In many cases the inefficiency of a proposed quantum algorithm is primarily due to the inefficiency of the required quantum measurement or post-processing within the group.

Infinite. What seems to be an obstacle for infinite groups is that the quantum computer should assume the state: $|G\rangle = |G|^{-1/2} \sum_{g \in G} |g\rangle$. The meaning of this is clear for finite groups.

The abelian infinite HSP was clearly first considered with Shor's algorithm, over $\mathbb{Z}$. In [46], infinite-dimensional HSP has been mentioned, particularly for infinite abelian groups $\mathbb{Z}^N$. Additionally, HSP has been defined and considered for infinite abelian groups of the form $\mathbb{R}^k \times \mathbb{Z}^l \times (\mathbb{R}/\mathbb{Z})^s \times H$ for some finite group H [47]. Other than the cases of $\mathbb{Z}^N$, $\mathbb{R}^N$, $\mathbb{T}^N$, and combinations of these in which an efficient algorithm exists, the infinite and continuous HSP has not been addressed within the literature for the *non-abelian* case.

References

1. Childs, A.: Lecture notes on quantum algorithms (2017)
2. Hart, D., et al.: A practical cryptanalysis of WalnutDSA™. In: Abdalla, M., Dahab, R. (eds.) PKC 2018. LNCS, vol. 10769, pp. 381–406. Springer, Cham (2018). https://doi.org/10.1007/978-3-319-76578-5_13

3. Anshel, I., Atkins, D., Goldfeld, D., Gunnells, P.: WalnutDSA(TM): a quantum resistant group theoretic digital signature algorithm. IACR Cryptology ePrint Archive (2017)

4. Wang, L., Wang, L.: Conjugate searching problem vs. hidden subgroup problem. In: The Third International Workshop on Post-Quantum Cryptography, Recent Results Session (2010)

5. Wang, L., Wang, L., Cao, Z., Yang, Y., Niu, X.: Conjugate adjoining problem in braid groups and new design of braid-based signatures. Sci. China Inf. Sci. **53**(3), 524–536 (2010)

6. Grover, L.: A fast quantum mechanical algorithm for database search. In: Proceedings of the Twenty-Eighth Annual ACM Symposium on Theory of Computing, pp. 212–219 (1996)

7. Wagner, N.R., Magyarik, M.R.: A public-key cryptosystem based on the word problem. In: Blakley, G.R., Chaum, D. (eds.) CRYPTO 1984. LNCS, vol. 196, pp. 19–36. Springer, Heidelberg (1985). https://doi.org/10.1007/3-540-39568-7_3

8. Flores, R., Kahrobaei, D.: Cryptography with right-angled artin groups. Theoret. Appl. Inform. **28**, 8–16 (2016)

9. Flores, R., Kahrobaei, D., Koberda, T.: Algorithmic problems in right-angled artin groups: complexity and applications. arXiv preprint arXiv:1802.04870 (2018)

10. Eick, B., Kahrobaei, D.: Polycyclic groups: a new platform for cryptology? arXiv preprint math/0411077 (2004)

11. Gryak, J., Kahrobaei, D.: The status of polycyclic group-based cryptography: a survey and open problems. Groups Complex. Cryptology **8**(2), 171–186 (2016)

12. Kahrobaei, D., Koupparis, C.: On-commutative digital signatures using non-commutative groups. Groups Complexity Cryptology, pp. 377–384 (2012)

13. Kahrobaei, D., Khan, B.: A non-commutative generalization of ELGamal key exchange using polycyclic groups. In: IEEE Global Telecommunications Conference 2006, pp. 1–5 (2006)

14. Habeeb, M., Kahrobaei, D., Koupparis, C., Shpilrain, V.: Public key exchange using semidirect product of (semi)groups. In: Jacobson, M., Locasto, M., Mohassel, P., Safavi-Naini, R. (eds.) ACNS 2013. LNCS, vol. 7954, pp. 475–486. Springer, Heidelberg (2013). https://doi.org/10.1007/978-3-642-38980-1_30

15. Shpilrain, V., Ushakov, A.: Thompson's group and public key cryptography. In: Ioannidis, J., Keromytis, A., Yung, M. (eds.) ACNS 2005. LNCS, vol. 3531, pp. 151–163. Springer, Heidelberg (2005). https://doi.org/10.1007/11496137_11

16. Chatterji, I., Kahrobaei, D., Lu, N.Y.: Cryptosystems using subgroup distortion. Theoret. Appl. Inform. **29**, 14–24 (2017)

17. Shpilrain, V., Zapata, G.: Combinatorial group theory and public key cryptography. Appl. Algebra Eng. Commun. Comput. **17**(3–4), 291–302 (2006)

18. Kahrobaei, D., Shpilrain, V.: Using semidirect product of (semi)groups in public key cryptography. In: Beckmann, A., Bienvenu, L., Jonoska, N. (eds.) CiE 2016. LNCS, vol. 9709, pp. 132–141. Springer, Cham (2016). https://doi.org/10.1007/978-3-319-40189-8_14

19. Baumslag, G., Fine, B., Xu, X.: Cryptosystems using linear groups. Appl. Algebra Eng. Commun. Comput. **17**(3–4), 205–217 (2006)

20. Petrides, G.: Cryptanalysis of the public key cryptosystem based on the word problem on the Grigorchuk groups. In: Paterson, K.G. (ed.) Cryptography and Coding 2003. LNCS, vol. 2898, pp. 234–244. Springer, Heidelberg (2003). https://doi.org/10.1007/978-3-540-40974-8_19

21. Grigoriev, D., Ponomarenko, I.: Homomorphic public-key cryptosystems over groups and rings. arXiv preprint cs/0309010 (2003)

22. Kobler, J., Schöning, U., Torán, J.: The Graph Isomorphism Problem: Its Structural Complexity. Springer Science & Business Media, New York (2012). https://doi.org/10.1007/978-1-4612-0333-9
23. Boneh, D., Lipton, R.J.: Quantum cryptanalysis of hidden linear functions. In: Coppersmith, D. (ed.) CRYPTO 1995. LNCS, vol. 963, pp. 424–437. Springer, Heidelberg (1995). https://doi.org/10.1007/3-540-44750-4_34
24. Grigoriev, D.: Testing shift-equivalence of polynomials by deterministic, probabilistic and quantum machines. Theoret. Comput. Sci. 180(1–2), 217–228 (1997)
25. Regev, O.: Quantum computation and lattice problems. SIAM J. Comput. 33(3), 738–760 (2004)
26. Alagic, G., Russell, A.: Quantum-secure symmetric-key cryptography based on hidden shifts. In: Coron, J.-S., Nielsen, J.B. (eds.) EUROCRYPT 2017. LNCS, vol. 10212, pp. 65–93. Springer, Cham (2017). https://doi.org/10.1007/978-3-319-56617-7_3
27. Simon, D.R.: On the power of quantum computation. SIAM J. Comput. 26(5), 1474–1483 (1997)
28. Shor, P.: Polynomial-time algorithms for prime factorization and discrete logarithms on a quantum computer. SIAM Rev. 41(2), 303–332 (1999)
29. Kitaev, A.: Quantum computations: algorithms and error correction. Russ. Math. Surv. 52(6), 1191–1249 (1997)
30. Brassard, G., Hoyer, P.: An exact quantum polynomial-time algorithm for Simon's problem. In: Proceedings of the Fifth Israeli Symposium on Theory of Computing and Systems, pp. 12–23. IEEE (1997)
31. Grigni, M., Schulman, L., Vazirani, M., Vazirani, U.: Quantum mechanical algorithms for the nonabelian hidden subgroup problem. In: Proceedings of the thirty-third annual ACM Symposium on Theory of Computing, pp. 68–74 (2001)
32. Gavinsky, D.: Quantum solution to the hidden subgroup problem for poly-near-hamiltonian groups. Quantum Inf. Comput. 4(3), 229–235 (2004)
33. Ivanyos, G., Magniez, F., Santha, M.: Efficient quantum algorithms for some instances of the non-abelian hidden subgroup problem. Int. J. Found. Comput. Sci. 14(05), 723–739 (2003)
34. Friedl, K., Ivanyos, G., Magniez, F., Santha, M., Sen, P.: Hidden translation and translating coset in quantum computing. SIAM J. Comput. 43(1), 1–24 (2014)
35. Hallgren, S., Russell, A., Ta-Shma, A.: Normal subgroup reconstruction and quantum computation using group representations. In: Proceedings of the Thirty-Second Annual ACM Symposium on Theory of Computing, pp. 627–635 (2000)
36. Childs, A., Ivanyos, G.: Quantum computation of discrete logarithms in semigroups. J. Math. Cryptology 8(4), 405–416 (2014)
37. Kuperberg, G.: A subexponential-time quantum algorithm for the dihedral hidden subgroup problem. SIAM J. Comput. 35(1), 170–188 (2005)
38. Regev, O.: A subexponential time algorithm for the dihedral hidden subgroup problem with polynomial space. arXiv preprint quant-ph/0406151 (2004)
39. Roetteler, M., Beth, T.: Polynomial-time solution to the hidden subgroup problem for a class of non-abelian groups. arXiv preprint quant-ph/9812070 (1998)
40. Friedl, K., Ivanyos, G., Magniez, F., Santha, M., Sen, P.: Hidden translation and orbit coset in quantum computing. In: Proceedings of the Thirty-Fifth Annual ACM Symposium on Theory of Computing, pp. 1–9 (2003)
41. Moore, C., Rockmore, D., Russell, A., Schulman, L.: The power of basis selection in Fourier sampling: hidden subgroup problems in affine groups. In: Proceedings of the Fifteenth Annual ACM-SIAM Symposium on Discrete Algorithms, pp. 1113–1122 (2004)

42. Inui, Y., Le Gall, F.: An efficient algorithm for the hidden subgroup problem over a class of semi-direct product groups. Technical report (2004)
43. Gonçalves, D., Portugal, R.: Solution to the hidden subgroup problem for a class of noncommutative groups. arXiv preprint arXiv:1104.1361 (2011)
44. Gonçalves, D., Fernandes, T., Cosme, C.: An efficient quantum algorithm for the hidden subgroup problem over some non-abelian groups. TEMA (São Carlos) **18**(2), 215–223 (2017)
45. Ettinger, M., Høyer, P., Knill, E.: The quantum query complexity of the hidden subgroup problem is polynomial. Inf. Process. Lett. **91**(1), 43–48 (2004)
46. Kissinger, A., Gogioso, S.: Fully graphical treatment of the quantum algorithm for the hidden subgroup problem. arXiv preprint quant-ph 1701.08669 (2017)
47. Eisenträger, K., Hallgren, S., Kitaev, A., Song, F.: A quantum algorithm for computing the unit group of an arbitrary degree number field. In: Proceedings of the Forty-Sixth Annual ACM Symposium on Theory of Computing, pp. 293–302 (2014)

Questions on Orbital Graphs

Paula Hähndel and Rebecca Waldecker$^{(\boxtimes)}$

Institut für Mathematik, Martin-Luther-Universität Halle-Wittenberg,
Theodor-Lieser-Straße 5, 06120 Halle, Germany
`paula.haehndel2@student.uni-halle.de`,
`rebecca.waldecker@mathematik.uni-halle.de`
`http://conway1.mathematik.uni-halle.de/~waldecker/index.html`

Abstract. Previous work on orbital graphs has shown that they are a powerful pruning tool in backtrack algorithms. In this article we consider a few questions that are relevant from this perspective, focussing on properties of orbital graphs that can be detected by an efficient algorithm. Roughly speaking, the challenge is to decide when to use orbital graphs and, possibly, how to choose a "best" orbital graph, and to make this decision early in the algorithm at low computational costs. In this note we discuss how to decide whether or not a given digraph is an orbital graph for some group and what groups are recognisable by their orbital graphs (or even just *one* orbital graph). We approach these problems from a theoretical point of view.

Keywords: Orbital graphs · Backtrack search · Permutation groups

1   Introduction

Orbital graphs can be used to significantly improve backtrack search algorithms for computing normalisers of subgroups in permutation groups. This has been demonstrated by Theißen in his PhD thesis (see [5]), where he uses these new methods for the construction of primitive permutation groups. He also mentions that improvements in backtrack search methods will substantially impact the performance of algorithms that compute the intersection of two permutation groups or the stabiliser of a set. This was confirmed recently: New refiners that use orbital graphs make it possible to skip large parts of the search tree in partition backtrack algorithms. Experiments in [3] show that for some typical problems (e.g. calculating set stabilisers), these new refiners improve the performance of partition backtrack by several orders of magnitude. These results lead to more questions on the usefulness of orbital graphs as a pruning tool. Therefore, in the present article we discuss some of these questions and indicate how hard it will be to give complete answers.

For the remainder of this note we let $n \in \mathbb{N}$, $\Omega := \{1, ..., n\}$ and $G := \mathcal{S}_n$ (the symmetric group on Ω). For all $\gamma \in \Omega$ and all $h \in H$ we write γ^h for the image of γ under h in the natural permutation action.

© Springer International Publishing AG, part of Springer Nature 2018
J. H. Davenport et al. (Eds.): ICMS 2018, LNCS 10931, pp. 227–234, 2018.
https://doi.org/10.1007/978-3-319-96418-8_27

Definition 1 (Orbital Graphs). *Let $H \leq G$ and let $\alpha, \beta \in \Omega$ be distinct elements, chosen in this order. We define a digraph $\Gamma = (\Omega, A)$ where the set of arcs A is defined as $A := \{(\alpha^h, \beta^h) \mid h \in H\}$. This digraph is called the **orbital graph of H with base-pair** (α, β), and is denoted by $\Gamma(H, \Omega, (\alpha, \beta))$.*

*Following [2] we say that an orbital graph is **self-paired** if and only if, for all $\gamma, \delta \in \Omega$, it is true that (γ, δ) is an arc if and only if (δ, γ) is an arc.*

We emphasise that this definition is directional (as it is in [1,2]), but that there also exist versions that are undirected.

Hypothesis 1. *Let $H \leq G$ and let $\alpha, \beta \in \Omega$ be distinct.*
Let $\Gamma := \Gamma(H, \Omega, (\alpha, \beta))$ and let A denote the set of arcs of Γ.

The statements in the next lemma are known, and a proof can be found for example on p. 7 in [3].

Lemma 2. *Suppose that Hypothesis 1 holds. Then we have the following:*

(i) $\Gamma = \Gamma(H, \Omega, (\gamma, \delta))$ if and only if $(\gamma, \delta) \in A$.
(ii) Γ is self-paired if and only if some $h \in H$ interchanges α and β.
(iii) α^H is precisely the set of vertices of Γ that are the starting point of some arc, and β^H is precisely the set of vertices of Γ that are the end point of some arc.
(iv) The number of arcs starting at α is $|\beta^{H_\alpha}|$ and the number of arcs going into β is $|\alpha^{H_\beta}|$.

Definition 2. *Given Hypothesis 1, we denote by $Aut_G(\Gamma)$ the group of graph automorphisms of Γ that are induced by elements of G.*

Remark 3. *Suppose that Hypothesis 1 holds. Then the definition of orbital graphs implies that H induces graph automorphisms on Γ and therefore $H \leq Aut_G(\Gamma)$. However, there are many examples where $H \neq Aut_G(\Gamma)$!*

Example 4. *Let $n = 4$ and $\Omega = \{1, 2, 3, 4\}$.*
The group $H := \langle (12)(34), (14)(23) \rangle$ is transitive on Ω and one of its orbital graphs is $\Gamma_1 := \Gamma(H, \Omega, (1, 2))$. The automorphism group $Aut_G(\Gamma_1)$ is isomorphic to D_8, hence strictly larger than H.

2   Recognising Graphs as Orbital Graphs

In this section we let Γ denote a directed graph with vertex set Ω.

How can we decide whether or not there is some subgroup H of G such that Γ is an orbital graph for H?

Definition 3. *Suppose that Hypothesis 1 holds. We define three **types** of vertices in Γ. A vertex $\omega \in \Omega$ is said to be*

- *-- **isolated** if and only if it is not on an arc in Γ,*
- *-- a **starting vertex** if and only if some arc in Γ starts at ω, and*
- *-- an **end vertex** if and only if some arc in Γ ends at ω.*

Being a starting vertex is equivalent to being in α^H and being an end vertex is equivalent to being in β^H (see Lemma 2).

Lemma 5. *Suppose that Γ is an orbital graph. Then all connected components with at least two vertices are isomorphic as digraphs. Moreover, all vertices of the same type have the same incoming and outgoing valences.*

Proof. We recall that H induces digraph automorphisms on Γ, so in particular all connected components with at least two vertices are isomorphic. The remaining statements are clear for isolated vertices, and for the other two types we consider α as a representative for starting vertices and β as a representative for end vertices. We start with the ingoing valency of end vertices: Let $\omega \in \Omega$ be an end vertex. Then there exist $\gamma \in \alpha^H$ and $h \in H$ such that $(\alpha^h, \beta^h) = (\gamma, \omega)$. Now (i) and (iv) of Lemma 2 imply that the ingoing valency of β and of ω is the same, namely $|\alpha^{H_\beta}|$.

A similar argument shows that all starting vertices have the same outgoing valency as α.

The following example illustrates why the problem of recognising a digraph as an orbital graph is difficult:

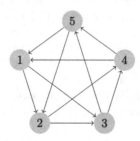

Example 6. *The digraph in the picture above, let us call it Γ, passes the obvious tests on symmetry as suggested by Lemma 5 above. It is connected and all vertices have ingoing valency 2 and outgoing valency 2. So it might well be an orbital graph!*

In fact it is not, and we will briefly argue why:

Assume that Γ is an orbital graph for a group H. Then $H \leq S_5$, as we see from the number of vertices. As $(1, 2)$ is an arc, but $(2, 1)$ is not, there is no element in H that interchanges 1 and 2. For the same reason there is no element in H that has a transposition in its cycle decomposition.

Next we consider the arcs $(1,2)$ and $(1,3)$. By definition of orbital graphs there is some $h \in H$ such that $1^h = 1$ and $2^h = 3$. As h stabilises 1 and cannot interchange 2 and 3, we conclude that h moves 3 to one of the numbers 4 or 5.

If $3^h = 4$, then the arc $(1,4)$ exists in Γ. However, it does not. If $3^h = 5$, then the arc $(1,5)$ exists in Γ. However, it does not.

Therefore Γ is not an orbital graph.

The idea in this example, namely restricting the structure of a potential group that has the given digraph as an orbital graph, can be refined and automated, at least partly. Typically, in our examples, we quickly find a lower bound for such a potential group by looking at the existing edges, and an upper bound by considering a point stabiliser in the automorphism group of the digraph. However, there is no general strategy, in particular because our current methods depend on whether or not the graph is self-paired.

As a first step towards understanding this problem better in theory, we prove a technical lemma characterising orbital graphs:

Lemma 7. *Let Π be a directed graph with vertex set Ω, with at least one arc and no loops. Let L be the group of digraph automorphisms of Π.*

Then Π is an orbital graph for some subgroup H of $Sym(\Omega)$ if and only if L acts arc-transitively on Π.

Proof. For the first direction let H be a subgroup of $Sym(\Omega)$ and let two $\alpha, \beta \in \Omega$ be distinct such that $\Pi = \Gamma(H, \Omega, (\alpha, \beta))$. We recall that H induces digraph automorphisms on its orbital graphs, so we may consider H to be a subgroup of L.

Let (γ, δ) be an arc in Π. By definition of orbital graphs there exists $h \in H \leq L$ such that $(\gamma, \delta) = (\alpha^h, \beta^h)$. This implies that L acts arc-transitively on Π.

Conversely suppose that L acts arc-transitively on Π and let H denote the subgroup of $Sym(\Omega)$ that consists of all the elements that induce digraph automorphisms on Π. By hypothesis there exists an arc (α, β) in Π. Let $\Gamma := \Gamma(H, \Omega, (\alpha, \beta))$ and let $\gamma, \delta \in \Omega$ be distinct. We prove that (γ, δ) is an arc in Γ if and only if it is an arc in Π.

If (γ, δ) is an arc in Γ, then there exists some $h \in H$ such that $(\gamma, \delta) = (\alpha^h, \beta^h)$. As h induces a digraph automorphism on Π, it follows that (α^h, β^h) is an arc of Π.

Conversely, if (γ, δ) is an arc of Π, then by the arc-transitive action there exists a digraph automorphism ϕ of Π that maps (α, β) to (γ, δ). As ϕ permutes the vertices in Ω, it is contained in H and therefore (γ, δ) is an arc of Γ. This finishes the proof.

Now we look at this lemma from a computational perspective. Given a digraph Γ, we would like to design an algorithm that checks whether or not there exists a subgroup of some symmetric group such that Γ is one of its orbital graphs.

A naive algorithm takes the number n of vertices of Γ and checks, for each subgroup H of S_n and each of its orbital graphs, respectively, whether or not this orbital graph is isomorphic to the given digraph Γ. This is very inefficient.

Using Lemma 7 we can draft a more efficient algorithm:

1. Check if the digraph has at least one arc and no loops.
2. Check if all the (incoming and outgoing) valances are the same.
3. Compute the automorphism group L of the digraph.
4. Check if L acts arc-transitively on the digraph.

An alternative for the last two steps is to check for each pair of arcs if there is a digraph automorphism that maps the first arc to the second one. As the number of arcs is substantially smaller than the number of subgroups of $\mathcal{S}_n$, this is a major improvement. This idea also allows for conclusions about the computational complexity of deciding whether or not a digraph is an orbital graph.

Following [4] we say a problem A is *polynomial-time many-one reducible* to a problem B if and only if there is a function f, computable in polynomial time, that transforms instances of A into instances of B such that an instance x fulfils the property stated in A if and only if $f(x)$ fulfils the property stated in B.

Lemma 8. *Deciding whether or not a given digraph Γ is an orbital graph has at most the same computational complexity as graph isomorphism. More precisely, there exists a polynomial-time many-one reduction from the problem of deciding whether or not a digraph is an orbital graph to the graph isomorphism problem.*

Proof. Let m be the number of arcs of Γ. If Γ has no arcs or at least one loop, then it is not an orbital graph. This is detectable in polynomial time. So suppose from now on that Γ has at least one arc and no loops. By Lemma 7 it is an orbital graph if and only if its group of digraph automorphisms acts arc-transitively. Hence let a_1, a_2 be two arcs of Γ. We only need to consider the $m \cdot (m - 1)$ possibilities where $a_1 \neq a_2$.

We construct two digraphs Π_1 and Π_2 such that each of them has exactly $m \cdot (m - 1)$ connected components. For each pair of different arcs a_1, a_2 of Γ a label $q = q(a_1, a_2)$ and two copies Δ_1, Δ_2 of the graph Γ are created: In Δ_1 the arc a_1 is labelled q and in Δ_2 the arc a_2 is labelled q. The connected components of the digraph Π_1 are exactly the graphs created as Δ_1 above, and the connected components of Π_2 are exactly the graphs created as Δ_2 above. This construction is possible in polynomial time.

We argue that Π_1 and Π_2 are graph isomorphic if and only if Γ is an orbital graph. This implies that the given construction is a polynomial-time many-one reduction, from the problem of deciding whether or not a digraph is an orbital graph, to the graph isomorphism problem.

First let Π_1 and Π_2 be isomorphic as digraphs. Let ϕ be a graph isomorphism from Π_1 to Π_2. Let a_1, a_2 be two arcs of Γ and let q be the corresponding label. As there is a unique arc b_1 in Π_1 with the label q (corresponding to a_1) and a unique arc b_2 in Π_2 with label q (corresponding to a_2), the isomorphism ϕ maps b_1 to b_2. Also, since ϕ is a digraph isomorphism, it maps the whole connected component Δ_1 of Π_1 containing the arc b_1 to the connected component Δ_2 of Π_2 containing the arc b_2. Now ϕ, restricted to Δ_1, is a digraph isomorphism from Δ_1 to Δ_2. By ignoring the labels, ϕ becomes a digraph isomorphism from one copy of Γ to another. Therefore there exists a digraph automorphism of Γ that maps a_1 to a_2. Then it follows that Γ is an orbital graph.

Now suppose, conversely, that Γ is an orbital graph. Then for each pair a_1, a_2 of distinct arcs in Γ there exists a digraph automorphism ϑ of Γ that maps a_1 to a_2. Let q be the label corresponding to a_1 in Δ_1 and a_2 in Δ_2, respectively, as described in the first paragraph. Then ϑ induces a digraph isomorphism from Δ_1 to Δ_2. As this construction is possible for all connected components of Π_1, there exists a digraph isomorphism from Π_1 to Π_2.

3   Recognising Groups from Their Orbital Graphs

In Example 4 we saw a group that could not be recovered from the orbital graph that we constructed. Let us briefly look at an example where it works and then phrase a definition that captures this phenomenon.

Example 9. *Let $n = 4$, $\Omega = \{1, 2, 3, 4\}$ and $H := \langle (1234) \rangle$.*

Then $\Gamma := \Gamma(H, \Omega, (1, 2))$ is just a directed cycle of length 4, and $Aut_G(\Gamma)$ is in fact H itself.

Definition 4. *The group H is **orbital graph recognisable (short: OGR)** if and only if*

$$H = \bigcap_{\alpha, \beta \in \Omega, \alpha \neq \beta} Aut_G(\Gamma(H, \Omega, (\alpha, \beta))).$$

*The group H is **strongly OGR** if and only if there are two distinct elements α, β of Ω such that $H = Aut_G(\Gamma(H, \Omega, (\alpha, \beta)))$ holds.*

*The group H is **absolutely OGR** if and only if for any pair α, β of distinct elements of Ω it is true that $H = Aut_G(\Gamma(H, \Omega, (\alpha, \beta)))$.*

This concept captures how many different orbital graphs we need in order to (possibly) recognise the group H. For example if a group is absolutely OGR, then for any randomly generated orbital graph Γ the group $Aut_G(\Gamma)$ is a perfect estimate for the group H. There are obvious implications in one direction, but we will give examples showing that the reverse implication is false in general.

Example 10

(a) *As in Example 4, let $n = 4$, $\Omega = \{1, 2, 3, 4\}$ and $H = \langle (12)(34), (14)(23) \rangle$. We have already seen $\Gamma_1 := \Gamma(H, \Omega, (1, 2))$. The other two possible orbital graphs are $\Gamma_2 := \Gamma(H, \Omega, (1, 3))$ and $\Gamma_3 := \Gamma(H, \Omega, (1, 4))$. These digraphs only differ by the number of arcs.*

Intersecting their automorphism groups induced by G yields the group H, so H is OGR, but in all three cases the automorphism group is isomorphic to D_8. Therefore H is not strongly OGR.

We would like to point out what this means: If in an algorithm we wish to replace the group H by a graph, because this is more efficient in computations, then one orbital graph is not sufficient.

(b) Next let $H := \langle(1234)\rangle$, let $\Gamma_1 := \Gamma(H, \Omega, (1,2))$ and $\Gamma_2 := \Gamma(H, \Omega, (1,3))$. Then $Aut_G(\Gamma_1) = H$, but $Aut_G(\Gamma_2)$ is isomorphic to D_8. Therefore H is strongly OGR and not absolutely OGR.

(c) Finally let $H := \langle(13), (12)(34)\rangle$ and let $\Gamma_1 := \Gamma(H, \Omega, (1,2))$ and $\Gamma_2 := \Gamma(H, \Omega, (1,3))$. These are the only orbital graphs of H and we see that both of them have H as its automorphism group in G. Hence H is absolutely OGR.

(d) All alternating group of degree at least 4 are not OGR because their orbital graphs are complete (see Lemma 11).

The last example does not occur accidentally, as the next lemma shows. It also illustrates that for small 2-transitive subgroups H of symmetric groups G of large degree, the group $Aut_G(\Gamma)$ is a very bad estimate for the group H.

Lemma 11. (a) If $n \geq 2$, then $G(= S_n)$ is absolutely OGR.
(b) If $H \leq G$, then H is 2-transitive if and only if $Aut_G(\Gamma) = G$. In particular, proper 2-transitive subgroups of G are never OGR.

Proof. If $H \leq G$ is 2-transitive and Γ is an orbital graph for H on Ω, then Γ is a complete graph because the 2-transitive action of H gives all possible arcs. In particular $Aut_G(\Gamma) = G$. This implies (a) and one direction of (b).

For the other direction of (b) suppose that $H \leq G$ and $Aut_G(\Gamma) = G$. Then every permutation of the vertices of Γ induces a digraph automorphism, and hence all possible arcs exist. Then Γ is a complete digraph and this implies that H acts 2-transitively on Ω. The last statement in (b) follows from this.

One last result before we list some open questions:

Lemma 12. Suppose that $H \leq G$ and that $\alpha \in \Omega$ is such that $H_\alpha = 1$. Then H is OGR.

Proof. Let $D := \bigcap_{\gamma, \delta \in \Omega, \gamma \neq \delta} Aut_G(\Gamma(H, \Omega, (\gamma, \delta)))$. Then $H \leq D$ because every element of H induces a graph automorphism on every orbital graph of H on Ω.

Assume for a contradiction that $D \neq H$ and let $g \in D \backslash H$. Let $\beta \in \Omega \backslash \{\alpha\}$ and set $\Gamma_1 := \Gamma(H, \Omega, (\alpha, \beta))$. Then g induces a graph automorphism on Γ_1 by choice, so (α^g, β^g) is an arc of Γ_1. By definition of orbital graphs let $h \in H \leq D$ be such that $(\alpha^h, \beta^h) = (\alpha^g, \beta^g)$ and set $d := g \cdot h^{-1}$. We note that $d \neq 1$ because $g \notin H$. Now $d \in D$ and it follows that

$$(\alpha^d, \beta^d) = ((\alpha^g)^{h^{-1}}, (\beta^g)^{h^{-1}}) = ((\alpha^h)^{h^{-1}}, (\beta^h)^{h^{-1}}) = (\alpha, \beta).$$

In particular d fixes the points α and β, so it lies in $G_\alpha \cap G_\beta$.

As $d \neq 1$, there exists some element of Ω that is not fixed by d. Let $\gamma \in \Omega$ be such that $\gamma^d \neq \gamma$ and let $\Gamma_2 := \Gamma(H, \Omega, (\alpha, \gamma))$.

We recall that $d \in D$, hence $d \in Aut_G(\Gamma_2)$ and therefore (α^d, γ^d) is an arc of Γ_2. Let $e \in H$ be such that $(\alpha^e, \gamma^e) = (\alpha^d, \gamma^d) = (\alpha, \gamma^d)$. Then $e \in H_\alpha = 1$.

Therefore $(\alpha^d, \gamma^d) = (\alpha^e, \gamma^e) = (\alpha, \gamma)$, which contradicts the fact that $\gamma^d \neq \gamma$. So it follows that $H = D$ and H is OGR.

We point out that H does not need to be transitive in this lemma!

Some final comments:

When using orbital graphs in partition backtrack search algorithms, performance is improved if only a small number of orbital graphs is built and if the ones that are built carry as much information as possible. Therefore, recognising groups that are OGR and using this theory in implementations of search algorithms makes the concept of orbital graph refiners even more useful.

Questions:

1. How can we characterise groups that are OGR (strongly OGR, absolutely OGR) in a way that can be checked quickly in an algorithm?
2. For a group that is strongly OGR: How can we quickly detect an orbital graph that is a witness for the strong OGR property?
3. For a group H that is OGR, but not strongly OGR: Can we efficiently find a small number of orbital graphs such that H is the intersection of their automorphism groups? (In Example 10(a) two orbital graphs already suffice.)
4. We know that 2-transitive groups do not provide useful orbital graphs for our pruning methods (see Lemma 27 in [3]). But primitive groups that are not 2-transitive are useful, by the same lemma, and it is known that all their orbital graphs are connected (Theorem 3.2A in [2]). How else does primitivity influence the structure of the orbital graphs? Here we need some specific experiments in order to phrase conjectures and develop more theory.
5. We already have results about groups that are absolutely OGR and act transitively, but not primitively. In general, how can we efficiently pick the most useful orbital graph for non-primitive groups? To begin with, what is the appropriate definition for a "most useful orbital graph"? Here we have some conjectures that we would like to test with experiments, implementing particular choices of orbital graphs into existing algorithms that use orbital graphs in partition backtrack.
6. We have seen (in Example 6) that basic checks for symmetry are not sufficient for deciding whether or not a digraph is an orbital graph. Currently we are working on further strategies that can be automated and that enable us to identify a graph as "not an orbital graph".

References

1. Cameron, P.: Permutation Groups. London Mathematical Society Student Texts. Cambridge University Press, Cambridge (1999)
2. Dixon, J.D., Mortimer, B.: Permutation Groups. Graduate Texts in Mathematics, vol. 163. Springer, New York (1996). https://doi.org/10.1007/978-1-4612-0731-3
3. Jefferson, Chr., Pfeiffer, M., Waldecker, R.: New refiners for permutation group search. J. Symb. Comput. (2018). https://doi.org/10.1016/j.jsc.2017.12.003
4. Köbler, J., Schöning, U., Torán, J.: The Graph Isomorphism Problem: Its Structural Complexity. Progress in Theoretical Computer Science. Birkhäuser, Boston (1993)
5. Theißen, H.: Eine Methode zur Normalisatorberechnung in Permutationsgruppen mit Anwendungen in der Konstruktion primitiver Gruppen, Ph.D. thesis, RWTH Aachen (1997)

Implementation of a Near-Optimal Complex Root Clustering Algorithm

Rémi Imbach[1](✉), Victor Y. Pan[2], and Chee Yap[3]

[1] TU Kaiserslautern, Kaiserslautern, Germany
imbach@mathematik.uni-kl.de
[2] City University of New York, New York, USA
victor.pan@lehman.cuny.edu
[3] Courant Institute of Mathematical Sciences,
New York University, New York, USA
yap@cs.nyu.edu
http://www.mathematik.uni-kl.de/en/agag/members/,
http://comet.lehman.cuny.edu/vpan/, http://www.cs.nyu.edu/yap/

Abstract. We describe Ccluster, a software for computing natural ε-clusters of complex roots in a given box of the complex plane. This algorithm from Becker et al. (2016) is near-optimal when applied to the benchmark problem of isolating all complex roots of an integer polynomial. It is one of the first implementations of a near-optimal algorithm for complex roots. We describe some low level techniques for speeding up the algorithm. Its performance is compared with the well-known MPSolve library and Maple.

1 Introduction

The problem of root finding for a polynomial $f(z)$ is a classical problem from antiquity, but remains the subject of active research to the present [6]. We consider a classic version of root finding:

> **Local root isolation problem:**
> **Given:** *a polynomial $f(z) \in \mathbb{C}[z]$, a box $B_0 \subseteq \mathbb{C}$, $\varepsilon > 0$.*
> **Output:** *a set $\{\Delta_1, \ldots, \Delta_k\}$ of pairwise-disjoint discs of radius $\leq \varepsilon$, each containing a unique root of $f(x)$ in B_0.*

It is local because we only look for roots in a locality, as specified by B_0. The local problem is useful in applications (especially in geometric computation) where we know where to look for the roots of interest. There are several variants of this

Rémi's work has received funding from the European Union's Horizon 2020 research and innovation programme under grant agreement No. 676541.
Victor's work is supported by NSF Grants # CCF-1116736 and # CCF-1563942 and by PSC CUNY Award 698130048.
Chee's work is supported by NSF Grants # CCF-1423228 and # CCF-1564132.

© Springer International Publishing AG, part of Springer Nature 2018
J. H. Davenport et al. (Eds.): ICMS 2018, LNCS 10931, pp. 235–244, 2018.
https://doi.org/10.1007/978-3-319-96418-8_28

problem: in the **global version**, we are not given B_0, signifying that we wish to find all the roots of f. The global version is easily reduced to the local one by specifying a B_0 that contains all roots of f. If we omit ε, it amounts to setting $\varepsilon = \infty$, representing the pure isolation problem.

Our main interest is a generalization of root isolation, to the lesser-studied problem of root clustering [8,10,12]. It is convenient to introduce two definitions: for any set $S \subseteq \mathbb{C}$, let $Z_f(S)$ denote the set of roots of f in S, and let $\#_f(S)$ count the total multiplicity of the roots in $Z_f(S)$. Typically, S is a disc or a box. For boxes and discs, we may write kS (for any $k > 0$) to denote the dilation of S by factor k, keeping the same center. The following problem was introduced in [16]:

Local root clustering problem:
Given: *a polynomial $f(z)$, a box $B_0 \subseteq \mathbb{C}$, $\varepsilon > 0$.*
Output: *a set of pairs $\{(\Delta_1, m_1), \ldots, (\Delta_k, m_k)\}$ where*

- Δ_i's are pairwise-disjoint discs of radius $\leq \varepsilon$,
- $m_i = \#_f(\Delta_i) = \#_f(3\Delta_i)$ for all i, and
- $Z_f(B_0) \subseteq \bigcup_{i=1}^{k} Z_f(\Delta_i)$.

This generalization of root isolation is necessary when we consider polynomials whose coefficients are non-algebraic (or when $f(z)$ is an analytic function, as in [16]). The requirement that $\#_f(\Delta_i) = \#_f(3\Delta_i)$ ensures that our output clusters are **natural** [1]; a polynomial of degree d has at most $2d - 1$ natural clusters (see [16, Lemma 1]). The local root clustering algorithm for analytic functions of [16] has termination proof, but no complexity analysis. By restricting $f(z)$ to a polynomial, Becker et al. [2] succeeded in giving an algorithm and also its complexity analysis based on the geometry of the roots. When applied to the **benchmark problem**, where $f(z)$ is an integer polynomial of degree d with L-bit coefficients, the algorithm can isolate all the roots of $f(z)$ with bit complexity $\widetilde{O}(d^2(L + d))$. Pan [13] calls such bounds **near-optimal** (at least when $L \geq d$). The clustering algorithm studied in this paper comes from [1], which in turn is based on [2]. Previously, the Pan-Schönhage algorithm has achieved near-optimal bounds with divide-and-conquer methods [13], but [1,2] was the first *subdivision* algorithm to achieve the near-optimal bound for complex roots. For real roots, Sagraloff-Mehlhorn [15] had earlier achieved near-optimal bound via subdivision.

Why the emphasis on "subdivision"? It is because such algorithms are implementable and quite practical (e.g., [14]). Thus the near-optimal real subdivision algorithm of [15] was implemented shortly after its discovery, and reported in [11] with excellent results. In contrast, all the asymptotically efficient root algorithms (not necessarily near-optimal) based on divide-and-conquer methods of the last 30 years have never been implemented; a proof-of-concept implementation of Schönhage's algorithm was reported in Gourdon's thesis [9]. Computer algebra systems mainly rely on algorithms with a priori guarantees of correctness. But in practice, algorithms without such guarantees are widely used. For complex root isolation, one of the most highly regarded multiprecision software is `MPSolve` [3]. The original algorithm in `MPSolve` was based on Erhlich-Aberth (EA) iteration; but since 2014, a "hybrid" algorithm [4] was introduced. It is based on the secular equation, and combines ideas from EA and `eigensolve` [7]. These algorithms are inherently global solvers (they must approximate *all* roots of a polynomial simultaneously). Another theoretical limitation is that the global convergence of these methods is not proven.

In this paper, we give a preliminary report about `Ccluster`, our[1] implementation of the root clustering algorithm from [1].

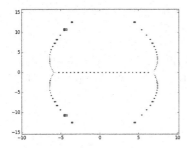 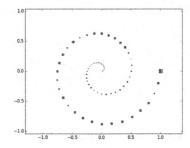

Fig. 1. Left: the connected components isolating all roots of the Bernoulli polynomial of degree 100. **Right:** the connected components isolating all roots of the Spiral polynomial of degree 64.

To illustrate the performance for the local versus global problem, consider the Bernoulli polynomials $\texttt{Bern}_d(z) := \sum_{k=0}^{d} \binom{d}{k} b_{d-k} z^k$ where b_i's are the Bernoulli numbers. Figure 1(Left) shows the graphical output of `Ccluster` for $\texttt{Bern}_{100}(z)$. Table 1 has four timings τ_X (for $X = \ell, g, u, s$) in seconds: τ_ℓ is the time for solving the local problem over a box $B_0 = [-1, 1]^2$; τ_g is the time for the global problem over the box $B_0 = [-150, 150]^2$ (which contains all the roots). The other two timings from `MPSolve` (τ_u for unisolve, τ_s for secsolve) will be explained later. For each instance, we also indicate the numbers of solutions (#Sols) and clusters (#Clus). When #Sols equals #Clus, we know the roots are isolated. Subdivision algorithms like ours naturally solve the local problem, but `MPSolve` can only solve the global problem. Table 1 shows that `MPSolve` remains unchallenged for the global problem. But in applications where locality can be exploited, local methods may win, as seen in the last two rows of the table. The corresponding time for `Maple`'s `fsolve` is also given; `fsolve` is not a guaranteed algorithm and may fail.

---

[1] Irina Voiculescu informed us that her student Dan-Andrei Gheorghe has independently implemented the same algorithm in a Masters Thesis Project (May 18, 2017) at Oxford University. Sewon Park and Martin Ziegler at KAIST, Korea, have implemented a modified version of Becker et al. (2016) for polynomials having only real roots being the eigenvalues of symmetric square matrices with real coefficients. See the technical report CS-TR-2018-415 at https://cs.kaist.ac.kr/research/techReport.

Table 1. Bernoulli polynomials with five timings: local (τ_ℓ), global (τ_ℓ), unisolve (τ_ℓ), secsolve (τ_ℓ) and Maple's `fsolve` (τ_f).

d	Ccluster local ($B_0 = [-1,1]^2$)			Ccluster global ($B_0 = [-150,150]^2$)			unisolve	secsolve	fsolve
	(#Sols:#Clus)	(depth:size)	τ_ℓ (s)	(#Sols:#Clus)	(depth:size)	τ_g (s)	τ_u (s)	τ_s (s)	τ_f (s)
64	(4:4)	(9:164)	0.12	(64:64)	(17:1948)	2.10	0.13	0.01	0.1
128	(4:4)	(9:164)	0.34	(128:128)	(16:3868)	9.90	0.55	0.05	6.84
191	(5:5)	(9:196)	0.69	(191:191)	(17:5436)	32.5	2.29	0.16	50.0
256	(4:4)	(9:164)	0.96	(256:256)	(17:7300)	60.6	3.80	0.37	>1000
383	(5:5)	(9:196)	2.06	(383:383)	(17:11188)	181	>1000	1.17	>1000
512	(4:4)	(9:164)	2.87	(512:512)	(16:14972)	456	>1000	3.63	>1000
767	(5:5)	(9:196)	6.09	(767:767)	(17:22332)	1413	>1000	10.38	>1000

Overview of Paper. In Sect. 2, we describe the experimental setup for Ccluster. Sections 3–5 describe some techniques for speeding up the basic algorithm. We conclude with Sect. 6.

2 Implementation and Experiments

The main implementation of Ccluster is in C language. We have an interface for Julia[2]. We based our big number computation on the arb[3] library. The arb library implements ball arithmetic for real numbers, complex numbers and polynomials with complex coefficients. Each arithmetic operation is carried out with error bounds.

Test Suite. We consider 7 families of polynomials, classic ones as well as some new ones constructed to have interesting clustering or multiple root structure.

(F1) The Bernoulli polynomial $\text{Bern}_d(z)$ of degree d is described in Sect. 1.
(F2) The Mignotte polynomial $\text{Mign}_d(z;a) := z^d - 2(2^a z - 1)^2$ for a positive integer a, has two roots whose separation is near the theoretical minimum separation bound.
(F3) The Wilkinson polynomials $\text{Wilk}_d(z) := \prod_{k=1}^{d}(z-k)$.
(F4) The Spiral Polynomial $\text{Spir}_d(z) := \prod_{k=1}^{d}\left(z - \frac{k}{d}e^{4ki\pi/n}\right)$. See Fig. 1(Right) for $\text{Spir}_{64}(z)$.
(F5) Wilkinson Multiple: $\text{WilkMul}_{(D)}(z) := \prod_{k=1}^{D}(z-k)^k$. $\text{WilkMul}_{(D)}(z)$ has degree $d = D(D+1)/2$ where the root $z = k$ has multiplicity k (for $k = 1,\ldots,D$).
(F6) Mignotte Cluster: $\text{MignClu}_d(z;a,k) := x^d - 2(2^a z - 1)^k(2^a z + 1)^k$. This polynomial has degree d (assuming $d \geq 2k$) and has a cluster of k roots near 2^{-a} and a cluster of k roots near -2^{-a}.

---

[2] https://julialang.org/. Download our code in https://github.com/rimbach/Ccluster.
[3] http://arblib.org/. Download our code in https://github.com/rimbach/Ccluster.jl.

(F7) Nested Cluster: $\mathtt{NestClu}_{(D)}(z)$ has degree $d = 3^D$ and is defined by induction on D: $\mathtt{NestClu}_{(1)}(z) := z^3 - 1$ with roots $\omega, \omega^2, \omega^3 = 1$ where $\omega = e^{2\pi i/3}$. Inductively, if the roots of $\mathtt{NestClu}_{(D)}(z)$ are $\{r_j : j = 1, \ldots, 3^D\}$, then we define $\mathtt{NestClu}_{(D+1)}(z) := \prod_{j=1}^{3^D} \left(z - r_j - \frac{\omega}{16^D}\right)\left(z - r_j - \frac{\omega^2}{16^D}\right)\left(z - r_j - \frac{1}{16^D}\right)$ See Fig. 2 for the natural ε-clusters of $\mathtt{NestClu}_{(3)}(z)$.

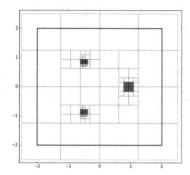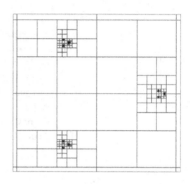

Fig. 2. Left: 3 clusters of $\mathtt{NestClu}_{(3)}$ found with $\varepsilon = 1$. **Right:** Zoomed view of 9 clusters of $\mathtt{NestClu}_{(3)}$ found with $\varepsilon = \frac{1}{10}$. **Note:** The initial box is in thick lines; the thin lines show the subdivisions tree.

Timing. Running times are sequential times on a Intel(R) Core(TM) i3 CPU 530 @ 2.93 GHz machine with linux. $\mathtt{Ccluster}$ implements the algorithm described in [1] with differences coming from the improvements described in Sects. 3–5 below. Unless explicitly specified, the value of ε for $\mathtt{Ccluster}$ is set to 2^{-53}; roughly speaking, it falls back to asking for 15 guaranteed decimal digits.

MPSolve. For external comparison, we use MPSolve. It was shown to be superior to major software such as Maple or Mathematica [3]. There are two root solvers in MPSolve: the original unisolve [3] which is based on the Ehrlich-Aberth iteration and the new hybrid algorithm called secsolve [4]. These are called with the commands $\mathtt{mpsolve\ -au\ -Gi\ -o\gamma\ -j1}$ and $\mathtt{mpsolve\ -as\ -Gi\ -o\gamma\ -j1}$ (respectively). $\mathtt{-Gi}$ means that MPSolve tries to find for each root a unique complex disc containing it, such that Newton iteration is guaranteed to converge quadratically toward the root starting from the center of the disc. $\mathtt{-o\gamma}$ means that $10^{-\gamma}$ is used as an escape bound, *i.e.*, the algorithm stops when the complex disc containing the root has radius less that $10^{-\gamma}$, regardless of whether it is isolating or not. Unless explicitly specified, we set $\gamma = 16$. $\mathtt{-j1}$ means that the process is not parallelized. Although MPSolve does not do general local search, it has an option to search only within the unit disc. This option does not seem to lead to much improvement.

$\widetilde{T}_k^G(\Delta, k)$ ◁ $f(z)$ is implicit argument
> Output: $r \in \{-1, 0 \dots, k\}$
>> ASSERT: if $r \geq 0$, then $\#_f(\Delta) = r$
>> $L \leftarrow 53$, $d \leftarrow \deg(f)$, $N \leftarrow 4 + \lceil \log_2(1 + \log_2(d)) \rceil$, $i \leftarrow 0$
>> $\tilde{f} \leftarrow$ getApproximation(f, L)
>> $\tilde{f} \leftarrow$ TaylorShift($\tilde{f}$, Δ)
>> While $i \leq N$
>>> Let $\tilde{f}$ be the i-th Graeffe iteration of $\tilde{f}$
>>> $r \leftarrow 0$
>>> While $r \leq k$
>>>> $j \leftarrow$ IntCompare($|\tilde{f}|_r$, $\sum_{k \neq r} |\tilde{f}|_k, 2^{-L}$) [a]
>>>> While $j =$ **unresolved**
>>>>> $L \leftarrow 2L$
>>>>> $\tilde{f} \leftarrow$ getApproximation(f, L)
>>>>> $\tilde{f} \leftarrow$ TaylorShift($\tilde{f}, \Delta$)
>>>>> Let $\tilde{f}$ be i-th Graeffe iteration of $\tilde{f}$
>>>>> $j \leftarrow$ IntCompare($|\tilde{f}|_r$, $\sum_{k \neq r} |\tilde{f}|_k, 2^{-L}$)
>>>> If $j =$ **true** then **Return** r
>>>> $r \leftarrow r + 1$
>>> $i \leftarrow i + 1$
>> **Return** -1

---

[a] IntCompare($\tilde{a}, \tilde{b}, 2^{-L}$) compares L-bit approximations of real numbers a and b. It returns **true** (resp. **false**) only if $a > b$ (resp. $a < b$). It returns **unresolved** when L is too small to conclude.

Fig. 3. $\widetilde{T}_k^G(\Delta, k)$. $|\tilde{f}|_i$ is the absolute value of the coefficient of the monomial of degree i of $\tilde{f}$, for $0 \leq i \leq d$.

3   Improved Soft Pellet Test

The key predicate in [1] is a form of Pellet test denoted $\widetilde{T}_k^G(\Delta, k)$ (with implicit $f(z)$). This is modified in Fig. 3 by adding an outer while-loop to control the number of Graeffe-Dandelin iterations. We try to get a definite decision (i.e., anything other than a **unresolved**) from the soft comparison for the current Graeffe iteration. This is done by increasing the precision L for approximating the coefficients of $\tilde{f}$ in the innermost while-loop. Thus we have two versions of our algorithm: (V1) uses the original $\widetilde{T}_k^G(\Delta, k)$ in [1], and (V2) uses the modified form in Fig. 3. Let τV1 and τV2 be timings for the 2 versions. Table 2 shows the time τV1 (in seconds) and the ratio τV1/τV2. We see that (V2) achieves a consistent 2.3 to 3-fold speed up.

In (V2), as in [1], we use $\widetilde{T}_0^G(\Delta)$ (defined as $\widetilde{T}_0^G(\Delta, 0)$) to prove that a box B has no root. We propose a new version (V3) that uses $\widetilde{T}_*^G(\Delta)$ (defined as $\widetilde{T}_k^G(\Delta, d)$, where d is the degree of f) instead of $\widetilde{T}_0^G(\Delta)$ to achieve this goal: instead of just showing that B has no root, it upper bounds $\#_f(B)$. Although counter-intuitive, this yields a substantial improvement because it led to fewer

Table 2. Solving within the initial box $[-50, 50]^2$ with $\varepsilon = 2^{-53}$ with versions (V1), (V2) and (V3) of `Ccluster`. n1: number of discarding tests. n2: number of discarding tests returning -1 (inconclusive). n3: total number of Graeffe iterations. τV1 (resp. τV2, τV3): sequential time for V1 (resp. V2, V3) in seconds.

	V1		V2		V3	
	(n1, n2, n3)	τV1	(n1, n2, n3)	τV1/τV2	(n1, n2, n3)	τV1/τV3
$\mathtt{Bern}_{64}(z)$	(2308, 686, 20223)	19.6	(2308, 686, 6028)	2.84	(2308, 8, 2291)	7.06
$\mathtt{Mign}_{64}(z; 14)$	(2060, 622, 18018)	17.3	(2060, 622, 5326)	3.03	(2060, 20, 2080)	7.68
$\mathtt{Wilk}_{64}(z)$	(2148, 674, 18053)	23.6	(2148, 674, 5692)	2.74	(2148, 0, 2140)	7.23
$\mathtt{Spir}_{64}(z)$	(2512, 728, 22176)	22.2	(2512, 728, 6596)	2.39	(2512, 15, 2670)	4.46
$\mathtt{WilkMul}_{(11)}(z)$	(724, 202, 6174)	9.69	(724, 202, 2684)	2.30	(724, 18, 2065)	3.37
$\mathtt{MignClu}_{64}(z; 14, 3)$	(2092, 618, 18515)	20.0	(2092, 618, 5600)	3.00	(2092, 12, 2481)	6.57
$\mathtt{NestClu}_{(4)}(z)$	(3532, 1001, 30961)	90.2	(3532, 1001, 9654)	3.09	(3532, 24, 4588)	6.81

Graeffe iterations overall. The timing for (V3) is τV3, but we display only the ratio τV1/τV3 in the last column of Table 2. This ratio shows that (V3) enjoys a 3.3-7.7 fold speedup. Comparing $n3$ for (V2) and (V3) explains this speedup.

4   Filtering

A technique for speeding up the evaluation of predicates is the idea of filters (e.g., [5]). The various Pellet tests can be viewed as a box predicate C that maps a box $B \subseteq \mathbb{C}$ to a value[4] in $\{\mathbf{true}, \mathbf{false}\}$. If C^- is another box predicate with property that $C^-(B) = \mathbf{false}$ implies $C(B) = \mathbf{false}$, we call C^- a **falsehood filter**. If C^- is efficient relatively to C, and "efficacious" (informally, $C(B) = \mathbf{false}$ is likely to yield $C^-(B) = \mathbf{false}$), then it is useful to first compute $C^-(B)$. If $C^-(B) = \mathbf{false}$, we do not need to compute $C(B)$. The predicate C_0 used in `Ccluster` is defined as follows: $C_0(B)$ is **true** if $\widetilde{T}^G_*(\Delta_B)$ returns 0 (then B contains no root of f) and is **false** if $\widetilde{T}^G_*(\Delta_B)$ returns -1 or $k > 0$ (then B may contain some roots of f). We next present the falsehood filter $C_0^-(B)$ for C_0.

Let f_Δ denote the Taylor shift of f in Δ, $f_\Delta^{[i]}$ its i-th Graeffe iterate, $(f_\Delta^{[i]})_j$ the j-th coefficient of $f_\Delta^{[i]}$, and $|f_\Delta^{[i]}|_j$ the absolute value of the j-th coefficient. Let d be the degree of f. The assertion below is a direct consequence of the classical test of Pellet (see [2, p. 12]) and justify the correctness of our filters:

(A) if $|f_\Delta^{[N]}|_0 \le |f_\Delta^{[N]}|_1 + |f_\Delta^{[N]}|_d$ then $\widetilde{T}^G_*(\Delta)$ returns -1 or $k > 0$.

Our C_0^- filter computes $|f_\Delta^{[N]}|_0$, $|f_\Delta^{[N]}|_1$ and $|f_\Delta^{[N]}|_d$ and checks hypothesis of (A) using `IntCompare`. $|f_\Delta^{[N]}|_0$ and $|f_\Delta^{[N]}|_d$ can respectively be computed as $(|f_\Delta|_0)^{2^N}$ and $(|f_\Delta|_d)^{2^N}$. $|f_\Delta^{[N]}|_1$ can be computed with the following well known formula:

$$(f_\Delta^{[i+1]})_k = (-1)^k((f_\Delta^{[i]})_k)^2 + 2\sum_{j=0}^{k-1}(-1)^j(f_\Delta^{[i]})_j(f_\Delta^{[i]})_{2k-j} \tag{1}$$

---

[4] We treat two-valued predicates for simplicity; the discussion could be extended to predicates (like $\widetilde{T}^G_*$) which returns a finite set of values.

Obtaining $|f_\Delta^{[N]}|_1$ with Eq. (1) requires to know $2^{N-1}+1$ coefficients of $f_\Delta^{[1]}$, $2^{N-2}+1$ coefficients of $f_\Delta^{[2]}, \ldots$, and finally $3 = 2^1 + 1$ coefficients of $f_\Delta^{[N-1]}$. In particular, it requires to compute entirely the iterations $f_\Delta^{[i]}$ such that $2^{N-i} \le d$, and it is possible to do it more efficiently that with Eq. (1) (for instance with the formula given in definition 2 of [2]).

Table 3. Solving within the initial box $[-50, 50]^2$ with $\varepsilon = 2^{-53}$ with versions (V3), (V4) of `Ccluster`. n3: number of Graeffe iterations. τV3 and τV4: sequential time in seconds.

		V3		V4	
		n3	τV3	n3	τV3/τV4
$\texttt{Bern}_d(z)$	$d = 64$	2291	2.61	2084	1.08
	$d = 128$	4496	14.5	3983	1.13
	$d = 256$	8847	94.5	7714	1.19
	$d = 512$	15983	620	11664	1.42
	$d = 767$	19804	1832	13863	1.53
$\texttt{Mign}_d(z; a)$	$(d, a) = (64, 14)$	2080	2.41	1808	1.22
	$(d, a) = (128, 14)$	3899	12.1	3257	1.21
	$(d, a) = (256, 14)$	7605	88.3	6339	1.33
	$(d, a) = (512, 14)$	15227	674	10405	1.57
$\texttt{Wilk}_d(z)$	$d = 64$	2140	3.27	1958	1.05
	$d = 128$	2240	10.0	1942	1.09
	$d = 256$	2414	36.6	2108	1.21
	$d = 512$	2557	129	1841	1.43
$\texttt{Spir}_d(z)$	$d = 64$	2670	4.43	2364	1.08
	$d = 128$	5090	28.8	4405	1.07
	$d = 256$	9746	182	8529	1.10
	$d = 512$	19159	1340	14786	1.19
$\texttt{WilkMul}_{(D)}(z)$	$(D, d) = (11, 66)$	2065	2.87	1818	1.14
	$(D, d) = (12, 78)$	2313	3.95	2053	1.12
	$(D, d) = (13, 91)$	2649	5.89	2336	1.18
	$(D, d) = (14, 105)$	2892	8.56	2537	1.29
$\texttt{MignClu}_d(z; a, k)$	$(d, a, k) = (64, 14, 3)$	2481	2.94	2145	1.13
	$(d, a, k) = (128, 14, 3)$	4166	14.4	3555	1.16
	$(d, a, k) = (256, 14, 3)$	7658	86.0	6523	1.27
	$(d, a, k) = (512, 14, 3)$	15044	650	10472	1.63
$\texttt{NestClu}_{(D)}(z)$	$(D, d) = (4, 27)$	1628	0.77	1459	1.07
	$(D, d) = (5, 81)$	4588	13.2	4085	1.12
	$(D, d) = (6, 243)$	13056	358	11824	1.26

Our C_0^- filter takes as input a precision L, the Taylor shift f_Δ of the L bit approximation of f and its i-th Graeffe iteration $f_\Delta^{[i]}$ such that $2^{N-i} \le \frac{d}{4}$ and $2^{N-(i+1)} > \frac{d}{4}$. It computes $|f_\Delta^{[N]}|_0$, $|f_\Delta^{[N]}|_d$ and the $2^{N-j} + 1$ first coefficients of $f_\Delta^{[j]}$ for $i < j \le N$ with Eq. (1). Then it checks the hypothesis of (A) using IntCompare, and returns **false** if it is verified, and **true** otherwise. In practice, it is implemented within the procedure implementing $\widetilde{T}_*^G(\Delta_B)$.

Incorporating C_0^- into Version (V3), we obtain (V4) and the speed up can be seen in Table 3. Filtering with C_0^- becomes more effective as degree grows and this is because one has $2^{N-i} \le \frac{d}{4}$ for smaller i (recall that $N = 4 + \lceil \log_2(1 + \log_2(d)) \rceil$).

5   Escape Bound

The ε parameter is usually understood as the precision desired for roots. But we can also view it as an escape bound for multiple roots as follows: we do not refine a disc that contains a simple root, even if its radius is $\ge \varepsilon$. But for clusters of size greater than one, we only stop when the radius is $< \varepsilon$. MPSolve has a similar option. This variant of (V4) is denoted (V4'). We see from Table 4 that (V4') gives a modest improvement (up to 25% speedup) over (V4) when $-\log \varepsilon = 53$. This improvement generally grows with $-\log \varepsilon$ (but $\mathtt{WilkMul}_{(11)}(z)$ shows no difference).

Table 4. Solving within the box $[-50, 50]^2$ with versions (V4) and (V4') of Ccluster with three values of ε. $\tau 53$ (resp. $\tau 530$, $\tau 5300$): sequential time for (V4) and (V4') in seconds.

ε:	(V4)			(V4')		
	2^{-53}	2^{-530}	2^{-5300}	2^{-53}	2^{-530}	2^{-5300}
	$\tau 53$ (s)	$\tau 530/\tau 53$	$\tau 5300/\tau 53$	$\tau 53$ (s)	$\tau 530/\tau 53$	$\tau 5300/\tau 53$
$\mathtt{Bern}_{64}(z)$	2.42	1.26	4.22	1.99	0.94	0.94
$\mathtt{Mign}_{64}(z; 14)$	1.97	1.63	4.56	1.61	1.45	1.38
$\mathtt{Wilk}_{64}(z)$	3.22	1.10	2.16	2.91	0.96	1.01
$\mathtt{Spir}_{64}(z)$	4.09	1.33	5.25	3.05	0.95	0.95
$\mathtt{WilkMul}_{(11)}(z)$	2.51	1.12	2.03	2.50	1.13	1.98
$\mathtt{MignClu}_{64}(z; 14, 3)$	2.60	1.89	4.15	2.20	1.70	1.80
$\mathtt{NestClu}_4(z)$	11.9	1.08	2.67	10.4	1.00	0.99

6   Conclusion

Implementing subdivision algorithms is relatively easy but achieving state-of-art performance requires much optimization and low-level development. This paper explores several such techniques. We do well compared to fsolve in Maple, but the performance of MPSolve is superior to the global version of Ccluster. But Ccluster can still shine when looking for local roots or when ε is large.

References

1. Becker, R., Sagraloff, M., Sharma, V., Xu, J., Yap, C.: Complexity analysis of root clustering for a complex polynomial. In: Proceedings of the ACM on International Symposium on Symbolic and Algebraic Computation, pp. 71–78. ACM (2016)
2. Becker, R., Sagraloff, M., Sharma, V., Yap, C.: A near-optimal subdivision algorithm for complex root isolation based on the pellet test and newton iteration. J. Symb. Comput. **86**, 51–96 (2018)
3. Bini, D.A., Fiorentino, G.: Design, analysis, and implementation of a multiprecision polynomial rootfinder. Numer. Algorithms **23**(2–3), 127–173 (2000)
4. Bini, D.A., Robol, L.: Solving secular and polynomial equations: a multiprecision algorithm. J. Comput. Appl. Math. **272**, 276–292 (2014)
5. Brönnimann, H., Burnikel, C., Pion, S.: Interval arithmetic yields efficient dynamic filters for computational geometry. Discrete Appl. Math. **109**(1–2), 25–47 (2001)
6. Emiris, I.Z., Pan, V.Y., Tsigaridas, E.P.: Algebraic algorithms. In: Computing Handbook, Third Edition: Computer Science and Software Engineering, pp. 10:1–10:30. Chapman and Hall/CRC (2014)
7. Fortune, S.: An iterated eigenvalue algorithm for approximating roots of univariate polynomials. J. Symb. Comput. **33**(5), 627–646 (2002)
8. Giusti, M., Lecerf, G., Salvy, B., Yakoubsohn, J.-C.: On location and approximation of clusters of zeros of analytic functions. Found. Comput. Math. **5**(3), 257–311 (2005)
9. Gourdon, X.: Combinatoire, Algorithmique et Géométrie des Polynomes. Ph.D. thesis, École Polytechnique (1996)
10. Hribernig, V., Stetter, H.J.: Detection and validation of clusters of polynomial zeros. J. Symb. Comput. **24**(6), 667–681 (1997)
11. Kobel, A., Rouillier, F., Sagraloff, M.: Computing real roots of real polynomials... and now for real! In: Proceedings of the ACM on International Symposium on Symbolic and Algebraic Computation, pp. 303–310. ACM (2016)
12. Niu, X.-M., Sakurai, T., Sugiura, H.: A verified method for bounding clusters of zeros of analytic functions. J. Comput. Appl. Math. **199**(2), 263–270 (2007)
13. Pan, V.Y.: Univariate polynomials: nearly optimal algorithms for numerical factorization and root-finding. J. Symb. Comput. **33**(5), 701–733 (2002)
14. Rouillier, F., Zimmermann, P.: Efficient isolation of polynomial's real roots. J. Comput. Appl. Math. **162**(1), 33–50 (2004)
15. Sagraloff, M., Mehlhorn, K.: Computing real roots of real polynomials. J. Symb. Comput. **73**, 46–86 (2016)
16. Yap, C., Sagraloff, M., Sharma, V.: Analytic root clustering: a complete algorithm using soft zero tests. In: Bonizzoni, P., Brattka, V., Löwe, B. (eds.) CiE 2013. LNCS, vol. 7921, pp. 434–444. Springer, Heidelberg (2013). https://doi.org/10.1007/978-3-642-39053-1_51

Towards a Unified Ordering for Superposition-Based Automated Reasoning

Jan Jakubův[1(✉)] and Cezary Kaliszyk[2]

[1] Czech Technical University in Prague, Prague, Czech Republic
jakubuv@gmail.com
[2] University of Innsbruck, Innsbruck, Austria
cezary.kaliszyk@uibk.ac.at

Abstract. We propose an extension of the automated theorem prover E by the weighted path ordering. Weighted path ordering is theoretically stronger than all the orderings used in E-prover, however its parametrization is more involved than those normally used in automated reasoning. In particular, it depends on a term algebra. We discuss how the parameters for the ordering can be proposed automatically for particular theorem proving problem strategies. We integrate the ordering in E-prover and perform an evaluation on the standard theorem proving benchmarks. The ordering is complementary to the ones used in E prover so far.

Keywords: Automated reasoning · Term orderings
Weighted path order · Superposition calculus

1 Introduction

In the last two decades the superposition calculus has become one of the main foundations of automated theorem provers for first-order logic. Indeed the systems regularly winning the yearly CADE ATP Systems Competition, such as E [7] and Vampire [2] are based on the superposition calculus. Also for the problems not previously solved by humans, superposition calculus based Prover9 has been most useful so far [5].

The use of powerful and efficient orderings is one of the major advantages of the superposition calculus for classical first-order theorem proving. Orderings allow provers to avoid redundant clauses, namely clauses which only differ in the order of literals, as well as permit orienting equations and therefore rewriting the clauses only in one direction. The three predominantly used orderings in automated theorem proving are LPO, KBO, and RPO. In fact for the former two optimized implementations are known [3,4].

Supported by the ERC Consolidator grant no. 649043 *AI4REASON*, ERC Starting grant no. 714034 *SMART*, and FWF grant P26201.

© Springer International Publishing AG, part of Springer Nature 2018
J. H. Davenport et al. (Eds.): ICMS 2018, LNCS 10931, pp. 245–254, 2018.
https://doi.org/10.1007/978-3-319-96418-8_29

However, term rewriting research has shown that there exist more powerful orderings, for example the *weighted path order* (WPO) [10] is one of the strongest known orderings. With carefully selected parameters is can subsume most known orderings including LPO, KBO, and RPO [11]. There are however two reasons, why such stronger orderings have not been tried for automated reasoning so far. First, they often rely on complicated parameters. For example WPO relies on an algebra on terms as an argument. Second, the efficiency of KBO, LPO, or even RPO has been optimized for the most common cases, whereas the more advanced orderings have been stated in a general manner, without optimizing their efficiency.

In this paper we attempt to overcome both of these obstacles and propose an efficient way to implement WPO as part of an automated reasoning system. We also propose parameters that allow WPO to function efficiently within a state-of-the-art automated theorem prover and help with actual theorem proving problems. After discussing the preliminaries on term orderings in Sect. 2 and on their use in the superposition calculus in Sect. 3, the particular contributions of this paper are:

- We propose algebras that can be used efficiently for first-order theorem proving (Sect. 4),
- We present an optimized pseudocode for WPO in terms of typical ATP structures and implement an extension of E-prover that supports WPO (Sect. 4).
- We evaluate WPO against existing orderings in E-prover on parts of the TPTP library, the proofs stemming from the AIM conjecture [9], and on the CoqHammer proofs [1] in Sect. 5.

2 Term Orderings and Rewriting

We work in first-order logic (FOL). A *signature* Σ is a collection of *symbols* with *arities*. The set of first-order *variables* is denoted $\mathcal{V}$, and $\mathcal{T}_\Sigma$ stands for the *terms* over signature Σ and variables $\mathcal{V}$. A *literal* is an atomic formula or its negation, and a *clause* is a disjunction of literals. In ATPs, clauses are used to describe both the input problem, and the knowledge inferred during the search. On occasion, *unit equality* clauses of the form $s = t$ are inferred. Such equalities can be used to simplify other clauses using $s \to t$ or $t \to s$ as a *rewriting rule*.

Rewriting systems, described by finite sets of rewriting rules, are often used inside ATPs to keep a set of clauses in *normal forms*. A crucial property for ATPs is the *termination* of every rewriting chain on any term. The termination of system $\mathcal{R}$ can be shown using a well-founded *term ordering* $>_T$ on terms $\mathcal{T}$, that orients every rule $(s \to t) \in \mathcal{R}$, meaning $s >_T t$. Terminating rewriting systems are called *reduction orders*. See [6,11] for details.

Reduction orders are successfully used in many state-of-the-art ATPs. Common orders [6,11] are lexicographic path order (LPO) and Knuth-Bendix order (KBO). LPO extends a *precedence* $>_\Sigma$ on symbols to a reduction order on $\mathcal{T}_\Sigma$ by a variety of subterm comparisons. KBO is generated by a precedence and symbol *weights*. Terms in KBO are first compared by weights and the subterm comparisons are necessary only if the weights differ. WPO further abstracts the idea of symbol weight comparisons to comparisons in *algebras* as follows.

Definition 1. *An* algebra $\mathcal{A}$ *over* Σ *consists of a well-ordered carrier set and of an interpretation* $f_{\mathcal{A}} : \mathbb{N}^n \rightarrow \mathbb{N}$ *for every n-ary function symbol f from* Σ. *An algebra* $\mathcal{A}$ *is* weakly monotone *iff* $a \geq b$ *implies* $f(\ldots, a, \ldots) \geq f(\ldots, b, \ldots)$, *and* weakly simple *iff* $f(\ldots, a, \ldots) \geq a$ *for every* $f \in \Sigma$.

In this work, we consider the carrier set always to be $\mathbb{N}$ with the standard order on $\mathbb{N}$. Given a variable assignment $\sigma : \mathcal{V} \rightarrow \mathbb{N}$, we can structurally interpret every term $t \in \mathcal{T}_{\Sigma}$ using interpretations from algebra $\mathcal{A}$ as the number $\sigma_{\mathcal{A}}(t) \in \mathbb{N}$, formally as follows.

$$\sigma_{\mathcal{A}}(x) = \sigma(x) \qquad \sigma_{\mathcal{A}}(f(s_1, \ldots, s_n)) = f_{\mathcal{A}}(\sigma_{\mathcal{A}}(s_1), \ldots, \sigma_{\mathcal{A}}(s_n)))$$

Thus the algebra $\mathcal{A}$ induces the following ordering $>_{\mathcal{A}}$ on terms: $s >_{\mathcal{A}} t$ iff $\sigma_{\mathcal{A}}(s) > \sigma_{\mathcal{A}}(t)$ for every variable assignment σ. Similarly, we write $s \geq_{\mathcal{A}} t$ iff $\sigma_{\mathcal{A}}(s) \geq \sigma_{\mathcal{A}}(t)$ for every σ. The following defines WPO induced by $\mathcal{A}$.

Definition 2 (WPO [11]). *Given a precedence* $>_{\Sigma}$ *and an algebra* $\mathcal{A}$ *over* Σ, *the* weighted path order $>_{\mathrm{wpo}}$ *on* $\mathcal{T}_{\Sigma}$ *is defined as follows:* $s = f(s_1, \ldots, s_n) >_{\mathrm{wpo}}$ *t iff (1)* $s >_{\mathcal{A}} t$, *or (2)* $s \geq_{\mathcal{A}} t$ *and one of the following holds:*

2a. $\exists i \in \{1, \ldots, n\}.\ s_i \geq_{\mathrm{wpo}} t$, *or*
2b. $t = g(t_1, \ldots, t_m), \forall j \in \{1, \ldots, m\}.\ s >_{\mathrm{wpo}} t_j$ *and either*
  (i) $f >_{\Sigma} g$, *or*
  (ii) $f = g$ *and* $(s_1, \ldots, s_n) >_{\mathrm{wpo}}^{lex} (t_1, \ldots, t_n)$.

Only terms comparable in $\mathcal{A}$ are comparable in $>_{\mathrm{wpo}}$. Strict order $s >_{\mathcal{A}} t$ alone implies $s >_{\mathrm{wpo}} t$. Otherwise $s \geq_{\mathcal{A}}$ must hold and various subterm conditions are checked. In (2a), $\geq_{\mathrm{wpo}}$ is the reflexive closure of $>_{\mathrm{wpo}}$, while $>_{\mathcal{A}}$ and $\geq_{\mathcal{A}}$ are separately defined orders induced by $\mathcal{A}$. In (2b/ii) the lexicographical extension $>_{\mathrm{wpo}}^{lex}$ of $>_{\mathrm{wpo}}$ to n-tuples is used when the compared terms have the same head symbol.

If the WPO algebra $\mathcal{A}$ is weakly monotone and weakly simple, then $>_{\mathrm{wpo}}$ is a reduction order [11, Theorem 13]. With different algebras, WPO is known to behave like LPO [11, Theorem 19], or like KBO [11, Theorem 16], or to subsume both [11, Theorem 20]. Instantiations of WPO with different algebras are discussed in Sect. 4.

3   Orderings in Superposition Calculus

Saturation based automated theorem provers, like E prover [7], attempt to prove a first-order goal conjecture G in a theory T, that is, $T \vdash G$. First, theory axioms with the negated conjecture $T \cup \{\neg G\}$ are translated to a logically equivalent set of clauses. Then, a saturation process is initiated, which selects an unprocessed clause C and computes all possible inferences of C with all the previously processed clauses. Clause C is then marked as processed and another unprocessed clause is selected. This process continues until an empty clause (contradiction) is derived, or there are no more unprocessed clauses (the set of processed clauses becomes *saturated*), or the prover runs out of resources.

The saturation process uses term orderings for various purposes depending on the selected inference rules. The classical *resolution* rule allows to infer the clause $(C_1 \vee C_2)\sigma$ from clauses $(L_1 \vee C_1)$ and $(\neg L_2 \vee C_2)$ provided L_1 and L_2 are unifiable with the unifier σ. The *ordered resolution* restricts the classical resolution rule to literals maximal in each clause (w.r.t. a fixed term ordering $>_T$). In *paramodulation*, inferred unit equality clauses of the form $s = t$, which can be oriented using the ordering (either $s >_T t$ or $t >_T s$), can be used as rewriting rules ($s \rightarrow t$ or $t \rightarrow s$, respectively). The processed clauses are then kept in their normal form with respect to the inferred rewriting rules (called *demodulators*). All these extensions restrict the number of possible inferences preserving completeness (that is, they do not prevent the inference of the empty clause). Clearly, the more terms are comparable, the more inferences are restricted, which leads to a more effective search space reduction.

E prover implements LPO and KBO. The desired term ordering can be selected using a command-line option. E implements approximately ten signature-independent methods to generate the precedence on the symbols. In this work, we shall consider the following.

(arity/iarity) Symbols are sorted by arity or reverse arity. Symbols with higher arity are larger/smaller.

(freq/ifreq) Symbols are sorted by the frequency of their occurrence in the input problem. Frequently occurring symbols are larger/smaller. In the case of the same frequency, symbols are sorted by arity.

(ufirst) Same as **arity** but unary symbols are smaller. In the case of the same arity, symbols are sorted by frequency.

(ufreq) Same as **ifreq** but unary symbols are always smaller.

KBO is additionally parametrized by a weight function (w, w_0). E implements several ways of generating weights for a given problem. We shall consider the following. All of these set the variable weight w_0 to 1 and only differ in w.

(const) The weights of all the symbols are set to the constant 1.

(arity/iarity) The weight of an n-ary function symbol is set to $n + 1$ (respectively to $m - n + 1$, where m is the largest symbol arity).

(prec/iprec) Given a symbol precedence $<$, the weight of symbol f is the number of symbols smaller/larger than f increased by 1.

(fcount/ifcount) The weight of symbol f is the number of occurrences of f in the input problem (respectively m minus the number of occurrences, where m is the frequency of the most occurring symbol).

(frank/ifrank) Sort all function symbols by frequency of occurrence (which induces a total quasi-ordering). The weight of a symbol is the rank of it's equivalence class, with less frequent symbols getting lower/higher weights.

Additionally, E allows user-defined weights for all constant symbols, which override the weight assigned by the above weight generation schemes. Finally, E allows both a specific user-defined precedence and specific symbol weights. We do not, however, consider these specific settings as they depends on a signature. Our implementation of WPO in E Prover is described in the next Sect. 4.

4   Implementation of WPO in E Prover

This section describes our implementation of WPO in E Prover. We introduce two specific algebras from the literature [11]. Both algebras are weakly monotone and simple, and hence instantiate WPO to a reduction order. We discuss the implementation of algebra comparisons and provide several coefficient generation schemes for WPO. We conclude by a brief description of our main WPO comparison method. First we introduce Sum-algebras which sum the arguments with a positive multiplier.

Definition 3 (Sum-algebra). *A Sum-algebra $\mathcal{A}$ over Σ induced by (w, c) is an algebra over Σ where an n-ary function symbol f is interpreted as*

$$f_{\mathcal{A}}(a_1, \ldots, a_n) = w(f) + \sum_{i=1}^{n} c(f, i) * a_i$$

where $w(f) > 0$ is the weight of f and $c(f, i) > 0$ is the coefficient of the i-th argument of f (called subterm coefficient*).*

Both the weights and subterm coefficients can be zero under certain additional conditions [11, Theorems 5 and 13]. All E weight generation schemes used in this work produce non-zero weights, and hence we consider only positive coefficients, mainly to simplify the implementation. Experimenting with non-zero values is left as future work. The carrier set of $\mathcal{A}$ can be instantiated by a subset of $\mathbb{N}$ ($\{n \in \mathbb{N} : n \geq w_0\}$ for some $w_0 \in \mathbb{N}$). Note, that a restriction of such a Sum-algebra to $w_0 > 0$ and $c(f, i) = 1$ is equivalent to KBO [11, Theorem 16].

Given a Sum-algebra $\mathcal{A}$ over Σ, every term $s \in \mathcal{T}_{\Sigma}$ can be interpreted in $\mathcal{A}$ as an expression of the grammar "$E ::= \mathbb{N} \mid \mathcal{V} \mid (E + E) \mid (\mathbb{N} * E)$". This expression contains variables $\text{vars}(s) = \{x_1, \ldots, x_n\}$. The expression can transformed to the equivalent expression $s_{\mathcal{A}}$ of the following form, which we say *interprets* s in $\mathcal{A}$ (for appropriate $c_i \in \mathbb{N}$).

$$s_{\mathcal{A}}(x_1, \ldots, x_n) = c_0 + c_1 * x_1 + \cdots + c_n * x_n$$

Since the definitions of $>_{\mathcal{A}}$ and $\geq_{\mathcal{A}}$ involve an infinite number of variable assignments, it is necessary to provide an efficient algorithm to check the algebra comparisons in WPO. The following lemma helps us to achieve that. Note that, we take the liberty of reordering variables so that shared variables come first.

Lemma 1. *Given* Sum*-algebra* $\mathcal{A}$ *over* Σ *and terms* $s, t \in \mathcal{T}_\Sigma$, *let* $\mathrm{vars}(t) \subseteq \mathrm{vars}(s) = \{x_1, \ldots, x_n\}$ *and let* $\mathrm{vars}(t) = \{x_1, \ldots, x_m\}$ *for some* $m \leq n$. *Let*

$$s_\mathcal{A}(x_1, \ldots, x_n) = c_0 + c_1 * x_1 + \cdots + c_n * x_n$$
$$t_\mathcal{A}(x_1, \ldots, x_m) = d_0 + d_1 * x_1 + \cdots + d_m * x_m$$

be the interpretations of s *and* t *in* $\mathcal{A}$. *Then the following holds.*

$$s >_\mathcal{A} t \quad \text{iff} \quad \forall i \in \{1, \ldots, m\}.\ c_i \geq d_i \ \text{and}\ c_0 > d_0$$
$$s \geq_\mathcal{A} t \quad \text{iff} \quad \forall i \in \{0, \ldots, m\}.\ c_i \geq d_i$$

Clearly, $s >_\mathcal{A} t$ (and also $s \geq_\mathcal{A} t$) implies $\mathrm{vars}(t) \subseteq \mathrm{vars}(s)$, hence the variable requirement is not a limitation. WPO requires algebras to be weakly monotone to generate a reduction order. Similarly, the notion of *strictly monotone* algebras can be defined (using strict comparisons instead of weak ones). Sum-algebras are strictly (and hence weakly) monotone. We next define the Max-algebras, which use max instead of addition, making them weakly monotone.

Definition 4 (Max-algebra). *A* Max*-algebra* $\mathcal{A}$ *over* Σ *induced by* (w, c) *is an algebra over* Σ *where an* n*-ary function symbol* f *is interpreted as*

$$f_\mathcal{A}(a_1, \ldots, a_n) = \max\left(w(f)\ ,\ \max_{i=1}^{n}(c(f, i) + a_i)\right)$$

where $w(f) > 0$ *is the weight of* f *and* $c(f, i) > 0$ *is the coefficient of the* i*-th argument of* f *(called* subterm penalty*).*

Again, zero weights and penalties are allowed under certain conditions, which we omit in this presentation. For example, setting all the weights and penalty coefficients to zeros makes WPO behave like LPO [11, Theorem 19]. Similarly to Sum-algebras, given a Max-algebra $\mathcal{A}$ over Σ, every term $s \in \mathcal{T}_\Sigma$ with $\mathrm{vars}(s) = \{x_1, \ldots, x_n\}$ can be interpreted by an expression $s_\mathcal{A}$ of the following form, which is said to interpret s in $\mathcal{A}$.

$$s_\mathcal{A}(x_1, \ldots, x_n) = \max(c_0, x_1 + c_1, \ldots, x_n + c_n)$$

The following allows efficiently comparing terms in Max-algebras.

Lemma 2. *Let* Max*-algebra* $\mathcal{A}$ *over* Σ *and terms* $s, t \in \mathcal{T}_\Sigma$ *be given. Let* $\mathrm{vars}(t) \subseteq \mathrm{vars}(s) = \{x_1, \ldots, x_n\}$ *and* $\mathrm{vars}(t) = \{x_1, \ldots, x_m\}$ *for some* $m \leq n$. *Let*

$$s_\mathcal{A}(x_1, \ldots, x_n) = \max(c_0, x_1 + c_1, \ldots, x_n + c_n)$$
$$t_\mathcal{A}(x_1, \ldots, x_m) = \max(d_0, x_1 + d_1, \ldots, x_m + d_m)$$

interpret s *and* t *in* $\mathcal{A}$. *Let* $c_{\max} = \max(c_0, \ldots, c_n)$ *and* $d_{\max} = \max(d_0, \ldots, d_m)$. *Then the following holds.*

$$s >_\mathcal{A} t \quad \text{iff}\ c_{\max} > d_{\max}\ \text{and}\ \forall i \in \{1, \ldots, m\}.\ c_i > d_i$$
$$s \geq_\mathcal{A} t \quad \text{iff}\ c_{\max} \geq d_{\max}\ \text{and}\ \forall i \in \{1, \ldots, m\}.\ c_i \geq d_i$$

Note that in $s >_{\mathcal{A}} t$, as opposed to Lemma 1, we require all the coefficients to be strictly greater. Otherwise $\max(x+2, y+1)$ would be strictly greater than $\max(x+1, y+1)$. We do not compare the constant coefficients c_0 and d_0, because, for example, $\max(1, x+3)$ is always greater than $\max(2, x+2)$ even though the constant coefficients are not. The proof of Lemma 2 follows from the observation that c_0 can be substituted by $c_{\max}$ without affecting the value of $s_{\mathcal{A}}$.

Inspired by precedence/weight generation schemes in E, we have implemented the following subterm coefficient generation schemes. These schemes generate coefficients $c(f, i)$ to be used both in $\mathcal{S}um$ and $\mathcal{M}ax$-algebras.

(constant) All coefficients are set to 1.
(arity) For an n-ary function symbol f we set $c(f, i) = n$.
(firstmax) For all f, the first coefficient $c(f, 1)$ is set 2 while the others to 1.
(firstmin) For all f, the first coefficient $c(f, 1)$ is set 1 while the others to 2.
(asc/desc) Set up ascending/descending coefficients. For an n-ary function symbol f we set $c(f, i) = i$ (respectively $c(f, i) = n - i + 1$).

To implement a new term ordering $>_{\mathcal{T}}$ in E, a term comparison method is required. The method takes two terms s and t as input and returns whether $s <_{\mathcal{T}} t$, or $s >_{\mathcal{T}} t$, or $s = t$, or the terms are incomparable. We have implemented the WPO comparison methods for $\mathcal{S}um$ and $\mathcal{M}ax$ algebras. Our implementation mostly follows Definition 2. At first we check strict algebra comparisons $>_{\mathcal{A}}$. To do that, we compute coefficients $\overline{c_i}$ and $\overline{d_i}$ from Lemma 1 or 2 by a traversal of s and t. If the coefficients are the same, we clearly have both $s \geq_{\mathcal{A}} t$ and $t \geq_{\mathcal{A}} s$. If $s >_{\mathcal{A}} t$, we return $s >_{\text{wpo}} t$ (and *vice versa*). For terms incomparable with $>_{\mathcal{A}}$, we proceed with the weak comparison $\geq_{\mathcal{A}}$. If they are weakly comparable, we proceed with the subterm checks.

5 Experimental Evaluation

We evaluate our experimental implementation[1] of WPO in E Prover on four complementary benchmarks with 200 problems each. Benchmark problems are from two TPTP [8] categories (LAT and REL), from the *Abelian Inner Mappings* project (AIM) [9], and from CoqHammer [1]. We evaluate all instances of LPO, KBO, and WPO induced by the generation schemes described above, in order to estimate the value of WPO for E. This gives us a collection of about 800 benchmark problems which we believe are reasonably orthogonal to allow us to perform an objective evaluation. As we evaluate around 1400 different ordering instances on all of the benchmark problems, it is important to limit the number of problems so that the evaluation can be done in a reasonable time.[2] The limit of 1000 processed clauses, instead of time limit, is used for an evaluation independent on implementation effectiveness. We use a single good-performing E strategy with the different term orders.

---

[1] https://github.com/ai4reason/eprover/tree/WPO.
[2] The evaluation took around 2 days employing 32 cores of Intel(R) Xeon(R) CPU E5-2698 v3 @ 2.30 GHz with 128 GB memory in total.

We have 6 instances of LPO, 108 instances KBO, and 1296 of WPO. The results for each benchmark are in Table 1. For each ordering, the column "*by*" shows the least number of instances necessary to solve the number in the column *solved*. Number of problems solved by E's automated term order selection is shown in column "*Auto*". The "*all*" columns show combined performance. Table 2 shows the best-performing instance for every order type, measuring number problems *solved* and the number of problems solved additionally to *Auto* mode (column "*E+*"). The parameters of the instances select the generation schemes for precedence, weights, algebra, and coefficients.

Table 1. Total number of problems solved by all LPO, KBO, and WPO instances.

	Auto	LPO		KBO		WPO		*all*	
	solved	*solved*	*by*	*solved*	*by*	*solved*	*by*	*solved*	*by*
TPTP/LAT	*27*	28	2	30	2	**34**	5	*36*	*5*
TPTP/REL	*49*	68	3	59	2	**75**	2	*77*	*3*
AIM	*35*	44	2	38	2	**54**	4	*54*	*4*
COQ	*22*	26	3	**27**	2	**27**	2	*27*	*2*

Table 2. Best instances of LPO, KBO, and WPO for each benchmark.

TPTP/REL	solved	E+
WPO(freq,prec,$\mathcal{S}$um,desc)	**63**	14
LPO(arity)	59	12
KBO(iarity,iarity)	57	8
E (Auto)	49	0

TPTP/LAT	solved	E+
KBO(iarity,iprec)	**29**	3
WPO(arity,iprec,$\mathcal{S}$um,const)	28	1
LPO(arity)	27	0
E (Auto)	27	0
WPO(ifreq,prec,$\mathcal{M}$ax,desc)	24	3

AIM	solved	E+
WPO(freq,fcount,$\mathcal{S}$um,desc)	**41**	5
LPO(arity)	41	4
KBO(freq,ifrank,c1)	37	1
E (Auto)	35	0

COQ	solved	E+
WPO(arity,fcount,$\mathcal{S}$um,desc)	**26**	4
KBO(arity,fcount)	25	3
LPO(ufreq)	24	4
E (Auto)	22	0

WPO helped to solve more problems for each benchmark. It also solved problems unsolved by *Auto*. Furthermore, the strongest WPO is usually equal or better than the strongest version of LPO and KBO. LPO(arity) is often the best of LPOs. As for WPO, $\mathcal{S}$um often performs better than $\mathcal{M}$ax overall but $\mathcal{M}$ax can solve unique problems. The algebra coefficients generated by **desc** often perform best.

As stated above, we used a limit on processed clauses rather than on runtime, in order to abstract from implementation details. In order to asses the effectiveness of our implementation, we have additionally evaluated the best performing ordering instances from Table 2 on the benchmark problems with runtime limit of 5 s. For each benchmark category (AIM, COQ, etc.) we have computed the average runtime on the problems solved by all the instances. The results vary on different categories but LPO is usually the fastest and KBO is in average from 10% to 40% slower. The speed of WPO varies, but in average it is from 40% to 140% slower than LPO. However, for example on TPTP/REL, our implementation of WPO is in average faster than both LPO and KPO. We conclude that our implementation can be definitely made more effective, but even in the current state, it can provide a valuable gain.

6   Conclusion

In this paper we proposed efficient implementations of algebras that allow integrating more powerful orderings in the superposition calculus. The resulting E strategies are more precise, resulting in complementary proofs on the various corpora and have a potential to benefit E prover and superposition calculus ATPs in general.

As future work, we would like to experiment with further algebras, additional coefficient settings, and with zero weights, as this might further reduce the number of derived clauses. We would also like to further optimize the efficiency of the algebra comparisons, as well as the computation of the ordering itself, as well as perform more thorough evaluations.

References

1. Czajka, Ł., Kaliszyk, C.: Hammer for Coq: automation for dependent type theory. J. Autom. Reason. **61**, 423–453 (2018)
2. Kovács, L., Voronkov, A.: First-order theorem proving and VAMPIRE. In: Sharygina, N., Veith, H. (eds.) CAV 2013. LNCS, vol. 8044, pp. 1–35. Springer, Heidelberg (2013). https://doi.org/10.1007/978-3-642-39799-8_1
3. Löchner, B.: Things to know when implementing KBO. J. Autom. Reason. **36**(4), 289–310 (2006)
4. Löchner, B.: Things to know when implementing LPO. Int. J. Artif. Intell. Tools **15**(1), 53–80 (2006)
5. McCune, W.: Solution of the Robbins problem. J. Autom. Reason. **19**(3), 263–276 (1997)
6. Middeldorp, A.: Term rewriting lecture notes. In: 9th International School on Rewriting (ISR 2017) (2017)
7. Schulz, S.: System description: E 1.8. In: McMillan, K., Middeldorp, A., Voronkov, A. (eds.) LPAR 2013. LNCS, vol. 8312, pp. 735–743. Springer, Heidelberg (2013). https://doi.org/10.1007/978-3-642-45221-5_49
8. Sutcliffe, G.: The TPTP problem library and associated infrastructure. from CNF to TH0, TPTP v6.4.0. J. Automa. Reason. **59**(4), 483–502 (2017)

9. Sutcliffe, G.: The 8th IJCAR automated theorem proving system competition - CASC-J8. AI Commun. **29**(5), 607–619 (2016)
10. Yamada, A., Kusakari, K., Sakabe, T.: Unifying the Knuth-Bendix, recursive path and polynomial orders. In: PPDP, pp. 181–192. ACM (2013)
11. Yamada, A., Kusakari, K., Sakabe, T.: A unified ordering for termination proving. Sci. Comput. Program. **111**, 110–134 (2015)

Numerical Integration in Arbitrary-Precision Ball Arithmetic

Fredrik Johansson[✉]

LFANT – INRIA – IMB, Bordeaux, France
fredrik.johansson@gmail.com
http://fredrikj.net

Abstract. We present an implementation of arbitrary-precision numerical integration with rigorous error bounds in the Arb library. Rapid convergence is ensured for piecewise complex analytic integrals by use of the Petras algorithm, which combines adaptive bisection with adaptive Gaussian quadrature where error bounds are determined via complex magnitudes without evaluating derivatives. The code is general, easy to use, and efficient, often outperforming existing non-rigorous software.

Keywords: Numerical integration · Interval arithmetic
Special functions

1 Introduction

Many users can attest that there is a non-negligible chance of getting an incorrect answer when asking a numerical package or computer algebra system for an approximation of a definite integral $\int_a^b f(x)dx$, as rapid variation, narrow peaks, non-smooth points, cancellation or ill-conditioned numerical evaluation of f are prone to break widely used heuristic numerical integration methods.

One remedy is to compute rigorous error bounds using interval arithmetic. However, little work has been done to date on efficient arbitrary-precision implementations. Here, we present a new implementation of rigorous numerical integration in Arb, a C library for ball arithmetic[1] [4]. The integration code is easy to use directly in C, or can be wrapped from high-level languages. For example, an interface in Sage [11] exists (thanks to Marc Mezzarobba and Vincent Delecroix), which we demonstrate by computing $\int_0^8 \sin(x + e^x) \, dx$:

```
sage: C = ComplexBallField(333)    # 333-bit precision
sage: C.integral(lambda x, d: sin(x+exp(x)), 0, 8)
[0.3474001726572478078795121591198931246574562548661801838854927136167648
2139887853205296851043466 +/- 5.97e-96]
```

---

[1] Arb (http://arblib.org) is open source (GNU LGPL) software. For documentation and example code related to this paper, see http://arblib.org/acb_calc.html.

© Springer International Publishing AG, part of Springer Nature 2018
J. H. Davenport et al. (Eds.): ICMS 2018, LNCS 10931, pp. 255–263, 2018.
https://doi.org/10.1007/978-3-319-96418-8_30

We obtain nearly 100 digits with a rigorous error bound in 0.04 s (0.02 s when using C directly). This relatively difficult test integral ($f(x)$ changes sign 950 times) was introduced by Rump [9] who observed that the quad function in Matlab took over a second only to return the erroneous 0.2511 (Rump's interval package Intlab computes 7 digits in about one second; see also [7]).

2   Algorithm and Implementation

We consider integration of a function $f : \mathbb{C} \to \mathbb{C}$ on a segment $[a, b]$, $a, b \in \mathbb{C}$. We represent real numbers as mid-rad intervals (balls) $[m \pm r]$ and complex numbers as rectangles $[m_1 \pm r_1] + [m_2 \pm r_2]i$ (which we also refer to as balls with slight abuse of terminology). True complex balls $B(m_1 + m_2 i, r)$ would sometimes provide slightly better bounds, but rectangles are usually more convenient.

The user supplies the integrand f as a pointer to a C function func implementing its evaluation (we refer to the documentation for the detailed API). In effect, func gets called with the argument z and an extra flag d. If $d = 0$, func is to evaluate $f(z)$ without any assumptions about regularity. If $d = 1$, func is to evaluate $f(z)$ and also check that f is analytic on z, returning a non-finite ball (e.g. NaN) otherwise. For meromorphic f, the user can ignore d since $f(z)$ automatically blows up at poles, but d needs to be handled for functions with branch cuts like $\sqrt{z}$ and $\log(z)$ (here by checking whether z overlaps $(-\infty, 0]$).

We use the Petras algorithm [8], which combines bisection with Gaussian quadrature of variable degree n. Error bounds for Gaussian quadrature use complex magnitudes. If f is analytic with $|f| \le M$ on an ellipse E with foci ± 1 and semiaxes X, Y, then $|\int_{-1}^{1} f(x)dx - \sum_{k=1}^{n} w_k f(x_k)| \le M\rho^{-2n} C_\rho$, $\rho = X + Y$, where e.g. $C_\rho < 50$ if $\rho > 1.1$. The tradeoff is that a larger E increases M, with $M = \infty$ if E hits a singularity of f, but also improves convergence as $n \to \infty$. Of course, the computed bound for M will not just depend on the function f but also on the stability of its evaluation in ball arithmetic if E is large.

Degree adaptivity ensures near-optimal complexity ($O(p)$ evaluations of f) for analytic f at high precision p, while space adaptivity (bisection) helps if there are singularities near $[a, b]$ or if the ball enclosures are not optimal. For piecewise analytic f with discontinuities on $[a, b]$ the complexity is typically $O(p^2)$, i.e. a bit worse but still polynomial in p. Degree or space adaptivity used alone would give $2^{O(p)}$ complexity or fail to converge for common types of integrals.

Our version of the integration algorithm can be described as follows:

- Initialize sum $S \leftarrow 0$, subinterval work queue $Q \leftarrow [(a, b)]$.
- While $Q = [(a_1, b_1), \ldots, (a_N, b_N)]$ is not empty:
  1. Pop $(\alpha, \beta) = (a_N, b_N)$ from Q.
  2. Compute the direct box enclosure $I = (\beta - \alpha)f([\alpha, \beta])$ (evaluating f on $z = [\alpha, \beta]$ with $d = 0$). If I meets the tolerance goal, if α, β overlap, or if evaluation limits have been exceeded, set $S \leftarrow S + I$ and go to 1.
  3. Try to find an ellipse E with foci (α, β) and an $n \le n_{\max}$ such that f is analytic on E (evaluating $f(E)$ with $d = 1$) and the error bound for n-point Gaussian quadrature determined via $|f(E)|$ meets the tolerance goal. If successful, compute this integral J, set $S \leftarrow S + J$ and go to 1.

4. Interval bisection: let $m = \frac{\alpha+\beta}{2}$ and extend Q with (α, m), (m, β).

Compared to Petras [8], there are minor differences. Our ρ is not fixed; we try several sizes of E in step 3 to reduce n. The handling of tolerances is slightly different. We also compute quadrature nodes (w_k, x_k) at runtime, without using pre-made tables. A key point is that generating nodes for high-precision Gaussian quadrature used to be considered too costly [1], but the recent work [6] solves this problem.[2] With default settings, computing nodes takes a few milliseconds for 100-digit precision and a few seconds for 1000 digits.[3] Nodes are automatically cached, so this cost is amortized for repeated integrations at the same or lower precision (possible n are restricted to a sparse sequence $\approx 2^{k/2}$ to avoid computing nodes for many nearby n). As an optional tuning parameter, the user can change the allowed range of n which defaults to $n_{max} = 0.5p + 60$.

2.1 Tolerances and Evaluation Limits

Besides the working precision p, the user specifies absolute and relative tolerances ε_{abs} and ε_{rel}. In effect, the algorithm attempts to achieve an error of $\max(\varepsilon_{abs}, V\varepsilon_{rel})$ where V is the magnitude of the integral. Reasonable values (used as defaults by the Sage wrapper) are $\varepsilon_{abs} = \varepsilon_{rel} = 2^{-p}$. Other values can be useful, e.g. if low accuracy is sufficient but a higher p must be used for numerical reasons. One might also set $\varepsilon_{abs} = 0$ to use relative tolerance only, though for efficiency, it is better to supply $\varepsilon_{abs} \approx V\varepsilon_{rel}$ if an estimate for V is known when $V \not\approx 1$. This Sage code shows computation of $\int_0^1 e^{-1000+x} \sin(10x)dx$:

```
sage: C = ComplexBallField(64); f = lambda x, _: exp(-1000+x)*sin(10*x)
sage: C.integral(f, 0, 1)
[+/- 4.09e-434]                                    # time 0.013 ms
sage: C.integral(f, 0, 1, abs_tol=0)
[1.574528586972758e-435 +/- 7.36e-451]            # time 1.1 ms
sage: C.integral(f, 0, 1, abs_tol=exp(-1000)/2^64)
[1.574528586972758e-435 +/- 7.27e-451]            # time 0.38 ms
```

Conversely, for a large integrand:

```
sage: f = lambda x, _: exp(1000+x)*sin(10*x)
sage: C.integral(f, 0, 1)
[6.11102916709322e+433 +/- 1.98e+418]             # time 1.1 ms
sage: C.integral(f, 0, 1, abs_tol=exp(1000)/2^64)
[6.11102916709322e+433 +/- 1.95e+418]             # time 0.39 ms
```

In reality, ε_{abs} and ε_{rel} are only *guidelines* and the algorithm does not strictly achieve the goal $\max(\varepsilon_{abs}, V\varepsilon_{rel})$. Indeed, due to the fixed working precision and

---

[2] Clenshaw-Curtis or double exponential quadrature could be used instead of Gaussian quadrature, but typically require more points for equivalent accuracy. We could also use Taylor series, but this makes supplying f more cumbersome for the user, and computing $f, f' \dots, f^{(n)}$ tends to be more costly than n evaluations of f.

[3] In benchmark results, we omit the first-time nodes precomputation overhead.

possibly inexact parameters, the goal cannot generally be achieved. It is implied that the user will work with some guard bits and if needed adjust $(p, \varepsilon_{abs}, \varepsilon_{rel})$ based on the reliable *a posteriori* information in the output ball radius.

Use of ε_{rel} further depends circularly on V (V is essentially what we are trying to compute!), so the algorithm must guess V. A too large guess means loss of accuracy and a too small guess means unnecessary work. Our approach is to start with the tolerance ε_{abs} and continuously update $\varepsilon_{abs} \leftarrow \max(\varepsilon_{abs}, I_a\varepsilon_{rel})$, $\varepsilon_{abs} \leftarrow \max(\varepsilon_{abs}, J_a\varepsilon_{rel})$ where $|I| = [I_a, I_b]$ and $|J| = [J_a, J_b]$ are intervals computed in steps 2 and 3; I_a and J_a will then be lower bounds for V (we err on the side of preserving accuracy), modulo global cancellation in the integral. As noted above, the user should exploit knowledge about V if possible since I_a and J_a may be pessimistic. More clever globally adaptive strategies are possible, but we settled for this simple approach in the present version.

To abort gracefully when convergence is too slow, *evaluation limits* include a bound on the number of calls to f (default $1000p + p^2$) and a bound on the size N of the work queue Q (default $2p$). By default, Q acts as a stack and step 4 puts the new subinterval with the larger error at the top; optionally, Q can be switched to a global priority queue, which may improve results if convergence is so slow that evaluation limits are exceeded. This has the downside of sometimes requiring N nearly as large as the number of calls to f (e.g. $N \sim p^2$), whereas we always have $N \lesssim p$ with the stack. A more clever algorithm might use a top-level priority queue down to some depth before switching to a stack locally.

3   Benchmarks

We test various integrals with precision between about 10 and 1000 digits. Timings were obtained on an Intel Core i5-4300U CPU. We compare Arb to the heuristic arbitrary-precision integration routines `intnum` in Pari/GP [10] and `quad` in mpmath [5]. Both use double exponential quadrature without adaptive subdivision, although `quad` is degree-adaptive. Further comparisons with other numerical and interval packages (as well as alternative methods in Pari/GP and mpmath[4]) would be useful, but out of scope for this brief overview.

We do not show the outputs, but note that in all cases, Arb computes correct balls with radius a small multiple of 2^{-p}. On some test cases, mpmath with default settings silently returns an inaccurate answer due to exceeding its limit on the quadrature degree, but it provides an optional mechanism to catch this. We increased the degree limit to let mpmath run to full accuracy in all cases, and have written (!) after a timing where the default is insufficient. Pari/GP is not adaptive and silently returns inaccurate answers without providing a catch mechanism or a way to increase the degree. It does provide an option to split the interval non-adaptively into 2^t parts, but it is up to the user to find a correct t. We have done so where necessary, which is also marked in the timings.

---

[4] For example, mpmath provides `quadgl` for Gaussian quadrature, which is 2–3 times faster on some examples, but its precomputations are prohibitive at high precision.

3.1   Integrals Without Singularities on the Path

Table 1 shows examples with smooth f on $[a, b]$. For meromorphic f, the number of subintervals largely depends on the location of the poles and does not change with p. The "spike integral" I_1 (Fig. 1) is a well known pathological example [2, 3]; all ordinary numerical integrators we have tested (Mathematica, GSL, SciPy, etc.) give inaccurate results with default settings. This integrand has poles near the real axis, forcing many local bisections. It is a piece of cake for the Petras algorithm, but Pari/GP and mpmath converge slowly unless the user manually splits the path at the peaks. I_2 could be sped up 40% in Arb by using $\cos^2(x) = \frac{1}{2}(1 + \cos(2x))$ for wide x to bound the denominator more tightly.

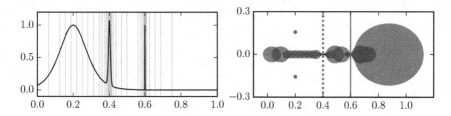

Fig. 1. Left: $f(x) = \operatorname{sech}^2(10(x - 0.2)) + \operatorname{sech}^4(100(x - 0.4)) + \operatorname{sech}^6(1000(x - 0.6))$, with subintervals used by Arb. Right: complex ellipses used. The dots show the poles of f.

For entire functions (I_4, I_5, I_6), the efficiency improves with larger p since arbitrarily large bounding ellipses can be used. I_5 is Rump's example again, and I_6 (whose graph has two sharp "bends") was provided by Silviu-Ioan Filip.

The code is seen to work well with special functions. In I_3 we integrate the Lambert W function, where we need to check for the branch cut on $(-\infty, -1/e]$ in the evaluation. I_7 also illustrates integration on a complex path.

Overall, Arb is faster than Pari/GP and mpmath, despite the fact that rigorous error bounds create extra work. The speedup is in part explained by faster arithmetic and transcendental functions in Arb and lower overhead due to using C, as well as the advantage of Gaussian quadrature over the double exponential method for smooth integrands. However, if these differences are accounted for, we can still conclude that the Petras algorithm in ball arithmetic holds up extremely well for high-precision integration, on top of giving rigorous bounds.

3.2   Endpoint Singularities and Infinite Intervals

The methods in Pari/GP and mpmath are designed to support typical integrals with infinite intervals or endpoint singularities, which often arise in applications. Arb requires finite a, b and a bounded f to return a finite result, but the user may provide a manual truncation (say $\int_0^\infty f(x)dx \approx \int_\varepsilon^N f(x)dx$) to work around this restriction. Tail bounds must then be added based on symbolic

Table 1. Integrals without singularities on $[a, b]$. Timings (Pari/GP, mpmath, Arb) are in seconds. Sub = number of terminal subintervals (requiring no further bisection) used by Arb, Eval = total number of integrand evaluations used by Arb.

p	Pari/GP	mpmath	Arb	Sub	Eval	Pari/GP	mpmath	Arb	Sub	Eval
	$I_0 = \int_0^1 1/(1+x^2)dx$					$I_1 = \int_0^1 \sum_{k=1}^3 \mathrm{sech}^{2k}(10^k(x-0.2k))\,dx$				
32	0.00039	0.00057	0.000025	2	32	0.54 $(t=8)$	1.9 (!)	0.0030	49	795
64	0.00039	0.0011	0.000036	2	52	0.54 $(t=8)$	5.0 (!)	0.0051	49	1299
333	0.0043	0.0058	0.00018	2	188	12 $(t=9)$	38 (!)	0.038	49	4891
3333	1.0	0.13	0.014	2	2056	3385 $(t=9)$	-	8.7	49	48907
	$I_2 = \int_0^\pi x\sin(x)/(1+\cos^2(x))dx$					$I_3 = \int_0^{1000} W_0(x)dx$				
32	0.00077	0.0021	0.00033	14	229	0.0037	0.012	0.00041	12	163
64	0.00077	0.0046	0.00054	14	373	0.0037	0.032	0.00093	12	273
333	0.0088	0.037	0.0040	14	1401	0.052 $(t=1)$	0.25	0.0099	12	1109
3333	2.2	4.4	1.0	14	14401	11 $(t=2)$	25	1.3	12	12043
	$I_4 = \int_0^{100}\sin(x)dx$					$I_5 = \int_0^8 \sin(x+e^x)\,dx$				
32	0.0012 $(t=1)$	0.0019	0.000047	1	53	0.063 $(t=6)$	0.23 (!)	0.0048	33	2115
64	0.0012 $(t=1)$	0.0014	0.000074	1	72	0.063 $(t=6)$	0.25 (!)	0.0055	27	2307
333	0.015 $(t=1)$	0.018	0.00030	1	139	0.22 $(t=4)$	0.58 (!)	0.017	22	4028
3333	2.0	0.71	0.032	1	526	14 $(t=2)$	12	1.1	8	10417
	$I_6 = \int_{-1}^1 e^{-x}\,\mathrm{erf}\left(\sqrt{1250}\,x + \frac{3}{2}\right)dx$					$I_7 = \int_1^{1+1000i}\Gamma(x)dx$				
32	0.024 $(t=3)$	0.018 (!)	0.0025	7	297	0.031 $(t=2)$	0.028	0.00076	11	103
64	0.024 $(t=3)$	0.057 (!)	0.0055	6	438	0.054 $(t=3)$	0.093	0.0035	12	280
333	0.50 $(t=3)$	0.22	0.047	4	791	0.65 $(t=3)$	1.1	0.081	14	1304
3333	173 $(t=2)$	466	5.7	2	2923	561 $(t=3)$	847	48	14	16535

knowledge about f. This is not ideal in terms of usability or efficiency, but since the Petras algorithm works well even with an endpoint very close to a singularity (or ∞), evaluating improper integrals to high precision in this way is at least feasible.[5]

In Table 2, E_0, E_1 and E_2 have algebraic or logarithmic singularities or decay, with E_1 requiring $N \approx 2^p$ and E_2 requiring $\varepsilon \approx 2^{-p}$ (no truncation is needed for E_0, as f is bounded at the algebraic branch point singularity $x = 1$). Here Arb needs $O(p)$ subintervals and $O(p^2)$ evaluations, while the double exponential algorithm in Pari/GP and mpmath only needs roughly $O(p)$ evaluations and therefore scales better.[6] For integrals with exponential decay (E_3, E_4 and E_5), a cutoff of $N \sim p$ is sufficient, and here Arb retains excellent performance.

In a future extension of this work, some reasonable class of improper integrals could be supported more efficiently and conveniently (e.g. with the user providing a symbolic bound like $|f(x)| < Cx^\alpha \exp(-\beta x^\gamma)$).

---

[5] An exception is when f has an essential singularity inducing oscillation combined with slow decay. Oscillation with exponential decay is not a problem (as in E_4, E_5), but integrals like $\int_0^1 \sin(1/x)dx = \int_1^\infty \sin(x)/x^2$ (not benchmarked here) require $2^{O(p)}$ work, so we can only hope for 5–10 digits without specialized oscillatory algorithms.

[6] As a means to improve performance, we note the standard trick of manually changing variables to turn algebraic growth or decay into exponential decay. Indeed, $x \to \sinh(x)$ gives $E_1 = E_3$. Similarly $x \to \tanh(x)$ and $x \to e^{-x}$ can be used in E_0, E_2.

Table 2. Improper integrals and integrals with endpoint singularities. For integration with Arb, all improper integrals (i.e. excluding E_0) have been truncated manually at a lower bound ε or upper bound N, chosen so that the omitted part is smaller than 2^{-p}.

p	Pari/GP	mpmath	Arb	Sub	Eval	Pari/GP	mpmath	Arb	Sub	Eval
	$E_0 = \int_0^1 \sqrt{1-x^2}\,dx$					$E_1 = \int_0^\infty 1/(1+x^2)\,dx$				
32	0.00041	0.00055	0.00022	22	234	0.00060	0.0010	0.00079	94	997
64	0.00041	0.00067	0.00057	44	674	0.00060	0.0012	0.0022	190	2887
333	0.0044	0.0060	0.015	223	12687	0.0068	0.011	0.048	997	51900
3333	0.94	0.18	6.6	2223	1187293	1.7	0.24	27	9997	4711128
	$E_2 = \int_0^1 \log(x)/(1+x)\,dx$					$E_3 = \int_0^\infty \operatorname{sech}(x)\,dx$				
32	0.00081	0.00080	0.00042	34	361	0.0011	0.0019	0.00017	9	144
64	0.00081	0.00094	0.0012	67	1026	0.0011	0.0043	0.00032	10	251
333	0.011	0.011	0.038	336	19254	0.013	0.098	0.0030	14	1277
3333	1.7	1.08	106	3336	1787191	3.5	3.3	0.95	17	16593
	$E_4 = \int_0^\infty e^{-x^2+ix}\,dx$					$E_5 = \int_0^\infty e^{-x}\operatorname{Ai}(-x)\,dx$				
32	0.0014	0.0067	0.00011	1	71	-	0.19	0.0028	4	269
64	0.0014	0.016	0.00018	1	98	-	0.91 (!)	0.012	9	842
333	0.017	0.13	0.0016	2	397	-	26 (!)	0.94	124	24548
3333	4.7	7.1	0.47	4	3894	-	10167 (!)	502	1205	709889

3.3 Piecewise and Discontinuous Functions

Piecewise real analytic functions can be integrated efficiently using piecewise complex analytic extensions. For example, $|x|$ on $\mathbb{R}$ extends to the function $\sqrt{z^2}$ of $z = x + yi$, which equals z in the right plane and $-z$ in the left plane with a branch cut on $\operatorname{Re}(z) = 0$.[7] We provide as library methods such extensions of $\operatorname{sgn}(x)$, $|x|$, $\lfloor x \rfloor$, $\lceil x \rceil$, $\max(x,y)$, $\min(x,y)$, with builtin branch cut detection.

Table 3. Integrals with point discontinuities in f or f'. Here $p(x) = x^4 + 10x^3 + 19x^2 - 6x - 6$ in D_0, and $u(x) = (x - \lfloor x \rfloor - \frac{1}{2})$, $v = \max(\sin(x), \cos(x))$ in D_3. For D_3, the function evaluation limit had to be increased for convergence at $p = 3333$.

p	Arb	Sub	Eval	Arb	Sub	Eval	Arb	Sub	Eval	Arb	Sub	Eval
	$D_0 = \int_0^1 |p(x)|\,e^x\,dx$			$D_1 = \int_0^{100} \lceil x \rceil\,dx$			$D_2 = \int_{-1-i}^{-1+i} \sqrt{x}\,dx$			$D_3 = \int_0^{10} u(x)v(x)dx$		
32	0.00058	38	412	0.0054	2208	6622	0.00064	68	506	0.011	699	5891
64	0.0016	70	1093	0.014	5536	16606	0.0021	132	1462	0.035	1437	19653
333	0.049	339	18137	0.12	33512	100534	0.067	670	28304	1.4	7576	436 K
3333	101	3339	1624951	1.6	345512	1036534	35	6670	2669940	2805	76101	42 M

---

[7] This works for integrating $|f|$ when f is real, but since $|\cdot|$ on $\mathbb{C}$ is not holomorphic, integrating $|f|$ for nonreal f must use direct enclosures, with $2^{O(p)}$ cost. In that case, the user should instead construct complex-extensible real and imaginary parts $f = g + hi$ (e.g. via Taylor polynomials if no closed forms exist) and integrate $\sqrt{g^2 + h^2}$.

Table 3 shows integrals with mid-interval jumps or kinks, including one complex integral crossing a branch cut discontinuity (D_2). The example D_0, where $p(x)$ changes sign once on $[0, 1]$, is due to Helfgott (see comments in [7]).

We see that a mid-interval singularity leads to use of $O(p)$ subintervals and $O(p^2)$ evaluations to isolate the problematic point by bisection. With k such points (D_1 and D_3), the cost simply increases by another factor k, and the user may have to raise the evaluation limits accordingly to let the algorithm complete (which we did for D_3). In contrast, Pari/GP and mpmath cope poorly with mid-interval singularities and cannot achieve high accuracy on these examples unless the user manually splits the interval precisely at the problematic points.

4   Complex Analysis

We conclude by illustrating integration as a tool for complex analysis. First, we consider computing derivatives via the Cauchy integral formula. Denote by $\wp(z; \tau) = \sum_{n=-2}^{\infty} a_n(\tau) z^n$ the Weierstrass elliptic function for the lattice $(1, \tau)$. We fix $\tau = i$ (placing the poles of $\wp$ at the Gaussian integers) and compute the Laurent coefficients $a_n = \frac{1}{2\pi i} \int_\gamma z^{-n-1} \wp(z) dz$ by integrating along the square connecting $\pm 0.5 \pm 0.5i$. We ignore symmetry and compute all four segments. With $p = 333$, some results are (note that $a_{-1} = a_{100} = 0$ and all a_n are real):

```
a[-2]  = [1.000000000000000000000000... +/- 3.57e-98] + [+/- 1.89e-98]*I
a[-1]  =                                [+/- 4.11e-98] + [+/- 2.57e-98]*I
a[2]   = [9.453636006461692614653069... +/- 4.44e-97] + [+/- 2.48e-97]*I
a[98]  = [395.999999999999648281345... +/- 2.90e-68] + [+/- 1.17e-68]*I
a[100] =                                [+/- 4.95e-68] + [+/- 4.95e-68]*I
```

We lose about n bits of precision to cancellation due to the integrand magnitude growing with n. Apart from this, the difficulty increases quite slowly with n: a_{-2} takes 0.67 s while a_{98} and a_{100} take 0.85 s at this precision.

As a second example, the number $N(T)$ of zeros ρ_k of the Riemann zeta function $\zeta(s)$ on the box $[0, 1] + [0, T]i$ can be computed via the argument principle

$$N(T) - 1 = \frac{1}{2\pi i} \int_\gamma \frac{\zeta'(s)}{\zeta(s)} ds = \frac{\theta(T)}{\pi} + \frac{1}{\pi} \operatorname{Im} \left[\int_{1+\varepsilon}^{1+\varepsilon+Ti} \frac{\zeta'(s)}{\zeta(s)} ds + \int_{1+\varepsilon+Ti}^{\frac{1}{2}+Ti} \frac{\zeta'(s)}{\zeta(s)} ds \right]$$

where γ traces the boundary of $[-\varepsilon, 1 + \varepsilon] + [0, T]i$ (plus an excursion for the pole at $s = 1$, whence the -1 term). The more numerically useful formula on the right, where $\varepsilon > 0$ now is arbitrary, is a well-known consequence of the functional equation, where $\theta(T)$ is the Hardy theta function. We set $\varepsilon = 99$ (!) so that only the horizontal segment is difficult, and evaluate the integrals with $\varepsilon_{\text{abs}} = 10^{-6}$:

T	p	Time (s)	Sub	Eval	$N(T)$
10^3	32	0.51	109	1219	[649.00000 +/- 7.78e-6]
10^5	32	12	353	4088	[138069.000 +/- 3.10e-4]
10^7	48	42	391	4500	[21136125.0000 +/- 5.53e-5]
10^9	48	1590	677	8070	[2846548032.000 +/- 1.95e-4]

We obtain balls that provably determine $N(T)$, and the method scales reasonably well. Unfortunately, the evaluation of $\zeta(s)$ in Arb is currently not well tuned for all s, which makes large T slower than necessary and can make this computation extremely slow with slightly different settings. In general, for complicated integrals, the user may need to customize the integrand evaluation to handle wide balls or large parameters optimally for a given path and precision.

References

1. Bailey, D.H., Borwein, J.M.: High-precision numerical integration: progress and challenges. J. Symbolic Comput. **46**(7), 741–754 (2011)
2. Cranley, R., Patterson, T.N.L.: On the automatic numerical evaluation of definite integrals. Comput. J. **14**(2), 189–198 (1971)
3. Hale, N.: Spike integral (2010). http://www.chebfun.org/examples/quad/SpikeIntegral.html
4. Johansson, F.: Arb: efficient arbitrary-precision midpoint-radius interval arithmetic. IEEE Trans. Comput. **66**, 1281–1292 (2017)
5. Johansson, F.: mpmath version 1.0 (2017). http://mpmath.org/
6. Johansson, F., Mezzarobba, M.: Fast and rigorous arbitrary-precision computation of Gauss-Legendre quadrature nodes and weights (2018). arXiv:1802.03948
7. Mahboubi, A., Melquiond, G., Sibut-Pinote, T.: Formally verified approximations of definite integrals. In: Blanchette, J.C., Merz, S. (eds.) ITP 2016. LNCS, vol. 9807, pp. 274–289. Springer, Cham (2016). https://doi.org/10.1007/978-3-319-43144-4_17
8. Petras, K.: Self-validating integration and approximation of piecewise analytic functions. J. Comp. Appl. Math. **145**(2), 345–359 (2002)
9. Rump, S.M.: Verification methods: rigorous results using floating-point arithmetic. Acta Numerica **19**, 287–449 (2010)
10. The Pari group. Pari/GP version 2.9.4 (2017). http://pari.math.u-bordeaux.fr/
11. The SageMath developers. SageMath version 8.2 (2018). http://sagemath.org/

New Counts for the Number
of Triangulations of Cyclic Polytopes

Michael Joswig$^{(\boxtimes)}$ and Lars Kastner

Institut für Mathematik, TU Berlin, Str. des 17. Juni 136, 10623 Berlin, Germany
{joswig,kastner}@math.tu-berlin.de
http://page.math.tu-berlin.de/~joswig/
http://page.math.tu-berlin.de/~kastner/

Abstract. We report on enumerating the triangulations of cyclic polytopes with the new software MPTOPCOM. This is relevant for its connection with higher Stasheff–Tamari orders, which occur in category theory and algebraic combinatorics.

1 Introduction

For an integer $d \geq 1$ the d-th *moment curve* is the map

$$\mu_d : \mathbb{R} \to \mathbb{R}^d, \ t \mapsto (t, \dots, t^d).$$

Picking n real numbers $t_1 < t_2 < \cdots < t_n$, where $n > d$, the convex hull

$$\mathcal{C}(n, d) \ = \ conv\{\mu_d(t_1), \mu_d(t_2), \dots, \mu_d(t_n)\} \tag{1}$$

is the d-dimensional *cyclic polytope* with n vertices. The combinatorics of $\mathcal{C}(n, d)$ is given by *Gale's evenness criterion*; cf. [16, Theorem 0.7]. In particular, the combinatorial type does not depend on the values $t_1, t_2, \dots, t_n$ but just on their number. The cyclic polytopes are neighborly, and hence their f-vectors attain McMullen's upper bound [16, Theorem 8.23]. The *higher Stasheff–Tamari orders* are certain poset structures on the set of all triangulations of $\mathcal{C}(n, d)$. Their study was initiated by Kapranov and Voevodsky [10] in the context of category theory; see also [6,12,14]. Here we address the problem raised in [10, Sect. 5.2], which asks for determining the number of triangulations of $\mathcal{C}(n, d)$. We report on new computational results, obtained via the new software MPTOPCOM [9]. This verifies and extends previous results of Rambau and Reiner [14, Table 1], which were obtained with TOPCOM [13]. The general question remains wide open. Notice that the planar case $d = 2$ gives the Catalan numbers.

Triangulations of polytopes and of finite point configurations are the subject of the monograph [5] by De Loera, Rambau and Santos. Cyclic polytopes are discussed in [5, Sect. 6.1]. For a combinatorial encoding of triangulations of cyclic polytopes see [11,15]. We are indebted to Jörg Rambau and Francisco Santos for suggesting to apply MPTOPCOM to cyclic polytopes and for many useful comments on an earlier version of this text.

© Springer International Publishing AG, part of Springer Nature 2018
J. H. Davenport et al. (Eds.): ICMS 2018, LNCS 10931, pp. 264–271, 2018.
https://doi.org/10.1007/978-3-319-96418-8_31

This research is carried out in the framework of Matheon supported by Einstein Foundation Berlin. Further partial support by DFG (SFB-TRR 109 and SFB-TRR 195) is gratefully acknowledged.

2   The First Higher Stasheff–Tamari Order

Let $P \subset \mathbb{R}^d$ be a finite point configuration. A *circuit* of P is a minimally affinely dependent subconfiguration. A *triangulation* of P is a subdivision of the convex hull $conv(P)$ whose vertices form a subset of the points in P. Two triangulations of P *differ by a flip* if they agree outside a circuit. Here we are interested in the triangulations of the point configuration given by the vertices of $\mathcal{C}(n, d)$ and their flips.

There is a canonical projection of the $(d+1)$-dimensional simplex $\mathcal{C}(d+2, d+1)$ onto $\mathcal{C}(d + 2, d)$ by forgetting the last coordinate. There are precisely two triangulations of $\mathcal{C}(d + 2, d)$, and these correspond to projecting the lower and the upper hull of $\mathcal{C}(d + 2, d + 1)$. Consequently, we call them the *lower* and the *upper triangulation* of $\mathcal{C}(d + 2, d)$, respectively. From the construction (1) it is immediate that each circuit of $\mathcal{C}(n, d)$ looks like $\mathcal{C}(d + 2, d)$. Combined with the observation on the two triangulations of $\mathcal{C}(d + 2, d)$, this has far reaching consequences for the structure of the triangulations of $\mathcal{C}(n, d)$ for arbitrary $n > d$.

Let Δ and Δ' be two triangulations of $\mathcal{C}(n, d)$ which differ by a flip. Then there is subset C of the vertices of cardinality $d+2$ such that Δ and Δ' restricted to C look like the upper and the lower triangulations of $\mathcal{C}(d + 2, d)$. If Δ is the lower and Δ' is the upper triangulation, then we call the flip $[\Delta \rightsquigarrow \Delta']$ from Δ to Δ' an *up-flip*. Conversely, the reverse flip $[\Delta' \rightsquigarrow \Delta]$ is a *down-flip*. In this case we write

$$\Delta \leq_1 \Delta'. \tag{2}$$

The partial ordering on the set of all triangulations of $\mathcal{C}(n, d)$ which is obtained as the transitive and reflexive closure of the relation (2) is the *first higher Stasheff-Tamari order*, denoted as $HST_1(n, d)$; cf. [5, Definition 6.1.18] and [12]. Figure 3 below shows $HST_1(6, 2)$ as an example.

On the same set of triangulations of $\mathcal{C}(n, d)$ there is a second natural partial ordering, the *second higher Stasheff-Tamari order*, $HST_2(n, d)$; cf. [5, Definition 6.1.16]. It is known that $HST_1(n, d)$ is a weaker partial order than $HST_2(n, d)$. Moreover, these two orders coincide for $d \leq 3$ and $n - d \in \{1, 2, 3\}$; cf. [14]. In general, it is open whether or not they agree.

3   GKZ-Vectors

For a triangulation Δ of an affine spanning point configuration $P \subset \mathbb{R}^d$ the *GKZ-vector* is

$$gkz_\Delta = (gkz_\Delta(p) \mid p \in P),$$

where $gkz_\Delta(p)$ is the sum of the normalized volumes of those simplices in Δ which contain p as a vertex. The *normalized volume* is the Euclidean volume multiplied by $d!$.

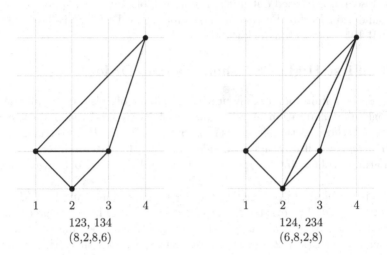

Fig. 1. Lower (left) and upper (right) triangulations of $C(4,2)$ with their GKZ-vectors. In the lower triangulation the gaps are 4 and 2, i.e., even; whereas in the upper triangulation the gaps are 3 and 1. Here $d = 2$ is even.

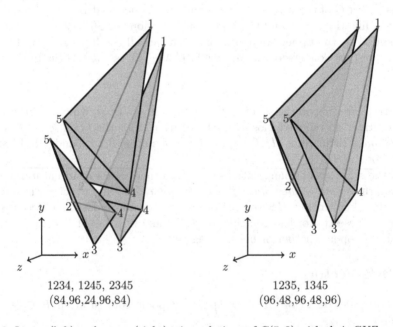

Fig. 2. Lower (left) and upper (right) triangulations of $C(5,3)$ with their GKZ-vectors. In the lower triangulation the gaps are 4 and 2, i.e., even; whereas in the upper triangulation the gaps are 5, 3 and 1. Here $d = 3$ is odd.

In order to determine the GKZ-vectors of triangulations of cyclic polytopes, we need to choose coordinates. To keep it simple we can take the lattice points $\mu_d(1), \mu_d(2), \ldots, \mu_d(n)$ as the vertices of $\mathcal{C}(n, d)$. The normalized volume of any d-simplex spanned by $\mu_d(i_1), \mu_d(i_2), \ldots, \mu_d(i_{d+1})$ with $i_1 < i_2 < \cdots < i_{d+1}$, is the Vandermonde determinant

$$\det \begin{pmatrix} 1 & i_1 & \cdots & i_1^d \\ 1 & i_2 & \cdots & i_2^d \\ \vdots & \vdots & \ddots & \vdots \\ 1 & i_{d+1} & \cdots & i_{d+1}^d \end{pmatrix} = \prod_{1 \leq k < \ell \leq d+1} (i_\ell - i_k). \tag{3}$$

In particular, these values do not change when we replace the standard parameters $1, 2, \ldots, n$ for the moment curve by any other set of n consecutive integers; cf. Fig. 1, where we chose the parameters $-1, 0, 1$ and 2, while we keep the labels $1, 2, 3, 4$. We fix the natural ordering of the vertices on the moment curve in order to identify GKZ-vectors of triangulations of $\mathcal{C}(n, d)$ with vectors in $\mathbb{R}^n$. The following basic observation is crucial.

Proposition 1. *Let Δ and Δ' be two triangulations of $\mathcal{C}(n, d)$ related by a flip $[\Delta \rightsquigarrow \Delta']$. Then we have*

$$\Delta \leq_1 \Delta' \iff \begin{cases} gkz_\Delta >_{lex} gkz_{\Delta'} & \text{if } d \text{ even,} \\ gkz_\Delta <_{lex} gkz_{\Delta'} & \text{if } d \text{ odd.} \end{cases}$$

Proof. Since each circuit looks like $\mathcal{C}(d + 2, d)$ it suffices to consider the case $n = d + 2$. We exploit the relationship of the triangulations of $\mathcal{C}(d + 2, d)$ with the upper and lower hull of $\mathcal{C}(d + 2, d + 1)$ previously explained.

The Oriented Gale's Evenness Criterion from [5, Corollary 6.1.9] describes the upper and lower facets of $\mathcal{C}(d+2, d+1)$. Let $F \subseteq \mathcal{C}(d+2, d+1)$ be a facet, then F can be written as a subset of $[d+2]$, the set of indices of vertices in F. The *gaps* of F are the elements of $[n] \backslash F$. A gap i of F is *even* if the number of elements in F that are larger than i is even. It is called *odd* otherwise. Correspondingly, a facet is called odd/even if all its gaps are odd/even. The odd facets correspond to the upper triangulation of $\mathcal{C}(d + 2, d)$ and the even facets give rise to the lower triangulation of $\mathcal{C}(d + 2, d)$.

Assume that 1 is a gap of F. Since every facet of $\mathcal{C}(d+2, d+1)$ is a simplex, F must be $\{2, 3, \ldots, d + 2\}$ and 1 is the only gap of F. Hence, if d is odd, then F is even. Conversely, if d is even, then F must be odd. We conclude that, if d is odd, then all odd facets contain 1. However, if d is even, then only the even facets contain 1.

Assume now that d is even. The odd case is similar.

Let Δ and Δ' be the lower and upper triangulations of $\mathcal{C}(d + 2, d)$, i.e. $\Delta \leq_1 \Delta'$. Then Δ contains all the even facets of $\mathcal{C}(d + 2, d + 1)$. But any even facet contains 1, thus the first entry of gkz_Δ is the entire normalized volume of $\mathcal{C}(d + 2, d)$. The facet $\{2, 3, \ldots, d + 2\}$ is odd, and hence it belongs to Δ'. Since it does not contain 1, we infer that $gkz_\Delta(1) > gkz_{\Delta'}(1)$. Hence we obtain $gkz_\Delta >_{lex} gkz_{\Delta'}$.

This argument can be reversed, and this completes the proof.   □

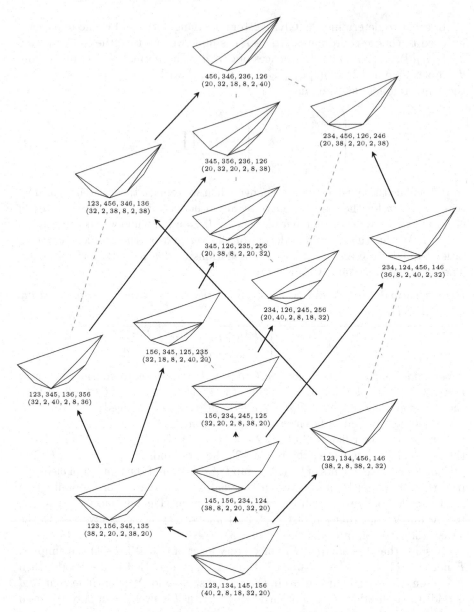

Fig. 3. First higher Stasheff-Tamari order $HST_1(6, 2)$ with reverse search tree marked. The lowest triangulation has the lexicographically largest GKZ-vector.

Figures 1 and 2 depict the situation considered in the proof above for $(n, d) = (4, 2)$ and $(n, d) = (5, 3)$, respectively. The interest in Proposition 1 comes from the following.

In [8] Imai et al. described an algorithm for computing all (regular) triangulations of a given point configurations, which is based on the reverse search enumeration scheme of Avis and Fukuda [2]. That algorithm, which we call *down-flip reverse search*, was improved and implemented by Skip Jordan with the authors of this extended abstract [9]. The basic idea is to orient each flip according to lexicographic ordering of the GKZ-vectors. Then down-flip reverse search produces a directed spanning tree of those triangulations which can be obtained from some seed triangulation by monotone flipping; cf. [5, Sect. 5.3.2]. For the cyclic polytopes we arrive at two choices for orienting the flips, one by GKZ-vectors, one according to the first higher Stasheff–Tamari order. Now Proposition 1 says that these two choices fortunately agree.

Corollary 1. *Down-flip reverse search computes a directed spanning tree of the first Stasheff-Tamari poset $HST_1(n, d)$, rooted at the triangulation with the lexicographically largest GKZ-vector. For d even, the root is the lowest triangulation of $\mathcal{C}(n, d)$, whereas, for d odd, the root is the highest triangulation. In particular, each triangulation of a cyclic polytope can be obtained by monotone flipping from the respective roots.*

The first higher Stasheff–Tamari order $HST_1(6, 2)$ with GKZ-vectors is shown in Fig. 3.

4   Computations with MPTOPCOM

The open source software MPTOPCOM is designed for computing triangulations in a massively parallel setup. Its algorithm is the down-flip reverse search method of Imai et al. [8] with several improvements as described in [9]. As its key feature reverse search is output sensitive, and this makes it attractive for extremely large enumeration problems. Our parallelization, based on the MPI protocol, employs *budgeting* for load balancing; cf. [1,3]. In this way MPTOPCOM can enumerate the (regular) triangulations of much larger point sets than other software before; extensive experiments are described in [9, Sect. 7]. MPTOPCOM uses linear algebra and basic data types from polymake [7], triangulations and flips from TOPCOM [13] and the budgeted parallel reverse search from mts [3].

The most recent census of triangulations of cyclic polytopes that we are aware of is by Rambau and Reiner [14, Table 1]; we use their notation and introduce the parameter $c := n - d$. Note that there are two rather obvious typos in the rows $c \in \{10, 11\}$ of the column $d = 1$ in [14, Table 1]. Apart from that we can confirm their results; cf. Table 1. Our new results are the values for $(c, d) \in \{(12, 3), (8, 5), (6, 8), (5, 14)\}$.

Our experiments used MPTOPCOM, version 1.0, on a cluster with four nodes, each of which comes with 2 x 8-Core Xeon E5-2630v3 (2.4 GHz) and 64 GB per node. We ran MPTOPCOM with 40 threads. The operating system is SMP Linux 4.4.121. For instance, the computation for $c = 5$ and $d = 14$, i.e., $n = 19$ took 71191 seconds, i.e., less than 20 h.

Azaola and Santos [4, p. 30] implicitly raised the following question.

Table 1. The number of triangulations of $C(c + d, d)$. The column $d = 2$ contains the Catalan numbers, while the row $c = 4$ is known by results of Azaola and Santos [4]. The rows $c \in \{1, 2, 3\}$ are trivial and only listed for completeness. The row $c = 5$ and the column $d = 2$ are marked for their relevance to Question 1. Our new results are written in blue; the rest of the table agrees with [14, Table 1].

c \ d:	2	3	4	5	6	7	8
1	1	1	1	1	1	1	1
2	2	2	2	2	2	2	2
3	5	6	7	8	9	10	11
4	14	25	40	67	102	165	244
5	42	138	357	1 233	3 278	12 589	35 789
6	132	972	4 824	51 676	340 560	6 429 428	68 007 706
7	429	8 477	96 426	5 049 932	132 943 239		
8	1 430	89 405	2 800 212	1 171 488 063			
9	4 862	1 119 280	116 447 760				
10	16 796	16 384 508					
11	58 786	276 961 252					
12	208 012	5 349 351 298					

c \ d:	9	10	11	12	13	14
1	1	1	1	1	1	1
2	2	2	2	2	2	2
3	12	13	14	15	16	17
4	387	562	881	1 264	1 967	2 798
5	159 613	499 900	2 677 865	9 421 400	62 226 044	247 567 074

Question 1. Is there an absolute constant $\beta > 1$ such that, for all $n \geq 7$:

$$\frac{1}{\beta} \leq \frac{\#\{\text{triangulations of } C(n, n - 5)\}}{\#\{\text{triangulations of } C(n, 2)\}} \leq \beta? \tag{4}$$

This relates the row $c = 5$ with the column $d = 2$; these are marked in Table 1. From MPTOPCOM's results we can derive the series (4) for $n \in \{7, 8, \ldots, 19\}$:

$$1,\ 1.045,\ 0.832,\ 0.862,\ 0.674,\ 0.750,\ 0.609,$$
$$0.767,\ 0.673,\ 1.001,\ 0.972,\ 1.760,\ 1.910.$$

Note that the sequence in [4, p. 30] lists the reciprocals of the above; moreover, that sequence contains two more (trivial) values for $n \in \{5, 6\}$, which we omit.

References

1. Avis, D., Devroye, L.: An analysis of budgeted parallel search on conditional Galton-Watson trees. Preprint arXiv 1703.10731 (2017)
2. Avis, D., Fukuda, K.: Reverse search for enumeration. Discrete Appl. Math. **65**(1–3), 21–46 (1996). (English)

3. Avis, D., Jordan, C.: A parallel framework for reverse search using mts. arXiv Preprint arXiv:1610.07735 (2016)

4. Azaola, M., Santos, F.: The number of triangulations of the cyclic polytope $C(n, n-4)$. Discrete Comput. Geom. **27**(1), 29–48 (2002). (English)

5. De Loera, J., Rambau, J., Santos, F.: Triangulations, Algorithms and Computation in Mathematics, vol. 25. Springer, Heidelberg (2010). https://doi.org/10.1007/978-3-642-12971-1. MR2743368 (2011j:52037)

6. Edelman, P.H., Reiner, V.: The higher Stasheff-Tamari posets. Mathematika **43**(1), 127–154 (1996). (English)

7. Gawrilow, E., Joswig, M.: polymake:a framework for analyzing convex polytopes, Polytopes–combinatorics and computation (Oberwolfach, 1997), DMV Sem., vol. 29, Birkhäuser, Basel, pp. 43–73 (2000). MRMR1785292 (2001f:52033)

8. Imai, H., Masada, T., Takeuchi, F., Imai, K.: Enumerating triangulations in general dimensions. Int. J. Comput. Geom. Appl. **12**(6), 455–480 (2002). MR1945594

9. Jordan, C., Joswig, M., Kastner, L.: Parallel enumeration of triangulations. Preprint arXiv:1709.04746 (2017)

10. Kapranov, M.M., Voevodsky, V.A.: Combinatorial-geometric aspects of polycategory theory: pasting schemes and higher Bruhat orders (list of results). Cahiers Topologie Géom. Différentielle Catég **32**(1), 11–27 (1991). International Category Theory Meeting (Bangor, 1989 and Cambridge, 1990). MR1130400

11. Oppermann, S., Thomas, H.: Higher-dimensional cluster combinatorics and representation theory. J. Eur. Math. Soc. (JEMS) **14**(6), 1679–1737 (2012). MR2984586

12. Rambau, J.: Triangulations of cyclic polytopes and higher Bruhat orders. Mathematika **44**(1), 162–194 (1997). MR1464385

13. Rambau, J.: TOPCOM: triangulations of point configurations and oriented matroids, Mathematical software (Beijing, 2002), pp. 330–340. World Sci. Publ., River Edge, NJ (2002). MR1932619

14. Rambau, J., Reiner, V.: A survey of the higher Stasheff-Tamari orders. In: Müller-Hoissen, F., Pallo, J., Stasheff, J. (eds.) Associahedra, Tamari Lattices and Related Structures. Tamari memorial Festschrift. PM, vol. 299, pp. 351–390. Birkhäuser, Basel (2012). https://doi.org/10.1007/978-3-0348-0405-9_18

15. Thomas, H.: New combinatorial descriptions of the triangulations of cyclic polytopes and the second higher Stasheff-Tamari posets. Order **19**(4), 327–342 (2002). MR1964443

16. Ziegler, G.M.: Lectures on Polytopes. Graduate Texts in Mathematics, vol. 152. Springer, New York (1995). https://doi.org/10.1007/978-1-4613-8431-1. MR1311028

Estimating Tropical Principal Components Using Metropolis Hasting Algorithm

Qiwen Kang[1] and Ruriko Yoshida[2(✉)]

[1] University of Kentucky, Lexington, USA
qiwen.kang@uky.edu
[2] Naval Postgraduate School, Monterey, USA
ryoshida@nps.edu
http://stat.as.uky.edu/users/qka222, http://polytopes.net/

Abstract. Principal component analysis is one of the most popular unsupervised learning methods for reducing the dimension of a given data set in a high-dimensional Euclidean space. However, computing principal components on a space of phylogenetic trees with fixed labels of leaves is a challenging task since a space of phylogenetic tree is not Euclidean. In 2017, Yoshida et al. defined a notion of tropical principal component analysis and they have applied it to a space of phylogenetic trees. The challenge, however, they encountered was a computational times.

In this paper we estimate tropical principal components in a space of phylogenetic trees using the Metropolis-Hasting algorithm. We have implemented an R software package to efficiently estimate tropical principal components and then we have applied it to African coelacanth genomes data set.

Keywords: Phylogenetic trees · Polytopes · Tropical geometry

1 Introduction

Principal component analysis (PCA) is one of the most popular and robust unsupervised learning methods for reducing the dimension of a high-dimensional data set in Euclidean spaces. PCA is a statistical method that takes data points in a high dimensional Euclidean space into a lower dimensional plane which minimizes the sum of squares between each point in the data set and their orthogonal projection onto the plane. It has been used for clustering high dimensional data points for statistical analysis and it is one of the simplest and most robust ways of doing such dimensionality reduction in a Euclidean vector space. However, it assumes the properties of a Euclidean vector space while the space of rooted equidistant trees on n leaves, a polyhedral complex of dimension $n-2$, realized as the set of all ultrametrics is not Euclidean.

This is a U.S. government work and its text is not subject to copyright protection
in the United States; however, its text may be subject to foreign copyright protection 2018
J. H. Davenport et al. (Eds.): ICMS 2018, LNCS 10931, pp. 272–279, 2018.
https://doi.org/10.1007/978-3-319-96418-8_32

One classical way to conduct a statistical analysis on phylogenetic trees with n leaves is to map each tree to a vector in $\mathbb{R}^{\binom{n}{2}}$, for example using the *dissimilarity map*. Given any tree T of n leaves with branch length information, one may produce a corresponding *distance matrix*, $D(T)$. The distance matrix is an $n \times n$ symmetric matrix of non-negative real numbers, with elements corresponding to $d_{ij}(T)$, the sum of the branch lengths between pairs of leaves in the tree. To calculate $d_{ij}(T)$, one simply determines which edges of the tree form the path from a leaf i to a leaf j, and then sums the lengths of these branches. Since $D(T)$ is symmetric and has zeros on the diagonal, the upper-triangular portion of the matrix contains all of the unique information found in the matrix. We can vectorize T by enumerating this unique portion of the distance matrix,

$$v_d(T) := (d_{12}(T), d_{13}(T), \ldots, d_{23}(T), \ldots, d_{n-1n}(T))$$

which is called the *dissimilarity map* of a tree T and is a vector in $\mathbb{R}^{\binom{n}{2}}$. If it is clear we simply abbreviate $D(T)$ with D.

Let D be a distance matrix computed from a phylogenetic tree, that is, a nonnegative symmetric $n \times n$-matrix $D = (d_{ij})$ with zero entries on the diagonal such that all triangle inequalities are satisfied:

$$d_{ik} \leq d_{ij} + d_{jk} \quad \text{for all } i, j, k \text{ in } [n] := \{1, 2, \ldots, n\}.$$

If a distance matrix D is computed from an equidistant tree, it is well-known that elements in D satisfy the following strengthening of the triangle inequalities:

$$d_{ik} \leq \max(d_{ij}, d_{jk}) \quad \text{for all } i, j, k \in [n]. \tag{1}$$

If (1) holds then the metric D is called an *ultrametric*. The set of all ultrametrics contains the ray $\mathbb{R}_{\geq 0}\mathbf{1} = (a, a, \ldots, a)$, where $s \in \mathbb{R}$, spanned by the metric $\mathbf{1} = (1, 1, \ldots, 1)$, which is defined by $d_{ij} = 1$ for $1 \leq i < j \leq n$. The image of the set of ultrametrics in the quotient space $\mathbb{R}^{\binom{n}{2}}/\mathbb{R}\mathbf{1}$ is denoted $\mathcal{U}_n$ and called the *space of ultrametrics*. Therefore, we can consider the space of ultrametrics as a treespace for all possible equidistant phylogenetic trees with n leaves.

However, the space of phylogenetic trees with n leaves is not an Euclidean space. In fact, it is a union of lower dimensional polyhedral cones in $\mathbb{R}^{\binom{n}{2}}$. Therefore we cannot directly apply classical PCA to a set of gene trees. Nye showed an algorithm in [11] to compute the first order principal component over the space of phylogenetic trees of n leaves using the unique shortest connecting paths, or geodesics, defined by the CAT(0)-metric introduced by Billera-Holmes-Vogtman (BHV) over the tree space of phylogenetic trees with fixed labeled leaves [3]. Nye in [11] used a convex hull of two points, i.e., the geodesic, on the tree space as the first order PCA. However, we could not generalize this idea for computing higher order principal components with the BHV metric because, in 2017, Lin et al. showed that the convex hull of three points with the BHV metric over the tree space has an arbitrary dimension [9]. On the other hand the tropical metric in tree space defined by the tropical convexity in the max-plus algebra is well studied [10].

Now we turn to *tropical mathematics* [13]. This furnishes a metric and a convexity structure on the tree space which is radically different from BHV. Let $e = \binom{n}{2}$. Tropical geometry gives an alternative geometric structure on $\mathcal{U}_n$, via the graphic matroid of the complete graph [10, Example 4.2.14], i.e., $\mathcal{U}_n$ can be written as a tropical linear space under the max-plus algebra. We mostly use the max-plus algebra, so our convention is opposite to that of [10,12]. The connection between phylogenetic trees and tropical lines, identifying tree space with a tropical Grassmannian, has been explained in many sources, including [10, Sect. 4.3], [12, Sect. 3.5], and [13, Fact 6]. However, the restriction to ultrametrics [2, Sect. 4] offers a fresh perspective.

In 2017, Yoshida et al. defined a notion of *tropical principal components* [14]: Tropical convex hull, i.e., tropical polytope, which minimizes the sum of squares between each point in the data set and their orthogonal projection onto the tropical polytope with the *tropical metric d_{tr}*. They have introduced a mathematical foundation on tropical principal components and they have applied it to computing tropical principal components in $\mathcal{U}_n$. However, it is not efficient to compute tropical principal components using their implementations even though the time complexity of computing tropical principal components is still unknown.

In this paper we have developed a method to estimate tropical principal components via Metropolis-Hasting algorithm and then we have applied it to coelacanths genome and transcriptome data from Liang et al. [8]. This paper is organized as follows: In Sect. 2 we discuss the basics of tropical geometry and review the interpretation of the space of equidistant trees as a tropical linear space. Then we review the tropical principal components introduced by Yoshida et al. In Sect. 3 we describe our algorithm and then in Sect. 4 we apply our method to the coelacanths genome data set.

2   Tropical Principal Components

In this section we review some basics of tropical geometry and then we review the tropical principal components developed by [14]. See [10] or [6] for more detail.

In the tropical semiring $(\mathbb{R} \cup \{-\infty\}, \oplus, \odot)$, the basic arithmetic operations of addition and multiplication are redefined as follows:

$$a \oplus b := \max\{a, b\}, \quad a \odot b := a + b \quad \text{where } a, b \in \mathbb{R}.$$

The element $-\infty$ is the identity element for addition and 0 is the identity element for multiplication: for all $a \in \mathbb{R} \cup \{-\infty\}$, we have $a \oplus -\infty = a$ and $a \odot 0 = a$.

With given scalars $a, b \in \mathbb{R} \cup \{\infty\}$ and vectors $v = (v_1, \ldots, v_e), w = (w_1, \ldots, w_e) \in (\mathbb{R} \cup \infty)^e$, we can define tropical scalar multiplication and tropical vector addition as

$$a \odot v = (a + v_1, a + v_2, \ldots, a + v_e)$$

$$a \odot v \oplus b \odot w = (\max\{a + v_1, b + w_1\}, \ldots, \max\{a + v_e, b + w_e\}).$$

In tropical geometry we often work in the *tropical projective torus* $\mathbb{R}^e/\mathbb{R}\mathbf{1}$, where $\mathbf{1}$ denotes the all-ones vector. Given two points v, w in the tropical projective torus, their *tropical distance* $d_{\mathrm{tr}}(v, w)$ is defined as follows:

$$d_{\mathrm{tr}}(v, w) \;=\; \max\{\, |v_i - w_i - v_j + w_j| \;:\; 1 \le i < j \le e \,\}, \tag{2}$$

where $v = (v_1, \ldots, v_e)$ and $w = (w_1, \ldots, w_e)$. This metric is also known as the *generalized Hilbert projective metric* [1, Sect. 2.2], [4, Sect. 3.3].

A subset $S \subset \mathbb{R}^e$ is said *tropically convex* if it contains the point $a \odot x \oplus b \odot y$ for all $x, y \in S$ and all $a, b \in \mathbb{R}$. The *tropical convex hull* or *tropical polytope* $\mathrm{tconv}(V)$ of a given subset $V \subset \mathbb{R}^e$ is the smallest tropically convex subset containing $V \subset \mathbb{R}^e$. The tropical convex hull of V can be also written as the set of all tropical linear combinations

$$\mathrm{tconv}(V) = \{a_1 \odot v_1 \oplus a_2 \odot v_2 \oplus \cdots \oplus a_r \odot v_r : v_1, \ldots, v_r \in V \text{ and } a_1, \ldots, a_r \in \mathbb{R}\}.$$

Any tropically convex subset S of $\mathbb{R}^e$ is closed under tropical scalar multiplication, $\mathbb{R} \odot S \subseteq S$.

Let $\mathcal{P}$ be a tropical polytope $\mathcal{P} = \mathrm{tconv}(D^{(1)}, D^{(2)}, \ldots, D^{(s)})$, where the $D^{(i)}$ are points in $\mathbb{R}^e/\mathbb{R}\mathbf{1}$. There is a projection map $\pi_{\mathcal{P}}$ sending any point D to a closest point in the tropical polytope $\mathcal{P}$ as

$$\pi_{\mathcal{P}}(D) \;=\; \lambda_1 \odot D^{(1)} \;\oplus\; \lambda_2 \odot D^{(2)} \;\oplus\; \cdots \;\oplus\; \lambda_s \odot D^{(s)}, \tag{3}$$

where $\lambda_k = \min(D - D^{(k)})$ for $k = 1, \ldots, s$. This formula appears as [10, Formula 5.2.3].

Now we review how tropical geometry connects to the space of phylogenetic trees. It is well known that all ultrametrics are tree metrics. In fact, all ultrametrics are derived from *equidistant trees*, where all leaves have the same distance to some distinguished root vertex. Furthermore, the tree metric of an equidistant tree is an ultrametric; hence ultrametrics and equidistant trees convey equivalent information.

Let L_n denote the subspace of $\mathbb{R}^e$ defined by the linear equations $x_{ij} - x_{ik} + x_{jk} = 0$ for $1 \le i < j < k \le n$. The tropicalization $\mathrm{Trop}(L_n) \subseteq \mathbb{R}^e/\mathbb{R}\mathbf{1}$ is the tropical linear space consisting of points $(v_{12}, v_{13}, \ldots, v_{n-1,n})$ such that $\max(v_{ij}, v_{ik}, v_{jk})$ is obtained at least twice for all triples $i, j, k \in [n]$.

Theorem 1. *[14] The image of $\mathcal{U}_n$ in the tropical projective torus $\mathbb{R}^e/\mathbb{R}\mathbf{1}$ coincides with $\mathrm{Trop}(L_n)$.*

A tropical principal component analysis defined in [14] is the tropical convex hull of s points in $\mathcal{U}_n$ minimizing the sum of distances between each point in the sample to its projection onto the convex hull. While we can generalize this to arbitrary s, here we focus on the second order principal components for simplification. The second order tropical principal components can be written as follows:

Problem 1. We seek a solution for the following optimization problem:

$$\min_{D^{(1)}, D^{(2)}, D^{(3)} \in \mathcal{U}_n} \sum_{i=1}^{n} d_{\mathrm{tr}}(d_i, d_i')$$

where

$$d_i' = \lambda_1^i \odot D^{(1)} \oplus \lambda_2^i \odot D^{(2)} \oplus \lambda_3^i \odot D^{(3)}, \quad \text{where } \lambda_k^i = \min(d_i - D^{(k)}), \quad (4)$$

and

$$d_{\mathrm{tr}}(d_i, d_i') = \max\{|d_i(k) - d_i'(k) - d_i(l) + d_i'(l)| : 1 \le k < l \le e\} \quad (5)$$

with

$$d_i = (d_i(1), \dots, d_i(e)) \text{ and } d_i' = (d_i'(1), \dots, d_i'(e)). \quad (6)$$

Even though we do not know the time complexity to solve the optimization problem in Problem 1, the implementation by [14] was not efficient in general. Therefore in this paper we have applied the Metropolis-Hasting algorithm to approximate the optimal solution for Problem 1.

3   Algorithm

We consider two different data sets: Apicomplexa data set and Lungfish data set. After reading these trees into R, we need to root each tree first and specify the outgroup. During the rooting process, some of the trees cannot be rooted with a specified outgroup. We remove these trees from the original data set and then do the sampling.

f is the function for calculating the sum of tropical distances; g is the function for calculating projected points.

Algorithm 2. (Markov Chain Monte Carlo sampling)

- **Input**: *Initial distance vectors D of trees T, the number of principal components P, the number of trees N.*
- **Output**: *The combination of trees comb, the projected points projPoints and the sum of tropical distances tropDist.*
- **Algorithm**:
    1. *Let $pcs = P$ random trees selected from T, $S = Z \setminus pcs$, $tropDist = 1000000$;*
    2. *For $i = 1, \dots, N$*
        (a) $a = $ Select one tree randomly from pcs;
        (b) $b = $ Select one tree randomly from S;
        (c) $p\hat{c}s = pcs \setminus a \cap b$;
        (d) $r = f(pcs)/f(p\hat{c}s)$;
        (e) Randomly select a number u from uniform(0,1) distribution;
            i. If $u \le \min(r, 1)$, $pcs = p\hat{c}s$;
                A. If $f(pcs) < tropDist$ $projPoints = g(p\hat{c}s)$, $tropDist = f(p\hat{c}s)$, $comb = p\hat{c}s$
        (f) $S = S \setminus b$;

We implement this method in R. All the code for this article can be found at https://github.com/QiwenKang/tropicalMCMC.git.

Before implementing our method, we should extract the distance matrix from each phylogenetic tree and transfer it to a vector format, `distVec_all`. This could be done using `disMat` function. Outgroups should be stated as well. `nr` is the number of repetition; `pcs` gives how many principle components would be considered; `to` is tip labels of the raw trees and ordered by names; `N` is the number of trees in whole data set.

```
distVec_all <- distMat(trees_ori, tipOrder = to,
                outgp = c("Leucoraja","Scyliorhin","Callorhinc"))
tropMCMC(distVec_all, N, pcs, nr = 100)
```

In lungfish data set, 1156 trees are included in analysis. It takes around 6 mins to finish a round. The running time could be reduced if we consider parallel computing. All the code is running on a computer with processor Intel Core i7-6700 3.40 GHz × 8, memory 15.6 GB and OS type Ubuntu 18.04 64-bit.

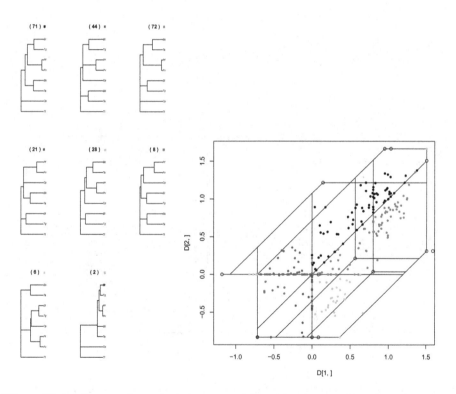

Fig. 1. Right: Projected topology frequencies from the Apicomplexa dataset: parenthesized numbers give the frequencies of each topology. Left: Projected points in the tropical polytope PCA.

4   Application to Empirical Datasets

4.1   Apicomplexa Data

The phylum Apicomplexa contains many important protozoan pathogens [7], including the mosquito-transmitted *Plasmodium* spp., the causative agents of malaria, *T. gondii*, which is one of the most prevalent zoonotic pathogens worldwide, and the water-born pathogen *Cryptosporidium* spp. Several members of the Apicomplexa also cause significant morbidity and mortality in both wildlife and domestic animals. These include *Theileria* spp. and *Babesia* spp., which are tickborne haemoprotozoan pathogens that infect and cause disease in ungulates, and several species of *Eimeria*, which are enteric parasites that are particularly detrimental to the poultry industry. Due to their medical and veterinary importance, whole genome sequencing projects have been completed for multiple prominent members of the Apicomplexa. The second order tropical principal components computed from the Apicomplexa data set is shown in Fig. 1.

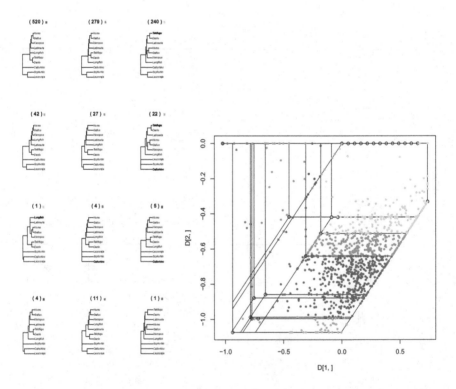

Fig. 2. Right: Projected topology frequencies from the Coelacanths genome dataset: parenthesized numbers give the frequencies of each topology. Left: Projected points in the tropical polytope PCA.

4.2   Coelacanths Genome and Transcriptome Data

We have applied the clustering methods to the data set comprising 1,290 nuclear genes encoding 690,838 amino acid residues obtained from genome and transcriptome data by [8]. Over the last decades, the phylogenetic relations between coelacanths, lungfishes, and tetrapods have been controversial despite there has been much work on the data set [5]. The dataset consisted of 1290 gene alignments for 10 species: lungfish, *Protopterus annectens*, and coelacanth, *Latimeria chalumnae*; three tetrapods, frog, *Xenopus tropicalis*, chicken, *Gallus gallus*, and human, *Homo sapiens*; two ray-finned fish, *Danio rerio* and *Takifugu rubripes*; and three cartilaginous fish included as an out-group, *Scyliorhinus canicula*, *Leucoraja erinacea* and *Callorhinchus milii*. The second order tropical principal components computed from the Coelacanths genome and transcriptome data set is shown in Fig. 2.

Acknowledgements. R. Y. is supported by NSF Division of Mathematical Sciences: CDS&E-MSS program. Proposal number:1622369.

References

1. Akian, M., Gaubert, S., Viorel, N., Singer, I.: Best approximation in max-plus semimodules. Linear Algebra Appl. **435**, 3261–3296 (2011)
2. Ardila, F., Klivans, C.: The Bergman complex of a matroid and phylogenetic trees. J. Comb. Theory Ser. B **96**, 38–49 (2006)
3. Billera, L., Holmes, S., Vogtman, K.: Geometry of the space of phylogenetic trees. Adv. Appl. Math. **27**, 733–767 (2001)
4. Cohen, G., Gaubert, S., Quadrat, J.P.: Duality and separation theorems in idempotent semimodules. Linear Algebra Appl. **379**, 395–422 (2004)
5. Hedges, S.B.: Vertebrates (Vertebrata). In: Hedges, S.B., Kumar, S. (eds.) The Timetree of Life, pp. 309–314. Oxford University Press (2009)
6. Joswig, M., Essentials of Tropical Combinatorics (2017). http://page.math.tu-berlin.de/~joswig/etc/index.html
7. Levine, N.D.: Progress in taxonomy of the Apicomplexan protozoa. J. Eukaryot Microbiol. **35**, 518–520 (1988)
8. Liang, D., Shen, X.X., Zhang, P.: One thousand two hundred ninety nuclear genes from a genome-wide survey support lungfishes as the sister group of tetrapods. Mol. Biol. Evol. **8**, 1803–1807 (2013)
9. Lin, B., Sturmfels, B., Tang, X., Yoshida, R.: Convexity in tree spaces. SIAM Discrete Math. **3**, 2015–2038 (2017)
10. Maclagan, D., Sturmfels, B.: Introduction to Tropical Geometry, Graduate Studies in Mathematics, vol. 161. American Mathematical Society, Providence (2015)
11. Nye, T.: Principal components analysis in the space of phylogenetic trees. Ann. Stat. **39**, 2716–2739 (2011)
12. Pachter, L., Sturmfels, B.: Algebraic Statistics for Computational Biology. Cambridge University Press, New York (2005)
13. Speyer, D., Sturmfels, B.: Tropical mathematics. Math. Mag. **82**, 163–173 (2009)
14. Yoshida, R., Zhang, L., Zhang, X.: Tropical principal component analysis and its application to phylogenetics (2017). https://arxiv.org/abs/1710.02682

Mathematics Classroom Collaborator (MC2): Technology for Democratizing the Classroom

Sohee Kang[1($\boxtimes$)], Marco Pollanen[2], Sotirios Damouras[1], and Bruce Cater[2]

[1] University of Toronto Scarborough, Toronto, Canada
soheekang@utsc.utoronto.ca
[2] Trent University, Peterborough, Canada

Abstract. In any classroom, different groups of students may have unequal voices. This "lack of democracy" may be particularly problematic in STEM fields. To promote a more inclusive classroom, we developed and tested an online, real-time communication tool: the Mathematics Classroom Collaborator (MC^2). MC^2 makes the entry of mathematics easy and intuitive, it includes an option for anonymity, and it works on a variety of platforms, including smart phones, tablets, and notebook computers. In this paper, we share our experience with employing MC^2 in a statistics service course and an introductory probability course. We describe how this tool creates new communication models for the technologically-enhanced class — models that may help overcome social barriers to create a more inclusive environment, and that may lead to further democratization of learning, including increased participation by women and/or English-language learners. The results of an experiment to measure the effectiveness of MC^2 compared to Microsoft Word Equation for novice users are also presented.

Keywords: Technology-enhanced classes
Mathematics Classroom Collaborator · Democracy in the classroom

1 Introduction

Students' interactions with instructors and classmates have long been a cornerstone of learning in any field of study. For instructors, one challenge, therefore, is to create a welcoming and inclusive classroom atmosphere by building rapport with students, and to encourage them to take full advantage of office hours. In STEM disciplines, however, a number of studies, including Eddy et al. (2015) and Krupnick (1985), have suggested that certain groups — namely, female students and English-language learners — may not have an equal voice, both in in-class and out-of-class engagements, relative to other groups. To overcome this barrier and build more "democratic" classrooms, we developed a novel online technology: the Mathematics Classroom Collaborator (MC^2). In this paper, we will first describe MC^2's full functionality for students and instructors. We will

© Springer International Publishing AG, part of Springer Nature 2018
J. H. Davenport et al. (Eds.): ICMS 2018, LNCS 10931, pp. 280–288, 2018.
https://doi.org/10.1007/978-3-319-96418-8_33

then share our experience of utilizing this tool in an introductory course in Probability and a second course in Statistics. Finally, we will present, and discuss the implications of, the results of a small-scale experiment designed to measure the effectiveness of MC^2, relative to Microsoft Word 2016 Equation.

2   Functionality

MC^2 (http://mc2.trentu.ca) (Pollanen et al. (2017)) is a Web-based application that works on a range of different hardware platforms — tablets, smartphones, and laptop computers — running on a variety of different operating systems, including Windows, MacOS, iOS, and Android. No installation or "sign-up" stages are needed for students. Once an instructor creates a classroom, students join the room with the name of the classroom and any self-selected anonymous username (Fig. 1).

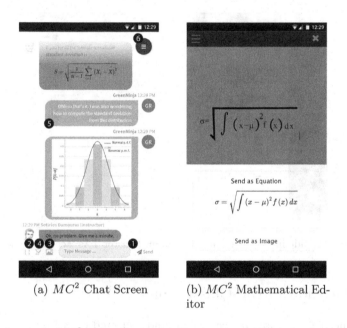

(a) MC^2 Chat Screen   (b) MC^2 Mathematical Editor

Fig. 1. Left panel: MC^2 Chat Screen (1) Text input field (2) Insert TEX into text field (3) Input image from camera or gallery (4) Launch the mathematical expression editor. (5) Dialogue Pane: clicking on an image or mathematical expression launches the expression editor for annotating or modification (6) Chat Option: tab to expand the options of On-line Users button and Logout button. Right Panel: MC^2 Mathematical Editor – non-TEX users can click on math symbols or Greek letters from the menu and make a diagrammatic equation that can be either converted to a TEX expression and sent to a text file, or sent directly to the dialogue pane as an image file.

The User-Side Interface: On its face, MC^2 is designed to be like a familiar texting application, with the addition of mathematical capabilities for both TEX and non-TEX users. TEX users can click the TEX icon and write TEX expressions with the extended chat window with the collection of commonly used TEX symbols for easy access for smart phone users. Non-TEX users can click the math icon to access the Math Editor where equations can be input using a diagrammatic equation editor. The symbols can be selected, moved, and resized based on the diagram editor UI. The expression is then recognized and converted to TEX which then be displayed by the messenger. Students can also take a picture from a smart phone, upload it, to share image of questions in the chat.

The Instructor-Side Interface: Instructors can do the following tasks: **create** a classroom; **announce (pin)** a specific message; **delete** the message; **zoom in/out** the message; **assign** TAs for the course; **access** the data of chat history; **upload and edit** the images from the built-in library and send it to chat for sharing with students (Fig. 2).

Fig. 2. Instructor side interface: create classroom, register TAs, access the data of chat history

3   Application in the Classroom

We introduced MC^2 into the classroom, and collected data from two different courses representing two different student audiences. The two courses were: (1) an introductory Probability course (STAB52, Fall 2017), consisting of 386 students, mostly majoring in Mathematics, Statistics and Computer Science and (2) a second service course in Statistics (STAB27, Winter 2018), consisting of 80 students, most of whom were majoring in the Social and Life Sciences. We now turn to a description of how the software was employed in each course, and of summary results from its use.

MC^2 in the Introductory Probability Course
The *Introduction to Probability* (STAB52) course followed a semi-inverted classroom format, where students were required to complete in-class worksheets

immediately following each lecture. The lecture and worksheet portions of the class were each one hour in length, occurring twice each week over a 12-week semester. For the worksheets, students could get help from the instructor and TAs and were allowed to work in groups, but they had to submit individual answers which were graded for credit. The course had three sections sharing the same instructor, and we chose one section per week to employ MC^2 during the worksheet activity, excluding the first two weeks, and the last week, of the semester. During the weeks when MC^2 was used in any given section, only the instructor answered student questions submitted through the tool, while the TAs provided face-to-face assistance. Students could post questions anonymously, and their questions and answers were visible to everyone.

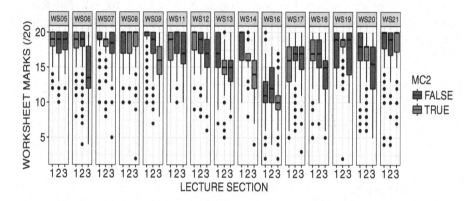

Fig. 3. Boxplots of worksheet marks distribution, grouped by worksheet number and lecture section.

To investigate the effect of MC^2 on student performance, we examined students' worksheet marks (graded out of 20). In total, the tool was used in 15 worksheets, and Fig. 3 presents side-by-side boxplots of the marks distribution by lecture section, worksheet number, and MC^2 usage. Considerable variability between worksheets and sections is evident, though the overall average was high (>17/20). During the first two weeks of classes, when MC^2 was not used, the average marks were 18.61, 18.04, and 17.85 for Sects. 1, 2, and 3, respectively. Figure 3 shows that, during the weeks of MC^2 use, average worksheet scores generally followed the same 1-2-3 ranking. There were, however, some exceptions to that rule that reveal mixed results. Most notably, on worksheet 19, when Sect. 2 was using MC^2, that section ranked first in average scores, for a 2-1-3 ranking. On the other hand, when Sect. 1 utilized MC^2 on worksheet 17, the section obtained the lowest average score (i.e., 2-3-1). So, at first glance, we cannot discern a consistent change in the relative ordering of average marks.

To help us isolate the effect of MC^2 on student performance, we ran an ANCOVA analysis on the data. Clearly, there are other factors besides MC^2 that affect worksheet marks. Variation both in the difficulty of topics across weeks and in individual student aptitude across sections could impact the marks. To control for these differences, we used the worksheet number and students' cumulative grade point averages (CGPA) at the start of the course as nuisance variables, with worksheet marks as the response variable. We also looked at the effect of MC^2 by section because of differences in the way sections were formed — Sect. 3, for example, was created only one week before the start of the semester from wait-listed students. The ANCOVA results are presented in Table 2, and show that all nuisance variables (LECTURE, WS, CGPA) are highly significant (p-value $\ll 10^{-10}$). The overall effect (across all sections) of MC^2 was not statistically significant, but it differs significantly across sections at the 5% level (p-value $= 0.01814$). In fact, looking at the fitted model parameters we find that MC^2 use in Sect. 3 leads on average to an extra $+0.49/20$ marks, relative to its effect in Sect. 1, but its use in Sect. 2 leads to $-.29/20$ fewer marks, relative to Sect. 1. These results, although not consistent across the board, suggest that MC^2 has a more pronounced positive effect when used by weaker students (Table 1).

Table 1. Analysis of Covariance Table: Average worksheet Marks by MC^2 use (MC2), lecture section (LECTURE), worksheet number (WS), student CGPA (CGPA), and MC2:LECTURE interaction.

Source	Df	SSE	MSE	F-value	P-value
MC2	1	1	0.69	0.0836	0.77244
LECTURE	2	2048	1023.88	123.9454	$<2e^{-16}$***
WS	14	14512	1036.60	125.4853	$<2e^{-16}$***
CGPA	1	1204	1203.85	145.7318	$<2e^{-16}$***
MC2:LECTURE	2	66	33.15	4.0131	0.01814*
Residuals	4764	39354	8.26		

Signif. codes: 0 '***' 0.001 '**' 0.01 '*' 0.05 '.' 0.1 ' ' 1

MC^2 in Second Course in Statistics

In the *Statistics II* (STAB27) course, students were from various disciplines and backgrounds. The class was composed of 49% female and 46% male students, with 5% undefined. We examined whether MC^2 usage depends on the gender or the status (international/domestic) of students. MC^2 was employed in weekly group work sessions and two sessions of online office hours before the final exam in the second half of the course.

We created a new dichotomous variable, Y, as follows: $Y = 0$, if MC^2 is not used; $Y = 1$, if MC^2 is used one or more times. A logistic regression was fitted to investigate which factors are statistically significant in terms of their impact on the engagement of students with MC^2.

$$Logit(\pi_i) = \beta_0 + \beta_1(GENDER) + \beta_2(LEG\_STATUS) + \beta_3(YEAR\_OF\_STUDY)$$

```
Coefficients:
                          Estimate Std. Error z value Pr(>|z|)
(Intercept)                0.06003    0.78618   0.076   0.9391
GENDER(M)                 -0.90656    0.51299  -1.767   0.0772.
LEG_STATUS(International) -0.51626    0.57096  -0.904   0.3659
YEAR_OF_STUDY            -0.05231    0.25646  -0.204   0.8384
---
Signif. codes:  0 '***' 0.001 '**' 0.01 '*' 0.05 '.' 0.1 ' ' 1
```

Controlling for the student's year of study and the legal status, the gender of the student was marginally significant factor (p-value < 0.1) in our estimation of the probability of being engaged in MC^2. Roughly 43.6% of female students used MC^2, while the figure was 24.3% for males, based on the Table 2.

Table 2. Two way table with MC^2 usage and Gender

Y = 0 (No, MC^2)		Y = 1 (Yes, MC^2)	
Female	Male	Female	Male
22	28	17	9

This partly supports Sankar et al. (2015) who found that, with the online bulletin board tool Piazza, female STEM students were more likely to use the anonymity feature than their male counterparts, although female undergraduate students ask 37% fewer questions than their male peers in computer science. There is also evidence that anonymity in the classroom boost the engagement (Jong et al. (2013)).

The instructor conducted two types of extended office hours before the final exam: face-to-face and online. During three hours of face-to-face office hours, only 4 students visited the instructor's office, while during one-hour online office hour, 26 students logged in the session and 14 students actively participated in conversations. Since students submit the questions simultaneously during online office hours, "**pinning a message**" is a very useful tool for the instructor to announce the message (it pins it to the top of the chat window) to which the instructor is responding. We have yet to assess the impact of MC^2 on students' learning in this course. Marginal evidence that it encouraged females students to engage in the course material through anonymous real-time communication does, however, suggest that the tool creates a more inclusive environment.

4   Experiment

To measure the efficiency of using the math editor in MC^2, relative to Microsoft Equation (ME), we conducted an experiment involving 14 first-year students (10 females and 4 males), none of whom had any previous experience in either software. We asked each student to enter each of the following 4 expressions into each of the two software tools:

1. $s = \sqrt{\frac{1}{n-1} \sum_{i=1}^{n} (x - \bar{x})^2}$ 2. $\int_0^\infty e^{-y} dy$ 3. $\int \frac{1}{\sqrt{2\pi}} e^{-\frac{x^2}{2}} dx$ 4. $\frac{\sqrt{\frac{\sqrt{x^2+2x+2}}{x}}}{x^2+1}$

A maximum time limit was set to allow students to submit incomplete equations. Figure 4 shows the completion time differences. Expression entry in MC^2 takes longer than ME for each expression. The differences are statistically significant.

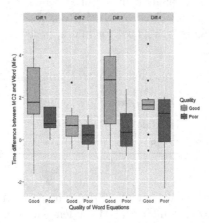

Fig. 4. Boxplots of completion time differences between MC^2 and Word of four equations, grouped by quality of completion by Word Equation

But students who submitted "poor" quality expressions, and who could thus be considered the most novice in the use of technology or mathematics, showed statistically insignificant differences for all 4 expressions, based on the non-parametric Wilcoxon signed-rank test. These results suggest that, for more novice users in particular, although its user-interface has room to improve, MC^2's usability could be comparable to Microsoft Equation (Table 3).

Table 3. Non-parametric test results by four equations of the poor quality group

Equation #	Eq. 1	Eq. 2	Eq. 3	Eq. 4
Non-parametric test statistic	9	7	7	6
P-value	0.1975	0.625	0.625	0.875

Potential errors in the entry of the square roots in expressions 1 and 4 in ME are illustrated by two examples:

$$s = \sqrt{\frac{1}{n-1}\sum_{i=1}^{n}(x - \bar{x})^2} \text{ and } \sqrt{\sqrt{\frac{\frac{x2+2x+2}{x}}{x2+1}}}$$

These entries show that students had difficulty understanding the nested concept in the software. A short survey was conducted to capture students' views on the usability of the two packages. On a scale of 1–5, the average response for MC^2 was 3 out of 5 for both quality groups, while the average usability score for ME was 2.75 for the poor quality group and 4.5 for the good quality group.

Students comments about the challenges of ME included: "having to find certain symbols and having to click on fractions or exponents (having to carefully manage when I press certain functions to get the equations I want)" and "You often end up deleting an equation when you're only trying to highlight it to add a function/square root on top". Comments on the challenges of MC^2 included: "Having to be precise on the position of each character", "at first, I did not know how to make the equation show up properly when converting, but then I got the hang of it", and "It was a little difficult moving things around".

There were also positive comments about MC^2, including: "I like how it is easy to move the symbols around to fit the equation", "The program was really cool. Really awesome how it can read the equations and send them in chat".

5   Conclusion

In STEM subjects, there appear to be barriers to communication that prevent certain groups, including women, from having an equal voice inside, and perhaps outside, the classroom. This is particularly apparent in the quantitative sciences, where communication may be suppressed by math or statistics anxiety. These barriers may then prevent women from achieving their full potential. One possibility for increasing communication is through the use of online technology and the anonymity that technology can provide. Mathematics communication, however, requires specialized user interfaces for input. To that end, we have developed the Web-based application, MC^2, that allows for easier entry and communication of mathematical expressions on all devices.

We introduced the use of MC^2 in a large first-year introductory probability course and a second course in Statistics. We found a statistically significant effect of MC^2 on average work sheet difference by sections, after controlling for other nuisance factors. We further found that females engaged at higher rates than males in MC^2 activities in the Statistics course. A small-scale experiment suggests that the functionality of the Math Editor in MC^2 is comparable to that in Microsoft Equation.

Research into the use of anonymity in mathematics and statistics courses is still in its infancy, and further research into how students write and perceive mathematical expressions is also required. The results presented here, however, suggest that the development of additional technologies to enable easy, anonymous communication and interaction in quantitative subjects is warranted.

Acknowledgment. This work was supported by a research grant from *eCampus Ontario*.

References

Eddy, S.L., Brownell, S.E., Thummaphan, P., Lan, M.C., Wenderoth, M.P.: Caution, student experience may vary: social identities impact a student's experience in peer discussions. CBE Life Sci. Educ. **14**(4), ar45 (2015)

Jong, B.-S., Lai, C.-H., Hsia, Y.-T., Lin, T.-W.: Effects of anonymity in group discussion on peer interaction and learning achievement. IEEE Trans. Educ. **56**, 292–299 (2013)

Krupnick, C.G.: Women and men in the classroom: inequality and its remedies. Teach. Learn. **1**(1), 18–25 (1985)

Pollanen, M., Kang, S., Cater, B., Chen, Y., Lee, K.: MC2: mathematics classroom collaborator. In: Proceedings of the Workshop on Mathematical User Interfaces, Edinburgh, UK (2017)

Sankar, P., Gilmartin, J., Sobel, M.: An examination of belongingness and confidence among female computer science students. ACM SIGCAS Comput. Soc. **45**(2), 7–10 (2015)

Software Citation in Theory and Practice

Daniel S. Katz[1](✉) and Neil P. Chue Hong[2]

[1] University of Illinois Urbana-Champaign, Urbana, USA
d.katz@ieee.org
[2] University of Edinburgh, Edinburgh, UK
n.chuehong@epcc.ed.ac.uk
https://www.software.ac.uk/

Abstract. In most fields, computational models and data analysis have become a significant part of how research is performed, in addition to the more traditional theory and experiment. Mathematics is no exception to this trend. While the system of publication and credit for theory and experiment (journals and books, often monographs) has developed and has become an expected part of the culture, how research is shared and how candidates for hiring, promotion are evaluated, software (and data) do not have the same history. A group working as part of the FORCE11 community developed a set of principles for software citation that fit software into the journal citation system, allow software to be published and then cited, and there are now over 50,000 DOIs that have been issued for software. However, some challenges remain, including: promoting the idea of software citation to developers and users; collaborating with publishers to ensure that systems collect and retain required metadata; ensuring that the rest of the scholarly infrastructure, particularly indexing sites, include software; working with communities so that software efforts "count"; and understanding how best to cite software that has not been published.

Keywords: Software citation · Credit · Software identifiers
Software metadata · Software repositories · Bibliometrics

1 Introduction

In most fields, computational models and data analysis have become a significant part of how research is performed, in addition to the more traditional theory and experiment. Evidence of the increased role and importance of software in today's research can be found in surveys and in papers, and while neither of these are specific to mathematics, it is likely no exception.

Two recent surveys, one of UK academics at Russell Group Universities [9, 10], and one of members of (US) National Postdoctoral Research Association [14, 15] asked researchers asked how important software is to them, and found that 67%/63% (UK/US respectively) of respondents said, "my research would not be possible without software." 21%/31% said, "my research would be possible

© Springer International Publishing AG, part of Springer Nature 2018
J. H. Davenport et al. (Eds.): ICMS 2018, LNCS 10931, pp. 289–296, 2018.
https://doi.org/10.1007/978-3-319-96418-8_34

but harder," while just 10%/6% said, "it would make no difference." A similar survey of mathematicians would be welcome.

One of the authors of this paper scanned six months of *Science* in mid-2013, and found that about half the papers were software-intensive projects, and most of the other papers also relied on some software. A formal study of 90 randomly selected papers in the biology literature in 2015 found that 80% mentioned software, and that those articles mentioned an average of 4.85 software packages [11]. A more recent study of *Nature* in Jan–Mar 2017 found software mentioned in 32 of 40 research articles, with an average of 6.5 software packages mentioned per article [16]. A similar study could be done of the mathematics literature. And while these studies have been manually performed by humans, natural language processing and machine learning could be used to expand their reach.

The system of publication and credit for theory and experiment (journals and books, often monographs) has developed and has become an expected part of the culture, how research is shared and how candidates for hiring, promotion are evaluated; software (and data) do not have the same history. In order to cite software, we could overload the current citation system to add software or alternatively, we could develop a new citation system that works for all kinds of products. As developing a new citation system would be very difficult, current efforts related to software citation have focused on the overloading approach.

2   Software Citation Principles

FORCE11[1] is a community of scholars, librarians, archivists, publishers and research funders that has arisen organically to help facilitate the change toward improved knowledge creation and sharing. In 2015 and 2016, a FORCE11 Software Citation working group developed a set of software citation principles [19]. The group grew to about 60 members, including researchers, developers, publishers, repository developer and maintainers, and librarians.

The group worked on GitHub[2] and on the FORCE11 web site[3]. It reviewed existing community practices and developed a set of use cases for software citation, and then drafted a software citation principles document. To do this, the group started with previously published data citation principles [5], updated them based on software use cases and related work, and further updated them based on working group discussions. This draft was then subjected to community feedback and review through a variety of channels, including a workshop at FORCE2016 in April 2016. In late 2016, the paper and its reviews were published [19]. The paper includes a set of six principles (general statements), use cases (where the principles should apply), and discussion (suggestions on how to apply the principles).

---

[1] https://www.force11.org.

[2] https://github.com/force11/force11-scwg.

[3] https://www.force11.org/group/software-citation-working-group.

The software citation principles, quoting from [19], are:

1. **Importance.** Software should be considered a legitimate and citable product of research. Software citations should be accorded the same importance in the scholarly record as citations of other research products, such as publications and data; they should be included in the metadata of the citing work, for example in the reference list of a journal article, and should not be omitted or separated. Software should be cited on the same basis as any other research product such as a paper or a book, that is, authors should cite the appropriate set of software products just as they cite the appropriate set of papers.
2. **Credit and Attribution.** Software citations should facilitate giving scholarly credit and normative, legal attribution to all contributors to the software, recognizing that a single style or mechanism of attribution may not be applicable to all software.
3. **Unique Identification.** A software citation should include a method for identification that is machine actionable, globally unique, interoperable, and recognized by at least a community of the corresponding domain experts, and preferably by general public researchers.
4. **Persistence.** Unique identifiers and metadata describing the software and its disposition should persist – even beyond the lifespan of the software they describe.
5. **Accessibility.** Software citations should facilitate access to the software itself and to its associated metadata, documentation, data, and other materials necessary for both humans and machines to make informed use of the referenced software.
6. **Specificity.** Software citations should facilitate identification of, and access to, the specific version of software that was used. Software identification should be as specific as necessary, such as using version numbers, revision numbers, or variants such as platforms.

There are now over 50,000 DOIs that have been issued for software, and more than 60% of them have been issued since the FORCE11 group published the first preprint of the principles paper [20].

3   Practices and Examples

In practice, the adoption of software citation depends on developing community guidelines that implement the software citation principles within the context of existing community scholarly communication and software development norms.

For some commonly used commercial software, there are mandatory citations, e.g. as specified by SAS [17] or Matlab [4]. In other cases, authors of research software may provide a recommended general citation referring to suite of related software, e.g. the HSL Mathematical Software Library [18]. However, in many of these cases, the citations do not provide enough information to allow crediting of the software authors (Principle 2), a machine actionable unique identifier (Principle 3) and persistent identifiers and metadata (Principle 4) or – in the case of HSL – an understanding of which version of the software was used (Principle 6).

Examples of mandatory and general software citations that do not fully implement the Software Citation Principles:

- The output for this paper was generated using SAS/STAT software, Version 14.1 of the SAS System for Unix. Copyright ©2018 SAS Institute Inc. SAS and all other SAS Institute Inc. product or service names are registered trademarks or trademarks of SAS Institute Inc., Cary, NC, USA.
- MATLAB and Statistics Toolbox Release 2012b, The MathWorks, Inc., Natick, Massachusetts, United States.
- HSL. A collection of Fortran codes for large scale scientific computation. http://www.hsl.rl.ac.uk/

Some software frameworks and platforms provide clear guidance on how to support particular versions or a specific citation for a package (Principle 6), e.g., by using the `citation()` function for R packages or the instructions for citing the GAP system for computational discrete algebra [23]. However these still do not provide persistent, machine actionable identifiers.

Examples of citations of specific packages as recommended by the software platform they are distributed with that mostly implement the principles:

- Maechler, M., Rousseeuw, P., Struyf, A., Hubert, M., Hornik, K. (2018). cluster: Cluster Analysis Basics and Extensions. R package version 2.0.7-1.
- Emma J. Moore, Christopher D. Wensley, groupoids - a GAP package, 1.54, 29/11/2017, https://gap-packages.github.io/groupoids/

However most software used in research does not provide guidance on how to cite it properly. If the software's website, or a CITATION file or README file with the source code, specifies how to cite the software, the author should use this information; this might be a reference to a software paper, or other publication. If the source code includes a codemeta.json [12] or Citation File Format (CFF) [7] file, the metadata in these files can be used with appropriate tooling to generate a citation automatically. Otherwise, the following guidance will help to construct a citation that implements the principles:

- For the authors, try to include all contributors to the software or, if this is not clear, name the project as the author. This may encourage some projects to make citation metadata available, including listing the authors.
- Include the name of the software, along with specific version/release information.

- Try to include a method for identification that is machine actionable, globally unique and interoperable. This ideally is a DOI but if there is no DOI, a URL pointing to a specific release might be the next best option.
- If there is a landing page that includes metadata, point to that, not directly to the software. Where you have the choice of pointing to a URL for general landing page including metadata, versus a specific URL (e.g. to a tag of a version) which does not contain sufficient metadata it is preferred to use the URL for the general landing page as the identifier, and clearly state the version.

Examples of citations for software using the suggested guidelines:

- Voevodsky, Vladimir and Ahrens, Benedikt and Grayson, Daniel and others. UniMath — a computer-checked library of univalent mathematics. https://github.com/UniMath/UniMath[accessed2018-04-27]
- Eigen Project. (2017). Eigen [software] version 3.3.4 Available from https://bitbucket.org/eigen/eigen/[accessed2018-04-27]

For developers of a piece of software, there are several things that can be done to make it easier for others to cite the software. At a minimum, the code should be published using a clear version number and license. If the code is in GitHub, the developer can make it easily citable using Github's integration with Zenodo [8]. Alternatively, the developer can manually deposit it in a digital repository such as Zenodo or Figshare – supplying metadata including the authors, title and version – and being provided with a Digital Object Identifier (DOI) and often a recommended citation that adheres to the Software Citation Principles. This information can be used to insert the citation that others should use into the software documentation, preferably as a CITATION file.

Example of a citation generated by Zenodo that implements the principles:

- Vince Knight, & Ria Baldevia. (2018, January 31). drvinceknight/ Nashpy: v0.0.13 (Version v0.0.13). Zenodo. http://doi.org/10.5281/ zenodo.1163694

Of course, the fact that swMath [21] exists means that citation should be integrated with it, providing suggested citations for software in it, and using it to track and understand citations of math software.

4   Challenges

In May 2017, the FORCE11 Software Citation Working Group ended, and a new Software Citation Implementation Working Group[4] started. This group has

---

[4] https://www.force11.org/group/software-citation-implementation-working-group.

the goal of moving the software citation principles to implementation. Those interested in following the new group can join it.

Many challenges remain, including:

- **Encouraging citation of software by authors**. Data citation is still not commonplace in many disciplines, let alone software citation. Author guidance for software citation is varied in the mathematical sciences. Both the Journal of Mathematical and Computer Simulation [22] and Journal of Statistical Software [13] provide guidance that follows the Software Citation Principles, but others - including the International Congress on Mathematical Software - do not. This will require the community to work with journals, conferences, and publishers to implement the Software Citation Principles in a way that they can be adopted by researchers in the area, similar to efforts in astronomy [2]. Tools such as CiteAs [1] may also help.
- **Promoting the idea of software citation to developers**. The benefits of making software more easily citable are not always obvious. The time taken to submit metadata can be reduced by the use of formats such as CodeMeta [12] and Citation File Format [7], particularly as they are adopted by repositories [3] and citation tools.
- **Citing unpublished software**. When authors do not publish their software, there is no archival link a citer can point to. The in-progress work to build a software archive for all source code by Software Heritage [6] may solve this problem.
- **Ensuring quality of information**. Even when information is provided, it may be discarded in the publication process. Collaboration with publishers, funders, and the identifier and citation infrastructure will be required to ensure that systems collect and retain required metadata, making it easier to discover and reuse software.
- **Giving credit for software through citation**. Ultimately, software citation will become widely practiced when the rest of the scholarly infrastructure, particularly indexing sites, includes software, and research communities recognize the value of software as a research output, thus providing an incentive for developers and authors to publish and reuse research software.

5   Conclusions

Although software citation is currently not standardized nor widely practiced, the publication of the Software Citation Principles has acted as a foundation on which to build community guidelines and improved tooling and infrastructure to support citation. The FORCE11 Software Citation Implementation Working Group is taking forward work to address the challenges standing in the way of software citation, and looks to the mathematical sciences community to work towards implementing the principles in the future.

References

1. Citeas. http://citeas.org/. Accessed 27 Apr 2018
2. American Astronomical Society: Citing repositories in AAS journals (AJ/ApJ) (2018). https://github.com/AASJournals/Tutorials/blob/master/Repositories/CitingRepositories.md. Accessed 27 Apr 2018
3. Caltech Library: Enhanced software preservation now available in CaltechDATA! (2018). https://www.library.caltech.edu/news/enhanced-software-preservation-now-available-caltechdata. Accessed 27 Apr 2018
4. Croucher, M.: How to cite MATLAB in research papers (2013). http://www.walkingrandomly.com/?p=4767. Accessed 27 Apr 2018
5. Data Citation Synthesis Group: Joint declaration of data citation principles. In: Martone, M. (ed.) FORCE 2011, San Diego, CA (2014). https://doi.org/10.25490/a97f-egyk
6. Di Cosmo, R., Zacchiroli, S.: Software heritage: why and how to preserve software source code. In: iPRES 2017: 14th International Conference on Digital Preservation, Kyoto, Japan (2017). https://hal.archives-ouvertes.fr/hal-01590958hal.archives-ouvertes.fr/hal-01590958/file/ipres-2017-software-heritage.pdf
7. Druskat, S.: Citation file format (CFF) (2017). https://doi.org/10.5281/zenodo.1003150. https://github.com/sdruskat/citation-file-format
8. GitHub: Making your code citable (2018). https://guides.github.com/activities/citable-code/. Accessed 27 Apr 2018
9. Hettrick, S.: It's impossible to conduct research without software, say 7 out of 10 UK researchers (2014). http://bit.ly/2B8y6Iz
10. Hettrick, S., Antonioletti, M., Carr, L., Chue Hong, N., Crouch, S., De Roure, D., Emsley, I., Goble, C., Hay, A., Inupakutika, D., Jackson, M., Nenadic, A., Parkinson, T., Parsons, M.I., Pawlik, A., Peru, G., Proeme, A., Robinson, J., Sufi, S.: UK research software survey 2014, December 2014. https://doi.org/10.5281/zenodo.14809
11. Howison, J., Bullard, J.: Software in the scientific literature: problems with seeing, finding, and using software mentioned in the biology literature. J. Assoc. Inf. Sci. Technol. **67**(9), 2137–2155 (2016). https://doi.org/10.1002/asi.23538
12. Jones, M.B., Boettiger, C., Mayes, A.C., Smith, A., Slaughter, P., Niemeyer, K., Gil, Y., Fenner, M., Nowak, K., Hahnel, M., Coy, L., Allen, A., Crosas, M., Sands, A., Chue Hong, N., Cruse, P., Katz, D.S., Goble, C.: CodeMeta: an exchange schema for software metadata. Version 2.0. (2017). https://doi.org/10.5063/schema/codemeta-2.0
13. Journal of Statistical Software: Journal of statistical software style guide. https://www.jstatsoft.org/pages/view/style. Accessed 27 Apr 2018
14. Nangia, U., Katz, D.S.: Survey of National Postdoctoral Association - dataset, August 2017. https://doi.org/10.5281/zenodo.843607
15. Nangia, U., Katz, D.S.: Track 1 paper: Surveying the U.S. National Postdoctoral Association Regarding Software use and Training in Research. Figshare (2017). https://doi.org/10.6084/m9.figshare.5328442.v3
16. Nangia, U., Katz, D.S.: Understanding software in research: initial results from examining nature and a call for collaboration. In: Proceedings of the 13th IEEE International Conference on eScience (eScience 2017) (2017). https://doi.org/10.1109/eScience.2017.78
17. SAS Institute Inc.: Referencing data analysis performed with SAS® software (2015). https://www.sas.com/en_us/legal/editorial-guidelines.html. Accessed 27 Apr 2018

18. Science & Technology Facilities Council: HSL a collection of Fortran codes for large scale scientific computation. http://www.hsl.rl.ac.uk/catalogue/. Accessed 27 Apr 2018

19. Smith, A.M., Katz, D.S., Niemeyer, K.E., FORCE11 Software Citation Working Group: Software citation principles. PeerJ Comput. Sci. **2**, e86 (2016). https://doi.org/10.7717/peerj-cs.86

20. Smith, A.M., Katz, D.S., Niemeyer, K.E., FORCE11 Software Citation Working Group: Software citation principles. PeerJ Prepr. **4**, e2169v1 (2016). https://doi.org/10.7287/peerj.preprints.2169v1

21. swMath: swMATH: an information service for mathematical software. http://www.swmath.org. Accessed 30 Apr 2018

22. Taylor & Francis: Taylor & Francis standard reference style—NLM. https://www.tandf.co.uk//journals/authors/style/reference/tf_NLM.pdf. Accessed 27 Apr 2018

23. The GAP Group: How to cite GAP (2018). https://www.gap-system.org/Contacts/cite.html. Accessed 27 Apr 2018

Identification of Errors in Mathematical Symbolism and Notation: Implications for Software Design

Seyeon Kim[1], Marco Pollanen[1], Michael G. Reynolds[2], and Wesley S. Burr[1(✉)]

[1] Department of Mathematics, Trent University, Peterborough, Canada
wesleyburr@trentu.ca
[2] Department of Psychology, Trent University, Peterborough, Canada

Abstract. Mathematical user interfaces for authoring, collaboration, problem-solving and reasoning invariably rely on the ability to read, write and manipulate complex mathematical expressions. However, very little research has been done on how people read mathematical expressions, let alone how they are understood by the mind. One technique which researchers use to gain insight into how people read and comprehend symbols and complex phenomena are studies using eye-tracking hardware: focus on, and tracking of, pupils in order to determine the reader's attention and fixation. In this paper we will explore the results of a study on two classes of students: mathematically "expert" (mathematical sciences students) and non-expert (Faculty of Science majors from outside the mathematical sciences). Each participant was presented with a series of mathematical problems (stimuli) and their eyes and attention/focus tracked as they worked through the problems mentally. We will discuss the differences in the two classes, both with respect to the correctness of responses to the problems and the structure of the scanning and identification of important components within the problem. This study has applications in mathematical software usability, accessibility, and design of interfaces, as comprehension of mathematical notation and formalism is assumed in the implementation of the modified symbolism inherent in structured mathematical software interfaces.

Keywords: Mathematical notation · Symbolism · Eye-tracking
Problem identification · Mathematical software interfaces

1 Introduction

Mathematics is limited and structured by the human brain and mental capacity. In terms of brain- and mind-based mathematics, eye movement research can reveal much about the working of the brain and mind, since our perception of the world is heavily influenced by our sense of sight. Given their importance for eye-mind research, surprisingly little is known about the cognitive basis of

© Springer International Publishing AG, part of Springer Nature 2018
J. H. Davenport et al. (Eds.): ICMS 2018, LNCS 10931, pp. 297–304, 2018.
https://doi.org/10.1007/978-3-319-96418-8_35

reading, understanding and solving mathematical questions. For this we investigate implicit mathematical practices by conducting an eye-tracking experiment
on the answering of mathematical questions. The eye-mind hypothesis [5] claims
a correlation between the cognitive processing of information and the person's
gaze at the specific location of the information. Implicit mathematical practices
can be made explicit by direct comparison of practices by expert (math familiar)
and non-expert (math unfamiliar) people.

2   Methods

Subjects: Twenty upper-year and graduate mathematics students (class: *expert*)
and eighteen science (non-mathematics) students (class: *non-expert*) volunteered
to participate in the present study. All provided written informed consent, and
were verbally debriefed at the end of the experiment. Trent University's ethics
committee approved all experiments presented here, which were carried out in
accordance with the provisions of the World Medical Association Declaration
of Helsinki [1]. All participants reported normal or corrected-to-normal visual
acuity.

Stimuli: The stimulus for the present study consisted of fourteen mathematical
questions, all at approximately an 11th to 12th grade (secondary school) level.
The questions ranged from True/False to "Find the Error", with one particular
question (the longest) showing particularly interesting findings, which we will
discuss below. Figures 1 and 2 show an example of the average question type
and the specific interesting problem (Question 07), respectively. Question 07 has
a sequence of mathematical equations, joined by logical statements (in words,
to the right).

$$\Omega = \{1, 2, 3, \cdots, 9, 10\},$$

$$A = \{1, 3, 6, 9\}, \text{ and}$$

$$B = \{2, 6, 8, 10\}.$$

True or False: if $x \in (A \cap B)$ then $x = 6$.

Fig. 1. Stimuli for Question 01 of 14. Stimuli was displayed full-screen on a 1400×900
pixel consumer monitor, as described in Procedure below.

Materials: Eye-movements were tracked using a Gazepoint G3 model eye-tracker. Calibration, stimulus presentation and response collection were controlled using Gazepoint Analysis Software, Standard Edition v4.2.0. Stimuli were presented to participants on an LG monitor attached to a Dell Optiplex (Intel i3) desktop running Windows 7 Personal service pack 1.

Find the error in the following.

1. $b = a$ We'll start by assuming this is true
2. $ab = a^2$ Multiply both sides by a
3. $ab - b^2 = a^2 - b^2$ Subtract b^2 from both sides
4. $b(a - b) = (a - b)(a + b)$ Factor both sides
5. $b = a + b$ Divide both sides by $a - b$
6. $b = 2b$ We started assuming $a = b$
7. $1 = 2$ Divide both sides by b.

Fig. 2. Stimuli for Question 07 of 14, the focus of this paper. Stimuli was displayed full-screen on a 1400×900 pixel consumer monitor, as described in Procedure below.

Procedure: Each subject was tested individually. The experimental session consisted of 2 phases. During the first phase of the experiment the eye-tracker was calibrated to the subject. A 9-point calibration was used. Calibration was not accepted until all calibration points were within default criterion. During the second phase of the experiment subjects were instructed to determine an answer to the posed question (either True/False or "Find the Error"). The problems included set theoretic, function, and matrix notation, as well as several common algebraic and arithmetic logical errors.

Each subject answered a total of 14 questions. The questions were presented in the same order for each subject. Only the particularly interesting case of Question 07, a "Find the Error" problem containing a logical fallacy, is analyzed in this paper. Each trial started by reminding the subject to read the problem carefully in order to determine what was being asked. A trial ended when the subject looked off-screen in the bottom right hand corner and telling the researcher. The researcher then asked them to indicate their response to the question, and the trial was coded as correct or incorrect. Participants had a short break in between trials, and at the end they were briefly interviewed to determine if any problems were encountered. No major occurrences were recorded for problems in the trials.

Data Analysis: Prior to conducting statistical analyses of Question 07, blind coders examined the fixation maps and identified two subjects that lost calibration during the study. The data from these participants was not analyzed.

One participant from the expert class was an outlier on all measures and was also excluded from all analyses. The remaining data were analyzed in two steps. First, a qualitative examination of individual scan paths led to the identification of behavioral sequences that capture how students answered the question. A second quantitative examination was also performed in order to assess whether there were any clear differences in performance using standard [2–4] eye-tracking techniques.

3   Quantitative Assessment

Four global measures of performance were examined: (1) total time until completion; (2) accuracy in identifying the source of the error; (3) the total number of eye-movements; and (4) the average fixation duration. Summaries of the numerical results can be found in Table 1. No statistical difference was found for total time to complete, $F(1, 33) = 0.818$, $p = 0.372$; total number of fixations/eye-movements, $F(1, 33) = 0.573$, $p = 0.454$; or average fixation duration, $F(1, 33) = 1.153$, $p = 0.291$. However, there was some suggestion that the novice group made more errors, $F(1, 33) = 3.513$, $p = 0.0698$.

As noted above the reading of the question can be broken down into two distinct periods: the first pass, where the initial attempt to solve the problem took place, and a second pass which either constituted a double check (for those who identified the error) or a second attempt to identify the error (for those who did not find it during their first attempt). For the first pass, the expert class completed their scan-through much more quickly than the novices, $F(1, 33) = 4.862$, $p = 0.0345$ (with significantly larger difference if three outliers are removed from consideration, $F(1, 30) = 10.8$, $p = 0.00259$). For the outlier-removed samples, the difference in the average time for the first pass is 14 s, with non-experts taking as much as 50% longer to scan the question. This difference was reflected in more fixations for the non-expert class, $F(1, 33) = 6.68$, $p = 0.015$.

As well, during the first pass, the expert class was much faster at reaching the location of the error (9 s, $F(1, 32) = 10.42$, $p = 0.00288$). Much of this time appeared to be due to the non-expert class participants spending more time reading the first three rows of the problem, seemingly in an attempt to understand the nature of the logical progression between stages.

4   Qualitative Assessment

For the qualitative analysis, videos were created from the fixation data for each subject. The videos were then examined by research assistants blind to the experimental conditions of the study ("blind coders"). The coders were instructed to watch the videos and identify and tag behavioural tokens that give insight into how people were solving the problems. Several behavioural markers were identified and are discussed below.

Many participants started by orienting themselves to the math problem. For some, this only consisted of reading the instructions, whereas others both

Table 1. Results for a variety of metrics across the two (expert and non-expert) classes of participants in the study.

			Expert (N = 18)	Novice (N = 17)
Overall performance	Total time (s)	Mean	64	71
		SD	50	57
	Total fixations	Mean	159	173
		SD	24	23
	Average fixation Duration (ms)	Mean	306	316
		SD	20	32
	Accuracy (%)		39	12
			50	33
First pass	Total time (s)	Mean	36	47
		SD	16	15
	Total fixations	Mean	82	102
		SD	36	34
	Average fixation Duration (ms)	Mean	306	323
		SD	27	38
Error fixation	Start time (s)	Mean	26	35
		SD	8	9
	Total time (s)	Mean	6	12
		SD	5	13
	Total fixations	Mean	16	28
		SD	12	29
	Average fixation Duration (ms)	Mean	303	327
		SD	36	78

read the instructions and did a quick scan of the problem before starting to read the mathematical statements. Surprisingly, some people neither read the instructions nor explored the problem space. Unfortunately, we have been unable to determine if this behaviour predicts performance at this time.

The initial exploration of the problem space was followed by a first pass through the problem. The first pass consisted of the period after the initial exploration and continued until there was clear evidence that the problem was being re-read from the beginning. Two broad solutions were used during the first pass. One, which we identified as *linear reading*, consisted of reading each statement in the problem left to right and then proceeding to the next statement in a top to bottom fashion. The other approach was more recursive and had participants moving both forwards and backwards between adjacent statements. For instance, the first statement would be read left to right before moving on to the second statement. After the second statement was read left to right, such a participant would return to the previous statement and begin a brief amount of time iterating between the two statements. The iterative approach appeared to be preferred by the experts though it was demonstrated by both groups.

$$3.\ ab - b^2 = a^2 - b^2 \quad \bullet \text{ Subtract } b^2 \text{ from both sides}$$
$$4.\ b(a - b) = (a - b)(a + b) \quad \circ \text{ Factor both sides}$$

Fig. 3. Example for Question 07 of a single expert participant's behaviour, showing the triangular pattern common with this class of participant: (starting at the black fixation) read left to right through a statement, then return to the previous statement's mathematical expression (grey fixation) for confirmation of logical linkage.

After the first pass through the problem, a second reading commonly took place. Examination of the scan path data suggested that multiple strategies were employed when rereading the problem. There was substantial variability in how much time was spent re-reading the problem, with some spending as little as 3 s and others spending over a minute. Some seemed to double check their answers, whereas others proceeded to reread the entire problem again from beginning to end. Future research will establish whether these differences are due to the type of question, or represent individual strategies.

Finally, additional signatures were observed when lines 6 and 7 of Question 07 were read. These lines occur after the *first* mathematical error (line 5) but contain evidence that a mathematical error has occurred ($b = 2b$ and $1 = 2$, respectively). Often when these lines were read, participants exhibited large eye-movements back and forth between the statements that $a = b$ (located on the first line) and line 6. We took these large movements as evidence that people were aware that an error was made (though not necessarily that they know what the error was). These large eye-movements appeared more common for the novice group than for the expert group suggesting either they were unprepared for them, or they were unaware that the error likely occurred on an immediately preceding line.

4.1   Non-expert Participants

In general, the non-expert participants in the study tended toward exhaustive, linear reading of the problem. They spent more time examining the associated text of each statement rather than the symbols, didn't recognize the error when they found it, and were more erratic after observing that an error had occurred (often a line or two after the actual error).

4.2   Expert Participants

In general, expert participants tended toward triangular (iterative) behaviour (example in Fig. 3, moving carefully back and forth between steps in a triangle pattern, starting at black and ending at grey). Such participants tended to be self-terminating (ending the problem once the error had been found), mostly recognized the error when it occurred, and performed a variety of re-visiting techniques for checking work.

5 Discussion

This paper examines the interplay between visual processing of mathematical symbolism and problem identification. There are a few obvious limitations to the study. First, while eye-tracking provides a "gaze into the mind's eye" [5], tracking eye movements does not actually give direct knowledge of thought processes (and problem solving technique). Secondly, although the "expert" class participants were recruited from the mathematics discipline (majors and joint majors) at Trent University, and the "non-expert" class from non-mathematical sciences, there remains a strong logical-mathematical intelligence in many non-mathematical science students, making the division less clear than would be preferred.

Previous eye movement research [2,3] has shown that the duration and frequency of fixations and number of regressions are related to the level of mental processing and effort needed to decipher the mathematical notation and symbolism [6,7] used in the problem. Thus, the longer and more often participants fixate on a particular component of notation, the more mental time and effort they are placing on trying to understand it. This implies that the observed differences in fixation number and transition time represent additional cognitive load on the part of non-expert participants, something also observed in recent studies of software interfaces [4] using similar techniques.

This study is relatively unique, as very few studies have looked at eye-tracking in the context of mathematical symbolism and error tracking. There are obvious applications to both mathematical software development and study and development of interfaces for the same. The expert participants understood and made use of the information logic paradigm, connecting the separate components of the derivation in seamless and integrated fashion, while the non-expert participants did not. This suggests that when presenting users with novel interface paradigms that *training* could be undertaken to demonstrate and exhibit proper logical pathways for understanding and mastery. This is particularly true when dealing with the necessary complexities of mathematical software, where symbols abound and, by necessity, interfaces tend toward the complex and hierarchical versus simple and flat.

In addition, there is a common trend in technology for devices to become smaller and smaller, as we move more functionality to smartphone-sized devices. The question for interface designers is then how to capture the key and essential functionality in such a small amount of screen real estate, especially for educationally oriented mathematical software. For example, software which allows students to compose complex mathematical formula on a smartphone using only touch requires a complex hierarchy of menus to encapsulate the required symbols

– we hope this study will assist in understanding and development of how users process and iterate through such symbolism.

In conclusion, this study of mathematical symbolism and error determination has applications in mathematical software usability, accessibility, and design of interfaces, as comprehension of mathematical notation and formalism is assumed in the implementation of the modified symbolism inherent in structured mathematical software interfaces. In particular, common user experience paradigms are used in many pieces of mathematical software, many of which are derived from the mathematics underlying their key functionality.

Acknowledgments. This work was supported by research grants from *eCampus Ontario* and Trent University's University Research Grants Program (URGP).

References

1. World Medical Association: Declaration of Helsinki: ethical principles for medical research involving human subjects. JAMA **310**(20), 2191–2194 (2013)
2. Clifton, C., Staub, A., Rayner, K.: Eye movements in reading words and sentences. In: Van Gompel, R.P.G., et al. (eds.) Eye Movements, pp. 341–371 (2007)
3. Holmqvist, K., Nyström, M., Andersson, R., Dewhurst, R., Jarodzka, H., Van de Weijer, J.: Eye Tracking: A Comprehensive Guide to Methods and Measures. OUP, Oxford (2011)
4. Kohlhase, A., Fürsich, M.: Understanding mathematical expressions: an eye-tracking study. In: FM4M/MathUI/ThEdu/DP/WIP@ CIKM, pp. 42–50 (2016)
5. Bone, M.: A gaze into the mind's eye: gaze as an indicator of cortical reinstatement during mental imagery. Unpublished doctoral dissertation, University of Toronto, Toronto, ON, Canada (2015)
6. Blostein, D. and Grbavec, A.: Recognition of mathematical notation. In: Handbook of Character Recognition and Document Image Analysis, pp. 557–582 (1997)
7. MacGregor, M., Stacey, K.: Cognitive models underlying students' formulation of simple linear equations. J. Res. Math. Educ. **24**(3), 217–232 (1993)

Image Analysis: Identification of Objects via Polynomial Systems

Robert H. Lewis(⊠)

Fordham University, New York, USA
rlewis@fordham.edu
https://fordham.academia.edu/RobertLewis

Abstract. The problem is to identify a movable object that is in some sense known, if it is encountered later. Suppose we have a sensor, on a fixed radar station or a moving platform. We have an object, say object A, previously measured, with certain distinct identifiable points p_i. We know the distances between these points. We later encounter a similar object B and want to know if it is A. We have a sensor that sends and receives electronic signals, and so we measure the distances t_i from the sensor to the distinguished points on B.

We first consider the two-dimensional case. Assume there are three distinct points on A. We have our measured distances t_1, t_2, t_3 and previously known distances between the points on A, d_1, d_2, d_3. We derive a polynomial system relating these quantities and show that it is easy to solve yielding a *resultant* that is the "signature" for A. Its use will eliminate B if B is not A.

The generalization to three dimensions is immediate. We need a fourth point. The polynomial system contains many parameters, but we solve it symbolically. We then discuss generalizations involving flexibility. In those cases we need five points and the systems are much more complex.

We compare solutions on *Magma*, *Maple*, and *Fermat* computer algebra systems.

Keywords: Image analysis · Polynomial system · Resultant
Parameters · Dixon · Gröbner basis

1 Introduction

Identifying an object if it moves or if the perspective changes is a classic problem in image recognition and computer vision [8,11]. In [8] we considered and solved the so-called "Six-Line Problem", in which a 3D object like a building has six distinguished lines. The object is considered "known" via these lines. The problem is to decide if a similar object encountered later from another viewpoint is in fact the same object. Via techniques of algebraic geometry, a system of polynomial equations was derived. The *variables* in these equations were the transformation coordinates, and the *parameters* specified the characteristics of the lines.

© Springer International Publishing AG, part of Springer Nature 2018
J. H. Davenport et al. (Eds.): ICMS 2018, LNCS 10931, pp. 305–309, 2018.
https://doi.org/10.1007/978-3-319-96418-8_36

There were four equations in three variables and 13 parameters. The equations were solved *symbolically*, i.e., the thirteen parameters are retained as symbolic names. The *resultant* [2,12], a single polynomial in 239 terms, was computed with the Dixon method [5,7]. To use this on a test object, numerical values for its 13 parameters would be substituted. If the result is not 0, this is not the original object. If it is 0, it is highly likely to be the original object.

In this work we consider a similar but actually simpler situation. Suppose we have a sensor, on a fixed radar station or a moving platform. We have an object, say object A, previously measured, with certain distinct identifiable points p_i. We know the distances between these points, d_i. We later encounter a similar object B and want to know if it is A. The sensor sends and receives electronic signals, and so we measure the distances t_i from the sensor to the distinguished points on B.

We first consider the two-dimensional case. Assume there are three distinct points on A. We have our measured distances t_1, t_2, t_3 and previously known distances between the points on A, d_1, d_2, d_3. We derive a polynomial system relating these quantities and show that it is easy to solve yielding a resultant that is the "signature" for A. Its use will eliminate B if B is not A.

The generalization to three dimensions is immediate. We need a fourth point. The polynomial system contains many parameters, but we solve it symbolically. We then discuss generalizations involving flexibility. In those cases we need five points and the systems are much more complex.

2   Polynomial Systems

The theory of eliminating variables from a system of equations has a long history, starting with Bezout around 1760. A key idea is the *resultant* of a system of polynomial equations [2,12]. Bezout did this for one-variable polynomials. Dixon [3] extended it to multivariate polynomials, and proved it would work in a certain ideal situation. However, for real problems the ideal situation rarely applies and often the method seems to fail. Kapur, Saxena, and Yang showed how to get around all those problems in 1994 [1,5]. Lewis refined and greatly improved the method in 2008 [7] to what is called Dixon-EDF. Gröbner bases can also be used to eliminate variables [12].

The Kapur-Saxena-Yang (KSY) method seems to have not been noticed in some research communities. Many researchers still try to use Gröbner bases even though that technique frequently fails, especially when there are parameters. In [6] it was shown than when there are parameters, Dixon-EDF is often enormously superior. We find that to be true in this work.

3   The Standard 2D and 3D Cases

In two dimensions suppose the sensor is at point (x, y). Assume there are three distinct points on A. We have our measured distances to B, t_1, t_2, t_3 and previously known distances between the points on A, d_1, d_2, d_3. Let the coordinates of the three points be $(x_1, y_1), (x_2, y_2), (x_3, y_3)$. If A = B we get six equations (set each to 0):

$$(x - x_1)^2 + (y - y_1)^2 - t_1^2, \quad (x - x_2)^2 + (y - y_2)^2 - t_2^2,$$
$$(x - x_3)^2 + (y - y_3)^2 - t_3^2, \quad (x_1 - x_2)^2 + (y_1 - y_2)^2 - d_1^2,$$
$$(x_1 - x_3)^2 + (y_1 - y_3)^2 - d_2^2, \quad (x_2 - x_3)^2 + (y_2 - y_3)^2 - d_3^2$$

We don't know the x_i and y_i so we have six equations in eight variables. However, by choice of coordinate system we can assume $x_1 = y_1 = 0$ and $y_2 = 0$, so there are really five variables. We compute the resultant very easily, in a fraction of a second with Dixon-EDF on Fermat [9,10]. Gröbner bases on Magma and Maple also succeed in a bit more time. The answer has 22 terms:

$$d_1^2 t_3^4 + d_3^2 t_2^2 t_3^2 - d_2^2 t_2^2 t_3^2 - d_1^2 t_2^2 t_3^2 - d_3^2 t_1^2 t_3^2 + d_2^2 t_1^2 t_3^2 - d_1^2 t_1^2 t_3^2 - d_1^2 d_3^2 t_3^2 - d_1^2 d_2^2 t_3^2 +$$
$$d_1^4 t_3^2 + d_2^2 t_2^4 - d_3^2 t_1^2 t_2^2 - d_2^2 t_1^2 t_2^2 + d_1^2 t_1^2 t_2^2 - d_2^2 d_3^2 t_2^2 + d_2^4 t_2^2 - d_1^2 d_2^2 t_2^2 + d_3^2 t_1^4 + d_3^4 t_1^2 -$$
$$d_2^2 d_3^2 t_1^2 - d_1^2 d_3^2 t_1^2 + d_1^2 d_2^2 d_3^2$$

In three dimensions we need four points; see Fig. 1. There are now six mutual distances and four t_i, yielding ten equations just like the six above. After assuming $x_1 = y_1 = z_1 = y_2 = z_2 = z_3 = 0$, there are nine variables to eliminate. Dixon-EDF takes 5.2 s and 20 MB of RAM. Maple running Faugere's FGb package [4] takes 17 s and 2.2 GB of RAM. Magma did not complete in 60 min. The resultant has 130 terms.

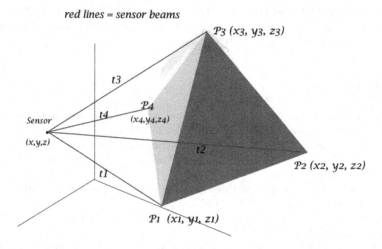

Fig. 1. Sensor and object in 3D, showing four points on yellow object. (Color figure online)

4   Flexible Objects

Motivated by airplanes that can flex their wings, we consider two cases: planes in which the wings rotate up and down, pivoting around the fuselage, and those in which the wings are swept back, maintaining the same altitude.

It is not hard to derive equations for these cases. A fifth point is needed, as not all of the six mutual distances among four points remain fixed. These are the ten equations for the first case:

$$z^2 + y^2 + x^2 - t_1^2, \ z^2 + y^2 + x^2 - 2x_2x + x_2^2 - t_2^2,$$
$$z^2 - 2z_3z + y^2 + x^2 - 2x_3x + z_3^2 + x_3^2 - t_3^2,$$
$$z^2 - 2z_4z + y^2 - 2y_4y + x^2 - 2x_4x + z_4^2 + y_4^2 + x_4^2 - t_4^2,$$
$$z^2 - 2z_4z + y^2 + 2y_4y + x^2 - 2x_4x + z_4^2 + y_4^2 + x_4^2 - t_5^2,$$
$$x_2^2 - d_1^2, \ z_3^2 + x_3^2 - 2x_2x_3 + x_2^2 - d_2^2, \ z_3^2 + x_3^2 - d_3^2,$$
$$z_4^2 + y_4^2 + x_4^2 - 2x_2x_4 + x_2^2 - d_4^2, \ z_4^2 + y_4^2 + x_4^2 - d_5^2$$

In the first case, Dixon-EDF finishes in 7 s, using 32 meg of RAM. The resultant has 2373 terms. Maple with FGb takes 9 min and 6.8 gig of RAM. Magma did not finish in 21 h.

In the second case, Dixon-EDF finishes in 14 s, using 100 meg of RAM. The resultant has 74668 terms. Maple with FGb crashed after 19 h and 62 gig of RAM. Magma did not finish in 7 h.

Further generalizations are possible. Another flexible configuration, not immediately identifiable as an airplane, takes Dixon-EDF 2 to 4 h to compute (depending on the variation of EDF used) and has 189441 terms. Maple with FGb crashed after 9 h and 60 gig of RAM. Magma was killed after 100 h.

One can also imagine scenarios in which five points are needed because one may not be visible to the sensor at times.

References

1. Buse, L., Elkadi, M., Mourrain, B.: Generalized resultants over unirational algebraic varieties. J. Symbolic Comp. **29**, 515–526 (2000)
2. Cox, D., Little, J., O'Shea, D.: Using algebraic geometry. In: Graduate Texts in Mathematics, vol. 185. Springer, New York (1998). https://doi.org/10.1007/b138611
3. Dixon, A.L.: The eliminant of three quantics in two independent variables. Proc. London Math. Soc. **6**, 468–478 (1908)
4. Faugere, J.-C.: A new efficient algorithm for computing Gröbner bases (F4). J. Pure Appl. Algebra **139**, 61–88 (1999)
5. Kapur, D., Saxena, T., Yang, L.: Algebraic and geometric reasoning using Dixon resultants. In: Proceedings of the International Symposium on Symbolic and Algebraic Computation. ACM Press (1994)

6. Lewis, R.H.: Dixon-EDF: The Premier Method for Solution of Parametric Polynomial Systems. In: Kotsireas, I.S., Martinez-Moro, E. (eds.) Applications of Computer Algebra, Proceedings in Mathematics & Statistics, Kalamata, Greece, vol.
   198. Springer, Cham (2017). https://doi.org/10.1007/978-3-319-56932-1_16

7. Lewis, R.H.: Heuristics to accelerate the Dixon resultant. Math. Comput. Simul.
   77(4), 400–407 (2008)

8. Lewis, R.H., Stiller, P.: Solving the recognition problem for six lines using the
   Dixon resultant. Math. Comput. Simul. **49**, 203–219 (1999)

9. Lewis, R.H.: Computer algebra system Fermat. http://home.bway.net/lewis/

10. Lewis, R.H.; Fermat code for Dixon-EDF. http://home.bway.net/lewis/dixon

11. Stiller, P.: Symbolic computation of object/image equations. In: Proceedings of the
    International Symposium on Symbolic and Algebraic Computation, pp. 359–364.
    ACM Press, New York (1997)

12. Sturmfels, B.: Solving systems of polynomial equations. In: CBMS Regional Con-
    ference Series in Mathematics, vol. 97. American Mathematical Society (2003)

Resultants, Implicit Parameterizations, and Intersections of Surfaces

Robert H. Lewis$^{(\boxtimes)}$

Fordham University, New York, USA
rlewis@fordham.edu
https://fordham.academia.edu/RobertLewis

Abstract. A fundamental problem in computer graphics and computer aided design is to convert between a parameterization of a surface and an implicit representation of it. Almost as fundamental is to derive a parameterization for the intersection of two surfaces.

In these problems, it seems that resultants, specifically the Dixon resultant, have been underappreciated. Indeed, several well known papers from ten to twenty years ago reported unsuitability of resultant techniques. To the contrary, we show that the Dixon resultant is an extremely effective and efficient method to compute an implicit representation.

To use resultants to compute a parameterization of an intersection, we introduce the concept of an "implicit parameterization." Unlike the conventional parameterization of a curve where x, y, and z are each explicitly given as functions of, say, t, we have three implicit functions, one each for (x, t), (y, t), and (z, t). This concept has rarely been mentioned before. We show that given a (conventional) parameterization for one surface and either an implicit equation for the second, or a parameterization for it, it is straightforward to compute an implicit parameterization for the intersection. Doing so is very easy for the Dixon resultant, but can be very daunting even for well respected Gröbner bases programs.

Further, we demonstrate that such implicit parameterizations are useful. We use builtin 3D plotting utilities of a computer algebra system to graph the intersection using our implicit parameterization. We do this for examples that are more complex than the quadric examples usually discussed in intersection papers.

Keywords: Surface · Polynomial system · Resultant · Dixon
Parameters · Intersection · Gröbner basis

1 Resultants and Implicitization

A classic problem in computer graphics and computer aided design is to derive an implicit equation for a surface given a parameterization of it. Since our surfaces are in three-dimensional space, we conventionally have three equations

$$x = f(s, t).$$
$$y = g(s, t).$$
$$z = h(s, t).$$

© Springer International Publishing AG, part of Springer Nature 2018
J. H. Davenport et al. (Eds.): ICMS 2018, LNCS 10931, pp. 310–318, 2018.
https://doi.org/10.1007/978-3-319-96418-8_37

If homogeneous coordinates are being used, there is a fourth equation for w.

The implicit equation is produced by eliminating the s and t. As a very simple two-dimensional example, for a circle of radius r, the parametric equations are $x = r\cos(\theta), y = r\sin(\theta)$. It is easy to eliminate θ by squaring and adding:

$$x^2 + y^2 = r^2 \cos^2(\theta) + r^2 \sin^2(\theta) = r^2$$

yielding the familiar equation for a circle. (r is not a variable, but a *parameter* in the other sense of the word "parameter.") Real examples of interest are much more complicated than this, and sophisticated elimination techniques are needed.

The simple example illustrates an important idea. Parametric systems frequently involve trig functions, usually sine and cosine. Elimination techniques usually require polynomial (or rational) functions. A system with sine and cosine is easily converted to a polynomial system by replacing cosine with, say, ct, sine with st, and adding a new equation $ct^2 + st^2 - 1 = 0$.

The theory of eliminating variables from a system of equations has a long history, starting with Bezout around 1760. A key idea is the *resultant* of a system of polynomial equations [3,18]. Bezout did this for one-variable polynomials. Dixon [4] extended it to multivariate polynomials, and proved it would work in a certain ideal situation. However, for real problems the ideal situation rarely applies and often the method seems to fail. Kapur, Saxena, and Yang showed how to get around all those problems in 1994 [1,5]. Lewis refined and greatly improved the method in 2008 [7] to what is called Dixon-EDF. Gröbner bases can also be used to eliminate variables [18].

In spite of the 1994 publication, the Kapur-Saxena-Yang (KSY) method seems to have not been noticed by the computer graphics community. In 2000 the authors of [2] explicitly reject resultants as unworkable. In 2004 Wang [19] was aware of the Bezout-Dixon method but not KSY. He develops a new method to implicitize surfaces and tests fifteen examples with his method, resultants, and Gröbner bases. As in [2] he reports that in many cases resultants will not work because the Dixon method returns 0. This is one of the situations that KSY overcomes!

In the following table we compare Wang's reported time using pre-KSY Dixon, Wang's method, and our solution today using Dixon-KSY-EDF. Our computer is 2.3 times faster than Wang's, so the final column is our time multiplied by 2.3. All times are in seconds. An asterisk $*$ means that Wang's Dixon failed after that many seconds. (The 47000 is not a typo.)

Example	Wang's Dixon	Wang's method	Dixon-EDF	2.3 Dixon-EDF
6	0.31*	0.019	0.004	0.009
9	1.21*	0.25	0.057	0.13
10	1830*	.051	0.09	.207
13	2673	47000	0.59	1.36

In 2017 Shen and Goldman [16] also report a new method for certain implicitizations. They also say that some resultant matrices have a 0 determinant and therefore resultants cannot be used. They do not refer to KSY.

In the following table we compare their reported times (in seconds) and our solutions today using Dixon-EDF working on some of their examples. We assume that our computers are comparable in speed.

Example	Shen-Goldman	Dixon-EDF
1	1.43	0.19
5	0.66	1.67
6	0.29	0.43
7	0.18	0.146
8	0.73	0.61
9	169.7	12.5
10	0.07	0.02
12	–	0.07

Example 5 has an answer with extremely large numerical coefficients.

Shen and Goldman describe their example 12 as a special case for which their method does not work directly and they need to "take remedial action." This example is completely straightforward with Dixon-EDF and finishes in 0.07 s.

Shen and Goldman try resultants in the generalized Sylvester form as found in [17] on their examples, and they also try Gröbner basis techniques. Gröbner bases failed in every case, meaning that nothing was returned within 10 min. Their resultants failed in the same way in every case except example 10.

Definition 1. *In the following,* Dixon *always denotes the complete combination Dixon-KSY-EDF.*

2   Resultants and Intersection of Surfaces

A very important problem is to compute the intersection of two surfaces. Many papers have addressed this question, such as [6, 10, 13, 15]. Virtually all the papers assume that the surfaces are *quadric*, i.e., degree 2. This means that the implicit equation is of the form

$$ax^2 + by^2 + cz^2 + dxy + exz + fyz + gx + hy + iz + j = 0$$

[10] asumes one is quadric and the other a torus (which is a degree 4 curve).

We describe here an apparently new way to compute intersections so long as at least one of the surfaces is given by a conventional parameterization, as in the previous section. There is no restriction on the degrees of the surfaces, at least theoretically. Suppose surface one is given by

$$x = f_1(s_1, t_1), \quad y = g_1(s_1, t_1), \quad z = h_1(s_1, t_1)$$

and surface two is

$$x = f_2(s_2, t_2), \quad y = g_2(s_2, t_2), \quad z = h_2(s_2, t_2)$$

For the intersection simply combine this to form a system of six equations. (N.B: it is the same x, y, z.) Use Dixon to eliminate five variables, say y, z, t_1, s_2, t_2. That yields one equation (resultant) involving x and s_1. If this is linear in x, solve for x and obtain the parametric equation for the x-coordinate of the intersection curve. Repeat for y and z. One could just as well express x in terms of s_2, t_1 or t_2. That might have computational advantages.

The process described above also works if one surface has a parameterization and the second has an implicit definition, say $p(x, y, z) = 0$. We then have four equations $x = f_1(s_1, t_1), y = g_1(s_1, t_1), z = h_1(s_1, t_1), p(x, y, z) = 0$ and we eliminate three variables, say y, z, t_1.

If the resultant is degree 2 in x, one can easily use the quadratic formula to get two possible expressions for x in terms of s_1. Numerical testing could determine which is correct. Of course, degree 3 or 4 could also be handled by formulas, but the expressions would no doubt become daunting.

What if the degrees are higher than 2 or we don't want to deal with messy formulas? This leads to a new concept:

Definition 2. *An* implicit parameterization *of a curve in 3-space is a set of three equations*

$$f(x, s) = 0, \quad g(y, s) = 0, \quad h(z, s) = 0$$

whose solution set includes the curve. s is called the curve parameter.

This is similar to the concept defined in [14]. f, g, and h could be any continuous functions, but will be rational functions in this work.

Theorem 1. *Given two surfaces defined as above with polynomial functions, the Dixon resultant will produce an implicit parameterization of their intersection.*

This follows immediately from the above discussion. The only possible flaw is if the set of six (or four) equations does not have a zero-dimensional solution space. That means for some values of the parameter s_1 there are infinitely many values of x. The authors of [1] show that Dixon can fail in that case.

3   Examples of Surface Intersections

We have found that when one of the surfaces is quadric and we use one of its parameters as curve parameter, f, g, and h in Definition 2 are quadratic in the variable. Explicit formulas for x, y, and z are then easy to compute.

In the examples below, we always use equations that are as general as possible. That means we use parameters, in the other sense of the word "parameter." The parametric equations of a torus are

$$x = (R + r\,cps)cth, \ \ y = (R + r\,cps)sth, \ \ z = r\,sps$$

in which there are six parameters: r and R are the radii, $cps = \cos(\psi)$, $sps = \sin(\psi)$, $cth = \cos(\theta)$, $sth = \sin(\theta)$. Similarly, below when we use spheres, ellipses, cones, etc., the equations contain parameters for radii, axes, etc. Polynomials using these parameters are fed to the Dixon resultant. That is *symbolic computation*. Only at the last step, when we want to produce a graph, do we substitute numerical values for these parameters into the symbolic resultants.

We compute Dixon resultants using Fermat [8,9]. Numerical values for the parameters were then substituted into the implicit resultants for x, y, and z. The ensuing functions were fed to Mathematica [12]. A fairly straightforward group of Mathematica commands [11] was used to produce the graphs.

3.1   Twisted Torus and Sphere

A twisted torus is defined by

$$x = (R + r\,cps)cth, \ \ y = (R + r\,cps)sth, \ \ z = r\,sps + 2\,sth\,cth$$

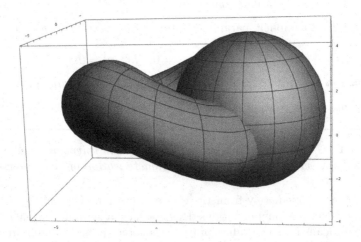

Fig. 1. Twisted torus and sphere

Each implicit resultant is quadratic in the main variable. Their intersection is in Fig. 2.

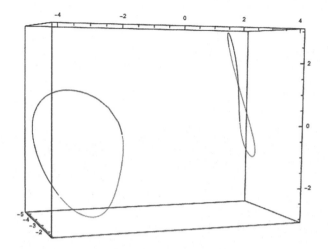

Fig. 2. Twisted torus and sphere intersection

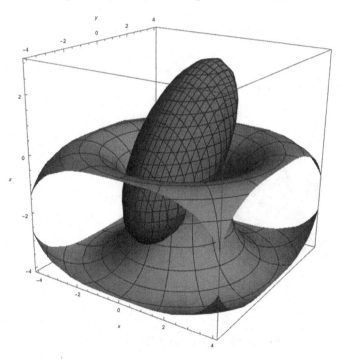

Fig. 3. Tilted ellipsoid and torus, cut-away view

3.2   Tilted Ellipsoid and Torus

Each implicit resultant is degree four in the main variable. Their intersection is
in Fig. 4

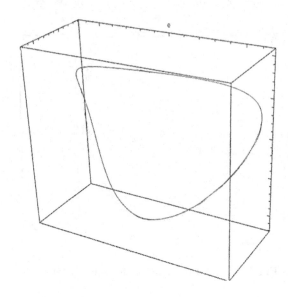

Fig. 4. Tilted ellipsoid and torus intersection

3.3 Other Examples

– arbitrarily oriented cones:

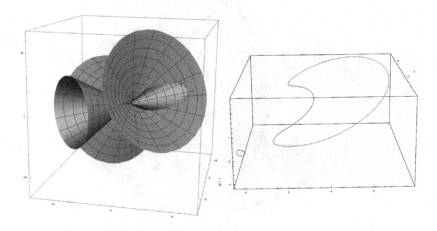

Fig. 5. Skew cones and their intersection

– distorted "bowl torus" and ordinary torus.
– two distorted "bowl tori"

The computation of the resultants for the arbitrarily oriented cones is easy for Dixon, but very hard for Gröbner basis implementations. Dixon running on Fermat takes 1.7 s for each resultant. Magma was killed after 7 h. Maple running FGb was killed after 8.5 h.

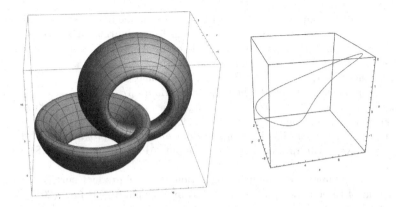

Fig. 6. Distorted tori intersection

4 Summary

Computing an implicitization with Dixon is straightforward and routine. No special conditions on the surfaces are needed.

The concept introduced here of "implicit parameterization" is easy to compute with Dixon. No special conditions on the surfaces are needed.

Implicit parameterizations can be dealt with in fairly straightforward ways with commercial software.

There is much room for further research. For example, how to decide which parameter to use for the curve parameter?

References

1. Buse, L., Elkadi, M., Mourrain, B.: Generalized resultants over unirational algebraic varieties. J. Symbolic Comp. **29**, 515–526 (2000)
2. Cox, D., Goldman, R., Zhang, M.: On the validity of implicitization by moving quadrics for rational surfaces with no base points. J. Symbolic Comput. **29**(3), 419–440 (2000)
3. Cox, D., Little, J., O'Shea, D.: Using Algebraic Geometry. In: Graduate Texts in Mathematics, vol. 185. Springer, New York (1998). https://doi.org/10.1007/b138611
4. Dixon, A.L.: The eliminant of three quantics in two independent variables. Proc. London Math. Soc. **6**, 468–478 (1908)
5. Kapur, D., Saxena, T., Yang, L.: Algebraic and geometric reasoning using Dixon resultants. In: Proceedings of the International Symposium on Symbolic and Algebraic Computation. ACM Press (1994)
6. Lazard, S., Peñaranda, L., Petitjean, S.: Intersecting quadrics: an efficient and exact implementation. Comput. Geom. **35**(1–2), 74–99 (2006)
7. Lewis, R.H.: Heuristics to accelerate the Dixon resultant. Math. Comput. Simul. **77**(4), 400–407 (2008)
8. Lewis, R.H.: Computer algebra system Fermat. http://home.bway.net/lewis/

9. Lewis, R.H.: Fermat code for Dixon-EDF. http://home.bway.net/lewis/dixon
10. Li, Q., Zhang, S., Ye, X.: Algebraic algorithms for computing intersections between torus and natural quadrics. Comput. Aided Des. Appl. **1**(1–4), 459–467 (2004)
11. Lichtblau, D.: Personal Communication (2018)
12. Mathematica. https://www.wolfram.com/mathematica/
13. Quadric intersections website. http://vegas.loria.fr/qi/
14. Sendra, J.R., Sevilla, D., Villarino, C.: Algebraic and algorithmic aspects of radical parameterizations. Comput. Aided Geom. Des. **55**, 1–14 (2017)
15. Shen, L.-Y., Cheng, J.-S., Xiaohong, J.: Homeomorphic approximation of the intersection curve of two rational surfaces. Comput. Aided Geom. Des. **29**(8), 613–625 (2012)
16. Shen, L., Goldman, R.: Implicitizing rational tensor product surfaces using the resultant of three moving planes. ACM Trans. Graph. **36**(5), 1–14 (2017)
17. Shi, X., Wang, X., Goldman, R.: Using μ-bases to implicitize rational surfaces with a pair of orthogonal directrices. Comput. Aided Geom. Des. **29**(7), 541–554 (2012)
18. Sturmfels, B.: Solving systems of polynomial equations. In: CBMS Regional Conference Series in Mathematics, vol. 97. American Mathematical Society (2003)
19. Wang, D.: A simple method for implicitizing rational curves and surfaces. J. Symbolic Comput. **38**, 899–914 (2004)

Fitting a Sphere via Gröbner Basis

Robert Lewis[1]([✉]), Béla Paláncz[2], and Joseph Awange[3]

[1] Department of Mathematics, Fordham University, NewYork, USA
rlewis@fordham.edu
[2] Department of Geoinformatics, Budapest Technical University,
Budapest, Hungary
palancz@epito.bme.hu
[3] Department of Spatial Sciences, Curtin University, Perth, Australia
J.awange@curtin.edu.au
https://fordham.academia.edu/RobertLewis
http://www.fmt.bme.hu/fmt/htdocs/dolgozok/
http://spatial.curtin.edu.au/people/index.cfm/J.Awange

Abstract. In indoor and outdoor navigation, finding the local position of a sphere in mapping space employing a laser scanning technique with low-cost sensors is a very challenging and daunting task. In this contribution, we illustrate how Gröbner basis techniques can be used to solve polynomial equations arising when algebraic and geometric measures for the error are used. The effectiveness of the suggested method is demonstrated, thanks to standard CAS software like *Mathematica*, using numerical examples of the real world.

Keywords: Point cloud · Outliers · SOM · Numerical Gröbner basis

1 Introduction

Laser scanning usually requires local reference points that are usually spherical in nature, from which the scanner registers coordinates of reflected points from objects in their own coordinate system (see Fig. 1). It is often desirable to know the object coordinates in the mapping space (also known as object space). To achieve this, the scanner's position should be known in mapping space, e.g. see [1]. This position can be computed if the position of the scanner relative to the reference point (center of a sphere) is known in the mapping space; see Fig. 1.

For outdoor and indoor navigation, low-cost sensors are generally employed, e.g. *Kinect* [2]. The problem, however, is that with low-cost sensors, not only are outliers present, but the data texture is not continuous and the resolution of the data points is rather low, see Fig. 2. Therefore, the usual sphere-fitting techniques provide low performance, e.g. [1,3–5].

Since the laser scanner is assumed to be moving in the mapping space, the coordinates of the center (a, b, c) and sometimes the radius R of the spheres

© Springer International Publishing AG, part of Springer Nature 2018
J. H. Davenport et al. (Eds.): ICMS 2018, LNCS 10931, pp. 319–327, 2018.
https://doi.org/10.1007/978-3-319-96418-8_38

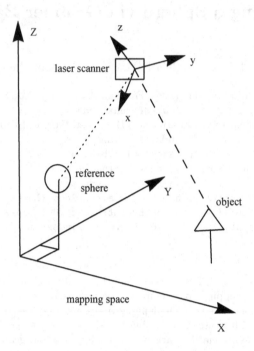

Fig. 1. Relative position of the laser scanner (x, y, z) in the mapping space (X,Y,Z).

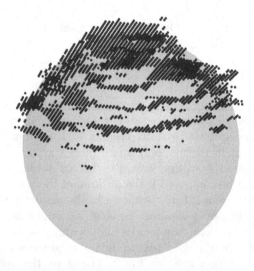

Fig. 2. Cloud of data points with outliers caused by reflections, quantized levels caused by round-off process, points cover only partly the spherical object, since the whole object is not accessible via laser rays.

having fixed position should be determined in real time, using scanner measurements. We compute the parameters a, b, c or R with least square techniques [6] using n data points (x_i, y_i, z_i). Then two objectives to be minimized are:

Algebraic error:

$$A(a, b, c, R) = \sum_{i=1}^{n} \delta_i^2 = \sum_{i=1}^{n} \left((x_i - a)^2 + (y_i - b)^2 + (z_i - c)^2 - R^2 \right)^2$$

Geometric error:

$$G(a, b, c, R) = \sum_{i=1}^{n} \delta_i^2 = \sum_{i=1}^{n} \left(\sqrt{(x_i - a)^2 + (y_i - b)^2 + (z_i - c)^2} - R \right)^2$$

From the computational point of view, employing algebraic error is the easier task; however the geometric error is more appropriate. Our aim is to use *Mathematica* to solve these computations employing Gröbner basis.

2   Algebraic Fitting

Sometimes the radius of the sphere is known, i.e., in the case of calibration, and we want to find only the position of the sphere. Therefore there are three parameters a, b and c to be estimated. Now, we have an overdetermined polynomial system to be solved,

$$(x_i - a)^2 + (y_i - b)^2 + (z_i - c)^2 - R^2 = 0 \qquad i = 1, 2, ...n$$

2.1   Determined Case

The solution of this problem is not unique since the reduced Gröbner basis for the determined polynomial subsystems ($n = 3$) consists of second order polynomials.

The prototype of the equations,

$$G = (x - a)^2 + (y - b)^2 + (z - c)^2 - R^2$$

proto = Table[(G/.$\{$x $\to \eta_i$, y $\to \xi_i$, z $\to \sigma_i\}$), $\{$i, 1, 3$\}$]

$\{-R^2 + (-a + \eta_1)^2 + (-b + \xi_1)^2 + (-c + \sigma_1)^2, -R^2 + (-a + \eta_2)^2 + (-b + \xi_2)^2 + (-c + \sigma_2)^2, -R^2 + (-a + \eta_3)^2 + (-b + \xi_3)^2 + (-c + \sigma_3)^2\}$

The polynomial for a can be computed via Gröbner basis as:

grba = GroebnerBasis[proto,$\{$a,b,c$\}$,$\{$b,c$\}$] // Simplify

The coefficient of the second order term is:

Coefficient[grba,a,2] // Simplify

$\{4(\eta_1^2\xi_2^2-2\eta_1^2\xi_2\xi_3+\eta_1^2\xi_3^2+\xi_2^2\sigma_1^2-2\xi_2\xi_3\sigma_1^2+\xi_3^2\sigma_1^2+\eta_3^2(\xi_1^2-2\xi_1\xi_2+\xi_2^2+(\sigma_1-\sigma_2)^2)-$
$2\xi_1\xi_2\sigma_1\sigma_2+2\xi_1\xi_3\sigma_1\sigma_2+2\xi_2\xi_3\sigma_1\sigma_2-2\xi_3^2\sigma_1\sigma_2+\eta_1^2\sigma_2^2+\xi_1^2\sigma_2^2-2\xi_1\xi_3\sigma_2^2+\xi_3^2\sigma_2^2+$
$\eta_2^2(\xi_1^2-2\xi_1\xi_3+\xi_3^2+(\sigma_1-\sigma_3)^2)+2\eta_1\eta_3(-\xi_2^2+\xi_1(\xi_2-\xi_3)+\xi_2\xi_3+(\sigma_1-\sigma_2)(\sigma_2-\sigma_3))+$
$2\xi_1\xi_2\sigma_1\sigma_3-2\xi_2^2\sigma_1\sigma_3-2\xi_1\xi_3\sigma_1\sigma_3+2\xi_2\xi_3\sigma_1\sigma_3-2\eta_1^2\sigma_2\sigma_3-2\xi_1^2\sigma_2\sigma_3+2\xi_1\xi_2\sigma_2\sigma_3+$
$2\xi_1\xi_3\sigma_2\sigma_3-2\xi_2\xi_3\sigma_2\sigma_3+\eta_1^2\sigma_3^2+\xi_1^2\sigma_3^2-2\xi_1\xi_2\sigma_3^2+\xi_2^2\sigma_3^2-2\eta_2(\eta_3(\xi_1^2+\xi_2\xi_3-\xi_1(\xi_2+$
$\xi_3)+(\sigma_1-\sigma_2)(\sigma_1-\sigma_3))+\eta_1(\xi_1(\xi_2-\xi_3)-\xi_2\xi_3+\xi_3^2+\sigma_1\sigma_2-\sigma_1\sigma_3-\sigma_2\sigma_3+\sigma_3^2)))\}$

And that of the third order term:

Coefficient[grba,a,3] // Simplify
{}

Thus a second order polynomial indicates a non-unique solution.

2.2   Over Determined Case

In case of n data points, the least square function to be minimized is

$$F = \sum_{i=1}^{n}\left((x_i-a)^2+(y_i-b)^2+(z_i-c)^2-R^2\right)^2$$

with the necessary conditions:

$eq_1 = \frac{d}{da}(F,a)$

$\sum_{i=1}^{n}(4a^3+4ab^2+4ac^2-4aR^2-12a^2x_i-4b^2x_i-4c^2x_i+4R^2x_i+12ax_i^2-4x_i^3-$
$8aby_i+8bx_iy_i+4ay_i^2-4x_iy_i^2-8acz_i+8cx_iz_i+4az_i^2-4x_iz_i^2)$

$eq_2 = \frac{d}{db}(F,b)$

$\sum_{i=1}^{n}(4a^2b+4b^3+4bc^2-4bR^2-8abx_i+4bx_i^2-4a^2y_i-12b^2y_i-4c^2y_i+4R^2y_i+$
$8ax_iy_i-4x_i^2y_i+12by_i^2-4y_i^3-8bcz_i+8cy_iz_i+4bz_i^2-4y_iz_i^2)$

$eq_3 = \frac{d}{dc}(F,c)$

$\sum_{i=1}^{n}(4a^2c+4b^2c+4c^3-4cR^2-8acx_i+4cx_i^2-8bcy_i+4cy_i^2-4a^2z_i-4b^2z_i-$
$12c^2z_i+4R^2z_i+8ax_iz_i-4x_i^2z_i+8by_iz_i-4y_i^2z_i+12cz_i^2-4z_i^3)$

Leading to a third order polynomial system with 18 parameters (i.e., $p_i, i = 1, ..., 18$). For the first equation, for example,

$eq_1 = 4a^3n+4ab^2n+4ac^2n-4aR^2n-12a^2p_1-4b^2p_1-4c^2p_1+4R^2p_1+12ap_2-$
$4p_3-8abp_4+8bp_5+4ap_6-4p_7-8acp_8+8cp_9+4ap_{10}-4p_{11}$

where

$$p_1 = \sum_{i=1}^{n}x_i, \ p_2 = \sum_{i=1}^{n}x_i^2, \ ..., \ p_{11} = \sum_{i=1}^{n}x_iz_i^2$$

Considering the additional two equations, we have a third order polynomial system with 18 parameters. For given parameter values numeric Gröbner basis works well and fast. We have three real solutions, and the proper one is that which provides the smallest objective value.

Numerical Example 1

Real field measurements adopted from [5]. The number of the measured data points is n = 2670 and the radius of the sphere is R = 0.152 m (see Fig. 2). Using these data, the 18 parameters are computed using numerical Gröbner basis built in NSolve function, giving the real solutions of the polynomial system as

$solabc = NSolve[\{eq1, eq2, eq3\}/.\{n \rightarrow 2670, R \rightarrow 0.152\}, \{a, b, c\}, Reals]$

$\{\{a \rightarrow -0.0431301, b \rightarrow -0.0122755, c \rightarrow 3.01883\}, \{a \rightarrow -0.0737348, b \rightarrow -0.0155016, c \rightarrow 2.89141\}, \{a \rightarrow -0.0536333, b \rightarrow -0.017366, c \rightarrow 2.83447\}\}$

From which the proper solution is selected by computing the value of the objective function for each solution to find the candidate that provides the smallest residual. This is the acceptable solution to the problem, i.e.,

$\{a \rightarrow -0.0431301, b \rightarrow -0.0122755, c \rightarrow 3.01883\}$

3   Geometric Fitting

For the geometrical fitting, we have an overdetermined *nonlinear equation system* that has to be solved for the parameter estimation problem even in case of *unknown R*. However, in this case, the solution of the system is unique.

$$\sqrt{(x_i - a)^2 + (y_i - b)^2 + (z_i - c)^2} - R = 0 \qquad i = 1, 2, ...n$$

3.1   Determined Case

Considering 4 exact noiseless measurement points, then the prototype system for these 4 points is,

$$G = \sqrt{(x - a)^2 + (y - b)^2 + (z - c)^2} - R$$

proto=Table[(G/.$\{x \rightarrow \eta_i, y \rightarrow \xi_i, z \rightarrow \sigma_i\}$), $\{i, 1, 4\}$]

$$\{-R + \sqrt{(-a + \eta_1)^2 + (-b + \xi_1)^2 + (-c + \sigma_1)^2},$$
$$- R + \sqrt{(-a + \eta_2)^2 + (-b + \xi_2)^2 + (-c + \sigma_2)^2},$$
$$- R + \sqrt{(-a + \eta_3)^2 + (-b + \xi_3)^2 + (-c + \sigma_3)^2},$$
$$- R + \sqrt{(-a + \eta_4)^2 + (-b + \xi_4)^2 + (-c + \sigma_4)^2}\}$$

One can realize that the elimination of R reduces the system to,

protoR=Take[proto,{2,4}]/.R $\rightarrow \sqrt{(-a+\eta_1)^2+(-b+\xi_1)^2+(-c+\sigma_1)^2}$

$\{-\sqrt{(-a+\eta_1)^2+(-b+\xi_1)^2+(-c+\sigma_1)^2}+\sqrt{(-a+\eta_2)^2+(-b+\xi_2)^2+(-c+\sigma_2)^2},$
$-\sqrt{(-a+\eta_1)^2+(-b+\xi_1)^2+(-c+\sigma_1)^2}+\sqrt{(-a+\eta_3)^2+(-b+\xi_3)^2+(-c+\sigma_3)^2},$
$-\sqrt{(-a+\eta_1)^2+(-b+\xi_1)^2+(-c+\sigma_1)^2}+\sqrt{(-a+\eta_4)^2+(-b+\xi_4)^2+(-c+\sigma_4)^2}\}$

To get univariate polynomial for parameter a let us eliminate b and c from the basis,

grba=GroebnerBasis[protoR,{a,b,c},{b,c}] // Simplify

The polynomial is:

grba[[1]]:

$2a\eta_2\xi_3\sigma_1-\eta_2^2\xi_3\sigma_1-\xi_2^2\xi_3\sigma_1+\xi_2\xi_3^2\sigma_1-2a\eta_2\xi_4\sigma_1+\eta_2^2\xi_4\sigma_1+\xi_2^2\xi_4\sigma_1-\xi_3^2\xi_4\sigma_1-$
$\xi_2\xi_4^2\sigma_1+\xi_3\xi_4^2\sigma_1-2a\eta_1\xi_3\sigma_2+\eta_1^2\xi_3\sigma_2+\xi_1^2\xi_3\sigma_2-\xi_1\xi_3^2\sigma_2+2a\eta_1\xi_4\sigma_2-\eta_1^2\xi_4\sigma_2-$
$\xi_1^2\xi_4\sigma_2+\xi_3^2\xi_4\sigma_2+\xi_1\xi_4^2\sigma_2-\xi_3\xi_4^2\sigma_2+\xi_3\sigma_1^2\sigma_2-\xi_4\sigma_1^2\sigma_2-\xi_3\sigma_1\sigma_2^2+\xi_4\sigma_1\sigma_2^2-2a\eta_2\xi_1\sigma_3+$
$\eta_2^2\xi_1\sigma_3+2a\eta_1\xi_2\sigma_3-\eta_1^2\xi_2\sigma_3-\xi_1^2\xi_2\sigma_3+\xi_1\xi_2^2\sigma_3-2a\eta_1\xi_4\sigma_3+\eta_1^2\xi_4\sigma_3+2a\eta_2\xi_4\sigma_3-$
$\eta_2^2\xi_4\sigma_3+\xi_1^2\xi_4\sigma_3-\xi_2^2\xi_4\sigma_3-\xi_1\xi_4^2\sigma_3+\xi_2\xi_4^2\sigma_3-\xi_2\sigma_1^2\sigma_3+\xi_4\sigma_1^2\sigma_3+\xi_1\sigma_2^2\sigma_3-\xi_4\sigma_2^2\sigma_3+$
$\xi_2\sigma_1\sigma_3^2-\xi_4\sigma_1\sigma_3^2-\xi_1\sigma_2\sigma_3^2+\xi_4\sigma_2\sigma_3^2+\eta_4^2(\xi_3(\sigma_1-\sigma_2)+\xi_1(\sigma_2-\sigma_3)+\xi_2(-\sigma_1+$
$\sigma_3))+2a\eta_4(\xi_3(-\sigma_1+\sigma_2)+\xi_2(\sigma_1-\sigma_3)+\xi_1(-\sigma_2+\sigma_3))+2a\eta_2\xi_1\sigma_4-\eta_2^2\xi_1\sigma_4-$
$2a\eta_1\xi_2\sigma_4+\eta_1^2\xi_2\sigma_4+\xi_1^2\xi_2\sigma_4-\xi_1\xi_2^2\sigma_4+2a\eta_1\xi_3\sigma_4-\eta_1^2\xi_3\sigma_4-2a\eta_2\xi_3\sigma_4+\eta_2^2\xi_3\sigma_4-$
$\xi_1^2\xi_3\sigma_4+\xi_2^2\xi_3\sigma_4+\xi_1\xi_3^2\sigma_4-\xi_2\xi_3^2\sigma_4+\xi_2\sigma_1^2\sigma_4-\xi_3\sigma_1^2\sigma_4-\xi_1\sigma_2^2\sigma_4+\xi_3\sigma_2^2\sigma_4+\xi_1\sigma_3^2\sigma_4-$
$\xi_2\sigma_3^2\sigma_4-\xi_2\sigma_1\sigma_4^2+\xi_3\sigma_1\sigma_4^2+\xi_1\sigma_2\sigma_4^2-\xi_3\sigma_2\sigma_4^2-\xi_1\sigma_3\sigma_4^2+\xi_2\sigma_3\sigma_4^2-2a\eta_3(\xi_4(-\sigma_1+$
$\sigma_2)+\xi_2(\sigma_1-\sigma_4)+\xi_1(-\sigma_2+\sigma_4))+\eta_3^2(\xi_4(-\sigma_1+\sigma_2)+\xi_2(\sigma_1-\sigma_4)+\xi_1(-\sigma_2+\sigma_4))$

It is a linear polynomial a; therefore the solution for parameter a is simple,

aG= $-\dfrac{\text{Coefficient[grba[[1]],a,0]}}{\text{Coefficient[grba[[1]],a,1]}}$

This means that from the coordinates of the corresponding 4 points, the parameter can be directly computed. Similar expressions can be developed for the other two parameters b and c.

3.2   Over Determined Case

Now, let us employ geometrical fitting to the data set used in *Numerical Example 1*. The approximation of the Gauss-Jacobi technique will be employed, but not for $\binom{2670}{4}$ subsets. We can reduce our data set and eliminate outliers at the same time, employing a neural network algorithm, the so called Self-Organizing Map (SOM), which uses a competitive learning technique to train itself in an unsupervised manner. SOMs are different from other artificial neural networks in the sense that they use a neighborhood function to preserve the topological properties of the input space. They have been used to create an ordered representation of multi-dimensional data, which simplifies complexity and reveals meaningful relationships.

In the last decade, a lot of research has been done to develop surface reconstruction methods, see e.g. [7]. A more recent approach to the problem of surface reconstruction is that of learning based methods. Learning algorithms are able to process very large or noisy data, such as point clouds obtained from 3D scanners and have been used to construct surfaces. Following this approach some studies have employed SOM and their variants for surface reconstruction. SOM is suitable for this problem because it can form topological maps and replicate the distribution of input data. In our case, this mapping occurs from 3D space to 2D.

We represent this cloud of data points by considerably fewer points using Kohonen-map with a lattice SOM of size 5×5 code-book vectors with symmetrical neighborhood. To carry out the computation the Neural Networks package of Mathematica was employed as follows, see e.g. [8].

<< NeuralNetworks'
{som, fitrecord} = UnsupervisedNetFit[dataQ, 25, 100, SOM → 5, 5];
// Quiet

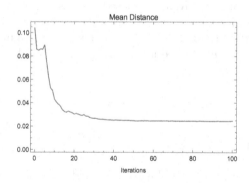

Fig. 3. The error of the Kohonen-map with a lattice of SOM of size 5×5 code-book vectors.

This computation, which required 100 iteration steps (see Fig. 3) took less than 5 seconds. The 25 representing code-book vectors are

dataSOM = som[[1]];

Let us display the resulting code-book vectors with the sphere in Fig. 4.
Numerical Example 2

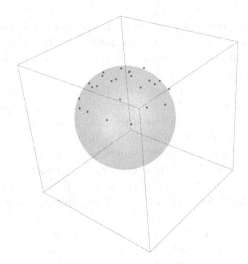

Fig. 4. Data points after elimination of outliers via Self-Organizing Map (SOM) neural network algorithm.

We can get a good approximation to compute the arithmetic average of the solution of the $\binom{25}{4} = 12650$ subsets, which is an approximation of the Gauss-Jacobi technique. In order to avoid ill - posed subsets, we computed the product of the distances of every 4-point subset, and selected the only subsets for which the product was higher than 2.5×10^{-5}. In this way the number of the equations to be considered could be reduced to 1020. Using parallel computation, we finally obtained.

$$\{a \rightarrow -0.0423085, b \rightarrow -0.0137875, c \rightarrow 3.02371, R \rightarrow 0.152695\}.$$

4   Conclusion

The two numerical examples employing real field data demonstrated the successfulness and effectiveness of Gröbner basis in solving algebraic as well as geometrical fitting in case of known or unknown sphere radius using standard Computer Algebra System software like Mathematica.

References

1. Franaszek, M., Cheok, G.S., Saidi, K.S., Witzgall, C.: Fitting spheres to range data from 3-D imaging systems. IEEE Trans. Instrum. Measur. **58**(10), 3544–3553 (2009)
2. Józków, G., Thoth, C., Koppanyi, Z., Grejner - Brzezinska, D.: Combined matching of 2D and 3D kinect data to support indoor mapping and navigation. In: ASPRS 2014 Annual Conference, Luisville, Kentucky, USA, 23–28 March 2014

3. Ogundana, O., Coggrave, C., Burguete, R.L., Huntley, J.M.: Fast hough transform for automated detection of spheres in three-dimensional point clouds. Opt. Eng. 0001 **46**(5), 051002–051002-11 (2007)
4. Zhou, Z., Guan, Y., Zhan, X., Lu, T.: Robust algorithm for fitting Sphere to 3D point clouds in terrestrial laser scanning. In: The International Archives of Photogrammetry and Spatial Information Science, vol. XXXVII Part B5, Beijing, pp. 519–522 (2008)
5. Molnár, B., Toth, C.K., Detrekõi, A.: Accuracy test of microsoft kinect for human morphological measurements. In: International Archives of the Photogrammetry, Remote Sensing and Spatial Information Sciences, vol. XXXIX - B3, 2012 XXII ISPRS Congress, Melbourne, Australia, 25 August – 01 September 2012
6. Awange, J., Palancz, B., Lewis, R.H., Völgyesi, L.: Mathematical Geosciences. Springer, Heidelberg (2018)
7. DalleMole, V.L., do Rego, R.L.M.E., Araujo, A.F.R.: The self - organizing approach for surface reconstruction from unstructured point clouds. In: Matsopoulos, G.K. (ed.) Self-Organizing Maps, INTECH, pp. 167–188. Rijeka, Croatia (2010)
8. Sjoberg, J.: Neural Networks 1.2, Mathematica Adds On (2018). https://www.wolfram.com/products/applications/neuralnetworks/

Homotopy Continuation in Macaulay2

Anton Leykin[✉]

Georgia Institute of Technology, Atlanta, USA
leykin@math.gatech.edu
http://people.math.gatech.edu/~aleykin3

Abstract. We describe the design and relationships of several Macaulay2 packages that use numerical polynomial homotopy continuation as their engine. Macaulay2 is a computer algebra system built around the classical symbolic computation tools such as Gröbner bases. However, recent Macaulay2 versions include its own fast implementation of homotopy continuation, interfaces to external numerical algebraic geometry software (Bertini and PHCpack), and a unified data structures design that allows the use of the internal and external capabilities interchangeably. The resulting numerical and hybrid tools are of general interest to Macaulay2 users interested in computational experimentation.

Keywords: Polynomial homotopy continuation
Numerical algebraic geometry · Macaulay2

1 Introduction

This extended abstract is written with a general attendee of ICMS in mind. To continue reading neither expertise in the subject nor familiarity with Macaulay2 is necessary.

The keywords *homotopy continuation* in our context refer to a technique that deforms one system of equations into another system while tracking the path that originates from an isolated solution of the former in hope that this path leads to a solution of the latter. When these systems consist of multivariate polynomials, under some mild assumptions, this hope may be upgraded to a guarantee of locating a solution. The extent of *polynomial homotopy continuation* is explained in a classic book by Morgan [14]. A currently booming area of *numerical algebraic geometry* develops these basic ideas further and delivers a powerful theoretical framework and computational machinery based on fast floating-point implementations of homotopy continuation algorithms.

Developing tools for homotopy continuation in Macaulay2 [8] pursues a dual goal. We not only build a fast implementation of basic homotopy continuation routines, but also provide the possibility of setting up problems and solving them using all other (not necessarily related to homotopy continuation) symbolic and numerical capabilities of Macaulay2.

© Springer International Publishing AG, part of Springer Nature 2018
J. H. Davenport et al. (Eds.): ICMS 2018, LNCS 10931, pp. 328–334, 2018.
https://doi.org/10.1007/978-3-319-96418-8_39

2   First Steps in Polynomial Homotopy Continuation

The basic problem to be solved is

Given polynomials $f_1, \ldots, f_n \in \mathbb{C}[x_1, \ldots, x_n]$ such that they generate a 0-dimensional ideal $I = (f_1, \ldots, f_n)$ find *numerical approximations* of all points of the underlying variety $\mathbb{V}(I) = \{\mathbf{x} \in \mathbb{C}^n \mid \boldsymbol{f}(\mathbf{x}) = 0\}$.

The main idea of homotopy continuation is to solve the *target* polynomial system

$$\boldsymbol{f} = (f_1, \ldots, f_n) = 0$$

by viewing it in a family of polynomial systems that also includes a *start* polynomial system $\boldsymbol{g} = (g_1, \ldots, g_n)$ possessing good properties; one typical property is that $\boldsymbol{g}$ has regular solutions that one can approximate inexpensively.

Define a *homotopy* from $\boldsymbol{g}$ to $\boldsymbol{f}$ as

$$\boldsymbol{h} = (1 - t)\boldsymbol{g} + \gamma t \boldsymbol{f} \in \mathbb{C}[\boldsymbol{x}, t], \quad \gamma \in \mathbb{C}^*, \, t \in [0, 1]. \tag{1}$$

The $\boldsymbol{h}(t) = 0, t \in \mathbb{C}$, defines a complex curve in the space of polynomial systems. One should imagine real segments of the branches of this curve projecting to the real interval $[0, 1]$. Those real segments are *continuation paths* leading from *start solutions* of $\boldsymbol{g} = \boldsymbol{h}|_{t=0}$ to the *target solutions* of $\boldsymbol{f} = \boldsymbol{h}|_{t=1}$.

As an example of a particular family giving rise to a homotopy technique one can take the space of square systems of polynomials with fixed degrees $d_1 = \deg f_1, \ldots, d_n = \deg f_n$. Then, according to Bézout's theorem, a generic system in this family has $d_1 \cdots d_n$ regular solutions. One can take $\boldsymbol{g} = (x_1^{d_1} - 1, \ldots, x_n^{d_n} - 1)$ as a start system; its solutions are easy to construct.

The following meta-statement holds for the construction above and other more special families that are complex spaces of polynomials.

- For a generic choice of $\gamma \in \mathbb{C}$ in (1) the homotopy continuation paths have no singularities with a possible exception of the endpoints corresponding to $t = 1$.
- Every isolated target solution is an endpoint, at $t = 1$, of some continuation path.

The exceptional set of parameters γ is contained in a proper Zariski closed subset of $\mathbb{R}^2 \simeq \mathbb{C}$. In practice, the exceptional set is hard to find and γ is chosen at random on the unit circle.

The function `solveSystem` of the package `NumericalAlgebraicGeometry` implements a basic homotopy continuation solver using the homotopy described above.

```
i1 : needsPackage "NumericalAlgebraicGeometry"

i2 : R = CC[x,y];

i3 : sols = solveSystem {x^2+y^2-1, x*y}

o3 = {{1.10617e-24-3.28085e-24*ii, -1}, {-1.10617e-24+3.28085e-24*ii, 1},
      {1,-3.59266e-24-6.15821e-24*ii}, {-1, 3.59266e-24+6.15821e-24*ii}}
```

Differentiating the homotopy equation $h = 0$, we get a system of ODEs

$$\frac{dx}{dt} = h_x^{-1} h_t, \tag{2}$$

where $h_x = \partial h / \partial x$ is the Jacobian and $h_t = \partial h / \partial t$. Following the solutions of (2) for $t \in [0, 1]$ with initial conditions given by the start solutions we can approximate the continuation paths iteratively predicting where the next approximation lies. As long as our approximation stays close to the tracked path one can correct an approximate solution to the polynomial system $h|_{t=t_0} (x) = 0$ using Newton's method.

One can imagine a tracking procedure alternates *predictor* and *corrector* steps as shown in Fig. 1.

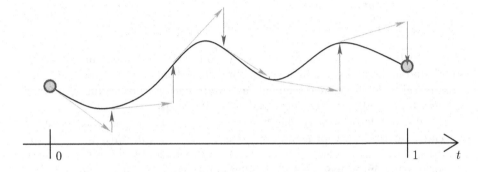

Fig. 1. Predictor steps (using tangent method) are followed by corrector steps.

3 Some Aspects of Software Design

The top-level language of Macaulay2 is an interpreted language. That is why the intensive part of homotopy continuation algorithm has been implemented in C++ in the core of the system. There are two major bottlenecks in the core functions track and trackHomotopy.

- Linear algebra subroutines: A linear system solver provided by LAPACK [1] is used for double precision and a custom one using MPFR [7] is used for arbitrary precision. The solvers are used at both predictor and corrector steps: e.g., Runge-Kutta 4th order method for predictor and Newton's method for corrector both call the solver several times.
- Evaluation of polynomial functions and their derivatives: The function h as well as its derivatives h_x and h_t have to be evaluated at all steps of the algorithm. To speed up the computation, these are passed to the core as straight line programs after the preprocessing step involving SLPexpressions supporting package.

Optional parameters allow users to experiment with various heuristic settings for the homotopy tracker and set numerical thresholds that control the accuracy of the computation. While the default precision is double precision, setting `Precision=>infinity` is `trackHomotopy` and `solveSystem`, the blackbox 0-dimensional system solver, instructs the tracker to *adapt* precision according to an estimate on the condition number of h_x at the current value of t.

One other notable option, `Software` gives a chance to use external homotopy continuation software instead of `Macaulay2` core routines: setting this to either `PHCPACK` or `BERTINI` will invoke PHCpack [15] or Bertini [3], respectively. This is done through the methods of corresponding `Macaulay2` packages `PHCpack.m2` [9] and `Bertini.m2` [2].

Given three main continuation `Software` choices — `M2engine`(default), `PHCPACK`, and `BERTINI` — one may ask: which one to choose? There are several answers to this question.

– Run them all: in many scenarios one is able to compute solutions to the same problem via different implementations. Given the overall heuristic nature of the continuation algorithms, having more than one option is a valuable feature.
– Pick the fastest: depending on the problem, each of the three may win the race. Currently, `M2engine` and `PHCPACK` are observed to be faster than `BERTINI` (on average) at solving 0-dimensional square systems with relatively well conditioned solutions. However, `BERTINI` could be faster and more reliable at the moment for dealing with overdetermined systems and numerical irreducible decomposition of positive-dimensional varieties.
– Use particular features absent in other implementations: e.g., `PHCPACK` is the only one that implements polyhedral homotopies.

4   A Higher Level Package Using Homotopy Continuation

Let us take `MonodromySolver` implementing a polynomial system solver described in detail in [6] as an example.

The essence of homotopy continuation can be formalized in the language outlined in Fig. 2.

The cartoon in Fig. 2 addresses a toy problem of computing all cubic roots. However, one can use the same cartoon to help imagine the setup of Sect. 2 following homotopy (1). There the base space B is the space of systems of polynomials of fixed degrees $d_1, \ldots, d_n$ with exactly $d_1 \cdots d_n$ solutions. These degrees come from the target system f, which may or may not be in the branch locus D, and a start system g, which is in B, is chosen in a particular way. The homotopy (1) corresponds to a path in B that depends on the parameter γ.

The algorithmic framework dubbed *monodromy solver* introduced in [6] provides another way to solve a system corresponding to a generic point in B for an arbitrary covering map $\pi : V \to B$ given that the total space (a.k.a. the *solution variety*) V is connected. The method starts by embedding a random graph G in B and "seeding" one vertex-system of G with *one* solution. It proceeds,

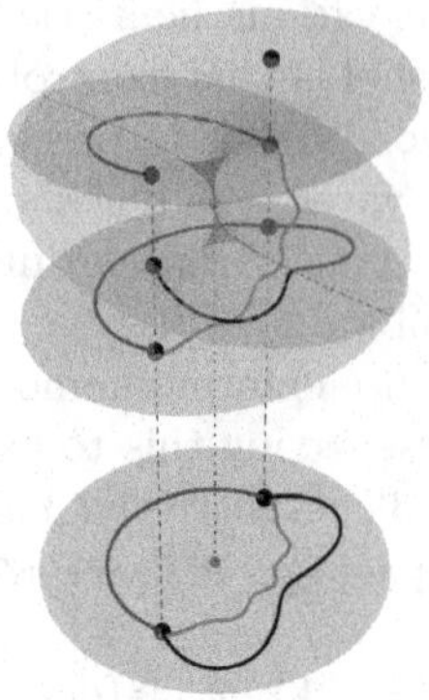

The *covering map*:

$$\pi : V \to B$$

Total space: the set of pairs (problem, solution),

$$V = \{(a, x) \in B \times \mathbb{C} \mid x^3 - a = 0\}$$

Base space: a parameterized space of problems,

$$B = \{a \in \mathbb{C} \mid x^3 - a = 0 \text{ has three solutions}\} = \mathbb{C} \backslash D$$

where $D = \{0\}$ is the *branch locus*.

Fig. 2. Homotopy continuation in a nutshell: lifting paths from B to V.

in a randomized fashion, to track edges-paths of $\pi^{-1}(G)$ starting from known solutions. The algorithm is guaranteed to succeed, i.e. discover all solutions for some vertex-system of G as long as $\pi^{-1}(G)$ is connected.

This machinery ultimately powers several blackbox solvers: one is implemented by the `solveFamily` method:

```
i1 : needsPackage "MonodromySolver";
i2 : R = CC[a,b,c,d,e,f][x,y];
i3 : (sys, sols) = solveFamily polySystem {a*x^2+b*y+c,d*x+e*y+f}

o3 = ({ (.23678 - .42551*ii)x^2  + (.536999 + .405563*ii)y + ... ,
        (- .352525 - .818762*ii)x + (.168395 - .624005*ii)y + ... },
      { {-.355531-.934665*ii, -.652688+.757627*ii},
        {1.32088+2.57812*ii, -5.47398-1.60337*ii} })
```

Another method, `sparseMonodromySolve`, provides an alternative to polyhedral homotopy solvers implemented in PHCpack [15] and HOM4PS [5].

```
i4 : R = CC[x,y,z];
i5 : # sparseMonodromySolve polySystem {
        1+11*x-3*y+30*x*y+55*x*y*z,
        3-5*x+7*y+2*x*y+9*x*y*z,
        6+13*x^2*y-5*y^2*z }

o5 = 5
```

The system above is an example from [10] with 5 solutions, while the Bézout bound is 27; i.e., the method of Sect. 2 is inefficient as it has to track 22 diverging paths.

5   Numerics in Macaulay2

A constellation of packages that grew around NumericalAlgebraicGeometry, whose early version was described in [12], has the following dependence structure at the time of writing (in Macaulay2 version 1.11).

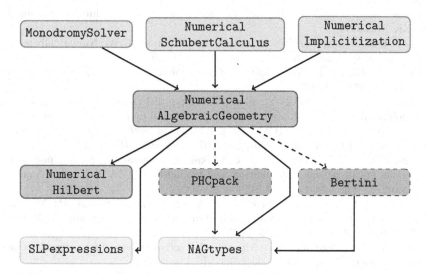

Note that the Bertini and PHCpack interfaces use the same Macaulay2 *types* defined for relevant data structures in NAGtypes facilitating the substitution of the corresponding external solver in the core routines if desired.

Some numerical scheme-theoretic devices, such as a local Hilbert function, are implemented in NumericalHilbert [11] and provided to the main package.

The package SLPexpressions provides a flexible way to create and manage arithmetic circuits in Macaulay2 together with the necessary functionality for automatic differentiation and routines for setting up straight line programs in the system's core. The efficient evaluation of SLPs is of particular importance for homotopy methods, since evaluation subroutines are the most frequently called ones in the main tracking algorithm.

Apart from MonodromySolver highlighted in Sect. 4, there are several packages that use homotopy continuation: NumericalImplicitization [4] supports user-friendly calculation of basic invariants of the image of a polynomial map and NumericalSchubertCalculus implements algorithms for solving instances of Schubert problems on Grassmannians; in particular, it implements the Littlewood-Richardson homotopy algorithm [13].

Acknowledgement. Research of AL is supported in part by DMS-1719968 award from NSF.

References

1. Anderson, E., Bai, Z., Bischof, C., Blackford, S., Demmel, J., Dongarra, J., Du Croz, J., Greenbaum, A., Hammarling, S., McKenney, A., Sorensen, D.: LAPACK Users' Guide, 3rd edn. Society for Industrial and Applied Mathematics, Philadelphia (1999)
2. Bates, D.J., Gross, E., Leykin, A., Rodriguez, J.I.: Bertini for Macaulay2. arXiv:1310.3297
3. Bates, D.J., Hauenstein, J.D., Sommese, A.J., Wampler, C.W.: Numerically solving polynomial systems with Bertini, vol. 25. SIAM (2013)
4. Chen, J., Kileel, J.: Numerical Implicitization for Macaulay2. arXiv preprint arXiv:1610.03034 (2016)
5. Chen, T., Lee, T.-L., Li, T.-Y.: Hom4PS-3: a parallel numerical solver for systems of polynomial equations based on polyhedral homotopy continuation methods. In: Hong, H., Yap, C. (eds.) ICMS 2014. LNCS, vol. 8592, pp. 183–190. Springer, Heidelberg (2014). https://doi.org/10.1007/978-3-662-44199-2_30
6. Duff, T., Hill, C., Jensen, A., Lee, K., Leykin, A., Sommars, J.: Solving polynomial systems via homotopy continuation and monodromy. IMA J. Numer. Anal. (to appear)
7. Fousse, L., Hanrot, G., Lefèvre, V., Pélissier, P., Zimmermann, P.: MPFR: a multiple-precision binary floating-point library with correct rounding. ACM Trans. Math. Softw. **33**(2), 13:1–13:15 (2007)
8. Grayson, D.R., Stillman, M.E.: Macaulay2, a software system for research in algebraic geometry. http://www.math.uiuc.edu/Macaulay2/
9. Gross, E., Petrović, S., Verschelde, J.: Interfacing with PHCpack. J. Softw. Algebra Geometry **5**(1), 20–25 (2013)
10. Huber, B., Sturmfels, B.: A polyhedral method for solving sparse polynomial systems. Math. Comp. **64**(212), 1541–1555 (1995)
11. Krone, R.: Numerical Hilbert functions for Macaulay2. arXiv preprint arXiv:1405.5293 (2014)
12. Leykin, A.: Numerical algebraic geometry. J. Softw. Algebra Geometry **3**, 5–10 (2011)
13. Leykin, A., del Campo, A.M., Sottile, F., Vakil, R., Verschelde, J.: Numerical schubert calculus via the littlewood-richardson homotopy algorithm. arXiv preprint arXiv:1802.00984 (2018)
14. Morgan, A.: Solving Polynomial Systems Using Continuation for Engineering and Scientific Problems. Prentice Hall Inc., Englewood Cliffs (1987)
15. Verschelde, J.: Algorithm 795: PHCpack: a general-purpose solver for polynomial systems by homotopy continuation. ACM Trans. Math. Softw. **25**(2), 251–276 (1999). http://www.math.uic.edu/~jan

Solving Polynomial Systems Using Numeric Gröbner Bases

Daniel Lichtblau$^{(\boxtimes)}$

Wolfram Research, Champaign, IL, USA
danl@wolfram.com

Abstract. Systems of polynomial or algebraic equations with finitely many solutions arise in many areas of applied mathematics. I will discuss the design and implementation of a hybrid symbolic-numeric method based on the endomorphism matrix approach pioneered by Stetter and others. It makes use of numeric Gröbner bases and arbitrary-precision eigensystem computations. I will describe how to assess accuracy, find and remove parasite solutions in the case of fractional degrees in the system, handle multiplicity, as well as some of the other finer points not usually covered in the literature. This work is one of the methods used in the Wolfram Language NSolve function.

Keywords: Polynomial systems
Numeric Gröbner bases · Endomorphism matrix · Eigensystem

1 Introduction

Systems of algebraic equations arise in many applied areas, including computational chemistry, computer-aided design, graphics, robotics, and elsewhere. Often one is interested in all solutions, or at least all solutions that satisfy some criterion, and in such cases a local solver such as Newton's method and its ilk simply do not suffice. As is well known, there are exact methods for handling such systems e.g. [3,8]. As these involve computation of exact Gröbner bases, and require a lexicographic term order, they tend to be quite expensive. Thus there is a need for faster methods that might use approximate arithmetic in a way that is reliable. Several global approximate methods are in current use. This reports on one such, and is intended as a long overdue update to [13]. The solver is based on prior literature spanning nearly 30 years (see e.g. [1,5,16]), and also utilizes several computational heuristics that will be described below. The key tools are numeric Gröbner bases, eigendecompositions in either machine arithmetic or higher, and numerous tactics for recognizing and appropriately handling multiplicity, spurious solutions, and other pathologies. The work described is implemented in the Wolfram Language [25] and is one of the main methods used by the function NSolve.

We start with a brief outline of the method. Given a polynomial system F in a set of variables X, the zero set is an ideal in $\mathbb{C}[X]$. If it has finitely many

© Springer International Publishing AG, part of Springer Nature 2018
J. H. Davenport et al. (Eds.): ICMS 2018, LNCS 10931, pp. 335–342, 2018.
https://doi.org/10.1007/978-3-319-96418-8_40

solutions then it is straightforward to show that the quotient module $\mathbb{C}[X]/F$ is a finite dimensional algebra [3,8]. The idea, first proposed in [1], is to obtain a set of monomials that generates this algebra as a vector space, and then use information about the eigensystems of certain endomorphisms to deduce solutions. An early version of an implementation is described in [13]. One way to obtain the vector space generators is to compute a Gröbner basis for F. This may be prohibitive if using exact arithmetic and a lexicographic term order, but there are two ways in which we can improve on matters. One is to use approximate arithmetic on the coefficients in order to avoid swell by limiting precision. The other is that, for purposes of a finding a generating set and computing a "good" endomorphism, we can use any term order. As the degree reverse lexicographic order is typically much faster to compute than lexicographic, we use that one. From here one typically finds an endomorphism matrix corresponding to multiplication by a "random" linear polynomial in the algebra. As shown in [1], eigenvalues of this matrix give the values of the solution set evaluated at that element. For example (and ignoring the use of a random element for simplicity), if we use multiplication by y then the eigenvalues give the solutions for y in the ideal under study. Various methods may then be used to solve for all variables. We say more about this presently.

The next sections discuss numeric Gröbner bases, the use of an eigensystem, handling of problematic cases such as multiplicity (in particular the derogatory case, wherein an eigenvalue has geometric multiplicity greater than one), and assessment of correctness by computing and gauging residuals. We finish with some brief and nontrivial examples. We point to some of the prior literature but make no pretense that this is complete. This work is intended as a practical guide to the method. We refer to the literature for most issues involving theoretical considerations, and focus instead on empirical results and concerns that arise in practice.

I thank Alexander Maletzky for a careful proof reading that caught numerous issues with the exposition.

2   Numeric Gröbner Bases

Given a system of polynomials with rational coefficients, there are various ways to go about computing a Gröbner basis using approximate arithmetic [2,9,11,12, 14,15,18–21,24]. The method used in the Wolfram Language, described in [12, 14], is most closely related to the original method of Shirayanagi [20,21]. Both use an approximation of interval arithmetic. In brief, each arithmetic computation retains a first order estimate of error, and coefficients whose error margin includes the origin are deemed to be zero. Further details will be provided in the full version of this paper.

Prior literature has placed some focus on what are called structural singularities [9,11,22]. In this situation, a perturbation of the coefficients of such a system will change the structure of the basis. Nonetheless, use of sufficiently high but finite precision arithmetic, with precision tracking, will uncover this fact; in practice the needed precision is typically modest [14]. If instead a system has coefficients that put it near but not at such a singularity, again a Gröbner basis computation at sufficient precision will allow the later steps of the numeric solver to proceed.

3   Eigendecomposition for a Multiplication Operator

A common method for solving polynomial systems sets up an eigendecomposition problem; this is done either with resultants [1], Gröbner bases [4,6,10,13,16], or more general normal forms [17]. The idea is as follows. Start with the polynomial system F for which we want to find the set of simultaneous zeros. As noted earlier, the assumption of a finite solution set implies that the polynomial algebra $\mathbb{C}[X]/F$ is a finite dimensional vector space. Moreover it is spanned by the "normal set" of monomials with respect to a given Gröbner basis. These are the monomials that are not reducible by the basis elements. One uses the canonical reduction modulo this basis to set up an endomorphism matrix that represents multiplication by a given polynomial, e.g. by the variable x [4]. Then, as noted in [1] and elsewhere, the eigenvalues of this matrix give the values of x on the solution set of F. We remark that all references above other than [13,17] were developed in the setting of exact coefficient arithmetic (since 1998 the NSolve function of Mathematica has used approximate coefficient arithmetic in its Gröbner basis computations, as noted in [13]).

The left eigenvectors also convey important information about the solutions of F. Assume such an eigenvector corresponds to an eigenvalue with geometric multiplicity of one. So it is the unique (up to scalar multiplication) eigenvector for this eigenvalue. Each component corresponds to an element in the normal set that spans the vector space $\mathbb{C}[X]/F$, and one such component corresponds to the monomial 1. The assumption of a one dimensional eigenspace implies that this component is nonzero. Normalizing the eigenvector so that this component has a value of 1, the values corresponding to the other generators are now the same as the values of those generators on the solution set of F [7,16]. For example, if the normal set is comprised of $\{1, y, y^2, x\}$ and an eigenvector is $(1, 7, 49, 3)$, then one solution has $x = 3$ and $y = 7$. Variables not in the normal set can be solved for in terms of the values of the remaining monomials.

Before proceeding with an example we discuss a modest but important technical detail. Most polynomial systems one encounters in practice do not have multiplicity (at least, such has been the authors experience). However it is not so uncommon for different solutions to have the same values for one or more coordinates even when there are no multiple solutions (in which case the ideal is said to be "not in general position" with respect to these coordinates). Were we to choose such a coordinate as our endomorphism generator, we would obtain one or more multiple eigenvalues. We want to avoid this apparent multiplicity since it complicates the later parts of the algorithm (more is said about this below), and in particular would fool the code into treating the corresponding eigenspaces as derogatory (since the geometric multiplicity would equal the algebraic multiplicity and hence be strictly larger than one). This avoidance is achieved by the simple tactic of choosing multiplication by a (pseudo)random linear polynomial as endomorphism (see e.g. [16]). Practically speaking, this will always suffice to separate solutions that are not multiple.

We now consider a simple example. It will be noticed that the input is already a Gröbner basis with respect to the variable order $y > x$, and in this setting the normal set is $(1, x, y, xy)$. We use a different order to show how NSolve handles this particular system.

$$\mathcal{F} = \left\{ x^2 - 3y + 12, y^2 - xy - 5y + 7 \right\};$$

One can readily check, e.g. by direct substitution, that one solution is in fact $x = 3$, $y = 7$. The Gröbner basis with variable order $x > y$ and degree reverse lexicographic term order is as below.

gb = GroebnerBasis[$N[\mathcal{F}, 100], \{x, y\}$,
  MonomialOrder → DegreeReverseLexicographic];
N[gb]
  $\left\{ -7. + 5.y + 1.xy - 1.y^2, 12. + x^2 - 3.y, -35. + 7.x + 44.y - 13.y^2 + 1.y^3 \right\}$

One will observe that the leading terms are $\left(xy, x^2, y^3 \right)$ and so a generating set of monomials for the algebra $\mathbb{C}[x, y] / < \mathcal{F} >$ is in fact $\left\{ 1, y, y^2, x \right\}$. We take as our (pseudo)random algebra element the linear polynomial $(-92291x)/87992 - (121001y)/175984$. The endomorphism matrix is as below.

$$\begin{pmatrix} 0. & -0.687568188017 & 0. & -1.04885671425 \\ -7.34199699973 & 5.24428357123 & -1.73642490226 & 0. \\ -60.7748715792 & 69.0606986999 & -17.3292401582 & 12.1549743158 \\ 7.77330325484 & 0.291270797345 & -0.687568188017 & 0. \end{pmatrix}$$

We form the eigensystem and normalize so that the first component, which corresponds to the monomial 1, has value of 1.

{vals, vecs} = Eigensystem[mat];
vecsNormalized = Map[#/First[#]&, vecs];
Chop[First[vecsNormalized]]

$\{1., 7., 49., 3.\}$

We see that the remaining components, corresponding to y, y^2, and x respectively, are 7, 49, and 3. So we recover the solution $x = 3$, $y = 7$. There are three other solutions, all obtained similarly from the three other eigenvectors. There is one other real solution, at around $x = 0.915, y = 4.279$.

The above discussion assumes that the eigenvector corresponds to an eigenvalue of geometric multiplicity one. As is well known (see e.g. [8]), this must happen if the ideal is radical and in general position with respect to the polynomial for which the endomorphism matrix was constructed $((-92291x)/87992 - (121001y)/175984$, in this discussion), since this forces every eigenvalue to be simple and that implies geometric multiplicity of one. Provided the ideal is radical, we can force that it be in general position simply by using for the endomorphism multiplication by a random linear combination of the polynomial variables e.g. $3/5x - 11/7y$. This is in fact the approach taken by NSolve. The remaining bad case is when the solution set is not radical. This is the topic of the next section.

4   Detection and Handling of Multiplicity

The case of algebraic multiplicity splits into two subcases. The easier one is where geometric multiplicity is one; in this case eigenvectors are repeated at exactly the algebraic multiplicity. This case is not too difficult to recognize, at least when multiplicity is modest. First one determines multiplicity by recognizing equal eigenvalues. This comparison is done at relatively lower precision than was used to compute them, to account for precision loss. Once multiple eigenvalues are grouped together, the corresponding eigenvectors are compared, again at lower precision. If they are deemed to be equal, after normalizing, then we are in the nonderogatory case. It has been shown [5, 22, 23] that a reliable way to handle this case is by taking the arithmetic average of the approximately equal eigenvectors, as the actual result is near the center.

In the derogatory case the eigenvectors that correspond to a multiple eigenvalue are not equal up to modest tolerance. The values cannot be used in this case because one cannot tell what linear combination of them gives the correct result. An empirical observation of the author is that these seem only to arise when there is both multiplicity and a dimensional component of solutions at infinity. But I have no proof as to whether this is either a necessary or a sufficient condition. Regardless, we proceed as follows. We know the eigenvalue is correct, hence we know the value of a particular linear combination of variables at the solution. We use that in a back substitution step, invoking the solver recursively. This new equation simplifies matters in that it both allows for elimination of one variable, and removes multiplicity by intersecting the solution set with a hyperplane. One back substitution is needed in order to fully determine the coordinate values at a given point of multiplicity (this is because in effect a polynomial with multiple solutions is replaced by a linear polynomial). As the new system is overdetermined, the recursive call uses lower precision so that the system does not become inconsistent for purposes of computing the new Gröbner basis. Further to this end we do not allow the term order to be changed. The benefit here is that the new basis computation will tend to not require too many operations, which carries the risks of both degrading precision and causing an artificial inconsistency if the eigenvalue computation was not sufficiently accurate.

5   Detection of Parasite Solutions and Other
Considerations

A system with radicals involving the variables can always be recast as a polynomial system, simply by adding a new variable for each radical and a defining equation for that radical. For example, $\sqrt{x^2 - y}$ can be rewritten as a new variable, call it r, with defining polynomial $r^2 - (x^2 - y)$. It is often the case that some solutions to this new system will not be valid solutions to the original one, since they may instead satisfy the system with the radical replaced by its conjugate. When radicals are present in the input, NSolve will attempt to detect and remove parasite solutions. The first step is to polish to higher precision using a local solver. After that the high precision solution is plugged into the system, and the residual is compared to zero. If the residual is not sufficiently small then the solution under scrutiny is discarded.

Another consideration is at what precision should the eigendecomposition be done. For best speed one would like to use Lapack library code at machine precision, but in some cases this is not adequate. The default behavior of NSolve is to compute the singular values of the endomorphism matrix and use these to assess its conditioning. If it is sufficiently ill conditioned as to give a bad result, the precision is artificially raised for the eigendecomposition step.

In practice such a precision increase is rarely needed. All the same, the method of assessment is of interest since it works quite well in cases where it is really needed. We note that the size of a residual is going to vary directly with the scale of the input equations, even though multiplying an equation by a constant has no effect on the solution set. We use a common convention to fix a scale for measuring the residual. The first step is to scale each polynomial so that its largest coefficient in norm is unity. As this still tends to give relatively larger residuals for polynomials with many terms, we then divide each by its number of terms. The solutions themselves are treated similarly since their size has an effect on residuals; see the full version of this paper for details.

Yet another consideration is that one would like "sparse" solutions, that is solutions with many components that are zero, to actually be sparse rather than having "small" (relative to the precision and conditioning) values in those components of the approximate solutions. Similarly, in the common case of inputs with only real coefficients, one wants real-valued solutions not to be approximated with small imaginary components. As eigensystem methods are not in general well able to distinguish between "small" values and zero, we use some filtering heuristics to decide when to set values to zero. In particular, vector components divided by the vector norm are deemed to be zero if they are zero to a number of decimal places that is a modest fraction of the precision that was used to compute them.

6   Examples

For brevity we show only two examples. First is the Caprasse system used frequently as a benchmark in the literature. It has multiplicity in the solution set. The second is a perturbation of this system. The original system is particularly problematic insofar as the endomorphism matrix is derogatory.

polysCaprasse $= \{-2x + 2txy - z + y^2z,$
$2 + 4x^2 - 10ty + 4tx^2y - 10y^2 + 2ty^3 + 4xz - x^3z + 4xy^2z,$
$-x + t^2x - 2z + 2tyz,$
$2 - 10t^2 - 10ty + 2t^3y + 4xz + 4t^2xz + 4z^2 + 4tyz^2 - xz^3\};$

Timing[Length[solnsCaprasse = NSolve[polysCaprasse,
  WorkingPrecision->200]]]

$\{3.676, 56\}$

Max[Abs[polysCaprasse/.solnsCaprasse]]

3.2530617312876992019313229$\overline{1}$.00556921532894*^-186

polysCaprasseModified = $\{-2x + 2txy - (100001z)/100000 + y^2z,$
    $2000001/1000000 + 4x^2 - 10ty + 4tx^2y - 10y^2 + 2ty^3 +$
    $4xz - x^3z + 4xy^2z, -x + t^2x - 2z + 2tyz,$
    $2 - 10t^2 - 10ty + 2t^3y + 4xz + 4t^2xz + 4z^2 + tyz^2 - xz^3\}$;

Timing[Length[solnsCaprasseModified = NSolve[polysCaprasseModified,
    WorkingPrecision → 200]]]

$\{2.016, 56\}$

Max[Abs[polysCaprasseModified/.solnsCaprasseModified]]

2.405961788477611232531519785$\overline{3}$.857117648856701*^-146

The perturbed system suffered somewhat more from precision loss in the Gröbner basis computation. This is not a surprise. A perturbation of a problem that gave rise to a derogatory endomorphism matrix leads to an ill conditioned system. Nonetheless we see that use of modest precision sufficed to give a reliable result.

References

1. Auzinger, W., Stetter, H.: An elimination algorithm for the computation of all zeros of a system of multivariate polynomial equations. Int. Ser. Numer. Math. **86**, 11–31 (1988)
2. Bodrato, M., Zanoni, A.: A numerical Gröbner bases and syzygies: an interval approach. In: Proceedings of the 6th International Symposium on Symbolic and Numeric Algorithms for Scientific Computing (SYNASC 2004), pp. 77–89 (2004)
3. Buchberger, B.: Gröbner-bases: an algorithmic method in polynomial ideal theory. In: Multidimensional Systems Theory - Progress, Directions and Open Problems in Multidimensional Systems, Chap. 6, pp. 184–232. Reidel Publishing Company, Dodrecht, Boston, Lancaster (1985)
4. Corless, R.: Editor's corner: Gröbner bases and matrix eigenproblems. ACM SIGSAM Bull. Commun. Comput. Algebra **30**, 26–32 (1996)
5. Corless, R., Gianni, P., Trager, B.: A reordered Schur factorization method for zero-dimensional polynomial systems with multiple roots. In: Proceedings of the International Symposium on Symbolic and Algebraic Computation (ISSAC 1997), pp. 133–140. ACM Press (1997)
6. Cox, D.: Introduction to Gröbner bases. In: Proceedings of Symposia in Applied Mathematics, pp. 1–24. ACM Press (1998)
7. Cox, D.A., Little, J., O'Shea, D.: Using Algebraic Geometry. Springer-Verlag New York, Inc., Secaucus (1998). https://doi.org/10.1007/b138611
8. Cox, D.A., Little, J., O'Shea, D.: Ideals, Varieties, and Algorithms: An Introduction to Computational Algebraic Geometry and Commutative Algebra. Springer-Verlag New York Inc., Secaucus (2007)

9. Faugère, J.-C., Liang, Y.: Pivoting in extended rings for computing approximate Gröbner bases. Math. Comput. Sci. **5**, 179–194 (2011)
10. Gianni, P., Mora, T.: Algebrric solution of systems of polynomirl equations using Groebher bases. In: Huguet, L., Poli, A. (eds.) AAECC 1987. LNCS, vol. 356, pp. 247–257. Springer, Heidelberg (1989). https://doi.org/10.1007/3-540-51082-6_83
11. Kondratyev, A., Stetter, H., Winkler, F.: Numerical computation of Gröbner bases. In: Proceedings of the 7th Workshop on Computer Algebra in Scientific Computation (CASC 2004), pp. 295–306 (2004)
12. Lichtblau, D.: Gröbner bases in mathematica 3.0. Math. J. **6**(4), 81–88 (1996). http://library.wolfram.com/infocenter/Articles/2179/
13. Lichtblau, D.: Solving finite algebraic systems using numeric Gröbner bases and eigenvalues. In: Proceedings of the World Conference on Systemics, Cybernetics, and Informatics (SCI 2000), vol. 10, pp. 555–560 (2000)
14. Lichtblau, D.: Polynomial GCD and factorization via approximate Gröbner bases. In: Proceedings of the 2010 12th International Symposium on Symbolic and Numeric Algorithms for Scientific Computing, SYNASC 2010, Washington, DC, USA, pp. 29–36. IEEE Computer Society (2010)
15. Lichtblau, D.: Approximate Gröbner bases overdetermined polynomial systems, and approximate GCDs. ISRN Comput. Math. **2013**, 13 (2013). http://www.hindawi.com/isrn/cm/2013/352806/
16. Möller, H.M.: Systems of algebraic equations solved by means of endomorphisms. In: Cohen, G., Mora, T., Moreno, O. (eds.) AAECC 1993. LNCS, vol. 673, pp. 43–56. Springer, Heidelberg (1993). https://doi.org/10.1007/3-540-56686-4_32
17. Mourrain, B., Trebuchet, P.: Generalized normal forms and polynomial system solving. In: Proceedings of the 2005 International Symposium on Symbolic and Algebraic Computation, ISSAC 2005, New York, NY, USA, pp. pages 253–260. ACM (2005)
18. Sasaki, T., Kako, F.: Computing floating-point Gröbner bases stably. In: Proceedings of the 2007 International Workshop on Symbolic-Numeric Computation, SNC 2007, New York, NY, USA, pp. 180–189. ACM (2007)
19. Sasaki, T., Kako, F.: Floating-point Gröbner basis computation with ill-conditionedness estimation. Comput. Math. **5081**, 278–292 (2008)
20. Shirayanagi, K.: An algorithm to compute floating point Groebner bases. In: Lee, T. (ed.) Mathematical Computation with Maple V, Ideas and Applications, pp. 95–106. Birkhäuser, Boston (1993). https://doi.org/10.1007/978-1-4612-0351-3_10
21. Shirayanagi, K.: Floating point Gröbner bases. Math. Comput. Simul. **42**(4–6), 509–528 (1996)
22. Stetter, H.: Stabilization of polynomial systems solving with Groebner bases. In: Proceedings of the 1997 International Symposium on Symbolic and Algebraic Computation (ISSAC 1997), New York, NY, USA, pp. 117–124. ACM (1997)
23. Stetter, H.: Numerical Polynomial Algebra. SIAM, Philadelphia (2004)
24. Traverso, C., Zanoni, A.: Numerical stability and stabilization of Groebner basis computation. In: Proceedings of the 2002 International Symposium on Symbolic and Algebraic Computation (ISSAC 2002), New York, NY, USA, pp. 262–269. ACM (2002)
25. Wolfram, I.: Research. Mathematica **11** (2018)

The Andrews-Curtis Conjecture, Term Rewriting and First-Order Proofs

A. Lisitsa$^{(\boxtimes)}$

Department of Computer Science, University of Liverpool, Liverpool, UK
A.Lisitsa@liverpool.ac.uk

Abstract. The Andrews-Curtis conjecture (ACC) remains one of the outstanding open problems in combinatorial group theory. In short, it states that every balanced presentation of the trivial group can be transformed into a trivial presentation by a sequence of simple transformations. It is generally believed that the conjecture may be false and there are several series of potential counterexamples for which required simplifications are not known. Finding simplifications poses a challenge for any computational approach - the search space is unbounded and the lower bound on the length of simplification sequences is known to be at least superexponential. Various specialised search algorithms have been used to eliminate some of the potential counterexamples. In this paper we present an alternative approach based on automated reasoning. We formulate a term rewriting system ACT for AC-transformations, and its translation(s) into the first-order logic. The problem of finding AC-simplifications is reduced to the problem of proving first-order formulae, which is then tackled by the available automated theorem provers. We report on the experiments demonstrating the efficiency of the proposed method by finding required simplifications for several new open cases.

1 Introduction

The topic of this paper can be described by two expressions: *applied automated reasoning* and *experimental mathematics*. We show how automated first-order theorem proving and disproving can be used to explore the Andrews-Curtis conjecture (ACC) [2]. This conjecture remains one of the outstanding open problems in combinatorial group theory. In short, it states that every balanced presentation of the trivial group can be transformed into a trivial presentation by a sequence of simple transformations. It is generally believed that the conjecture may be false and there are several series of potential counterexamples for which required simplifications are not known.

For a group presentation $\langle x_1, \ldots, x_n; r_1, \ldots r_m \rangle$ with generators x_i, and relators r_j, consider the following transformations.

AC1 Replace some r_i by r_i^{-1}.
AC2 Replace some r_i by $r_i \cdot r_j$, $j \neq i$.
AC3 Replace some r_i by $w \cdot r_i \cdot w^{-1}$ where w is any word in the generators.

© Springer International Publishing AG, part of Springer Nature 2018
J. H. Davenport et al. (Eds.): ICMS 2018, LNCS 10931, pp. 343–351, 2018.
https://doi.org/10.1007/978-3-319-96418-8_41

AC4 Re-order the relators.

AC5 Introduce a new generator y and relator y or delete a generator y and relator y.

We notice that AC4 rule is redundant in a sense that its effect can be achieved by an application of a sequence of AC1 and AC2 rules. Indeed, for any two relators r_i and r_j their transposition $\ldots r_i \ldots r_j \ldots \mapsto \ldots r_j, \ldots r_i \ldots$ is the result of the application of the sequence of rules $AC2_{ij}$ $AC1_i$ $AC2_{ji}$ $AC1_j$ $AC2_{ij}$ $AC1_i$. As any permutation is a composition of transpositions the statement follows.

Two presentations g and g' are called *Andrews-Curtis equivalent (AC-equivalent)* if one of them can be obtained from the other by applying a finite sequence of transformations of the types (AC1)–(AC4). Two presentations are *stably AC-equivalent* if one of them can be obtained from the other by applying a finite sequence of transformations of the types (AC1)–(AC5).

A group presentation $g = \langle x_1, \ldots, x_n; r_1, \ldots r_m \rangle$ is called *balanced* if $n = m$, that is a number of generators is the same as a number of relators. Such n we call a *dimension* of g and denote by $Dim(g)$.

Conjecture 1 (Andrews-Curtis [2])
If $\langle x_1, \ldots, x_n; r_1, \ldots r_n \rangle$ is a balanced presentation of the trivial group it is AC-equivalent to the trivial presentation $\langle x_1, \ldots, x_n; x_1, \ldots x_n \rangle$.

The *weak form* of the conjecture states that every balanced presentation for a trivial group is stably AC-equivalent (i.e. transformations AC5 are allowed) to the trivial presentation.

In what follows we will assume that we are dealing with the strong form of the conjecture unless stated otherwise.

Both variants of the conjecture remain open and challenging problems. According to [4] the prevalent opinion is that the conjecture is false, but no counterexamples have been found so far. There are, however, potential counterexamples and even infinite series of potential counterexamples, which provide an opportunity to use a computational approach to explore the conjecture. Notice, that if the statement of the conjecture holds for a particular presentation this fact can be established, at least in principle, by enumeration and application of all possible sequences of transformations until the trivial presentation is obtained. Then, in principle, one may attack potential counterexamples for AC-conjecture by the automated search of the AC-sequences leading to the trivial presentations (AC-simplifying sequences). Such a search is a computationally difficult and the search space grows exponentially with the length of the sequences. As it was noticed in [18], neither total enumeration, nor random search can be effectively applied here. More efficient search procedure using *genetic algorithms* has been proposed in [18] and it was used to show that a well-known potential counterexample $\langle x, y | xyxy^{-1}x^{-1}y^{-1}, x^2 y^{-3} \rangle$ is, in fact, AC-equivalent to the trivial presentation, and by that it is not a counterexample. Further exploration and improvement of genetic approach can be found in [13, 20] where many new simplifications are presented as well.

In [12] it was shown that a systematic breadth-first search of the tree of equivalent presentations is a viable alternative to genetic algorithms of [18] which allowed to show, in particular, that the potential counterexample

$$\langle x, y | xyxy^{-1}x^{-1}y^{-1}, x^3y^{-4} \rangle$$

is unique up to the AC-equivalence among all balanced presentations of trivial groups with two generators up to the length 13. This counterexample (AK-3) is one of the infinite series of presentations proposed by Akbulut and Kirby [1] and is the smallest for which it is not known whether it is AC-equivalent to trivial presentation. The paper [16] discusses the implementation aspects of the breadth-first search for AC-simplifications on high-performance computer platform using disk-based hash tables. The approach is illustrated by successful search of AC-simplifications for some known non-trivial cases. In [11] an alternative approach for refuting the potential counterexamples based on the methods from computational group theory was proposed. In this approach AC-simplifications are extracted from the results produced by Todd-Coxeter coset enumeration algorithm, by application of ad hoc techniques. The approach has been used to find some non-trivial AC-simplifications.

Lower bound on the length of simplifications is known to be superexponential [7,14]. So the failure to deal AK-3 example by any known computational approach should not be overestimated, we are still exploring very small part of the huge search space.

In this paper we propose an alternative approach for testing the groups presentations as to whether they satisfy the Andrews-Curtis conjecture which is based on use of term-rewriting systems and first-order logic. We formulate the term rewriting system ACT for AC-transformations, and its translations into the first-order logic. The problem of finding AC-simplifications is reduced to the problem of proving first-order formulas, which is then tackled by the available automated theorem provers. We show that the approach is competitive by demonstrating simplifications for a few open cases. An abstract with an announcement of the proposed method and simplifications of known cases has appeared in [15].

2   ACT Term Rewriting Systems

Let T_G be the equational theory of groups defined by the following equations in a vocabulary $(\cdot, r, e)$:

- $(x \cdot y) \cdot z = x \cdot (y \cdot z)$
- $x \cdot e = x$
- $x \cdot r(x) = e$

For each $n \geq 2$ we formulate a term rewriting system modulo T_G, which captures AC-transformations of presentations of dimension n. We start with dimension $n = 2$.

For an alphabet $A = \{a_1, a_2\}$ a term rewriting system ACT_2 consists the following rules:

R1L $f(x,y) \rightarrow f(r(x), y))$
R1R $f(x,y) \rightarrow f(x, r(y))$
R2L $f(x,y) \rightarrow f(x \cdot y, y)$
R2R $f(x,y) \rightarrow f(x, y \cdot x)$
R3L$_i$ $f(x,y) \rightarrow f((a_i \cdot x) \cdot r(a_i), y)$ for $a_i \in A, i = 1,2$
R3R$_i$ $f(x,y) \rightarrow f(x, (a_i \cdot y) \cdot r(a_i))$ for $a_i \in A, i = 1,2$

The term rewriting system ACT_2 gives rise to the rewrite relation $\rightarrow_{ACT}$ on the set of all terms defined in the standard way [3]. For terms t_1, t_2 in groups vocabulary we write $t_1 =_G t_2$ if equality $t_1 = t_2$ is derivable in T_G. We extend $=_G$ homomorphically by defining $f(t_1, t_2) =_G f(s_1, s_2)$ iff $t_1 =_G s_1$ and $t_2 =_G s_2$. Denote by $[t]_G$ the equivalence class of t wrt $=_G$, that is $[t]_G = \{t' \mid t =_G t'\}$.

Then rewrite relation $\rightarrow_{ACT/G}$ for ACT modulo theory T_G is defined [3] as follows: $t \rightarrow_{ACT/G} s$ iff there exist $t' \in [t]_G$ and $s' \in [s]_G$ such that $t' \rightarrow_{ACT} s'$.

Claim (on formalization). The notion of rewrite relation $\rightarrow_{ACT/G}$ captures adequately the notion of AC-rewriting, as defined in Sect. 1 that is for presentations p_1 and p_2 we have $p_1 \rightarrow^*_{AC} p_2$ iff $t_{p_1} \rightarrow^*_{ACT/G}$. Here t_p denotes a term encoding of a presentation p, that is for $p = \langle a_1, a_2 \mid t_1.t_2 \rangle$ we have $t_p = f(t_1, t_2)$.

The term rewriting system ACT_2 can be simplified without changing the transitive closure of the rewriting relation. Reduced term rewriting system $rACT_2$ consists of the following rules:

R1L $f(x,y) \rightarrow f(r(x), y))$
R2L $f(x,y) \rightarrow f(x \cdot y, y)$
R2R $f(x,y) \rightarrow f(x, y \cdot x)$
R3L$_i$ $f(x,y) \rightarrow f((a_i \cdot x) \cdot r(a_i), y)$ for $a_i \in A, i = 1,2$

Proposition 1. *Term rewriting systems ACT_2 and $rACT_2$ considered modulo T_G are equivalent, that is $\rightarrow^*_{ACT_2/G}$ and $\rightarrow^*_{rACT_2/G}$ coincide.*

Proposition 2. *For ground t_1 and t_2 we have $t_1 \rightarrow^*_{ACT_2/G} t_2 \Leftrightarrow t_2 \rightarrow^*_{ACT_2/G} t_1$, that is $\rightarrow^*_{ACT_2/G}$ is symmetric.*

Now we present two variants of translations of ACT_2 into first-order logic with an intention to use automated theorem proving to show AC-equivalence.

2.1   Equational Translation

Denote by E_{ACT_2} an equational theory $T_G \cup rACT^=$ where $rACT^=$ includes the following axioms (equality variants of the above rewriting rules):

E-R1L $f(x,y) = f(r(x), y))$
E-R2L $f(x,y) = f(x \cdot y, y)$
E-R2R $f(x,y) = f(x, y \cdot x)$

E-R3L$_i$ $f(x,y) = f((a_i \cdot x) \cdot r(a_i), y)$ for $a_i \in A, i = 1, 2$

Proposition 3. *For ground terms t_1 and t_2 $t_1 \rightarrow^*_{ACT_2/G} t_2$ iff $E_{ACT_2} \vdash t_1 = t_2$*

Proof (sketch). By Proposition 2 $t_1 \rightarrow^*_{ACT_2/G} t_2 \Leftrightarrow t_2 \leftrightarrow^*_{ACT_2/G} t_1$. By Birkhoff's theorem [5, 8, 19] the latter condition is equivalent to $E_{ACT_2} \models t_1 = t_2$ and therefore $E_{ACT_2} \vdash t_1 = t_2$.

In a variant of the equational translation the axioms **E − R3L$_i$** are replaced by "non-ground" axiom **E − RLZ** : $f(x,y) = f((z \cdot x) \cdot r(z), y)$ and the corresponding analogue of Proposition 3 holds true.

2.2   Implicational Translation

Denote by I_{ACT_2} the first-order theory $T_G \cup rACT_2^{\rightarrow}$ where $rACT_2^{\rightarrow}$ includes the following axioms:

I-R1L $R(f(x,y)) \rightarrow R(f(r(x), y))$
I-R2L $R(f(x,y)) \rightarrow R(f(x \cdot y, y))$
I-R2R $R(f(x,y)) \rightarrow R(f(x, y \cdot x))$
I-R3L$_i$ $R(f(x,y)) \rightarrow R(f((a_i \cdot x) \cdot r(a_i), y))$ for $a_i \in A, i = 1, 2$

Proposition 4. *For ground terms t_1 and t_2 $t_1 \rightarrow^*_{ACT_2/G} t_2$ iff $I_{ACT_2} \vdash R(t_1) \rightarrow R(t_2)$.*

Similarly to the case of equational translation "non-ground" axiom **I-R3Z**: $R(f(x,y)) \rightarrow R(f((z \cdot x) \cdot r(z), y))$ can be used instead of **I-R3L$_i$** with a corresponding analogue of Proposition 4 holding true.

2.3   Higher Dimensions

For dimensions $n > 2$ the rewriting systems ACT_n, their reduced versions $rACT_n$, their equational and implicational translations can be formulated such that the analogues of Propositions 3 and 4 hold true. To cut a long story short we show here only an equational translation $rACT_3^{=}$ ("non-ground" variant):

$$f(x,y,z) = f(r(x), y, z) \qquad f(x,y,z) = f(x, r(y), z)$$
$$f(x,y,z) = f(x, y, r(z)) \qquad f(x,y,z) = f(x \cdot y, y, z)$$
$$f(x,y,z) = f(x \cdot z, y, z) \qquad f(x,y,z) = f(x, y \cdot x, z)$$
$$f(x,y,z) = f(x, y \cdot z, z) \qquad f(x,y,z) = f(x, y, z \cdot x)$$
$$f(x,y,z) = f(x, y, z \cdot y) \qquad f(x,y,z) = f((v \cdot x) \cdot r(v), y, z)$$
$$f(x,y,z) = f(x, (v \cdot y) \cdot r(v), z) \qquad f(x,y,z) = f(x, y, (v \cdot z) \cdot r(v)).$$

3   Automated Proving and Disproving for ACC Exploration

Propositions 3 and 4 (and their analogues) suggest a way of using automated reasoning for exploration of ACC. For any concrete pair of presentations p_1 and

p_2, to establish whether they are AC-equivalent one can formulate a theorem proving/disproving tasks of the form $E_{ACT_n} \vdash t_{p_1} = t_{p_2}$, or $I_{ACT_n} \vdash R(t_{p_1}) \rightarrow R(t_{p_2})$ ($E_{ACT_n} \nvdash t_{p_1} = t_{p_2}$, or $I_{ACT_n} \nvdash R(t_{p_1}) \rightarrow R(t_{p_2})$).

Unfortunately disproving by finite countermodel model finding has its fundamental limitations in the context of ACC. Based on the results of [6] it cannot be used to disprove ACC. At the same time one can get some non-trivial results on necessity of some of the rules for simplification, both in solved cases and non-solved cases. For example we have:

Proposition 5. *To simplify AK-3 (if at all it is possible) one really needs conjugation with both generators a and b.*

We have used finite model builder Mace4 [17] to build countermodels of sizes 12 and 6 respectively for the cases where either of the conjugation rules was missing.

3.1 Theorem Proving for Simplification

Known Cases. We have applied automated theorem proving using Prover9 prover [17] to confirm that all cases eliminated as potential counterexamples in [11–13,16,18] can be eliminated by our method too.

New Cases. Using automated theorem proving we were able to eliminate the following potential counterexamples for ACC, which are all *irreducible cyclically presented groups* [10] whose status was open to the best of our knowledge [9,10,13]. We use notation of [9] to refer to these examples. We also follow the standard convention to use capital letters $A, B, C \ldots$ to denote inverse of $a, b, c, \ldots$ respectively.

Dim = 2
  T14 $\langle a, b \mid ababABB, babaBAA \rangle$
  T28 $\langle a, b \mid aabbbbABBBB, bbaaaaBAAAA \rangle$
  T36 $\langle a, b \mid aababAABB, bbabaBBAA \rangle$
  T62 $\langle a, b \mid aaabbAbABBB, bbbaaBaBAAA \rangle$
  T74 $\langle a, b \mid aabaabAAABB, bbabbaBBBAA \rangle$
Dim = 3
  T16 $\langle a, b, c \mid ABCacbb, BCAbacc, CABcbaa \rangle$
  T21 $\langle a, b, c \mid ABCabac, BCAbcba, CABcacb \rangle$
  T48 $\langle a, b, c \mid aacbcABCC, bbacaBCAA, ccbabCABB \rangle$
  T88 $\langle a, b, c \mid aacbAbCAB, bbacBcABC, ccbaCaBCA \rangle$
  T89 $\langle a, b, c \mid aacbcACAB, bbacBABC, ccbaCBCA \rangle$
Dim = 4
  T96 $\langle a, b, c, d \mid adCADbc, baDBAcd, cbACBda, dcBDCab \rangle$
  T97 $\langle a, b, c, d \mid adCAbDc, baDBcAd, cbACdBa, dcBDaCb \rangle$

We were able to prove corresponding formulas in both equational and (variants of) implicational translations. The proofs for implicational translations are more transparent and more amenable for simplifying transformations extractions. The proofs generated by Prover9 for implicational translations are essentially sequences of atomic formulas of the from $R(r_1, r_2)$ (for Dim $= 2$) which encompass simplification sequences of presentations $\langle a, b \mid r_1, r_2 \rangle$. All such atomic formulas produced with the references to the applied clauses which encode particular rules from (AC1)–(AC3). In the Appendix we show a simplification extracted manually from the proof for T16 (Dim $= 3$) presentation.

4 Conclusion

As it was noticed in [18] neither total enumeration, nor random search can be effectively applied to disproving the Andrews-Curtis conjecture. We have shown in this paper that systematic, goal-oriented search implemented in automated theorem proving procedures provides an interesting and viable alternative.

Furthermore, although finite model finding can not be used directly to disprove AC-conjecture, it can be a tool for establishing non-derivability for subsystems of transformations.

We have published all computer-generated proofs online[1].

Appendix

5.1 Technical Details

We used Prover9 and Mace4 version 0.5 (December 2007) [17] and one of the two following system configurations:

(A) AMD A6-3410MX APU 1.60 Ghz, RAM 4 GB, Windows 7 Enterprise
(B) Intel(R) Core(TM) i7-4790 CPU 3.60 Ghz, RAM 32 GB, Windows 7 Enterprise

Table 1. Time to prove simplifications for system configuration (B)

	T14	T28	T36	T62	T74	T16	T21	T48	T88	T89	T96	97
Dim	2	2	2	2	2	3	3	3	3	3	4	4
Equational	6.02 s	6.50 s	7.18 s	24.34 s	57.17 s	12.87 s	11.98 s	34.63 s	57.69 s	17.50 s	114.05 s	115.10 s
Implicational	1.57 s	2.46 s	1.34 s	22.50 s	6.29 s	1.61 s	1.45 s	2.17 s	1.97 s	2.14 s	102.34 s	89.65 s
Implicational GC	t/o	t/o	t/o	t/o	t/o	3.76 s	1.61 s	t/o	0.86 s	0.75 s	t/o	t/o

"t/o" stands for timeout in 200s; "GC" means encoding with ground conjugation rules; all other encodings are with non-ground conjugation rules.

---

[1] https://zenodo.org/record/1248986 DOI: 10.5281/zenodo.1248986.

5.2 AC-Trivialization for T16

Initial presentation:

$\langle a, b, c \mid ABCacbb, BCAbacc, CABcbaa \rangle$

Simplification: $\langle ABCacbb, BCAbacc, CABcbaa \rangle$

$\xrightarrow{x,y,z \to x,y,azA} \langle ABCacbb, BCAbacc, aCABcba \rangle$

$\xrightarrow{x,y,z \to x,y,zx} \langle ABCacbb, BCAbacc, aCABacbb \rangle$

$\xrightarrow{x,y,z \to x,y,bzB} \langle ABCacbb, BCAbacc, baCABacb \rangle$

$\xrightarrow{x,y,z \to x,y,zy} \langle ABCacbb, BCAbacc, bac \rangle$

$\xrightarrow{x,y,z \to x,y,czC} \langle ABCacbb, BCAbacc, cba \rangle$

$\xrightarrow{x,y,z \to x',y,z} \langle BBCAcba, BCAbacc, cba \rangle$

$\xrightarrow{x,y,z \to x,y,z'} \langle BBCAcba, BCAbacc, ABC \rangle$

$\xrightarrow{x,y,z \to xz,y,z} \langle BBCA, BCAbacc, ABC \rangle$

$\xrightarrow{x,y,z \to x',y,z} \langle acbb, BCAbacc, ABC \rangle \xrightarrow{x,y,z \to x,y,z'} \langle acbb, BCAbacc, cba \rangle$

$\xrightarrow{x,y,z \to x,y,azA} \langle acbb, BCAbacc, acb \rangle \xrightarrow{x,y,z \to x,y,z'} \langle acbb, BCAbacc, BCA \rangle$

$\xrightarrow{x,y,z \to x,y,zx} \langle acbb, BCAbacc, b \rangle \xrightarrow{x,y,z \to x,y,z'} \langle acbb, BCAbacc, B \rangle$

$\xrightarrow{x,y,z \to xz,y,z} \langle acb, BCAbacc, B \rangle \xrightarrow{x,y,z \to xz,y,z} \langle ac, BCAbacc, B \rangle$

$\xrightarrow{x,y,z \to x,y',z} \langle ac, CCABacb, B \rangle \xrightarrow{x,y,z \to x,yz,z} \langle ac, CCABac, B \rangle$

$\xrightarrow{x,y,z \to x,y',z} \langle ac, CAbacc, B \rangle \xrightarrow{x,y,z \to x,y,z'} \langle ac, CAbacc, b \rangle$

$\xrightarrow{x,y,z \to x',y,z} \langle CA, CAbacc, b \rangle$

$\xrightarrow{x,y,z \to x,yx,z} \langle CA, CAbacA, b \rangle \xrightarrow{x,y,z \to x,y',z} \langle CA, aCABac, b \rangle$

$\xrightarrow{x,y,z \to x,yx,z} \langle CA, aCAB, b \rangle \xrightarrow{x,y,z \to x,yz,z} \langle CA, aCA, b \rangle$

$\xrightarrow{x,y,z \to x',y,z} \langle ac, aCA, b \rangle \xrightarrow{x,y,z \to x,yx,z} \langle ac, a, b \rangle$

$\xrightarrow{x,y,z \to x,y',z} \langle ac, A, b \rangle \xrightarrow{x,y,z \to x,yx,z} \langle ac, c, b \rangle$

$\xrightarrow{x,y,z \to x,y',z} \langle ac, C, b \rangle \xrightarrow{x,y,z \to xy,y,z} \langle a, C, b \rangle$

$\xrightarrow{x,y,z \to x,yz,z} \langle a, Cb, b \rangle \xrightarrow{x,y,z \to x,y',z} \langle a, Bc, b \rangle$

$\xrightarrow{x,y,z \to x,y,zy} \langle a, Bc, c \rangle \xrightarrow{x,y,z \to x,y,z'} \langle a, Bc, C \rangle$

$\xrightarrow{x,y,z \to x,yz,z} \langle a, B, C \rangle \xrightarrow{x,y,z \to x,y,z'} \langle a, B, c \rangle$

$\xrightarrow{x,y,z \to x,y',z} \langle a, b, c \rangle$

References

1. Akbulut, S., Kirby, R.: A potential smooth counterexample in dimension 4 to the Poincare conjecture, the Schoenflies conjecture, and the Andrews-Curtis conjecture. Topology **24**(4), 375–390 (1985)
2. Andrews, J., Curtis, M.L.: Free groups and handlebodies. Proc. Amer. Math. Soc. **16**, 192–195 (1965)
3. Baader, F., Nipkow, T.: Term Rewriting and All That. Cambridge University Press, New York (1998)

4. Baumslag, G., Myasnikov, A.G., Shpilrain, V.: Open Problems in Combinatorial Group Theory, 2nd edn., vol. 296, pp. 1–38. Amer. Math. Soc., Providence, RI (2002)
5. Birkhoff, G.: On the structure of abstract algebras. In: Mathematical Proceedings of the Cambridge Philosophical Society, vol. 31, pp. 433–454. Cambridge University Press (1935)
6. Borovik, A.V., Lubotzky, A., Myasnikov, A.G.: The finitary Andrews-Curtis conjecture. In: Bartholdi, L., Ceccherini-Silberstein, T., Smirnova-Nagnibeda, T., Zuk, A. (eds.) Infinite Groups: Geometric, Combinatorial and Dynamical Aspects. Progress in Mathematics, vol. 248, pp. 15–30. Birkhäuser Basel (2005). https://doi.org/10.1007/3-7643-7447-0_2
7. Bridson, M.R.: The complexity of balanced presentations and the Andrews-Curtis conjecture. ArXiv e-prints, April 2015
8. Dershowitz, N., Jouannaud, J.-P.: Rewrite systems. In: van Leeuwen, J. (ed.) Formal Models and Semantics. Handbook of Theoretical Computer Science, vol. B, pp. 243–320. Elsevier, Amsterdam (1990)
9. Edjvet, M., Swan, J.: Irreducible cyclically presented groups 2005–2010. https://www.maths.nottingham.ac.uk/personal/pmzme/Irreducible-Cyclically-Presented-Groups.pdf
10. Edjvet, M., Swan, J.: On irreducible cyclic presentations of the trivial group. Exp. Math. 23(2), 181–189 (2014)
11. Havas, G., Ramsay, C.: Andrews-Curtis and Todd-Coxeter proof words. Technical report, in Oxford, vol. I, London Math. Soc. Lecture Note Ser (2001)
12. Havas, G., Ramsay, C.: Breadth-first search and the Andrews-Curtis conjecture. Int. J. Algebra Comput. 13(01), 61–68 (2003)
13. Krawiec, K., Swan, J.: Ac-trivialization proofs eliminating some potential counterexamples to the Andrews-curtis conjecture (2015). www.cs.put.poznan.pl/kkrawiec/wiki/uploads/Site/ACsequences.pdf
14. Lishak, B.: Balanced finite presentations of the trivial group. ArXiv e-prints, April 2015
15. Lisitsa, A.: First-order theorem proving in the exploration of Andrews-Curtis conjecture. TinyToCS 2 (2013)
16. McCaul, S.B., Bowman, R.S.: Fast searching for Andrews-Curtis trivializations. Exp. Math. 193–197, 2006 (2006)
17. McCune, W.: Prover9 and mace4, 2005–2010. http://www.cs.unm.edu/~mccune/prover9/
18. Miasnikov, A.D.: Genetic algorithms and the Andrews-curtis conjecture. Int. J. Algebra Comput. 09(06), 671–686 (1999)
19. Plaisted, D.A.: Equational reasoning and term rewriting systems. In: Gabbay, D.M., Hogger, C.J., Robinson, J.A. (eds.) Handbook of Logic in Artificial Intelligence and Logic Programming, vol. 1, pp. 274–364. Oxford University Press Inc, New York (1993)
20. Swan, J., Ochoa, G., Kendall, G., Edjvet, M.: Fitness landscapes and the Andrews-Curtis conjecture. IJAC 22(2) (2012)

Francy - An Interactive Discrete Mathematics Framework for GAP

Manuel Machado Martins[1(✉)] and Markus Pfeiffer[2]

[1] Universidade Aberta, Lisbon, Portugal
manuelmachadomartins@gmail.com
[2] University of St Andrews, St Andrews, Scotland
markus.pfeiffer@st-andrews.ac.uk
https://github.com/mcmartins, https://markusp.morphism.de

Abstract. Data visualization and interaction with large data sets is known to be essential and critical in many businesses today, and the same applies to research and teaching, in this case, when exploring large and complex mathematical objects. GAP is a computer algebra system for computational discrete algebra with an emphasis on computational group theory. The existing XGAP package for GAP works exclusively on the X Window System. It lacks abstraction between its mathematical and graphical cores, making it difficult to extend, maintain, or port. In this paper, we present Francy, a graphical semantics package for GAP. Francy is responsible for creating a representational structure that can be rendered using many GUI frameworks independent from any particular programming language or operating system. Building on this, we use state of the art web technologies that take advantage of an improved REPL environment, which is currently under development for GAP. The integration of this project with Jupyter provides a rich graphical environment full of features enhancing the usability and accessibility of GAP.

Keywords: Visualization · Interaction · Graphics · Mathematics
GAP · Jupyter

1 Introduction

By providing a mechanism for quickly demonstrating a topic, or result, visual learning has been proven effective and advantageous. It helps with engagement and allows students to look at problems in a different way [1].

In mathematics, especially in group theory, having a graphical representation of certain structures is invaluable when formulating conjectures and counterexamples, and when analyzing data. GAP is a computer algebra system (CAS) focused on computational group theory and it helps to explore algebraic structures and solve a variety of problems [2]. The primary existing package for GAP, which provides facilities displaying graphics and visualization of mathematical data structures, is XGAP. This package is integrated with the Unix X-Window

© Springer International Publishing AG, part of Springer Nature 2018
J. H. Davenport et al. (Eds.): ICMS 2018, LNCS 10931, pp. 352–358, 2018.
https://doi.org/10.1007/978-3-319-96418-8_42

System, which provides a basic framework for a Graphical User Interface (GUI), and includes a wide range of mathematical functionality focused on the lattice of subgroup of a group [3]. Further such packages include Interactive Todd Coxeter (ITC) which was developed using XGAP and provides an interactive environment for exploring coset enumerations [4]. GAP.APP is another project based on XGAP and it provides a native Macintosh interface for GAP [5]. All of these projects enable GAP to be used as a tool to visualize objects with computer graphics.

Technology evolves quickly and today multiple web platforms allow users to experience, learn, and share in a simple and fast paced environment. Jupyter is one of these projects and, as mentioned on the official website [6], aims *"to develop open-source software, open-standards, and services for interactive computing across dozens of programming languages"*, leveraging learning processes and the way people share their work. The purpose of the OpenDreamKit project is to provide a framework for the advancement of mathematics in Europe, as part of the Horizon2020 European Research Infrastructure, and Jupyter is a core component of OpenDreamKit [7]. Jupyter allows a centralized system for the dissemination of content and uses an intuitive interface where users interact with notebooks containing live code, equations, visualizations and narrative text.

Francy [8] arose from the necessity of having a lightweight framework for building interactive graphics, generated from GAP, running primarily on the web, primarily in a Jupyter Notebook. An initial attempt to re-use XGAP and port it was made, but the lack of a standardized data exchange format between GAP and the graphics renderer, and the simplistic initial requirements of the project were the basis for the creation of a new GAP package.

2   Functionality

Francy provides an interface to draw graphics using objects. This interface is based on simple concepts of drawing and graph theory, allowing the creation of directed and undirected graphs, trees, line charts, bar charts and scatter charts. These graphical objects are drawn inside a canvas that includes a space for menus and to display informative messages. Within the canvas it is possible to interact with the graphical objects by clicking, selecting, dragging and zooming.

In terms of interaction with the kernel, we use callbacks which allow the execution of functions in GAP from the graphical objects. A callback holds the function signature and any arguments that it requires. If a callback requires user input, a modal window will be shown before the execution of the function.

3   Applications

Francy does not provide any mathematical functionality as it is intended to be used by other mathematical software packages. Existing GAP packages can be easily ported to use it. Francy has potentially many applications and can be used to provide a graphical representation of data structures, allowing one to navigate through and explore properties or relations of these structures. In this way, Francy can be used to enrich a learning environment where GAP provides a library of thousands of functions implementing algebraic algorithms as well as large data libraries of algebraic objects.

In the following code we show a simple usage of Francy to display interactively the directed graph of all subgroups of the Symmetric Group S_3, using the GAP package Digraphs [9]:

```
LoadPackage("digraphs"); LoadPackage("francy");

G := SymmetricGroup(3); as := AllSubgroups(G); nodes := [];
d := Digraph(as, {H, K} -> IsSubgroup(H, K));

vertices := DigraphVertices(d); edges := DigraphEdges(d);

canvas := Canvas(Concatenation("Subgroups Digraph of ",
  String(G)));
graph := Graph(GraphType.DIRECTED);
Add(canvas, graph);

customMessage := FrancyMessage(FrancyMessageType.INFO,
  "Simple Groups", "A group is simple if it is nontrivial
  and has no nontrivial normal subgroups.");

IsGroupSimple := function(i)
  Add(canvas, simpleGroupMessage);
  if IsSimpleGroup(as[i]) then
    Add(canvas, FrancyMessage("Simple",
      Concatenation("The vertex ", String(i),
        ", representing the subgroup ", String(as[i]),
        ", is simple.")));
  else
    Add(canvas, FrancyMessage("Not Simple",
      Concatenation("The vertex ", String(i),
        ", representing the subgroup ", String(as[i]),
        ", is not simple.")));
  fi;
  return Draw(canvas);
end;

for i in vertices do
  nodes[i] := Shape(ShapeType.CIRCLE, String(i));
  Add(nodes[i], Menu("Is this subgroup simple?",
      Callback(IsGroupSimple, [i])));
  Add(graph, nodes[i]);
```

```
od;

for i in edges do
  Add(graph, Link(nodes[i[1]], nodes[i[2]]));
od;

Draw(canvas);
```

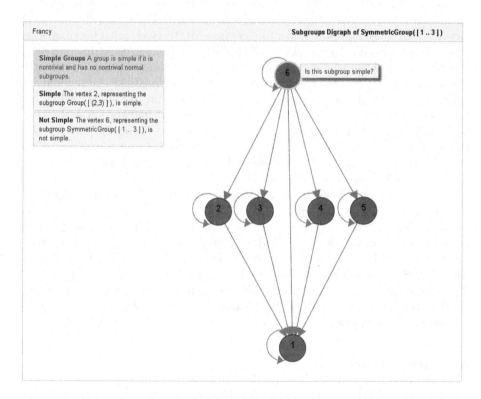

Fig. 1. The graphics produced by the code listing above.

4 Technical Contribution

In terms of software design, Francy follows some principles such as Separation
of Concerns and Modularity. These principles are perfectly articulated in the
Computer Science Handbook [10] *"Any domain or application can be divided and
decomposed into major building blocks and components (separation of concerns).
This decomposition allows the application requirements to be further defined and
refined, while partitioning these requirements into a set of interacting components
(modularity). Changes to the application are (it is hoped) localized. In addition,*

team-oriented design and development can proceed with different team members concentrating on particular components" (Fig. 1).

Francy consists of two main components, a GAP package that is responsible for the semantic representation of graphics, and a second component, a GUI library that is responsible for generating the actual interactive graphical representation.

The GAP package creates a semantic representation of graphics, providing a thin layer between GAP objects and graphical objects to be rendered. This is done using JSON, a lightweight, text-based, language-independent data interchange format [11]. The semantic model follows a JSON Schema [12,13], and is identified with the *application/vnd.francy+json* MIME type [14]. This creates an abstraction and allows the development of new GUI libraries, using different data rendering dependencies or even different programming languages, independently of the GAP package. This package is somehow based in XGAP throughout its application programming interface (API), but avoiding any non-GAP code. This has been the main concern, in order to allow a smooth integration with other GAP packages. In fact, Francy has only one dependency, the JSON package [15], that is distributed with GAP by default, and it is needed to communicate with Jupyter. Access to the GAP language shell (ReadEvalPrint Loop or REPL) is abstracted and managed by a kernel [16,17].

At the moment, Francy has a JavaScript GUI library, based on d3.js [18], for rendering the semantic representation produced by the GAP package. This library is distributed both as a browser module and as a Jupyter extension. The browser module can be used for displaying graphics outside a Jupyter environment or to build applications that can be integrated with GAP, for instance, using WebSockets [19] and a web-based terminal emulator such as tty.js [20]. The Jupyter extension can be used in Jupyter Notebooks or Jupyter Lab, using the Jupyter GAP Kernel [17] and the MIME type *application/vnd.francy+json* to render the document.

5   Future Work

Many other interactive features can be implemented providing a richer learning environment. Features such as rendering multiple topological graphs on the same canvas would allow, for instance, easier comparison of data structures. Other ways for users to input data would provide a more intuitive user experience.

Packages such as the Francy-Monoids [21], Subgroup Lattice [22] and the Interactive Todd-Coxeter [23] still need to be polished and finished.

At the moment, the semantic model produced is not being validated against the JSON Schema [12]. This can be addressed in the future by extending the actual JSON package and implement the JSON Schema specification for validation of documents.

It would also be beneficial moving some of the processing JavaScript code into Web Workers [24], such that rendering of huge structures does not block the web page.

In some cases, having a local installation of **Francy** could be a requirement, and porting it to a desktop application is also possible as there are many tools to help on this process, for instance with ElectronJS [25].

Acknowledgements. We are grateful to James D. Mitchell, Pedro A. García-Sánchez, João Araújo and Francesca Fusco for their suggestions that led to a much improved version of the paper.

We are also very grateful to the anonymous referees for their careful reviews and helpful suggestions.

The first author is grateful to CoDiMa (CCP in the area of Computational Discrete Mathematics - EPSRC EP/M022641/1, 01/03/2015-29/02/2020) for supporting the attendance at the event Computational Mathematics with Jupyter 2017 in Edinburgh, in which some of this research was done. The second author has received funding from the European Union project Open Digital Research Environment Toolkit for the Advancement of Mathematics (EC Horizon 2020 project 676541, 01/09/2015-31/08/2019).

References

1. Berg, J.: Visual Leap: A Step-by-Step Guide to Visual Learning for Teachers and Students. Bibliomotion Incorporated, San Francsico (2015)
2. The GAP Group: Gap - groups, algorithms, and programming, version 4.9.1 (2018). https://www.gap-system.org/
3. Neunhöffer, M., Celler, F.: XGAP documentation, what is XGAP? (2018). https://www.gap-system.org/Manuals/pkg/xgap-4.26/htm/CHAP002.htm
4. Neubüser, J., Felsch, V., Hippe, L.: ITC documentation, what is ITC? (2018). https://www.gap-system.org/Manuals/pkg/itc/htm/CHAP001.htm
5. Woodroofe, R.: Introducing gap.app (2018). https://cocoagap.sourceforge.io/
6. Jupyter Community: Project jupyter (2018). http://jupyter.org/
7. OpenDreamKit Community: Project opendreamkit (2018). http://opendreamkit.org/
8. Machado Martins, M.: Francy github page (2018). https://github.com/mcmartins/francy/
9. De Beule, J., Jonušas, J., Mitchell, J.D., Torpey, M., Wilson, W.A.: Digraphs - gap package, version 0.12.1, April 2018. https://doi.org/10.5281/zenodo.596465
10. Tucker, A.B.: Computer Science Handbook, 2nd edn. CRC Press, New York (2004)
11. Crockford, D.: The Javascript object notation (JSON) data interchange format (2018). https://tools.ietf.org/html/rfc8259
12. Machado Martins, M.: Francy schema github page (2018). https://github.com/mcmartins/francy/blob/master/gap/schema/francy.json
13. JSON Schema Community: JSON schema (2018). http://json-schema.org/
14. Mozilla and individual contributors: Mime types (2018). https://developer.mozilla.org/en-US/docs/Web/HTTP/Basics_of_HTTP/MIME_types
15. Jefferson, C.: JSON - reading and writing JSON (2018). https://www.gap-system.org/Manuals/pkg/json-1.2.0/doc/chap0.html
16. Jupyter Community: Jupyter documentation, jupyter kernel gateway (2018). http://jupyter-kernel-gateway.readthedocs.io/en/latest/
17. Pfeiffer, M.: Native Jupyter kernel for gap github page (2018). https://github.com/gap-packages/JupyterKernel

18. Bostock, M.: Data-driven documents, d3 (2018). https://d3js.org/
19. Mozilla and individual contributors: Websockets (2018). https://developer.mozilla. org/en-US/docs/Web/API/WebSockets_API
20. Jeffrey, C.: tty.js github page (2018). https://github.com/chjj/tty.js
21. García-Sánchez, P.A.: Francy monoids github page (2018). https://github.com/ pedritomelenas/francy-monoids
22. Machado Martins, M.: Subgroup lattice github page (2018). https://github.com/ mcmartins/subgroup-lattice
23. Machado Martins, M.: Interactive todd-coxeter github page (2018). https://github. com/mcmartins/interactive-todd-coxeter
24. Mozilla and individual contributors: Web workers API (2018). https://developer. mozilla.org/en-US/docs/Web/API/Web_Workers_API
25. Electron Community: Build cross platform desktop apps with Javascript, HTML, and CSS (2018). https://electronjs.org/

Sparse Multivariate Hensel Lifting: A High-Performance Design and Implementation

Michael Monagan$^{(\boxtimes)}$ and Baris Tuncer

Department of Mathematics, Simon Fraser University, Vancouver, Canada
{mmonagan,ytuncer}@sfu.ca

Abstract. Our goal is to develop a high-performance code for factoring a multivariate polynomial in n variables with integer coefficients which is polynomial time in the sparse case and efficient in the dense case. Maple, Magma, Macsyma, Singular and Mathematica all implement Wang's multivariate Hensel lifting, which, for sparse polynomials, can be exponential in n. Wang's algorithm is also highly sequential.

In this work we reorganize multivariate Hensel lifting to facilitate a high-performance parallel implementation. We identify multivariate polynomial evaluation and bivariate Hensel lifting as two core components. We have also developed a library of algorithms for polynomial arithmetic which allow us to assign each core an independent task with all the memory it needs in advance so that memory management is eliminated and all important operations operate on dense arrays of 64 bit integers. We have implemented our algorithm and library using Cilk C for the case of two monic factors. We discuss details of the implementation and present experimental timing results.

Keywords: Hensel lifting · Polynomial factorization · Cilk C

1 Introduction

Let $a = fg$ where f and g are two irreducible polynomials in $\mathbb{Z}[x_1, x_2, \ldots, x_n]$. Let $\alpha := (\alpha_2, \alpha_3, \ldots, \alpha_n) \in \mathbb{Z}^{n-1}$ be an evaluation point. For a given polynomial $h \in \mathbb{Z}[x_1, x_2, \ldots, x_n]$ let us use the notation $h_j = h(x_1, \ldots, x_j, \alpha_{j+1}, \ldots, \alpha_n)$ so that $a_1 = a(x_1, \alpha_2, \ldots, \alpha_n)$. To factor a we first factor the image a_1 over $\mathbb{Z}$. With high probability $f(x_1, \alpha)$ and $g(x_1, \alpha)$ will be irreducible so we obtain f_1 and g_1. Next we use a process known as Multivariate Hensel Lifting (MHL) to recover f and g from a, f_1, g_1. Maple, Magma, Macsyma, Singular and Mathematica all implement Wang's MHL from [7,8]. A complete description of Wang's MHL may be found in Chap. 6 of Geddes et al. [3].

The input to Wang's MHL is a, α, f_1, g_1 and a lifting prime p. The evaluation point α and prime p must satisfy $\gcd(f_1, g_1) = 1$ in $\mathbb{Z}_p[x_1]$. The algorithm lifts the factors f_1, g_1 to f_2, g_2 then f_2, g_2 to f_3, g_3 etc. until we obtain f_n, g_n. At the

© Springer International Publishing AG, part of Springer Nature 2018
J. H. Davenport et al. (Eds.): ICMS 2018, LNCS 10931, pp. 359–368, 2018.
https://doi.org/10.1007/978-3-319-96418-8_43

jth step we have $a_j - f_j g_j \mod p = 0$. At the end of this iteration we have $a - f_n g_n \mod p = 0$. Thus for sufficiently large p we obtain the factorization $a = fg$ over $\mathbb{Z}$. The reason Hensel lifting is done modulo a prime p is to avoid an expression swell that would otherwise occur over $\mathbb{Q}$.

Throughout the paper we restrict our presentation to two factors f and g both monic in x_1. We refer the reader to [3] for how to modify MHL for the non-monic and multi-factor cases. Algorithm 1 below shows the jth step of MHL.

Algorithm 1. j^{th} step of Multivariate Hensel Lifting for $j > 1$: Monic Case.

Input : p, $\alpha_j \in \mathbb{Z}_p, a_j \in \mathbb{Z}_p[x_1,\ldots,x_j]$, $f_{j-1}, g_{j-1} \in \mathbb{Z}_p[x_1,\ldots,x_{j-1}]$ where a_j, f_{j-1}, g_{j-1} are monic in x_1 and $a_j(x_j = \alpha_j) = f_{j-1} g_{j-1}$.
Output : $f_j, g_j \in \mathbb{Z}_p[x_1,\ldots,x_j]$ such that $a_j = f_j g_j$ or FAIL.

1: $f_j \leftarrow f_{j-1}$; $g_j \leftarrow g_{j-1}$.
2: $error \leftarrow a_j - f_j g_j$.
3: **for** $i = 1, 2, 3, \ldots$ **while** $\deg(f_j, x_j) + \deg(g_j, x_j) < \deg(a_j, x_j)$ **do**
4:     $c_i \leftarrow$ Taylor coefficient of $(x_j - \alpha_j)^i$ of $error$
5:     **if** $c_i \neq 0$ **then**
6:         Solve the MDP $\sigma_i g_{j-1} + \tau_i f_{j-1} = c_i$ in $\mathbb{Z}_p[x_1,\ldots,x_{j-1}]$ for σ_i and τ_i.
7:         $(f_j, g_j) \leftarrow (f_j + \sigma_i \times (x_j - \alpha_j)^i, g_j + \tau_i \times (x_j - \alpha_j)^i)$
8:         $error \leftarrow a_j - f_j g_j$
9:     **end if**
10: **end for**
11: **if** $error = 0$ **then return** f_j, g_j **else return** FAIL **end if**

There are two main computations in Algorithm 1, namely, the multivariate polynomial diophantine equation (MDP) in Step 6, which typically dominates the cost, and the multivariate multiplication of $f_j \times g_j$ in Steps 2 and 8. Wang's method for solving an MDP resembles his Hensel lifting. He first solves the univariate polynomial diophantine equation

$$\sigma_{i1} g_{j-1}(x_1, \alpha_2, \ldots, \alpha_{j-1}) + \tau_{i1} f_{j-1}(x_1, \alpha_2, \ldots, \alpha_{j-1}) = c_i(x_1, \alpha_2, \ldots, \alpha_{j-1})$$

using the Euclidean algorithm then recovers x_2 then x_3 etc. in σ_i and τ_i. For each x_j there is an iteration on the degree of x_j similar to Algorithm 1. This results in a highly serial algorithm which precludes a parallel implementation.

Wang's solution to the MDP is exponential in j when the evaluation points $\alpha_2, \alpha_3, \ldots, \alpha_{j-1}$ are non-zero. This makes the whole Hensel lifting process exponential for sparse f and g. Polynomial time algorithms were developed by Zippel in 1981 [9], Kaltofen in 1985 [5], and Monagan and Tuncer in 2016 [6].

Let us use the notation $\text{supp}(h)$ to denote the set of monomials appearing in the polynomial h. Monagan and Tuncer [6] solved this exponential problem by observing that if α_j in Algorithm 1 is chosen at random from a sufficiently large set then with high probability the monomials in σ_i for $i \geq 1$ will be contained in the monomials in f_{j-1}, that is $\text{supp}(\sigma_i) \subseteq \text{supp}(f_{j-1})$. Similarly, $\text{supp}(\tau_i) \subseteq \text{supp}(g_{j-1})$ with high probability. They interpolate σ_i and τ_i by picking $\beta_2, \ldots, \beta_{j-1}$ at random from $\mathbb{Z}_p$, computing sufficiently many images of $\sigma_{ik} = \sigma_i(x_1, \beta_2^k, \beta_3^k, \ldots, \beta_{j-1}^k)$ and $\tau_{ik} = \tau_i(x_1, \beta_2^k, \ldots, \beta_{j-1}^k)$ for $1 \leq k$ by solving univariate diophantine equations

$$\sigma_{ik}\, g_{j-1}(x_1, \beta_2^k, \ldots, \beta_{j-1}^k) + \tau_{ik}\, f_{j-1}(x_1, \beta_2^k, \ldots, \beta_{j-1}^k) = c_i(x_1, \beta_2^k, \ldots, \beta_{j-1}^k)$$

for σ_{ik} and τ_{ik} in $\mathbb{Z}_p[x_1]$. Equating coefficients we obtain linear systems. The linear systems are Vandermonde systems which can be solved efficiently in quadratic time and linear space – see Zippel [10]. This improves on Kaltofen's solution to the MDP which results in large unstructured linear systems. The second author has installed this new approach in Maple. It will be available in Maple 2019.

2   High Performance Considerations

Following Bernardin [1] we first reorganize the computation of c_i in Algorithm 1 to avoid recomputing the entire product $f_j \times g_j$. At the ith iteration of the loop we have $f_j = f_{j-1} + \sum_{k=1}^{i-1} \sigma_k (x_j - \alpha_j)^k$ and $g_j = g_{j-1} + \sum_{k=1}^{i-1} \tau_k (x_j - \alpha_j)^k$ and

$$c_i = \operatorname{coeff}(a_j - f_j g_j, (x_j - \alpha_j)^i) = \frac{a_j^{(i)}(\alpha_j)}{i!} - \sum_{k=1}^{i-1} \sigma_k \tau_{i-k}$$

where $a_j^{(i)}$ is the ith derivative of a_j wrt x_j. So we may write the loop in Algorithm 1 as follows.

1: $f_j \leftarrow f_{j-1};\ g_j \leftarrow g_{j-1};\ da \leftarrow \deg(a_j, x_j);\ df \leftarrow 0;\ dg \leftarrow 0.$
2: **for** $i = 1, 2, 3, \ldots$ **while** $df + dg < da$ **do**
3:      $a_j \leftarrow \partial a_j / \partial x_j$
4:      $c_i \leftarrow a_j(\alpha_j)/i! - \sum_{k=1}^{i-1} \sigma_k\, \tau_{i-k}$
5:      Solve the MDP $\sigma_i g_{j-1} + \tau_i f_{j-1} = c_i$ in $\mathbb{Z}_p[x_1, \ldots, x_{j-1}]$ for σ_i and τ_i.
6:      **if** $\sigma_i \neq 0$ set $df \leftarrow i$ **end if**
7:      **if** $\tau_i \neq 0$ set $dg \leftarrow i$ **end if**
8: **end for**
9: $f_j \leftarrow f_{j-1} + \sum_{k=1}^{df} \sigma_k \times (x_j - \alpha_j)^k$
10: $g_j \leftarrow g_{j-1} + \sum_{k=1}^{dg} \tau_k \times (x_j - \alpha_j)^k$

How can we parallelize this for a multi-core computer? We are using Cilk C (see [2]), a parallel extension of C available with the gcc compiler. Because of the time needed to start a Cilk process, the units of work should be of size at least 10^4 clock cycles, equivalently, at least 10^3 multiplications in $\mathbb{Z}_p$. Also, small units of work must require no memory allocations, otherwise memory management will become a parallel bottleneck. We propose to reduce the multivariate Hensel lifting in $\mathbb{Z}_p[x_1, \ldots, x_j]$ to Hensel lift bivariate images in $\mathbb{Z}_p[x_1, x_j]$. That is we will Hensel lift x_j in

$$a_j(x_1, \beta_2^k, \ldots, \beta_{j-1}^k, x_j),\ f_{j-1}(x_1, \beta_2^k, \ldots, \beta_{j-1}^k),\ \text{and } g_{j-1}(x_1, \beta_2^k, \ldots, \beta_{j-1}^k).$$

Algorithm 2. HenselLift1: Bivariate Hensel Lift of x_j for $j > 1$.

Input: p, $\alpha_j \in \mathbb{Z}_p$, $a \in \mathbb{Z}_p[x_1, x_j]$, $f_0, g_0 \in \mathbb{Z}_p[x_1]$ where a, f_0, g_0 are monic in x_1,
$a(x_1, \alpha_j) = f_0 g_0$, $\gcd(f_0, g_0) = 1$ and $p > \deg(a_j, x_j)$.
Output : $f_j, g_j \in \mathbb{Z}_p[x_1, x_j]$ such that $a_j = f_j g_j$.

1: $da \leftarrow \deg(a, x_j)$; $df \leftarrow 0$; $dg \leftarrow 0$;
2: Solve $sg_0 + tf_0 = 1$ for $s, t \in \mathbb{Z}_p[x_1]$ using the Euclidean Alg. . . . . $O(d_1^2)$
3: **for** $i = 1, 2, 3, \ldots$ **while** $df + dg < da$ **do**
4:     $a \leftarrow \partial a/\partial x_j$ . . . . $O(d_1 d_j)$
5:     $c_i \leftarrow a(x_1, x_j = \alpha_j)/i! - \sum_{k=1}^{i-1} \sigma_k(x_1)\, \tau_{i-k}(x_1)$ . . . . $O(d_1 d_j) + O(i d_1^2)$
6:     Solve $\sigma_i g_0 + \tau_i f_0 = c_i$ for $\sigma_i, \tau_i \in \mathbb{Z}_p[x_1]$ via
7:         $\sigma_i \leftarrow (c_i s)$ rem f_0; $\tau_i \leftarrow (c_i - \sigma_i g_0)$ quo f_0 . . . . . . $O(d_1 \deg(f_0, x_1)) \subset O(d_1^2)$
8:     **if** $\sigma_i \neq 0$ **then** $df \leftarrow i$ **end if**
9:     **if** $\tau_i \neq 0$ **then** $dg \leftarrow i$ **end if**
10: **end for**
11: We have $f_j = f_0 + \sum_{i=1}^{df} \sigma_i(x_1)(x_j - \alpha_j)^i$ and $g_j = g_0 + \sum_{i=1}^{dg} \tau_i(x_1)(x_j - \alpha_j)^i$.
12: **return** $[f_0, \sigma_1, \ldots, \sigma_{df}]$ and $[g_0, \tau_1, \ldots, \tau_{dg}]$

Algorithm 2 is our main unit of work. The complexity estimates on the right count arithmetic operations in $\mathbb{Z}_p$. Here $d_1 = \deg(a, x_1)$ and $d_j = \deg(a, x_j)$.

In Algorithm 2 the loop runs to either $df = \deg(f_j, x_j)$ or $dg = \deg(g_j, x_j)$, whichever is greater. Now since $d_f + d_g = d_j$ the most expensive step is the sum of products $\Sigma = \sum_{k=1}^{i-1} \sigma_i(x_1) \tau_{k-i}(x_1)$ in Step 5 which costs $\sum_{i=1}^{d_j} O(i d_1^2) \in O(d_j^2 d_1^2)$ in total. This is the same cost as Bernardin obtains in [1] for two factors. To reduce the cost of Step 5 consider evaluating then interpolating x_1 as follows.

5 Evaluate $\sigma_{il} \leftarrow \sigma_{i-1}(l)$ and $\tau_{il} \leftarrow \tau_{i-1}(l)$ for $0 \leq l \leq d_1$.
  for $l = 0$ to d_1 **do** $c_{il} \leftarrow \sum_{k=1}^{i-1} \sigma_{kl} \times \tau_{(i-k)l}$ **end for**
  Interpolate $\Sigma(x_1)$ from values $\{(l, c_{il}) : 0 \leq l \leq d_1\}$.
  $c_i \leftarrow a(x_1, x_j = \alpha_j)/i! - \Sigma(x_1)$.

Notice that the values σ_{il} and τ_{il} are reused in subsequent iterations. Using Horner's method for evaluation and Newton interpolation the cost of Step 5 becomes $O(d_1^2) + O(i d_1) + O(d_1^2) + O(d_1 d_j)$ and the total cost of Algorithm 2 is now $O(d_1^2 d_j + d_1 d_j^2)$. The only new requirement is that $p > d_1$. Note, if $\deg(f_0, x_1) > \deg(g_0, x_1)$ then one may either interchange f_0 and g_0 or use $\tau_i \leftarrow (c_i t)$ rem g_0 and $\sigma_i \leftarrow (c_i - \tau_i f_0)$ quo g_0 to minimize the cost of Step 7.

2.1   Implementation of HenselLift1

In Algorithm 2 there are univariate operations in $\mathbb{Z}_p[x_1]$ and bivariate operations in $\mathbb{Z}_p[x_1, x_j]$. For a high performance implementation we have designed a library of polynomial arithmetic for $\mathbb{Z}_p[x_1]$ and $\mathbb{Z}_p[x_1, \ldots, x_n]$. The data structure for $\mathbb{Z}_p[x_1]$ is just a dense array of coefficients. For $\mathbb{Z}_p[x_1, \ldots, x_n]$ we use a sparse representation. We encode, e.g., the trivariate polynomial $\sum_{i=1}^{t} a_i M_i(x_1, x_2, x_3)$ as two arrays of integers $A = [a_1|a_2|\ldots|a_t]$ and the monomials $X = [M_1|M_2|\ldots|M_t]$ also stored as an array of integers, that is, each monomial $x_1^i x_2^j x_3^k$ in X is stored as the 64 bit integer $2^{42}i + 2^{21}j + k$. Each subroutine in our library has inputs which are either integers or arrays of integers or arrays of arrays of integers. The arrays may be for inputs, outputs, and, if needed, temporary storage. For example, in Step 7 the multiplications $c_i s$ and $\sigma_i g_0$ are done using the C routine

```
# define LONG long long int // 64 bit signed C integer
int polmul64s( LONG *A, LONG *B, LONG *C, int da, int db, LONG p );
```

Here $da = \deg(a, x)$, $db = \deg(b, x)$, the coefficients of $a(x)$ and $b(x)$ are stored in the arrays A and B. The product $c(x) = a(x)b(x) \bmod p$ is computed in the array C which must be an array of size at least $da + db + 1$.

As a second example, in Step 4 we differentiate $a(x_1, x_j)$ with respect to x_j by calling the routine poldiff64s($A, X, t, 2, 2, p$) below. Here $a(x_1, x_j)$ is input in the arrays (A, X) and the routine overwrites (A, X) with the derivative $\partial a / \partial x_j$.

```
int poldiff64s( LONG *A, LONG *X, int t, int n, int j, LONG p ) {
// diff(a,x[j]): a is stored as pair (A,X) with t terms in n variables
// compute result in (A,X) and return the number of terms
```

To implement Algorithm 2 we first coded it by allocating space for the polynomials in Algorithm 2, so including space for $\sigma_1, \ldots, \sigma_{df}$ for example. Then we make all polynomials parameters of HenselLift1 so that Algorithm 2 does not allocate any new memory. This is possible because all polynomials have bounded degree. The resulting code will be called on many inputs in parallel and the temporary space can be reused.

2.2   Reduction from Multivariate to Bivariate Hensel Lifting

Algorithm 3 describes how we reduce Hensel lifting of x_j in f_{j-1}, g_{j-1} to many bivariate Hensel lifts of x_j. When we implemented Algorithm 3 we tested it on polynomials f and g with $100 - 8000$ terms in $n = 6 - 15$ variables of degree 7. We observed that almost all the time was spent evaluating a_j at Y_k in step 8. In the

$a_j(x_1, x_2, \ldots, x_j)$	$f_j(x_1, x_2, x_3, \ldots, x_j)$
$f_{j-1}(x_1, x_2, \ldots, x_{j-1})$	$g_j(x_1, x_2, x_3, \ldots, x_j)$
evaluate $x_3, \ldots, x_{j-1}$ for $1 \le k \le s$	$\uparrow$
$\downarrow$	sparse interpolate $x_3, \ldots, x_{j-1}$
$a_j(x_1, x_2, \beta_3^k, \ldots, \beta_{j-1}^k, x_j)$ Hensel	$f_j(x_1, x_2, \beta_3^k, \ldots, \beta_{j-1}^k, x_j)$
$f_{j-1}(x_1, x_2, \beta_3^k, \ldots, \beta_{j-1}^k)$ lift x_j	$g_j(x_1, x_2, \beta_3^k, \ldots, \beta_{j-1}^k, x_j)$
evaluate x_2 for $1 \le l \le \deg(a_j, x_2)$	$\uparrow$
$\downarrow$	dense interpolate x_2
$a_j(x_1, \gamma_l, \beta_3^k, \ldots, \beta_{j-1}^k, x_j)$ Hensel	$f_j(x_1, \gamma_l, \beta_3^k, \ldots, \beta_{j-1}^k, x_j)$
$f_{j-1}(x_1, \gamma_l, \beta_3^k, \ldots, \beta_{j-1}^k)$ lift x_j	$g_j(x_1, \gamma_l, \beta_3^k, \ldots, \beta_{j-1}^k, x_j)$

Fig. 1. Homomorphism diagram depicting our evaluation/interpolation strategy

next section we discuss how we implement evaluation and how we parallelized it. Here we point out that if instead of evaluating out $x_2, \ldots, x_{j-1}$ we evaluate out $x_3, \ldots, x_{j-1}$, and thus interpolate the σ_i and τ_i from bivariate images in x_1, x_2, then we likely reduce the number of evaluations s thus leading to a speedup. We describe what we have implemented using a homomorphism diagram.

Algorithm 3. Hensel Lift x_j

Input: Prime p, $\alpha_j \in \mathbb{Z}_p$, Monic polynomials $a_j \in \mathbb{Z}_p[x_1, \ldots, x_j]$ $f_{j-1}, g_{j-1} \in \mathbb{Z}_p[x_1, \ldots, x_{j-1}]$ with $j > 2$, s.t. $a_j(x_1, \ldots, x_{j-1}, \alpha_j) = f_{j-1}g_{j-1}$.

1: Let $f_j = \sum_{i=0}^{df} \sigma_i(x_2, \ldots, x_{j-1}) x_1^i$ where $\sigma_i = \sum_{k=1}^{s_i} a_{ik} M_{ik}$ where $x_1^i M_{ik}$ are the monomials in $\text{supp}(f_{j-1})$ and $df = \deg(f_{j-1}, x_1)$.

2: Let $g_j = \sum_{i=0}^{dg} \tau_i(x_2, \ldots, x_{j-1}) x_1^i$ where $\tau_i = \sum_{k=1}^{t_i} b_{ik} N_{ik}$ where $x_1^i N_{ik}$ are the monomials in $\text{supp}(g_{j-1})$ and $dg = \deg(g_{j-1}, x_1)$.

3: Set $s = \max(s_i, t_i)$.

4: Pick $(\beta_2, \ldots \beta_{j-1}) \in \mathbb{Z}_p$ at random.

5: Compute monomial evaluation sets
$$\{S_i = \{m_{ik} = M_{ik}(\beta_2, \ldots, \beta_{j-1}) : 1 \le k \le s_i\} : 0 \le i \le df\} \text{ and}$$
$$\{T_i = \{n_{ik} = N_{ik}(\beta_2, \ldots, \beta_{j-1}) : 1 \le k \le t_i\} : 0 \le i \le dg\}.$$
If any $|S_i| \ne s_i$ or any $|T_i| \ne t_i$ try a different choice for $(\beta_2, \ldots, \beta_{j-1})$.
If this fails **return** FAIL(1). (*p is not big enough*)

6: **for** k from 1 to s **in parallel do** (*Compute univariate images of σ_i and τ_i*)

7:     Let $Y_k = (x_2 = \beta_2^k, \ldots, x_{j-1} = \beta_{j-1}^k)$.

8:     Evaluate: $a_k, f_0, g_0 \leftarrow a_j(x_1, Y_k, x_j), f_{j-1}(x_1, Y_k), g_{j-1}(x_1, Y_k)$.

9:     **if** $\gcd(f_0, g_0) \ne 1$ **return** FAIL(2) (*an unlucky evaluation*)

10:     Call HenselLift1($p, \alpha_j, a_k, f_0, g_0$) to compute $\sigma_{ik}(x_1)$ and $\tau_{ik}(x_1)$ such that $a_k - f_k g_k = 0$ where $f_k = \sum_{i=0}^{df} \sigma_{ik}(x_j - \alpha_j)^i$ and $g_k = \sum_{i=0}^{dg} \tau_{ik}(x_j - \alpha_j)^i$.

11: **end for**

12: **for** i from 0 to df **do**

13:     Construct and solve the $s_i \times s_i$ linear system
$$\left\{ \sum_{k=1}^{s_i} a_{ik} m_{ik}^n = \text{coefficient of } x_1^i \text{ in } \sigma_{in}(x_1) \text{ for } 1 \le n \le s_i \right\}$$
for the coefficients a_{ik} of $\sigma_i(x_2, \ldots, x_{j-1})$. Because it is a Vandermonde system in m_{ik} which are distinct by Step 5 it has a unique solution.

14: **end for**

15: Do the same for the $t_i \times t_i$ linear systems to solve for the coefficients b_{ik} of the τ_i.

16: Substitute the solutions for a_{ik} into f_j and b_{ik} into g_j and return(f_j, g_j).

In Fig. 1 the reader will see two Hensel lifting steps which represent two possible ways of computing $f_j(x_1, x_2, \beta^k, x_j)$ and $g_j(x_1, x_2, \beta^k, x_j)$. In the first way (the top Hensel lift) the diophantine equations $\sigma_i g_0 + \tau_i f_0 = c_i$ in Step 6 of Algorithm 2 are in $\mathbb{Z}_p[x_1, x_2]$ thus bivariate. One can solve these using dense evaluation and interpolation of x_2 in $O(d_1^2 d_2 + d_1 d_2^2)$ arithmetic operations in $\mathbb{Z}_p$ where $d_2 = \deg(a_j, x_2)$. See Monagan and Tuncer [6].

We coded this approach in Maple as an experiment and found that the most expensive computation is the sum of products $\Sigma = \sum_{k=1}^{i-1} \sigma_k(x_1, x_2) \tau_{i-k}(x_1, x_2)$

in Step 5 of Algorithm 2 which are now bivariate multiplications which cost $O(id_1^2 d_2^2)$. To reduce this cost, we experimented with evaluating and interpolating x_2 which is described by the bottom Hensel lift in Fig. 1. So the number of univariate images $f_j(x_1, \gamma_l, \beta^k, x_j)$, $g_j(x_1, \gamma_l, \beta^k, x_j)$ needed to interpolate x_2 is $\max(\deg(f_j, x_2), \deg(g_j, x_2)) < \deg(a_j, x_2) = d_2$. In our current implementation we have parallelized the computation of the Hensel lifts of these images.

2.3 Parallelizing Evaluation

We describe how we parallelize the evaluations in Step 8 of Algorithm 3. Let $a_j = (A, X)$ where the monomials in X are sorted in lexicographical order with $x_1 > x_2 > \cdots > x_j$. We first sort the monomials into $x_1 > x_j > x_2 > \cdots > x_{j-1}$. Now when we evaluate $a_j(x_1, x_j, \beta_2^k, \ldots, \beta_{j-1}^k)$ the evaluated monomials will be sorted on $x_1 > x_j$. Let $a_j = \sum_{i=1}^{t} a_i x_1^{d_i} x_j^{e_i} M_i(x_2, \ldots, x_{j-1})$. Let $A = [a_1, a_2, \ldots, a_t]$ be the array of coefficients, $m_i = M_i(\beta_2, \ldots, \beta_{j-1})$ and $B = [m_1, m_2, \ldots, m_t]$ be the array of monomial evaluations and let Y be the array of monomials $[x_1^{d_1} x_j^{e_1}, \ldots, x_1^{d_t} x_j^{e_t}]$. If we initialize $C_0 := A = [a_1, \ldots, a_t]$ and define $C_k = [a_1 m_1^k, \ldots, a_t m_t^k]$ then we have

$$a_j(x_1, x_j, \beta_2^k, \ldots, \beta_{j-1}^k) = \sum_{i=1}^{t} a_i m_i^k x_1^{d_i} x_2^{e_i} = \sum_{i=1}^{t} C k_i Y_i$$

and we can compute C_{k+1} from C_k and B using t multiplications with

$$C_{k+1} \leftarrow [B_1 \times C_{k1}, \ldots, B_t \times C_{kt}] = [a_1 m_1^{k+1}, \ldots, a_t m_t^{k+1}]$$

Then we assemble the result from $\sum_{i=1}^{t} C_{k+1,i} Y_i$ which requires adding coefficients of equal monomials in x_1, x_j. Since the monomials in Y_i are already sorted on $x_1 > x_j$ this is $O(t)$. Thus the total number of multiplications needed is st plus those needed to compute $m_1, \ldots, m_t$.

Our first attempt to parallelize this for N cores was to do N evaluations at a time as done by Hu and Monagan in [4]. First compute $C_1, C_2, \ldots, C_N$ and the array $\Gamma = [m_1^N, m_2^N, \ldots, m_t^N]$. To obtain the next N evaluations in parallel, on the kth core compute $C_{k+N} \leftarrow [C_{k1} \times \Gamma_1, \ldots, C_{kt} \times \Gamma_t] = [a_1 m_1^{k+N}, \ldots, a_t m_1^{k+N}]$. One problem with this approach is that we require $\#a$ words of memory for each $C_1, \ldots, C_N$. For one of our benchmark problems where $\#a = 64,000,000$ this is about a half a gigabyte per core. Another problem is that we did not obtain full parallel speedup on our 16 core computer as the computation becomes memory bound when $\#a$ is this large. The following works.

Split $a_j = (A, X)$ into N blocks of size t/N terms. Each core evaluates a block of a_j at β^{k+1} which must be combined later. Numbering the cores $0, 1, \ldots, N-1$ core c computes $C_{kl} \times B_l$ for $c\lfloor t/N \rfloor < l \le (c+1)\lfloor t/N \rfloor$. We found that we obtained a 20% improvement by also computing the evaluation β^{k+2} at the same time so that we compute two evaluations at a time.

3   Experimental Results

We give two sets of experimental results. The first set (see Table 1) is for polynomials in many variables with relatively low degree. Here, evaluation of a_j is the bottleneck in our method – the time spent Hensel lifting images is negligible. The second set (see Table 2) is for polynomials with higher degree where Hensel lifting becomes the bottleneck. All experiments were performed on a server with two Intel E5-2660 8 core CPUs running at 2.2 GHz (base) and 3.0 GHz (turbo) hence the maximum theoretical parallel speedup is a factor of $16.2/3.0 = 11.7$.

In Tables 1 and 2 the factors f and g are of the form $x_1^d + \sum_{i=2}^{t} a_i \prod_{j=1}^{n} x_j^{e_{ji}}$ where the coefficients a_i are chosen randomly from $[1, 999]$ and the exponents e_{ji} randomly from $[0, d-1]$. The timings are for Hensel lifting x_n the last variable only, which is always most of the time. The quantity s in column 4 is the number of images needed to interpolate $x_3, \ldots, x_n$ in Fig. 1. Table 1 shows we achieve very good parallel speedup for evaluations. To obtain the parallel speedups for the Hensel lifting in Table 2 we needed to parallelize the evaluations and interpolations of x_2 in Fig. 1 as well as the Hensel Lifts.

For Maple we report two timings. The first is the best case of Wang's method where the evaluation points $\alpha_2, \ldots, \alpha_n$ are all 0. To obtain this timing we forced Maple to use x_1 as the main variable (by default, it chooses a variable of least degree) and we added a constant to f and g as Maple requires that the leading and trailing coefficient in x_1 not vanish at α. The second timing is the worst case for Wang's method where all evaluation points are non-zero. It is the actual timing for Maple on these inputs.

Table 1. Timings (real time in seconds) for increasing n and t. NA = not attempted.

n	d	t	s	New times (1 core)			New times (16 cores)			Maple 2018	
				Total	(hensel)	(eval)	Total	(hensel)	(eval)	Best	Worst
6	7	500	18	0.098	(0.015)	(0.042)	0.074	(0.019)	(0.008 – 5.2x)	0.411	28.84
6	7	1000	30	0.414	(0.025)	(0.247)	0.180	(0.027)	(0.030 – 8.2x)	1.140	58.46
6	7	2000	47	1.593	(0.041)	(1.132)	0.285	(0.042)	(0.121 – 9.4x)	3.066	99.88
6	7	4000	81	5.072	(0.069)	(4.070)	0.814	(0.074)	(0.380 – 10.7x)	7.173	162.49
6	7	8000	145	12.75	(0.122)	(10.95)	1.896	(0.130)	(0.939 – 11.7x)	15.61	NA
9	7	500	16	0.105	(0.013)	(0.040)	0.101	(0.024)	(0.010 – 4.0x)	1.171	7564.9
9	7	1000	29	0.524	(0.025)	(0.297)	0.233	(0.026)	(0.030 – 11.4x)	3.704	10010.4
9	7	2000	50	2.838	(0.042)	(1.973)	0.483	(0.045)	(0.193 – 10.2x)	13.43	NA
9	7	4000	93	18.35	(0.078)	(14.84)	2.325	(0.083)	(1.350 – 11.0x)	51.77	NA
9	7	8000	164	116.6	(0.139)	(102.5)	11.50	(0.145)	(7.947 – 12.9x)	NA	NA

Table 2. Timings (real time in seconds) for increasing degree.

n	d	t	s	New time (1 core)			New time (16 cores)			Maple 2018	
				Total	(hensel)	(eval)	Total	(hensel)	(eval)	Best	Worst
6	10	500	10	0.099	(0.029)	(0.025)	0.081	(0.024 – 1.2x)	(0.006)	0.571	92.49
6	15	500	6	0.134	(0.070)	(0.016)	0.093	(0.034 – 2.1x)	(0.006)	0.751	7956.5
6	20	500	5	0.238	(0.168)	(0.017)	0.130	(0.065 – 2.6x)	(0.005)	0.919	48610.1
6	40	500	3	1.207	(1.128)	(0.015)	0.282	(0.203 – 5.6x)	(0.006)	1.615	NA
6	60	500	3	4.580	(4.486)	(0.015)	0.732	(0.631 – 7.1x)	(0.011)	3.343	NA
6	80	500	3	13.76	(13.65)	(0.016)	1.674	(1.554 – 8.8x)	(0.012)	4.485	NA
6	10	2000	30	1.775	(0.089)	(1.067)	0.374	(0.055 – 1.6x)	(0.121)	5.237	976.94
6	15	2000	18	1.616	(0.221)	(0.706)	0.413	(0.107 – 2.1x)	(0.061)	7.166	23128.5
6	20	2000	12	1.635	(0.451)	(0.480)	0.431	(0.150 – 3.0x)	(0.040)	9.195	NA
6	40	2000	6	4.008	(2.993)	(0.260)	0.854	(0.505 – 5.9x)	(0.038)	15.98	NA
6	60	2000	6	14.25	(13.15)	(0.292)	1.926	(1.500 – 8.8x)	(0.052)	42.32	NA
6	80	2000	4	26.34	(25.25)	(0.217)	3.340	(2.839 – 8.9x)	(0.050)	57.33	NA

4   Implementation Notes and Cilk C

We end with some comments about programming in Cilk C. Cilk has a very simple task model. One starts a new task using the **spawn** directive. Typically one creates several tasks in a C for loop inside a C function. One may wait for all the tasks started inside the function to complete using the Cilk **sync;** directive. And that's essentially it! We had few problems with Cilk. But ...

Coding in Cilk C basically means we are coding in C where we must manage the memory needed for every polynomial operation. Naively calling **malloc** and **free** in every subroutine will ruin parallel performance and degrade serial performance. Having to manage memory greatly increases coding effort. To reduce memory allocations we re-designed many algorithms to run in-place, that is, to require no additional memory.

It was very hard work getting an algorithm that took about two days to code in Maple to work in Cilk C. In C there is no array bounds checking. Incorrect memory references result in corrupted data which is difficult to track down. Maintaining an identical version of the code in Maple is helpful here. What we would find helpful is to code in C++ using the array data type, which does not support bounds checking, but have some tool for automatically converting arrays to C++ vectors where array bounds checking is available.

The data structure we use for multivariate polynomials assumes monomials can be packed into a 64 bit integer which limits the degree and number of variables that our software can handle. To accommodate more variables we plan to use the 128 bit integer type available in gcc, thus doubling the number of variables of a given degree that we can handle.

References

1. Bernardin, L.: On bivariate Hensel lifting and its parallelization. In: Proceedings of ISSAC 1998, pp. 96–100. ACM Press (1998)
2. Frigo M., Leiserson C.E., Randall K.H.: The implementation of the Cilk-5 multi-threaded language. In: Proceedings of PLDI 1998, pp. 212–223. ACM (1998)
3. Geddes, K.O., Czapor, S.R., Labahn, G.: Algorithms for Computer Algebra. Kluwer Academic, Boston (1992). ISBN: 0-7923-9259-0
4. Hu J., Monagan M.: A fast parallel sparse polynomial GCD algorithm. In: Proceedings of ISSAC 2016, pp. 271–278. ACM (2016)
5. Kaltofen, E.: Sparse Hensel lifting. In: Caviness, B.F. (ed.) EUROCAL 1985. LNCS, vol. 204, pp. 4–17. Springer, Heidelberg (1985). https://doi.org/10.1007/3-540-15984-3_230
6. Monagan, M., Tuncer, B.: Using sparse interpolation in Hensel lifting. In: Gerdt, V.P., Koepf, W., Seiler, W.M., Vorozhtsov, E.V. (eds.) CASC 2016. LNCS, vol. 9890, pp. 381–400. Springer, Cham (2016). https://doi.org/10.1007/978-3-319-45641-6_25
7. Wang, P.S., Rothschild, L.P.: Factoring multivariate polynomials over the integers. Math. Comput. **29**(131), 935–950 (1975)
8. Wang, P.S.: An improved multivariate polynomial factoring algorithm. Math. Comput. **32**(144), 1215–1231 (1978)
9. Zippel, R.E.: Newton's iteration and the sparse Hensel algorithm. In: Proceedings of SYMSAC 1981, pp. 68–72. ACM (1981)
10. Zippel, R.E.: Interpolating polynomials from their values. J. Symbolic Comput. **9**(3), 375–403 (1990)

TheoryGuru: A Mathematica Package to Apply Quantifier Elimination Technology to Economics

Casey B. Mulligan[1]([✉]), James H. Davenport[2], and Matthew England[3]

[1] University of Chicago, Chicago, USA
c-mulligan@uchicago.edu
[2] University of Bath, Bath, UK
J.H.Davenport@bath.ac.uk
[3] Coventry University, Coventry, UK
Matthew.England@coventry.ac.uk

Abstract. We consider the use of Quantifier Elimination (QE) technology for automated reasoning in economics. There is a great body of work considering QE applications in science and engineering but we demonstrate here that it also has use in the social sciences. We explain how many suggested theorems in economics could either be proven, or even have their hypotheses shown to be inconsistent, automatically via QE.

However, economists who this technology could benefit are usually unfamiliar with QE, and the use of mathematical software generally. This motivated the development of a MATHEMATICA Package TheoryGuru, whose purpose is to lower the costs of applying QE to economics. We describe the package's functionality and give examples of its use.

Keywords: Quantifier elimination · Economic reasoning

1 Introduction

A general task in economic reasoning is to determine whether, with variables $v = (v_1, \ldots, v_n)$, a hypothesis $H(v)$ follows from assumptions $A(v)$, i.e. is it the case that $\forall v \, . \, A \Rightarrow H$? Ideally the answer would be `True` or `False`, but in practice life is more complicated: the answer could differ depending on the value of v; or the assumptions could even be contradictory, i.e. $A(v)$ alone is `False`.

We can categorise these possibilities via the outcome of a pair of quantified statements (Table 1). Should technology provide any one automatically then an economist gains important information: either a proof or a disproof of her theory; or an identification of where her theory may be true (a description of $\{v : A(v) \Rightarrow H(v)\}$); or the knowledge that her assumptions contradict.

Such technology could also allow for exploration. An economist could vary the question: the assumptions generating a `True` result can be weakened, or those generating a `Mixed` result strengthened, by quantifying more or less of v.

© Springer International Publishing AG, part of Springer Nature 2018
J. H. Davenport et al. (Eds.): ICMS 2018, LNCS 10931, pp. 369–378, 2018.
https://doi.org/10.1007/978-3-319-96418-8_44

Table 1. Possible outcomes from a potential theorem $\forall v . A \Rightarrow H$

	$\neg\exists v[A \wedge \neg H]$	$\exists v[A \wedge \neg H]$
$\exists v[A \wedge H]$	True	Mixed
$\neg\exists v[A \wedge H]$	Contradictory assumptions	False

For example, we can partition v into v_1, v_2 and ask for $\{v_1 : \forall v_2 . A(v_1, v_2) \Rightarrow H(v_1, v_2)\}$. The result is a formula in the free variables v_1 that weakens or strengthens the assumptions. If generated automatically the economist gains information about how to reformulate assumptions that justify her hypothesis.

1.1 Quantifier Elimination

Such problems fall within the framework of *Quantifier Elimination* (QE): the generation of an equivalent quantifier-free formula from one that contains quantifiers. QE is known to be possible over real closed fields thanks to the seminal work of Tarski [17]. Practical implementations followed with Collins' Cylindrical Algebraic Decomposition [4] and Weispfenning's Virtual Substitution [18]. There are modern implementations of QE in many computer algebra systems.

QE has found many applications within engineering and the life sciences. Recent examples include: the derivation of optimal numerical schemes [5], weight minimisation for truss design [3], and biological network analysis [1]. However, applications in the social sciences are lacking (the nearest we can find is [8]). On the few occasions when QE has been mentioned in economics it has been dismissed as infeasible, e.g. "something that is do-able in principle, but not by any computer that you and I are ever likely to see" [12]. But that dismissal is based on theoretical complexity results rather than experience with actual software applied to actual economic reasoning. Many meaningful economics problems can be studied with modern QE implementations[1], with the barrier to further use acceptance by the community, and experience with the software.

1.2 A New Mathematica Package TheoryGuru

This motivated the development of a new tool to aid the application of QE to economics: a package called TheoryGuru to run in the MATHEMATICA computer algebra system [19]. This is able to parse input from economists, run some error and sanity checks, and then utilise MATHEMATICA's QE tools and offer interpretations of the results. These QE tools are accessed by the MATHEMATICA Resolve command with some of the underlying algorithms described in [13–16]. The paper proceeds in Sect. 2 by introducing the functionality of TheoryGuru. Then in the remaining sections we describe examples of its use.

---

[1] A dataset of 45 economic reasoning examples that may be tackled with QE technology is available here: https://doi.org/10.5281/zenodo.1226892.

2   Functionality

2.1   Main Functionality

The purpose of `TheoryGuru` is to lower the costs of applying QE to economics. Hence it assumes the expression of reasoning in the format traditional to the field: as a conclusion to be possibly deduced from a set of assumptions.
The core functionality of `TheoryGuru` is then as follows:

Check for errors: Provide warnings on likely typographical errors in variables (e.g., when a variable appears only once in the entire formula) or formula structure (e.g. the user may have confused = with ==).

Parse input: This includes identifying from context whether a variable is a vector, scalar, or boolean; processing input given in a *pretty* mathematical notation (e.g. derivatives) into a format accepted by `Resolve`; standardizing dot products and integrals (e.g., distribute plus and alphabetically sort arguments of commutative operators).

Adding standard assumptions: If dot products are present, then add to user assumptions the necessary and sufficient conditions for the Gramian matrix (representing dot products for all pairs of vectors) to be positive semi-definite. This rules out vectors with imaginary elements.

Check assumptions: The package will next check that the assumptions provided are not mutually contradictory: the situation of the bottom left entry in Table 1. This is done via a call to `Resolve` to check there is at least one solution to $A(v)$ – a fully existentially quantified QE sub-problem.

Form main QE input: Automatically assemble the two Tarski formulae for the main calls (as given in Table 1).

Make algorithm choices: Currently this refers to (a) whether to process a universal or existential sentence and (b) the variable ordering determining the sequence for eliminating quantifiers. It is well known that the choice (b), while not affecting the correctness of the output, can have a large effect on computational resources required [6].

Output interpretation: Then after making the two calls to MATHEMATICA's `Resolve`, the package interprets the results by identifying the relevant cell from Table 1. The package also suggests what to do next: e.g., when applicable, show a counterexample, solve simultaneous equations appearing in the assumptions, or redo the QE with some free variables.

2.2   Access and Documentation

To access `TheoryGuru` the reader will need a modern version of MATHEMATICA[2] and then installation follows from simply running the command:
`Get["http://economicreasoning.com"]` which produces an interface as in Fig. 1. Not only does this install the underlying code, it also provides links to tutorials,

---

[2] The `Resolve` function has evolved and improved over the years and so the performance of `TheoryGuru` will alter correspondingly.

tips, help and a large bank of examples (as shown in Fig. 1). The examples are also available online at http://examples.economicreasoning.com/ as both interactive MATHEMATICA notebooks and static pdfs.

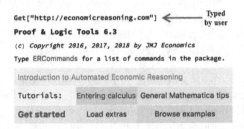

Fig. 1. Initial load screen of TheoryGuru

The main functionality is accessed via the function TheoryGuru which requires two arguments: a collection of assumptions and a hypothesis.

3 Examples of TheoryGuru Use

3.1 Tax Incidence

Our first, admittedly simple, *Tax Incidence* example is about the effect of a tax on buyers and sellers in a market. Each transaction involves the buyer paying price to the seller in addition to paying tax to the government. The symbolic functions demand(.) and supply(.) represent the quantities that buyers purchase and sellers sell, respectively, as a function of the price they pay or receive (so for the buyer that includes the tax). A market equilibrium price has the quantity demanded equal to the quantity supplied. The equilibrium condition can be input to MATHEMATICA as shown in the top cell of Fig. 2, which assigns the condition the natural language name Equilibrium.

In[2]:= Equilibrium = demand[price + tax] == supply[price];

In[3]:= TheoryGuru$\left[\left\{\dfrac{d\,\text{Equilibrium}}{d\,\text{tax}}, \text{demand}'[\text{price} + \text{tax}] < 0, \text{supply}'[\text{price}] > 0\right\}, \dfrac{d\,\text{price}}{d\,\text{tax}} \le 0\right]$

Out[3]= True

Fig. 2. Tax Incidence example in TheoryGuru

The first argument of the call to TheoryGuru in the second cell of Fig. 2 is a set of assumptions. The first of these is that changing the tax changes the market from one equilibrium to another[3]. The remaining two constrain the slope of the

---

[3] The notation is consistent with an economist saying that she "totally differentiates the equilibrium condition". This differentiation is automatically performed by MATHEMATICA when the TheoryGuru function is evaluated.

demand and supply curves in the neighborhood of the market equilibrium. The second argument is the hypothesis the user wishes to test, in this case whether the price impact is negative or zero.

The call causes `TheoryGuru` to automatically assemble Tarski formulae, in which it recognizes `demand'(price+tax)` and `supply'(price)` as partially interpreted functions [7, p. 73]. Following the generic format presented above, there are two QE problems for `TheoryGuru` to consider: the existence of an example and the existence of a counterexample. Here the output is simply `True` because there is no counterexample: i.e., no way to have a positive price impact while satisfying the assumptions.

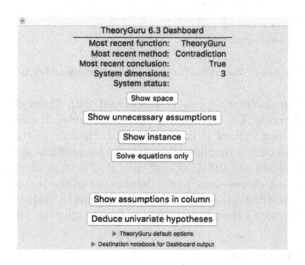

Fig. 3. The `TheoryGuru` dashboard

When `TheoryGuru` evaluates, a dashboard (Fig. 3) appears summarizing the calculation and offering the user possible next steps. In the tax incidence example, the user may be wondering what else can be concluded about the price impact. The button labelled "Deduce univariate hypotheses" on the dashboard serves this purpose. Pressing it automatically generates a call to the function `TheoryPossibilities` as shown in Fig. 4. Here, one or more free variables are provided by the user, or else variables are chosen by the software (giving priority to total derivative variables and alphabetical order). The assumptions are

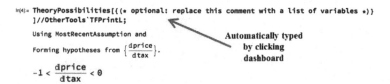

Fig. 4. `TheoryPossibilities` call from dashboard button to propose new hypothesis

then projected on each of the free variables separately (eliminating existential quantifiers from all variables except that one), with the resulting formulae simplified. In this example we discover the price impact must be strictly between -1 and 0.

Users can be forgetful or have an imperfect understanding of an economic model. In the top cell of Fig. 5 no definitive conclusion about price impact is reached because the user has forgotten to constrain the supply curve's slope.

In[5]:= TheoryGuru $\left[\left\{\dfrac{\text{dEquilibrium}}{\text{dtax}}, \text{demand'[price + tax]} < 0\right\}, \dfrac{\text{dprice}}{\text{dtax}} \le 0\right]$

Out[5]= True for some, False for others

In[6]:= TheorySufficient[]

Out[6]= supply'[price] $\ge$ 0

Fig. 5. Example use of `TheorySufficient` to recommend additional assumptions

The forgotten assumption can be discovered with `TheorySufficient`. It assembles the formula $A \wedge \neg H$ defining counterexamples. It then projects that set on each of the axes (three in this example). It then shows the disjunction of each formula, after each is simplified based on the assumptions and then negated. Here two formulae are discarded because they are `False` or identical to H. The third is the missing supply-slope restriction output in the second cell of Fig. 5. Note that, by construction, any of `TheorySufficient`'s disjunction branches, together with the user's (insufficient) assumptions, imply the user's hypothesis.

3.2 Gender Wage Gap

We now look at a more involved *Gender Wage Gap* example that studies the effect of wage inequality on women's supply of human capital to the market. Women are assumed to have (possibly correlated) skills h and r in market work and non-market activities, respectively. These skills have a population distribution modelled with the joint density function $f(h, r)$, which is normalized to have unconditional means of zero. Women work if and only if their non-market log wage $r + \mu_r$ is less than $\sigma h + \mu_w$, their market log wage. It follows that mean non-market and market log wages are μ_r and μ_w, respectively. The employment

In[7]:= p[$\sigma\_$, meangap_] = $\displaystyle\int_{-\infty}^{\infty}\int_{-\infty}^{\sigma h - meangap} f[h, r]\, dr\, dh;$

In[8]:= S[$\sigma\_$, meangap_] = $\dfrac{\displaystyle\int_{-\infty}^{\infty}\int_{-\infty}^{\sigma h - meangap} h\, f[h, r]\, dr\, dh}{p[\sigma, meangap]};$

In[9]:= DefineExperiment = $\left\{\dfrac{dp[\sigma, \mu R - \mu w]}{dz} == \dfrac{d\mu w}{dz} == 0, \dfrac{d\sigma}{dz} > 0\right\};$

Fig. 6. Defining variables for the Gender Wage Gap example

rate of women is $p(\sigma, \mu_w - \mu_r)$ as defined in the top cell of Fig. 6, and the average skill in the market is $S(\sigma, \mu_w - \mu_r)$ as defined in the next cell.

In [11] the Gaussian model is used to show how wage inequality, as modelled by σ, affects the average skill in the market. However, the *selection rule* result – that is, the effect of σ on S holding constant p by varying μ_r – does not require the Gaussian assumption. To show this, we define z to be any change in σ and μ_r that increases σ and holds p constant, as defined in the third cell of Fig. 6 (DefineExperiment). Figure 7 then assigns natural language to two definitions (top and bottom cells) as well as restrictions on partially interpreted functions (middle cell). The top cell of Fig. 8 shows that a positive skill impact can be deduced from the assumed properties of the partially interpreted functions. In economics terms: inequality increases the average skill that women supply to the market, thereby narrowing the measured wage gap with men.

In[10]:= `PositiveEmployment = p[σ, μR - μw] > 0;`

In[11]:= `InequalityOntheMargin =` $\dfrac{\int_{-\infty}^{\infty} h^2\, f[h,\ \mu w - \mu R + h\,\sigma]\ dh}{\int_{-\infty}^{\infty} f[h,\ \mu w - \mu R + h\,\sigma]\ dh} > \left(\dfrac{\int_{-\infty}^{\infty} h\, f[h,\ \mu w - \mu R + h\,\sigma]\ dh}{\int_{-\infty}^{\infty} f[h,\ \mu w - \mu R + h\,\sigma]\ dh}\right)^2 \wedge \int_{-\infty}^{\infty} f[h,\ \mu w - \mu R + h\,\sigma]\ dh > 0;$

In[12]:= `SkillImpact =` $\dfrac{dS[\sigma,\ \mu R - \mu w]}{dz};$

Fig. 7. Further definitions and restrictions to the Gender Wage Gap example.

In[13]:= `TheoryGuru[{DefineExperiment, PositiveEmployment, InequalityOntheMargin}, SkillImpact > 0]`

Out[13]= `True`

In[14]:= `TheorySpace[]//OtherTools`TFPrintL;`

Using MostRecentTheory.

$$\left\{ \dfrac{d\mu R}{dz},\ \dfrac{d\mu w}{dz},\ \dfrac{d\sigma}{dz},\ \int_{-\infty}^{\infty} f(h,\ h\,\sigma - \mu R + \mu w)\ dh,\ \int_{-\infty}^{\infty} h\, f(h,\ h\,\sigma - \mu R + \mu w)\ dh, \right.$$
$$\left. \int_{-\infty}^{\infty} h^2\, f(h,\ h\,\sigma - \mu R + \mu w)\ dh,\ \int_{-\infty}^{\infty}\int_{-\infty}^{h\,\sigma - \mu R + \mu w} f(h,\ r)\ dr\,dh,\ \int_{-\infty}^{\infty}\int_{-\infty}^{h\,\sigma - \mu R + \mu w} h\, f(h,\ r)\ dr\,dh \right\}$$

Fig. 8. Evaluating the Gender Wage Gap example

At first glance, the gender wage gap example appears to involve integrable probability density functions rather than the scalar variables required by the QE algorithms employed by MATHEMATICA's Resolve function. But the reasoning in this and many other examples depends on the probability density functions only as they are summarized by various scalars. TheoryGuru automatically discovers these scalar variables, which can be viewed by the user who clicks "Show space" on TheoryGuru's dashboard. The result of that click is the last cell of Fig. 8.

4   Run Times

Figure 9 shows the run times for several of the function calls shown above. As explained, each evaluation involves preparation of a QE problem for MATHE-MATICA's Resolve function, followed by that QE call. The cell numbers refer to those used by MATHEMATICA in the screen shots above. The figure's first run time column is for the entire evaluation of the TheoryGuru command. The next column shows, when applicable, the amount of time it took for just the "universal" QE regarding the existence of a counterexample. The final column is the amount of time it took for just the "existential" QE regarding the existence of an example (the faster QE for calls that have no counterexamples).

Function call	Dimensions represented	Run times in seconds		
		Entire evaluation	QE only	
			Universal	Existential
TheoryGuru (cell 3)	3	0.10	< 0.01	< 0.01
TheoryPossibilities (cell 4)	3	0.19	N/A	< 0.01
TheorySufficient (cell 6)	3	0.15	N/A	< 0.01
TheoryGuru (cell 13)	8	3.31	< 0.01	< 0.01
Addenda: TheoryGuru queries about production function concavity				
3 inputs, delineated individually	12	1.56	0.41	< 0.01
4 inputs, delineated individually	18	469.68	8.89	0.03
Any number of inputs (vector mode)	10	0.41	0.13	< 0.01

Notes: Run time was calculated with Mathematica 11.2 on a Macbook Pro Mid 2014 2.8GHz Intel i7 Quadcore processor, running TheoryGuru in Parallel Mode with popup dashboard already open and using the default tuning for variable search. Universal (existential) QE refers to deciding the existence of a counterexample (example), respectively, with the Tarski formula already assembled and variable-elimination sequence already determined.

Fig. 9. Run times for several examples

In order to explore the limits of the software, we consider queries regarding the concavity of quasiconcave production functions, whose economics and algebra we discuss in [10]. The three-input version evaluates in less than two seconds. The four-input version is considerably more complicated, and evaluates in about eight minutes primarily because of a long search to find a relatively efficient order for eliminating quantifiers[4]. The problem can be solved more elegantly with vectors, with a quicker run time as shown in the final row (see also [9]).

5   Final Thoughts

We have demonstrated how economic reasoning may be automated using QE procedures and how the TheoryGuru tools greatly reduce the costs to an economist of accessing that technology. We note that a set of benchmark examples that

---

[4] It is once that order is obtained that the corresponding QE needs only 8.89 s.

originate from economics and may be tackled with QE is now available at https://doi.org/10.5281/zenodo.1226892 and described in [10]. Future work will involve considering how the underlying QE technology could be optimised for such examples, whose structure is often not well represented in the broader QE literature.

References

1. Bradford, R., Davenport, J., England, M., Errami, H., Gerdt, V., Grigoriev, D., Hoyt, C., Košta, M., Radulescu, O., Sturm, T., Weber, A.: A case study on the parametric occurrence of multiple steady states. In: Proceedings of the ISSAC 2017, pp. 45–52. ACM (2017). https://doi.org/10.1145/3087604.3087622
2. Caviness, B., Johnson, J.: Quantifier Elimination and Cylindrical Algebraic Decomposition. Texts & Monographs in Symbolic Computation. Springer, Wien (1998). https://doi.org/10.1007/978-3-7091-9459-1
3. Charalampakis, A., Chatzigiannelis, I.: Analytical solutions for the minimum weight design of trusses by cylindrical algebraic decomposition. Arch. Appl. Mech. **88**(1), 39–49 (2018). https://doi.org/10.1007/s00419-017-1271-8
4. Collins, G.E.: Quantifier elimination for real closed fields by cylindrical algebraic decompostion. In: Brakhage, H. (ed.) GI-Fachtagung 1975. LNCS, vol. 33, pp. 134–183. Springer, Heidelberg (1975). https://doi.org/10.1007/3-540-07407-4_17. (reprinted in [2])
5. Erascu, M., Hong, H.: Real quantifier elimination for the synthesis of optimal numerical algorithms (Case study: square root computation). J. Symbolic Comput. **75**, 110–126 (2016). https://doi.org/10.1016/j.jsc.2015.11.010
6. Huang, Z., et al.: Applying machine learning to the problem of choosing a heuristic to select the variable ordering for cylindrical algebraic decomposition. In: Watt, S.M., Davenport, J.H., Sexton, A.P., Sojka, P., Urban, J. (eds.) CICM 2014. LNCS (LNAI), vol. 8543, pp. 92–107. Springer, Cham (2014). https://doi.org/10.1007/978-3-319-08434-3_8
7. Kroening, D., Strichman, O.: Decision Procedures: An Algorithmic Point of View. Springer, New York (2013). https://doi.org/10.1007/978-3-540-74105-3
8. Li, X., Wang, D.: Computing equilibria of semi-algebraic economies using triangular decomposition and real solution classification. J. Math. Econ. **54**, 48–58 (2014). https://doi.org/10.1016/j.jmateco.2014.08.007
9. Mulligan, C.: Automated Economic Reasoning with Quantifier Elimination. NBER Working Paper No. 22922 (2016). https://doi.org/10.3386/w22922
10. Mulligan, C., Bradford, R., Davenport, J.H., England, M., Tonks, Z.: Non-linear real arithmetic benchmarks derived from automated reasoning in economics. In: Proceedings of the 3rd International Workshop on Satisfiability Checking and Symbolic Computation (SC²) (2018, to appear). Preprint: https://arxiv.org/abs/1806.11447
11. Mulligan, C., Rubinstein, Y.: Selection, investment, and women's relative wages over time. Q. J. Econ. **123**(3), 1061–1110 (2008). https://doi.org/10.1162/qjec.2008.123.3.1061
12. Steinhorn, C.: Tame topology and O-minimal structures. In: Brown, D., Kubler, F. (eds.) Computational Aspects of General Equilibrium Theory, pp. 165–191. Springer, Heidelberg (2008). https://doi.org/10.1007/978-3-540-76591-2_11

13. Strzeboński, A.: Cylindrical algebraic decomposition using validated numerics. J. Symbolic Comput. **41**(9), 1021–1038 (2006). https://doi.org/10.1016/j.jsc.2006.06.004
14. Strzeboński, A.: Computation with semialgebraic sets represented by cylindrical algebraic formulas. In: Proceedings of the ISSAC 2010, pp. 61–68. ACM (2010). https://doi.org/10.1145/1837934.1837952
15. Strzeboński, A.: Solving polynomial systems over semialgebraic sets represented by cylindrical algebraic formulas. In: Proceedings of the ISSAC 2012, pp. 335–342. ACM (2012). https://doi.org/10.1145/2442829.2442877
16. Strzeboński, A.: Cylindrical algebraic decomposition using local projections. J. Symbolic Comput. **76**, 36–64 (2016). https://doi.org/10.1016/j.jsc.2015.11.018
17. Tarski, A.: A Decision Method For Elementary Algebra And Geometry. RAND Corporation, Santa Monica (1948). (reprinted in [2])
18. Weispfenning, V.: The complexity of linear problems in fields. J. Symbolic Comput. **5**(1/2), 3–27 (1988). https://doi.org/10.1016/S0747-7171(88)80003-8
19. Wolfram, S.: The Mathematica Book. Wolfram Research Inc., San Francisco (2000)

Collaborative Use of Mathematical Content Generated by CindyJS on Tablets

Takeo Noda$^{(\boxtimes)}$ and Masataka Kaneko

Toho University, Funabashi, Japan
noda@c.sci.toho-u.ac.jp, masataka.kaneko@phar.toho-u.ac.jp

Abstract. CindyJS is a system which enables researchers and learners to interactively handle mathematical models on browsers. Though the enhancement of mathematical user interfaces is widely anticipated and promoted so that mathematical models can be handled via familiar web and mobile devices, it seems that the resulting realities and benefits have not yet been fully investigated. In particular, for educational use, more precise knowledge about them will likely help maximize the effect of using newly developed systems. This research is mainly concerned with the comparison between individual use and group use of CindyJS content on tablets. It can be assumed that, in the case of group use, communication between members would give some influence on the strategy of their handling of mathematical models. To investigate how members of a group influence each other's handling of mathematical models when using CindyJS, we tracked some characteristic quantities from the recorded processes of users' operations. Through statistical analysis (approximation with finite mixture of beta distributions) of the quantities derived from the cases of individual use and group use respectively, it can be shown that the difference between these two cases is visualized and the above mentioned influence is illustrated.

Keywords: Cinderella · CindyJS · Web and mobile devices
Finite mixture of beta distributions · Influence of users' communication

1 Introduction

Some previous researches illustrate that interactive operations on mathematical models can lead human beings to be creatively engaged in mathematical reasoning [1,2]. For instance, though almost all of the science-major students are taught the knowledge about Maclaughlin expansion,

$$f(x) = f(0) + f'(0)x + \frac{f''(0)}{2}x^2 + \frac{f'''(0)}{6}x^3 + \cdots$$

it is not necessarily easy for them to fully and accurately grasp the concept of an infinitesimal of higher order without interactively moving the graphs of approximating functions. Dynamic geometry software is the most appropriate tools for

© Springer International Publishing AG, part of Springer Nature 2018
J. H. Davenport et al. (Eds.): ICMS 2018, LNCS 10931, pp. 379–388, 2018.
https://doi.org/10.1007/978-3-319-96418-8_45

such purposes since they enable us to generate mathematical objects (like points, lines, circles, function graphs, and areas) and move them by interactively controlling the variables on a PC screen by clicking on them or dragging them with a mouse. Using Cinderella (https://cinderella.de) [3] which is one of the most popular dynamic geometry software systems, we can visualize those graphs and move them on a graphical user interface (Cinderella screen) as in Fig. 1. Here the blue curve is the graph of the function $y = \sqrt{x+1}$ and the red curve is the graph of the function $y = a + bx + cx^2 + dx^3$. The coefficients a, b, c, d are interactively determined by dragging the red points in the blue segments. Also, as the red point for each coefficient is dragged along the slider, their values at each position are displayed on the screen. The shape of the red curve changes according to the specified values of a, b, c, d. When we click the button "RESET" on the screen, the graph returns to the initial state in which $a = b = c = d = 0$.

While the underlying technology of Cinderella is Java, the extension named CindyJS (https://cindyjs.org) is being developed so that the resulting mathematical content can be exported in the format of plugin-less web technology like JavaScript, HTML5 and WebGL [4–6]. Using CindyJS, we can generate mathematical content which can be used not only on PCs but also on familiar web and mobile devices. Once Cinderella content is generated, we can easily export it to CindyJS. In fact, we only need to choose "Export to CondyJS" from the file menu in the Cinderella screen and the program does the rest. Figure 2 shows the image of content generated by CindyJS as displayed via Google Chrome.

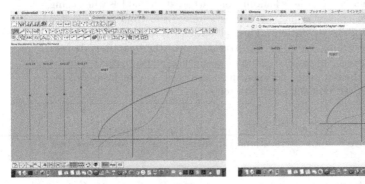
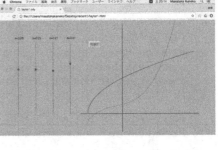

Fig. 1. Cinderella screen (Color figure online)

Fig. 2. Exported CindyJS content (Color figure online)

The content can also be displayed via any ordinary web browser like Safari or Internet Explorer. The movement on the screen is the same as that of Cinderella content except that objects are moved by users' finger strokes instead of their mouse operations. A more detailed description of the procedure to generate and use CindyJS content has been given in [7].

From the viewpoint of mathematical user interfaces, not only the mathematical notions which users can understand but also the procedure in which those notions are understood through their handling mathematical models is crucial.

In the case when a group of users moves CindyJS content on a single tablet, the way they handle the model is most likely influenced by the communication between members of the group which cannot occur in the case of individual use. Therefore, the answers to the following two questions should be the fundamental knowledge for further developments of mathematical user interfaces.

1. How can the difference in the procedure of users' operations between group use and individual use be visualized?
2. How can that difference be analyzed so that the above mentioned influence can be illustrated?

In this paper, the authors show the methods and results of their attempts to give some answers to these questions by conducting an experiment with 25 individual and 25 group subjects. In this experiment, the subjects moved the CindyJS content in Fig. 2 on iPads and their operating procedures were detected through ethnomethodological study. The relative frequencies of their moving lower order coefficients during some specific intervals were tracked. These ratios were used as the characteristic quantities for visualizing the difference between individual use and group use. The resulting distributions of these quantities were statistically summarized in terms of the approximation with finite mixture of beta distributions [8,9].

2   Methods

The authors recruited first grade (19 years old) students from their university as subjects. Their major is not mathematical science but pharmaceutical science. Twenty five students were chosen randomly as subjects for individual use. The remaining 75 students were divided into 25 groups. At the time of the experiment, three months had passed since the subjects studied this formula in the calculus classroom. So, it could be assumed that most of them had forgotten the precise form of the formula. All subjects were asked to use the material in Fig. 2 to find the values a, b, c, d with which the red curve best fits into the blue curve near the point $(0, 1)$. No advice was given and they were allowed to touch the "RESET" button as many times as they liked. Also the physical and verbal behaviors of the subjects were videotaped as shown in Figs. 3 and 4.

To detect the changes in the subjects' strategies for searching for an optimal approximation, the videotaped images taken as they worked were imported into SportsCode system (https://www.hudl.com/elite/sportscode) and each time interval in which they moved the four red points respectively was coded chronologically on one time lines. The typical workflow is demonstrated at the following URL.

https://drive.google.com/file/d/0B200rbx3ihxGSjhjSFJIZFlRUFE/view?usp=sharing

It was anticipated that the subjects would first fix the points corresponding to a and b so that the graph of $y = a + bx$ is tangent to the graph of $y = \sqrt{x+1}$

Fig. 3. Individual use   **Fig. 4.** Group use

at the point $(0, 1)$ and then move the points corresponding to c and d. Thus the anticipated result is shown in Fig. 5 in which each time line contains rows corresponding to a, b, c, d and "RESET" arranged from top to bottom.

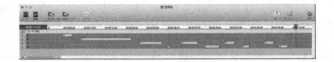

Fig. 5. Model result of the coding

When a subject leaves an interval between two successive movements of the same point, it is very difficult to infer the reasoning process of the subject during that interval. Therefore the authors regarded such successive movements as a single movement. Since it is also very difficult to precisely evaluate the length of each movement, the authors transformed the coded result on a timeline into a string of characters **a**, **b**, **c**, **d**. For example, the coded result in Fig. 5 is transformed into the following string [**a b c d c d c d c d**]. Moreover, the authors evaluated the strings derived from the coded time lines of each individuals and groups with equal weights.

3   Results

Figures 6 and 7 show the resulting coded time lines derived from 25 individual use and 25 group use respectively. As seen there, the actual results are largely different from the anticipation shown in Fig. 5. In fact, the subjects moved the points far more often than anticipated. This can be regarded as the processes by which they empirically studied the fact that the functions cx^2 and dx^3 are infinitesimals of higher order than the function bx near $x = 0$ [7].

In line with the plan stated in the end of Sect. 2, the authors visualized the resulting strings as in Fig. 8. Here, the right pointing horizontal axis represents the progress of the subjects' operating processes. Red and yellow color correspond to the character **a** and **b** respectively. Rows of the upper half part are the

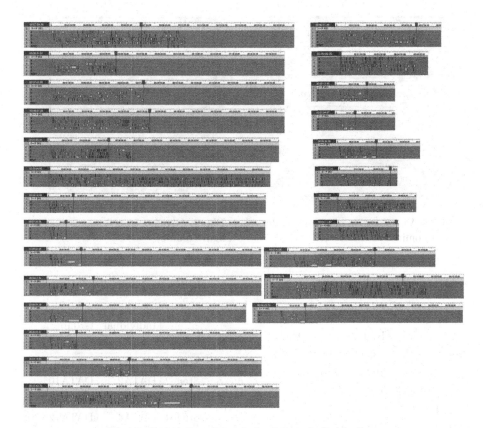

Fig. 6. Coded time lines derived from individual use

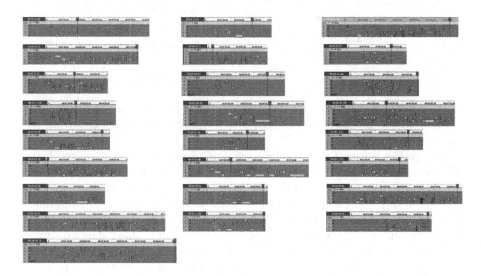

Fig. 7. Coded time lines derived from group use

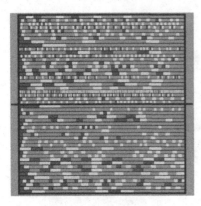

Fig. 8. Visualization of strings (Color figure online)	**Fig. 9.** Division into 8 stages (Color figure online)

Table 1. Frequencies at each stages

	Individual		Group	
	Low	High	Low	High
Stage1	101.375	86.25	51.125	52.75
Stage2	80.375	107.25	25.625	78.25
Stage3	69.875	117.75	26.5	77.375
Stage4	69.875	117.75	33.75	70.125
Stage5	71.125	116.5	29.625	74.25
Stage6	68.375	119.25	22.375	81.5
Stage7	70.75	116.875	25.25	78.625
Stage8	51.25	136.375	24.25	79.625

Table 2. Names assigned to each stages

	Individual		Group	
	Low	High	Low	High
Stage1	DATA1		DATA9	
Stage2	DATA2		DATA10	
Stage3	DATA3		DATA11	
Stage4	DATA4		DATA12	
Stage5	DATA5		DATA13	
Stage6	DATA6		DATA14	
Stage7	DATA7		DATA15	
Stage8	DATA8		DATA16	

results of individual use and rows of the lower half part are those of group use. In Fig. 9, all the processes are divided into 8 equally weighted stages from the beginning to the end.

Though Fig. 9 shows that the frequency in which the subjects moved lower order term points a and b decreased at stage 8, the decrease of frequency is not so apparent at the stages from 3 to 7. Not only that, the frequency increased at stages 4 and 5 in the case of group users as opposed to our anticipations. Moreover, remarkable difference in the frequencies at each stage can be observed between the individual users and the group users. In fact, Table 1 shows the total numbers of operations in which the subjects moved lower order term points a, b and higher order term points c, d respectively at each stage. When one movement belongs to two successive stages, its frequency is allocated to those stages proportionally. Applying pairwise Fisher's exact test to this data with the assigned numbers as shown in Table 2, we obtain the resulting p-values in Table 3.

Table 3. Result of pairwise Fisher's exact test

	1	2	3	4	5	6	7	8	9	10	11	12	13	14	15
2	0.08526	–													
3	0.00821	0.43919	–												
4	0.00821	0.43919	1.00000	–											
5	0.01343	0.57029	0.97201	0.97201	–										
6	0.00613	0.39684	0.97201	0.97201	0.94059	–									
7	0.01068	0.50371	1.00000	1.00000	1.00000	0.94059	–								
8	1.2e–05	0.01278	0.09830	0.09830	0.08526	0.14374	0.08526	–							
9	0.61736	0.48410	0.12521	0.12521	0.15473	0.09830	0.12693	0.00306	–						
10	3.4e–05	0.01607	0.08526	0.08526	0.07642	0.10469	0.07642	0.90713	0.00452	–					
11	4.5e–05	0.01607	0.08526	0.08526	0.08526	0.12816	0.08526	0.90713	0.00452	1.00000	–				
12	0.00459	0.18878	0.60346	0.60346	0.60346	0.76843	0.60346	0.50371	0.07310	0.43689	0.43689	–			
13	0.00067	0.07271	0.27190	0.27190	0.22383	0.33188	0.27190	0.90713	0.01851	0.78285	0.78285	0.79078	–		
14	4.6e–06	0.00235	0.02234	0.02234	0.01712	0.03116	0.01712	0.41484	0.00067	0.77726	0.66998	0.15695	0.41484	–	
15	2.0e–05	0.00872	0.07531	0.07531	0.06228	0.08526	0.06228	0.74051	0.00306	1.00000	0.96099	0.35888	0.68343	0.88801	–
16	1.2e–05	0.00613	0.04582	0.04582	0.03532	0.07508	0.04582	0.64043	0.00208	0.96099	0.88801	0.27675	0.60346	0.96099	1.00000

While the frequency is almost constant throughout the stages from 3 to 7 in the case of individual users, discrepancy can be observed in the case of group users as seen in Table 1. In fact, among the elements of the 9th column in Table 3 which are surrounded by a red rectangle, only those corresponding to the stages 4 and 5 are not statistically significant. Also, among the diagonal elements of the matrix which is surrounded by a yellow rectangle, those corresponding to the stages 2, 3, 6, 7 are statistically significant. Therefore it can be concluded that the significant difference between individual use and group use arose mainly from the difference at these four stages. In particular, the relative frequency of the movement of points a and b decreased more rapidly in the case of group users compared to that in the case of individual users. Contrarily, group users moved these points more frequently again at stages 4 and 5. Together with the fact that the frequency is always lower in the case of group users compared to the case of individual users, the results stated above indicate that communications between users made their search processes far more efficient.

4   Discussions

Though the overall difference in the pattern of users' operations between individual use and group use has been illustrated in Sect. 3, discrepancy in the data obtained from all samples has not been considered. Since the comparison of this discrepancy between individual use and group use may lead to tracking some quantities which illustrate the influence of users' communications more clearly, the authors turned their attention to the distribution of the relative frequencies of users' moving lower order term points in each stage. Regarding to the fact that, in Table 1, almost no difference in relative frequency was observed throughout the stages from 3 to 7 in the case of individual use, the authors counted up those relative frequencies at the stages from 3 to 7 in the case of individual use and group use respectively. Figures 10 and 11 are the histograms which show the distributions of the above accumulated data.

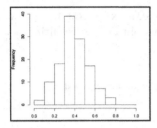

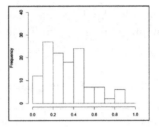

Fig. 10. Individual use   **Fig. 11.** Group use

As assumed from Fig. 11, the distributions are sparse at many stages and completely different from normal distribution in the case of group use. Therefore, we cannot apply traditional analysis of variance to compare the distributions between individual use and group use. To make this comparison, the authors summarized these two distributions by approximating them with finite mixture of beta distributions since it is known that any prior density on $(0,1)$ can be arbitrarily approximated by a finite mixture of betas [8]. Generating large-scale random samples from the specific beta distributions with R, the authors searched two mixtures of betas so that the generated data would give the histograms which are similar to Figs. 10 and 11 respectively. In fact, χ^2 test of goodness-of-fit using R tells us that these two distributions are well approximated with the mixtures shown in Figs. 12 and 13.

As seen in Figs. 12 and 13, there are three and four components in the distributions corresponding to individual use and group use respectively. However, identifiability can hardly be expected since there is no estimation of the number of beta components which can be definitely presumed from the data [9]. Therefore it cannot be claimed that the obtained mixture is unique. Nevertheless, the inferred index and weight of each beta component can provide valuable information about the users' operations belonging to that "cluster", since they may indicate the extent to which users persisted in moving lower order term points and the rate of incidence at which users' search processes belongs to that pattern.

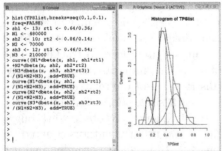

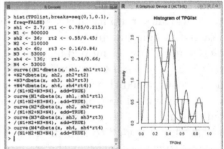

Fig. 12. Individual use   **Fig. 13.** Group use

Regarding to this point, one of the most remarkable differences between individual use and group use is the larger variance of data in the latter case. In fact, there are some stages in which the incidence rate of users' moving points corresponding to a and b is high (more than 0.8). Figure 14 shows the gesture of some subject at the time near to one of those stages. She moved her finger over the iPad screen tracing the shape of the blue curve.

Fig. 14. The gesture of one female subject (Color figure online)

This behavior indicates that the subjects working in her group moved the points in order to make the red curve overlap with the blue curve not in the neighbourhood of $x = 0$ but on the whole interval $[-1, 2]$. In fact, when the authors asked the subjects working in her group, "why did you move points corresponding to a and b at these middle and final stages of your search process?" while displaying the recorded video image, they answered as follows.

Though we were aware that the line $y = ax + b$ coincides with the tangent line, she began to worry about the difficulty in finding suitable coefficients c and d so that the shape of the red curve became the same as that of the blue curve. So we tried to find appropriate values c and d first, then we adjusted the position of the red curve by moving a and b.

Similar gestures and statements can be found in many cases of this cluster. Without her worries being communicated to the other members, the search process of this group should get closer to the model case as seen in Fig. 5. Even though the search process of this group is different from the model case, they could understand the concept of infinitesimal of higher order through observing that only one choice of a and b will lead to the desired approximation. Therefore, it can be seen that the communication between users gave them more opportunities to understand the target concept.

5   Concluding Remarks

Based on the observations described at the end of Sect. 4, it can be concluded that the larger variance of beta components in the case of group use resulted from the fluctuation of the users' thinking pattern caused by the communications between them. The result of this research strongly indicates that CindyJS can serve one of the most effective interfaces on which users' interactive operations can trigger the above mentioned communications directly connected to mathematical models. To extract the illustrative signal tracking users' reasoning processes from their operating pattern, comparison of that pattern with the discourse between users has proven to be helpful. In future, the system for automatically recording users' operating processes would enable us to obtain firm knowledge about the effective combination of various mathematical activities.

Acknowledgements. The authors are grateful to Professor Jürgen Richter-Gebert, Professor Ulrich Kortenkamp, and their colleagues for their great efforts to develop Cinderella and CindyJS. They are also grateful to Professor Jinfang Wang for his helpful advices with the statistical analysis of this research.

This work was supported by JSPS KAKENHI (15K01037).

References

1. Piaget, J. : The role of action in the development of thinking. In: Overton, W.F., Gallagher, J.M. (eds.) Knowledge and Development, pp. 17–43 (1977)
2. Pappert, S.: Teaching children to be mathematicians versus teaching about mathematics. Int. J. Math. Educ. Sci. Technol. **3**, 249–262 (1972)
3. Richter-Gebert, J., Kortenkamp, U.: Interactive Geometry with Cinderella. In: The Cinderella.2 Manual. Springer, Heidelberg (2012). https://doi.org/10.1007/978-3-540-34926-6_4
4. von Gagern, M., Kortenkamp, U., Richter-Gebert, J., Strobel, M.: CindyJS. In: Greuel, G.-M., Koch, T., Paule, P., Sommese, A. (eds.) ICMS 2016. LNCS, vol. 9725, pp. 319–326. Springer, Cham (2016). https://doi.org/10.1007/978-3-319-42432-3_39
5. von Gagern, M., Richter-Gebert, J.: CindyJS Plugins. In: Greuel, G.-M., Koch, T., Paule, P., Sommese, A. (eds.) ICMS 2016. LNCS, vol. 9725, pp. 327–334. Springer, Cham (2016). https://doi.org/10.1007/978-3-319-42432-3_40
6. Montag, A., Richter-Gebert, J.: CindyGL: authoring GPU-based interactive mathematical content. In: Greuel, G.-M., Koch, T., Paule, P., Sommese, A. (eds.) ICMS 2016. LNCS, vol. 9725, pp. 359–365. Springer, Cham (2016). https://doi.org/10.1007/978-3-319-42432-3_44
7. Kaneko, M.: Using tangible contents generated by CindyJS and its influence on mathematical cognition. In: Gervasi, O., Murgante, B., Misra, S., Borruso, G., Torre, C.M., Rocha, A.M.A.C., Taniar, D., Apduhan, B.O., Stankova, E., Cuzzocrea, A. (eds.) ICCSA 2017. LNCS, vol. 10407, pp. 199–215. Springer, Cham (2017). https://doi.org/10.1007/978-3-319-62401-3_15
8. Titterlington, D.M., Smith, A.F.M., Makov, U.E.: Statistical Analysis of Finite Mixture Distributions. Wiley, New York (1985)
9. Frühwirth-Schnatter, S. (ed.): Finite Mixture and Markov Switching Models. Springer Series in Statistics, 1st edn. Springer, New York (2006)

A Novel Dynamic Mathematics System Based on the Internet

Yongsheng Rao[1], Hao Guan[2,3], Ruxian Chen[1], Yu Zuo[4], and Ying Wang[5(✉)]

[1] Institute of Computing Science and Technology, Guangzhou University,
Guangzhou, China
rysheng@gzhu.edu.cn, chrxian@qq.com
[2] University of Chinese Academy of Sciences, Beijing, China
gd.ucas@qq.com
[3] Chengdu Institute of Computer Application, Chinese Academy of Sciences,
Chengdu, China
[4] Institute of Mathematics and Computer Science, Guizhou Normal College, ·
Guiyang, China
[5] South China Institute of Software Engineering, Guangzhou University,
Guangzhou, China
wying@sise.com.cn

Abstract. In this paper, we introduce a novel dynamic mathematics system called NetPad for teaching and learning mathematics in elementary and secondary school. NetPad is a product of Internet Plus Education and can be launched directly from the internet using a web browser. It combines the Internet with dynamic geometry, computer algebra, and automated reasoning technology. NetPad distinguishes itself from other dynamic geometry systems by being an open, internet-based and sharing oriented intelligent system. NetPad is not only a tool but also a cloud platform for creating and sharing. Since NetPad is developed in HTML5, it is platform independent, runs on every operating system and intelligent device, and can be seamlessly integrated into other websites, PowerPoint and other software. The resources of NetPad can be shared to various social networks directly. The functions of NetPad include dynamic geometry drawing, symbolic computation, programming, automated reasoning in geometry, and so on. NetPad was published in March, 2016. Nowadays, there are more than 100,000 users and 30,000 mathematical resources on the NetPad website.

Keywords: Dynamic mathematics · The Internet
Mathematics education · NetPad

This work was supported in part by the National Natural Science Foundation of China (11701118 and U1201252), the Guangdong Provincial Engineering and Technology Research Center ([2015]1487), the Specialized Fund for Science and Technology Platform and Talent Team Project of Guizhou Province (QianKeHePing-TaiRenCai[2016]5609), the Guangdong Provincial Key Platform and Major Scientific Research Projects (2016KQNCX238).

© Springer International Publishing AG, part of Springer Nature 2018
J. H. Davenport et al. (Eds.): ICMS 2018, LNCS 10931, pp. 389–396, 2018.
https://doi.org/10.1007/978-3-319-96418-8_46

1   Introduction

The international education community has reached a consensus that the impact of dynamic geometry on education is positive [1]. Many dynamic geometry systems (DGSs) are used in mathematics education. The popular DGSs include The Geometer's Sketchpad (GSP) [2], Cabri [3], and Cinderella [4]. However, there are not only geometry, but arithmetic, algebra, analysis, programming, and proof systems used in mathematics education at elementary and secondary school. In order to satisfy the demands of mathematics education, our dynamic geometry system has been developed into a dynamic mathematics system (DMS) by integrating computer algebra and other technology. GeoGebra(GGB) [5] and Super Sketchpad (SSP) [6] are popular DMSs in elementary and secondary school.

With the development of mobile Internet, various devices and operating systems are widely used in education. For better integration of the Internet with mathematics education, we developed a novel dynamic mathematics system based on SSP, called NetPad. NetPad combines the Internet with dynamic geometry, computer algebra, and automated reasoning technology. Therefore it is not only a dynamic mathematics teaching tool, but also a cloud platform for creating and sharing educational resources. Since NetPad is developed in HTML5, it is platform independent. NetPad can be launched directly from internet and is freely available at www.netpad.net.cn. Figure 1 is its homepage, and Fig. 2 is its user interface.

Fig. 1. Homepage of NetPad

Fig. 2. User interface of NetPad

2   Functions of NetPad

The functions of NetPad are listed below.

2.1   Geometric Drawing

The geometric drawing function of NetPad includes geometric figures, conic curves and function graphs, geometric transformations, customized coordinate systems, animation, graphic loci, iterations, variables, built-in functions, and so on. For example, Fig. 3(a) contains two iterative figures, the square and triangle. B_1 is generated rotating point A by 150° around point B. We make a further iteration from $(A; B)$ to $(A_1; B_1)$. The iterations of the two iterative figures are shown in Fig. 3(b) where the iteration depth is 5. This resource can be found via ID 27858 at www.netpad.net.cn.

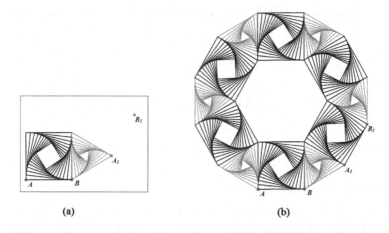

Fig. 3. The iteration of the iteration figures

The Intelligent Pen is a drawing tool based on context-aware technology. With it, users can construct about 20 kinds of dynamic geometric figures with the mouse alone, without toolbar buttons or menus, allowing the construction of dynamic geometric figures in an accurate and efficient manner. Further details are explained in [7]. With the Intelligent Pen, NetPad only needs 8 mouse operations (clicks or movements) to draw the orthocenter of a triangle (Fig. 4) and

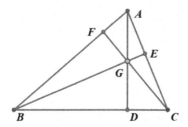

Fig. 4. The orthocenter of a triangle

requires no switching between the graphic window and menu. But it takes 22 mouse operations and 8 window switches in GSP, or 19 mouse operations and 2 window switches in Cinderella.

NetPad can conveniently construct intersection points of almost all linear geometric figures. The types of intersection points that can be constructed in NetPad(NPD), GSP, GGB and SSP are shown in Table 1. "Y" means this type of intersection point can be constructed, "-" means this type of intersection cannot be constructed.

Table 1. Comparison of intersection points function

	S - S	S - C	S - P	S - V	C - C	C - P	C - V	P - P	P - V	V - V	V - L	L - L
GSP	Y	Y	-	Y	Y	-	Y	-	-	Y	-	-
GGB	Y	Y	-	Y	Y	-	Y	-	-	Y	-	-
SSP	Y	Y	-	Y	Y	-	Y	-	-	-	-	-
NPD	Y	Y	Y	Y	Y	Y	Y	Y	Y	Y	Y	Y

S means Straight Line, C means Circle, P means Polygon, V means Curve, L means Locus.

2.2 Symbolic Computation

NetPad supports symbolic computation. The feature can assist students in learning some basic concepts about algebraic operations as well as in carrying out mathematical calculus and finding the results of computations, as shown in Fig. 5.

```
Factor(80501181299);
>>(7)^2*(11)^5*(101)^2

Factor(x^6*y^6-1);
>>(x^2*y^2-x*y+1)*(x^2*y^2+x*y+1)*(x*y+1)*(x*y-1)

Diff(a*x^2+ln(x),x);
>>(2*x^2*a+1)/(x)

(18*a^5*b^4)^(1/2);
>>3*a^2*b^2*(2*a)^(1/2)

(x^3-1)/(x^2-1);
>>(x^2+x+1)/(x+1)
```

```
Dichotomy(a, b, d) {
    u = f(a),v = f(b);
    while( u * v < 0 && b - a > d ) {
        c=(a+ b) / 2;
        w= f(c);
        if( w == 0 ) {    u=w; }
        else{
            if( u * w < 0) { v = w; b = c;}
            else{ u = w; a = c; }
        }}
        (a + b)/ 2;
}
>>Dichotomy(a, b, d)

f(x) { e^x + x; }
Dichotomy(-1, 0, 0.000001);
>>f(x) -(1134285926818847)/(2000000000000000)=-0.567143
```

Fig. 5. Symbolic computation   **Fig. 6.** Programming

2.3   Programming

NetPad provides a simple interpretive language for programming whose syntax is similar to C. It includes assignment, conditional branching, loops, and definition of functions. Figure 6 shows the definition of a program for finding the zero of a function $f(x)$, x in (a, b) using the dichotomy method, where the error of the zero is less than d.

2.4   Automated Reasoning in Geometry

Automated reasoning in geometry is highly useful in mathematics education [8]. Based on our automated reasoning technologies [9–11], NetPad can solve most elementary geometric problems. Figure 7 shows the five-circle theorem. A readable proof of the theorem is generated and shown in the text box. The reasoning system generates 2276 pieces of information via the detailed derivation process, including 30 pieces of information about similar triangles, and 480 pieces of information about equal angles.

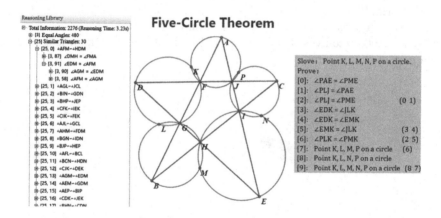

Fig. 7. The automated reasoning of the Five-Circle Theorem.

3   Features of NetPad

3.1   Convenient Platform to Share and Communicate

One notable feature of NetPad is webpage hyperlinks, which can be shared on various social networks such as Twitter, Facebook, WeChat, Microblog, etc. via the resource link or a QR code. For example, via its QR code (Fig. 8), a resource can be viewed in WeChat on a cell phone (Fig. 9). Furthermore, users can download resources from NetPad and the desktop application at https://www.netpad.net.cn/en/index.html, then use the resources offline without accessing the Internet.

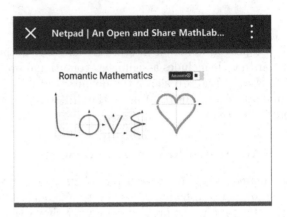

Fig. 8. The QR code   **Fig. 9.** NetPad in a cell phone

NetPad is also a community with users and resources. All resources created by users are stored in NetPad's resource cloud. Users can also ask for help through this platform. Through the time of publication of NetPad (March, 2016) to the time of writing this article, NetPad has acquired more than 100,000 registered users and 30,000 resources. The resources of Fig. 10 classified by knowledge topics are a part of the resource cloud.

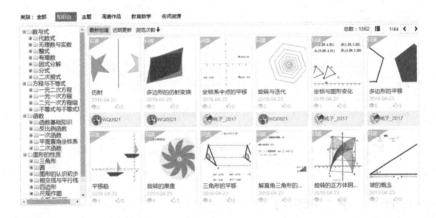

Fig. 10. Resources classified by knowledge topics

3.2   Seamless Integration with Other Systems

Microsoft SharePoint is a collection of web-based tools and technologies that help people store, share, and manage digital information. With SharePoint Add-ins Technology [12], NetPad becomes available in the Microsoft Office Store, such that NetPad can be embedded into PowerPoint in only two steps. The first step is searching for NetPad in the Office Store with the keyword "NetPad" and adding it into a PowerPoint document, as shown in Fig. 11. Because the resources are stored in the cloud in the web-archive format, the second step is searching for a resource in the resource cloud by keyword, author or resource ID. In Fig. 12, the resource is embedded using the key words "The Properties of the Linear Equation", then users can change the variables k and b by dragging the sliders dynamically in the slide.

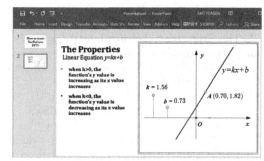

Fig. 11. Search for NetPad in the Office Store

Fig. 12. Insert NetPad into PPT

Because each resource of NetPad has a webpage hyperlink associated with it, it can be seamlessly integrated into other websites using its ID and NetPad APIs. For example, the resource "Romantic Mathematics" can be embedded into any website with the following code, in which 42406 is its ID.

```
<iframe src="https://www.netpad.net.cn/thirdInnerPad.
   html#posts/42406" style="width:800px;height:600px
   ;"></iframe>
```

4   Conclusion

NetPad is a novel dynamic mathematics system for teaching and learning mathematics in elementary and secondary school. It is a product of Internet Plus Education and is not only a teaching tool, but also a cloud platform for creating and sharing resources. The most important characteristics of NetPad are that it is open, sharing oriented, intelligent, internet-based, and that it features rich functions. The resources of NetPad can be shared to various social networks directly, seamlessly integrated into other websites, PowerPoint and other software. The functions of NetPad include dynamic geometry drawing, symbolic computation, programming, automated reasoning in geometry, and so on. NetPad was published in March 2016. Nowadays, there are more than 100,000 users and 30,000 resources on the NetPad website.

In future work, we will enrich and optimize the functions of NetPad, improve its running efficiency and stability, and release a 3D version.

Acknowledgements. We are grateful to Masataka Kaneko for proposing many good suggestions and Zak Tonks for improving the English.

References

1. Zhang, J., Xiong, H., Peng, X.: Free software SSP for teaching mathematics. In: Symbolic Computation and Education, pp. 115–135 (2014)
2. The Geometer's Sketchpad. http://www.dynamicgeometry.com
3. Cabri. http://www.cabri.com
4. Richter-Gebert, J., Kortenkamp, U.: The interactive geometry software Cinderella 2. Am. Math. Mon. **107**(8) (1999)
5. GeoGebra. http://www.geogebra.org
6. Super Sketchpad. http://ssp.gzhu.edu.cn
7. Wang, Y., Rao, Y., Zou, Y., Huang, Y.: An algorithm for dynamic geometric intelligent drawing based on context awareness. In: Proceedings of IEEE the 2nd International Conference on Computational Intelligence and Applications, pp. 547–550 (2017)
8. McCharen, J.D., Overbeek, R.A., Wos, L.A.: Problems and experiments for and with automated theorem-proving programs. IEEE Trans. Comput. **25**(8), 773–782 (1976)
9. Jingzhong Zhang, L., Yang, X.G., Chou, S.: Automated generation of readable proofs in geometry. Chin. J. Comput. **18**(5), 380–394 (1995)
10. Zhang, J., Gao, X., Chou, S.: The geometry information search system by forward reasoning. Chin. J. Comput. **19**(10), 722–727 (1996)
11. Zhang, J., Gao, X., Chou, S.: Geometric Invariant Methods of Geometric Theorem Proving. The Science Publishing Company, Beijing (2015)
12. SharePoint Add-ins. https://docs.microsoft.com/en-us/sharepoint/dev/sp-add-ins/sharepoint-add-ins

polyTop: Software for Computing Topology of Smooth Real Surfaces

Danielle A. Brake[1], Jonathan D. Hauenstein[2], and Margaret H. Regan[2(✉)]

[1] Department of Mathematics, University of Wisconsin-Eau Claire, Eau Claire, USA
brakeda@uwec.edu
[2] Department of Applied and Computational Mathematics and Statistics,
University of Notre Dame, Notre Dame, USA
{hauenstein,mregan9}@nd.edu
http://www.danibrake.org, http://www.nd.edu/~jhauenst,
http://www.nd.edu/~mregan9

Abstract. A common computational problem is to compute topological information about a real surface defined by a system of polynomial equations. Our software, called `polyTop`, leverages numerical algebraic geometry computations from `Bertini` and `Bertini_real` with topological computations in `javaPlex` to compute the Euler characteristic, genus, Betti numbers, and generators of the fundamental group of a smooth real surface. Several examples are used to demonstrate this new software.

Keywords: Numerical algebraic geometry · Topology
Cell decomposition · Graphs · Euler characteristic
Betti numbers · Fundamental group

1 Introduction

Let $X \subset \mathbb{R}^N$ be a smooth, closed, and orientable surface defined by the vanishing of a system of polynomial equations. Common topological quantities of interest regarding X include the Euler characteristic, genus, Betti numbers, and generators of the fundamental group [11,12]. This paper presents an approach to compute these quantities that combines numerical algebraic geometry with computational topology, and is implemented in the new software `polyTop`.[1]

The input to `polyTop` is a cell decomposition of X, which is computed from the polynomial system f as follows. First, `Bertini` [2,3] is used to compute a numerical irreducible decomposition of the solution set of $f = 0$ over the complex numbers. From this numerical irreducible decomposition, the software `Bertini_real` [6,7] computes a cell decomposition of the real surface X.

Using the cell decomposition as input, `polyTop` computes a topologically equivalent simplicial complex that immediately yields the Euler characteristic, genus, and Betti numbers. Interfacing with the computational topology software

---

[1] Available at http://dx.doi.org/10.7274/R0PV6HF4.

© Springer International Publishing AG, part of Springer Nature 2018
J. H. Davenport et al. (Eds.): ICMS 2018, LNCS 10931, pp. 397–404, 2018.
https://doi.org/10.1007/978-3-319-96418-8_47

javaPlex [15] yields confirmation of the Betti numbers and generators of the fundamental group.

There are several alternatives to our numerical approach. One could compute topological data from numerical sampling X, e.g., [4,9,10,13]. Another approach is to utilize symbolic computations to perform similar computations for a Riemann surface arising from a complex curve [16]. For example, similar topological computations are implemented in the software package algcurves in Maple.

The remainder is organized as follows. In Sect. 2, a method to move from a cell decomposition computed by Bertini_real to a topologically equivalent simplicial complex is presented. Section 3 explains the use of MATLAB and javaPlex in order to compute the Euler characteristic, genus, Betti numbers, and generators of the fundamental group. We demonstrate the software with various examples in Sect. 4 and conclude in Sect. 5.

2 Cell Decomposition and Simplicial Complex

For a smooth, closed, and orientable surface $X \subset \mathbb{R}^N$, we compute a simplicial complex $S(X)$ that is topologically equivalent to X. In our case, the simplicial complex $S(X)$ is a set composed of 0-, 1-, and 2-simplices, called vertices, edges, and faces, respectively and visually represented in Fig. 1. The key aspect is that such a topologically equivalent simplicial complex $S(X)$ for X can be constructed from a cell decomposition of X computed by Bertini_real.

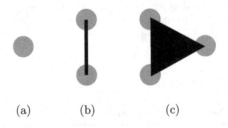

(a)   (b)   (c)

Fig. 1. A visual representation of (a) a 0-simplex or vertex, (b) a 1-simplex or edge, and (c) a 2-simplex or face.

Following [1,5], the real surface X can be decomposed into a finite union of cells that mirror a simplicial complex. Each 2-cell, called a face, of X is a subset of X that has a generic interior point and a boundary consisting of 1-cells. Each 1-cell, called an edge, of X is a subset of X that has a generic interior point and a 0-cell (vertex) at each end. Figure 2(a) provides an illustration of a 2-cell.

The software polyTop constructs a topologically equivalent simplicial complex $S(X)$ of X by looping over each cell of the cell decomposition and constructing a corresponding simplicial complex as follows. The vertices of the simplicial complex consist of the generic interior point of the 2-cell, each generic interior point of the 1-cells, and the vertices at the end of each 1-cell. The edges of the simplicial complex consist of "interior" edges connecting the generic interior point of the 2-cell with each vertex on the boundary of the 2-cell and "boundary" edges connecting the generic interior point of each 1-cell with its vertices. The faces consist of the generic interior point of the 2-cell and two vertices connected by a "boundary" edge. This construction is illustrated in Fig. 2(b). Naturally connecting the simplicial complexes along neighboring cells of the decomposition yields a simplicial complex $S(X)$ that is topologically equivalent to X.

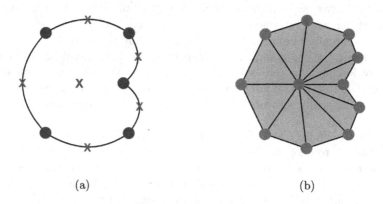

(a)   (b)

Fig. 2. A visual representation of a 2-cell and corresponding simplicial complex.

Let V, E, and F denote the number of vertices, edges, and faces, respectively, of the simplicial complex $S(X)$. The Euler characteristic χ of X is computed from $S(X)$ via

$$\chi = V - E + F. \tag{1}$$

Suppose that X is connected. Then, the Euler characteristic χ and genus g are related via

$$\chi = 2 - 2g \tag{2}$$

and the Betti numbers of X are $\beta_0 = \beta_2 = 1$ and $\beta_1 = 2g$. A basis for the fundamental group $\pi_1(X)$, which consists of $2g$ loops, is computed by javaPlex from the simplicial complex $S(X)$.

In order to test for connectivity, we take the transitive closure A^+ of the adjacency matrix A of edges of $S(X)$. Since $S(X)$ has V vertices, then A is a $V \times V$ symmetric matrix where $A_{ij} = A_{ji}$ is 1 if there is an edge in $S(X)$ between vertex i and vertex j and 0 otherwise. The transitive closure A^+ of A describes which vertices are connected, i.e., A_{ij}^+ is 1 if there is a path connecting vertex i and vertex j and 0 otherwise. The transitive closure can be computed using Boolean matrix multiplication and addition via the following:

$$A^+ = A + A^2 + A^3 + \cdots + A^V. \tag{3}$$

In particular, X is connected if and only if every entry of A^+ is 1. If X is not connected, A^+ can be used to decompose X into connected components.

3   Software

The software `polyTop` is written in MATLAB to utilize the preexisting MATLAB interfaces of both `Bertini_real` and `javaPlex`.

Given a polynomial system f, one first uses `Bertini` to compute a numerical irreducible decomposition which is then used by `Bertini_real` to compute a cell decomposition. The data for the cell decomposition is loaded into MATLAB utilizing the command `gather_br_samples` from `Bertini_real` which creates a file called `BRinfo#.mat` which can be used by all of the other MATLAB functions in the `Bertini_real` interface. For example, one can plot the surface using this file via the command `bertini_real_plotter` within MATLAB.

After using the command `load_javaplex` to load the `javaPlex` library and separately loading the cell decomposition data in MATLAB, `polyTop` can be executed. The first task of `polyTop` is to organize the cell decomposition data to create a topologically equivalent simplicial complex using the method described in Sect. 2.

Next, a stream is created in `javaPlex` that organizes the simplicial complex data for use in topological computations within `javaPlex`. Vertices are added using the command `stream.addVertex(i,0)`. Edges between vertices a and b are added via the command `stream.addElement([a, b])` while faces consisting of vertices a, b, and c are added via `stream.addElement([a, b, c])`.

Finally, a call to `javaPlex` performs homology computations on the simplex stream. The homology is computed with $\mathbb{Z}/2\mathbb{Z}$ coefficients.

4   Examples

The following summarizes several computations using `polyTop`. The input is computed via a numerical irreducible decomposition using `Bertini` followed by a cell decomposition using `Bertini_real`. A topologically equivalent simplicial complex is then constructed yielding the Euler characteristic, genus, and Betti numbers. The software `javaPlex` is then used for confirming the Betti numbers and generators of the fundamental group. The following timings are based on using a 2.4 GHz Intel Core i5 processor: the sphere and torus examples ran in under 0.1 s while the tanglecube and Crixxi examples completed in under 15 s.

4.1   Sphere

The unit sphere $X \subset \mathbb{R}^3$, defined by $x^2 + y^2 + z^2 = 1$, is a simply connected real surface. That is, the fundamental group of X is trivial, the Euler characteristic is $\chi = 2$, genus is $g = 0$, and the Betti numbers are $\beta_0 = \beta_2 = 1$ with $\beta_1 = 0$. A topologically equivalent simplicial complex derived from a cell decomposition computed by `Bertini_real` is shown in Fig. 3 consisting of $V = 6$ vertices, $E = 12$ edges, and $F = 8$ faces in agreement with (1).

Fig. 3. Topologically equivalent simplicial complex for the unit sphere.

4.2 Torus

For an illustrative real surface with a nontrivial fundamental group, we consider the torus $X \subset \mathbb{R}^3$ defined by

$$(x^2 + y^2 + z^2 + 15/4)^2 - 16(x^2 + y^2) = 0.$$

Figure. 4(a) shows a topologically equivalent simplicial complex derived from a cell decomposition computed by `Bertini_real`. In particular, there is a single hole which is commensurate with the fact that the genus is $g = 1$, Euler characteristic is $\chi = 0$, and the Betti numbers are $\beta_0 = \beta_2 = 1$ with $\beta_1 = 2$. This simplicial complex consists of $V = 32$ vertices, $E = 96$ edges, and $F = 64$ faces in agreement with (1).

Interfacing with `javaPlex` using this simplicial complex yields representatives for the two generators of the fundamental group. The output is

Dimension : 1

$[0.0, \text{infinity}) : [1, 14] + [2, 9] + [2, 14] + [1, 9]$

$[0.0, \text{infinity}) : [3, 18] + [3, 17] + [1, 17] + [1, 18]$

In this notation, an edge connecting vertices v and w is represented by $[v, w]$ and a loop is a sum of edges. Hence, this shows that each of the two generating loops consists of 4 edges which we can equivalently write as

$$1 \rightarrow 14 \rightarrow 2 \rightarrow 9 \rightarrow 1 \quad \text{and} \quad 3 \rightarrow 18 \rightarrow 1 \rightarrow 17 \rightarrow 3$$

and are visually represented in Fig. 4(b).

4.3 Tanglecube

The tanglecube is a degree four surface in $\mathbb{R}^3$ of genus $g = 5$ defined by

$$x^4 - 5x^2 + y^4 - 5y^2 + z^4 - 5z^2 + 11.8 = 0.$$

This surface, shown in Fig. 5(a), was used, for example, in [14] to demonstrate creating a meshing of the volume inside of the tanglecube surface using approximately 50,000 vertices and 200,000 tetrahedra. Using a cell decomposition computed by `Bertini_real`, a topologically equivalent simplicial complex of just the

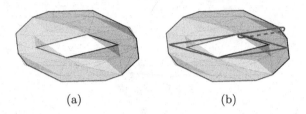

Fig. 4. (a) Simplicial complex for a torus with (b) visualizing two generating loops of the fundamental group.

tanglecube surface consists of $V = 296$ vertices, $E = 912$ edges, and $F = 608$ faces. This confirms that the surface has genus $g = 5$ with Euler characteristic $\chi = -8$ via (1) and (2).

Passing the simplicial complex to `javaPlex` confirms that the Betti numbers are $\beta_0 = \beta_2 = 1$ with $\beta_1 = 10$ and computes ten loops that generate the fundamental group. Figure 5(b) shows two representatives of these ten loops.

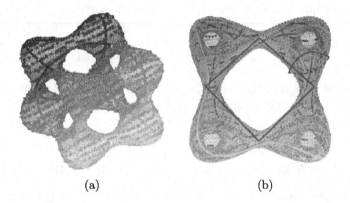

Fig. 5. (a) Simplicial complex for the tanglecube with (b) visualizing two representative loops of the ten generating loops of the fundamental group.

4.4   Crixxi

The Crixxi surface defined by

$$\left(\frac{1}{25}x^2 + \frac{1}{25}y^2 - 1\right)^3 + \left(\frac{1}{25}y^2 + \frac{1}{25}z^2 - 1\right)^2 = 0$$

is singular. By perturbing the right-hand side, say by replacing 0 with $1/10$, the real surface becomes smooth and orientable [8, p. 110] as shown in Fig. 6(a). This visualization suggests that the genus is $g = 2$ so that the Euler characteristic is

$\chi = -2$ and Betti numbers are $\beta_0 = \beta_2 = 1$ with $\beta_1 = 4$. This is confirmed by (1) and (2) after computing a topologically equivalent simplicial complex from a cell decomposition computed by `Bertini_real` having $V = 346$ vertices, $E = 1044$ edges, and $F = 696$ faces.

Passing the simplicial complex to `javaPlex` yields a confirmation of the Betti numbers above and computes four loops that generate the fundamental group. Two of the loops consist of eight edges while the other two loops consist of 32 and 40 edges. A visualization of these four loops is shown in Fig. 6(b).

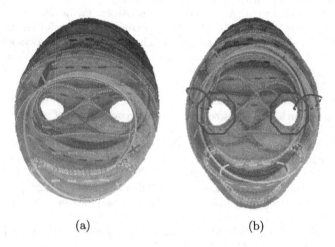

(a)   (b)

Fig. 6. (a) Simplicial complex for the perturbed Crixxi surface with (b) visualizing four generating loops of the fundamental group.

5   Conclusion

Computing topological information about a real surface defined by a system of polynomial equations is a recurrent problem within computational algebraic geometry. Using numerical algebraic geometry computations from `Bertini` and `Bertini_real` and topological computations in `javaPlex`, `polyTop` computes the Euler characteristic, genus, Betti numbers, and generators of the fundamental group of a real surface.

Acknowledgments. The authors thank Mikael Vejdemo-Johansson for input regarding `javaPlex`. All authors acknowledge support from NSF ACI-1440607/1460032. Additional support for JDH was provided by Sloan Research Fellowship BR2014-110 TR14 and for MHR by Schmitt Leadership Fellowship in Science and Engineering.

References

1. Bates, D.J., Brake, D.A., Hauenstein, J.D., Sommese, A.J., Wampler, C.W.: On computing a cell decomposition of a real surface containing infinitely many singularities. In: Hong, H., Yap, C. (eds.) ICMS 2014. LNCS, vol. 8592, pp. 246–252. Springer, Heidelberg (2014). https://doi.org/10.1007/978-3-662-44199-2_39
2. Bates, D.J., Hauenstein, J.D., Sommese, A.J., Wampler, C.W.: Bertini: Software for numerical algebraic geometry. bertini.nd.edu
3. Bates, D.J., Hauenstein, J.D., Sommese, A.J., Wampler, C.W.: Numerically solving polynomial systems with Bertini. In: SIAM (2013)
4. Berger, M., Tagliasacchi, A., Seversky, L.M., Alliez, P., Levine, J.A., Sharf, A., Silva, C.T.: State of the art in surface reconstruction from point clouds. In: Eurographics 2014 - State of the Art Reports. The Eurographics Association (2014)
5. Besana, G.M., Di Rocco, S., Hauenstein, J.D., Sommese, A.J., Wampler, C.W.: Cell decomposition of almost smooth real algebraic surfaces. Num. Alg. **63**(4), 645–678 (2013)
6. Brake, D.A., Bates, D.J., Hao, W., Hauenstein, J.D., Sommese, A.J., Wampler, C.W.: Algorithm 976: Bertini_real: numerical decomposition of real algebraic curves and surfaces. ACM Trans. Math. Softw. **44**(1), 10 (2017). bertinireal.com
7. Brake, D.A., Bates, D.J., Hao, W., Hauenstein, J.D., Sommese, A.J., Wampler, C.W.: Bertini_real: software for one- and two-dimensional real algebraic sets. In: Hong, H., Yap, C. (eds.) ICMS 2014. LNCS, vol. 8592, pp. 175–182. Springer, Heidelberg (2014). https://doi.org/10.1007/978-3-662-44199-2_29
8. do Carmo, M.P.: Differential Geometry of Curves and Surfaces. Prentice Hall, New Jersey (1976)
9. Cucker, F., Krick, T., Shub, M.: Computing the homology of real projective sets. Found. Comput. Math. **15**, 281–312 (2015)
10. Dufresne, E., Edwards, P.B., Harrington, H.A., Hauenstein, J.D.: Sampling real algebraic varieties for topological data analysis. arXiv:1802.07716 (2018)
11. Hatcher, A.: Algebraic Topology. Cambridge University Press, Cambridge (2002)
12. Munkres, J.R.: Topology. Prentice Hall, New Jersey (2000)
13. Niyogi, P., Smale, S., Weinberger, S.: Finding the homology of submanifolds with high confidence from random samples. Disc. & Comput. Geom. **389**(1–3), 419–441 (2008)
14. Oudot, S., Rineau, L., Yvinec, M.: Meshing volumes bounded by smooth surfaces. In: Hanks, B.W. (eds.) Proceedings of the 14th International Meshing Roundtable, Sandia National Laboratories. Springer, Heidelberg, pp. 203–220 (2005). https://doi.org/10.1007/3-540-29090-7_12
15. Adams, H., Tausz, A., Vejdemo-Johansson, M.: javaPlex: a research software package for persistent (co)homology. In: Hong, H., Yap, C. (eds.) ICMS 2014. LNCS, vol. 8592, pp. 129–136. Springer, Heidelberg (2014). https://doi.org/10.1007/978-3-662-44199-2_23
16. Tretkoff, C.L., Tretkoff, M.D.: Combinatorial group theory, Riemann surfaces and differential equations. Contemp. Math.: Contrib. Group Theory **33**, 467–519 (1984)

Solving the Likelihood Equations
to Compute Euler Obstruction Functions

Jose Israel Rodriguez$^{(\boxtimes)}$ (iD)

Department of Statistics, University of Chicago,
Chicago, IL 60637, USA
JoisRo@uchicago.edu
http://home.uchicago.edu/~joisro

Abstract. Macpherson defined Chern-Schwartz-Macpherson classes by introducing the (local) Euler obstruction function, which is an integer valued function on the variety that is constant on each stratum of a Whitney stratification. By understanding the Euler obstruction, one gains insights about a singular algebraic variety. It was recently shown by the author and B. Wang, how to compute these functions using maximum likelihood degrees. This paper discusses a symbolic and a numerical implementation of algorithms to compute the Euler obstruction at a point.

Keywords: Euler obstructions · Maximum likelihood degrees

1 Introduction

Studying singularities of algebraic varieties is of great interest in applied and computational algebraic geometry. For example, in applications the singular locus is important when finding the closest point to a variety, while in computations it can lead to bottlenecks. One way to understand a singular algebraic variety is by stratifying it into locally closed sets called Whitney strata. Then, for each stratum one considers the local information at a point.

To study how the closures of these strata interact with one another, the Euler obstruction function, defined in [19] by Macpherson, is considered. Informally, this function gives a measure of the singularity of a stratum. For an equivalent definition using Euler characteristics of complex links see [4,6]. In [20], it is shown that the Euler obstruction function at a point is given by an alternating sum of maximum likelihood degrees, which will be reviewed in the next section. It is in this framework, where we will develop our algorithms. The first sections of this paper recall the definition of ML degree, likelihood equations, and removal ML degrees to provide the statement of Theorem 1 [20], which enables us to compute the Euler obstruction. The main results are in Sect. 4, where the Macaulay2 [9] package MaximumLikelihoodObstructionFunction is described. This package implements the algorithms of [20]. The goal is to provide the research community

© Springer International Publishing AG, part of Springer Nature 2018
J. H. Davenport et al. (Eds.): ICMS 2018, LNCS 10931, pp. 405–413, 2018.
https://doi.org/10.1007/978-3-319-96418-8_48

tools to compute interesting examples to drive new areas of study.[1] The package is available at:

https://github.com/JoseMath/MaximumLikelihoodObstructionFunction

2   Maximum Likelihood Degrees

2.1   Maximum Likelihood Degrees of Very Affine Varieties

Consider an irreducible affine variety Z of $\mathbb{C}^n$. The set of points of Z with nonzero coordinates is denoted by Z^o and said to be the *underlying very affine variety* of Z. The underlying very affine variety of Z is a subvariety of $(\mathbb{C}^*)^n$. Consider the logarithmic 1-form on $(\mathbb{C}^*)^n$ given by

$$\ell_{\boldsymbol{\mu}}(\mathbf{z}) := \mu_1 \log z_1 + \mu_2 \log z_2 + \cdots + \mu_n \log z_n,$$

where $\boldsymbol{\mu} = (\mu_1, \ldots, \mu_n) \in \mathbb{C}^n$. The gradient of this one form is $\nabla \ell_{\boldsymbol{\mu}}(\mathbf{z}) := [\mu_1/z_1 \ldots \mu_n/z_n]$. Let $\mathbf{z}$ denote a regular point of Z^o. Then, the one form restricted to Z^o is said to have a *critical point* $\mathbf{z}$ if $\nabla \ell_{\boldsymbol{\mu}}(\mathbf{z})$ is orthogonal to the tangent space of Z^o at $\mathbf{z}$.

Definition 1. *The* maximum likelihood degree *(ML degree) of the very affine variety* Z^o, *is defined to be the number of critical points of* $\ell_{\boldsymbol{\mu}}(\mathbf{z})$ *on* Z^o *for general* $\boldsymbol{\mu}$ *and is denoted by* MLdegree(Z^o). *The ML degree of an affine variety* Z *is defined to be the ML degree of the underlying very affine variety* Z^o.

The notion of ML degree was first introduced in [5,14]. The name "maximum likelihood" comes from statistics, where the log likelihood function is the 1-form $\ell_{\boldsymbol{\mu}}(\mathbf{z})$, with μ_i denoting the number of times event i is observed. For a more geometric interpretation of Definition 1 see [16], and for a survey of results see [17]. Moreover, the *Gaussian degree* [7] is in some cases equivalent to the ML degree, and the *data singular locus* for maximum likelihood estimation is studied in [13]. In addition, the ML degree appears in other contexts, including Gaussian graphical models [22], variance component models [10], and in missing data [15].

Our convention is that the ML degree of an empty set is zero.

2.2   Likelihood Equations

The critical points of $\ell_{\boldsymbol{\mu}}(\mathbf{z})$ are defined by a zero dimensional variety and can be found by solving a system of polynomial equations. We determine the ML degree by determining the degree of this zero dimensional variety. Let $F \subset \mathbb{C}[z_1^{\pm}, \ldots, z_n^{\pm}]$ denote a set of generators of the ideal of Z^o, J denote the ideal of the singular locus of Z^o, and M denote the ideal of the $1 + \mathrm{codim}(Z^o)$ minors of

$$\begin{bmatrix} \nabla \ell_{\boldsymbol{\mu}}(\mathbf{z}) \\ \nabla F \end{bmatrix}, \tag{2.1}$$

where ∇F is the matrix of partial derivatives of F. The variety of $\mathtt{saturate}(M + F, J)$ is the set of critical points.

---

[1] The author is thankful for the helpful comments of Botong Wang and Xiping Zhang.

Definition 2. *The* likelihood equations *of Z^o (with respect to $\ell_{\boldsymbol{\mu}}(\mathbf{z})$) are defined by setting each element of a set of generators of the ideal* $\mathtt{saturate}(M + F, J)$ *to be zero.*

Proposition 1. *For generic $\boldsymbol{\mu}$, the degree of the variety of the ideal of the likelihood equations is the ML degree of Z^o.*

Often, it is easier to work with the affine variety Z rather than Z^o. Let $\hat{F}$ denote a set of generators of the ideal of Z, $\hat{J}$ denote a set of generators of the ideal of the singular locus of Z, and $\hat{M}$ denote a set of generators of the (codim $Z^o + 1$)-minors of

$$\begin{bmatrix} \nabla \ell_{\boldsymbol{\mu}}(\mathbf{z}) \\ \nabla \hat{F} \end{bmatrix} \mathrm{diag}([z_1 \ \ldots \ z_n]). \tag{2.2}$$

Lemma 1. *A set of generators of the ideal*

$$\mathtt{saturate}(\mathtt{ideal}\ \hat{M} + \mathtt{ideal}\ \hat{F}, \mathtt{ideal}\ \hat{J} * \mathtt{ideal}\ (z_1 z_2 \cdots z_n))$$

set to zero define likelihood equations of Z^o.

Remark 1. The definition of the ML degree of a projective variety in [14] agrees with the definition of the ML degree of a very affine variety when we restrict the projective variety to the affine chart where the coordinates sum to one.

The likelihood equations are often an overdetermined system of equations. For numerical computation we prefer for the number of equations to equal the dimension of the ambient space. Suppose $\hat{F}$ is a set of $c := \mathrm{codim}\ Z^o$ generators of the ideal of the ideal Z. Let $\{\lambda_0, \ldots, \lambda_c\}$ denote a set of indeterminants called *Lagrange multipliers*. Let $\hat{M}_{\mathrm{Lag}}$ denote the following set of n polynomials

$$[\lambda_0 \ \lambda_1 \ \ldots \ \lambda_c] \begin{bmatrix} \nabla \ell_{\boldsymbol{\mu}}(\mathbf{z}) \\ \nabla F \end{bmatrix} \mathrm{diag}([z_1 \ \ldots \ z_n]). \tag{2.3}$$

Definition 3. *The* Lagrange likelihood equations *are a system of equations defined by $\hat{F} = 0$ and $\hat{M}_{\mathrm{Lag}} = 0$.*

The variety of $\mathtt{ideal}\ \hat{F} + \mathtt{ideal}\ \hat{M}_{\mathrm{Lag}}$ is in the product space $\mathbb{C}^n \times \mathbb{P}^c$. Let $(\mathbb{C}^*)^n \times (\mathbb{C}^*)^c$ be the dense Zariski open set of $\mathbb{C}^n \times \mathbb{P}^c$ defined by $\lambda_0 = 1, \prod_{i=1}^n z_i \prod_{j=1}^c \lambda_j \neq 0$. These Lagrange likelihood equations were studied in [11].

Proposition 2. *For generic $\boldsymbol{\mu}$, the projection to $(\mathbb{C}^*)^n$ of $(\mathbb{C}^*)^n \times (\mathbb{C}^*)^c$ intersected with the variety of the Lagrange likelihood equations is the variety of the likelihood equations of Z^o.*

2.3   Computing Removal ML Degrees and Euler Obstructions

Let X^o denote a very affine variety of $(\mathbb{C}^*)^n$, and let $\mathbf{p}$ denote a point in $(\mathbb{C}^*)^n$. Let $\mathcal{H}_1, \ldots, \mathcal{H}_n$ denote general hyperplanes containing the point $\mathbf{p}$, and let $H_1(\mathbf{x}), \ldots, H_n(\mathbf{x})$ denote affine linear polynomials defining these hyperplanes. We denote the embedding of X^o to $(\mathbb{C}^*)^{n+1}$ via $y = H_1(\mathbf{x})$ by $X^o \setminus \mathcal{H}_1$.

Definition 4. *For $k = 0$, let $Z^o = X^o$, otherwise let $Z^o = (X^o \setminus \mathcal{H}_1) \cap \left(\cap_{i=2}^k \mathcal{H}_i \right)$. The k-th removal ML degree of X^o with respect to the point $\mathbf{p}$ is defined to be the ML degree of Z^o and is denoted by $r_k(X^o, \mathbf{p})$.*

Using removal ML degrees, we can compute the Euler obstruction by the following theorem.

Theorem 1 ([20]). *The signed alternating sum of removal ML degrees with respect to the point $\mathbf{p}$ in X^o of dimension d equals the value of the Euler obstruction function at a point $\mathbf{p}$, i.e.,*

$$(-1)^d \operatorname{Eu}_{X^o}(\mathbf{p}) = \sum_{k=0}^{d+1} (-1)^k r_k(\mathbf{p}, X^o).$$

With the symbolic implementation, we compute the removal ML degrees by solving the likelihood equations using Grobner basis. With the numerical implementation, we will use *coefficient parameter homotopies* [21, Theorem 7.1.1]. For fixed k, we choose a generic $k \times n$ matrix $[\gamma_{ij}]_{k \times n}$, and let $\mathcal{L}_b$ denote the following family of linear spaces defined by

$$\begin{bmatrix} H_1(\mathbf{x}) \\ \vdots \\ H_k(\mathbf{x}) \end{bmatrix} = [\gamma_{ij}]_{k \times n} \begin{bmatrix} x_1 \\ \vdots \\ x_n \end{bmatrix} - \begin{bmatrix} b_1 \\ \vdots \\ b_k \end{bmatrix}. \tag{2.4}$$

Let $Z = X \cap \mathcal{L}_b$. For general $[\gamma_{ij}]_{k \times n}$ in (2.4), the Lagrange likelihood equations of Z define a coefficient parameter homotopy with *parameters* $b = (b_1, \ldots, b_k)$, Lagrange multipliers $\lambda_0, \ldots, \lambda_c$, and *primal variables* $x_1, \ldots, x_n, y$ for $k > 0$.

Theorem 2 (Corollary 4.6 [20]). *If γ_{ij} are general and $Z = X \cap \mathcal{L}_b$, then for a parameter homotopy with target parameters $([\gamma_{ij}]_{k \times n}[\mathbf{p}])$, the number of regular endpoints not in the coordinate hyperplanes equals the k-th removal ML degree of X with respect to the point $\mathbf{p}$.*

In summary, we solve the Lagrange likelihood equations for a general choice of parameters, thereby determining the removal ML degrees with respect to a general point. Then, we use a parameter homotopy to determine the removal ML degrees for any other point $\mathbf{p}$ of interest.

3 Whitney Sombrilla

The illustrative example in this section comes from a Whitney umbrella in $\mathbb{C}^3$ defined by $x_1^2 - x_2^2 x_3 = 0$ that has been translated by $(1,1,1)$. We call this translation the *Whitney sombrilla* and denote it by X. The defining equation is $f := (x_1 - 1)^2 - (x_2 - 1)^2(x_3 - 1) = 0$. Note that the original Whitney umbrella and the Whitney sombrilla have different underlying very affine varieties. Indeed, the underlying very affine variety of the Whitney umbrella is smooth while the underlying very affine variety of the Whitney sombrilla is not. A Whitney stratification of X is given by the regular points; the singular points with $\{(1,1,1)\}$ removed; and $\{(1,1,1)\}$. We denote these strata by S_1, S_2, S_3 respectively. Let S_0 denote $\mathbb{C}^3 \setminus X$. Let $\mathbf{p}_i$ be a point in $S_i \cap (\mathbb{C}^*)^3$. The removal ML degrees of these points and their respective Euler obstructions are below. Note that the defining equations of $X \setminus \mathcal{H}_1$, $(X \setminus \mathcal{H}_1) \cap \mathcal{H}_2$, $(X \setminus \mathcal{H}_1) \cap (\mathcal{H}_2 \cap \mathcal{H}_3)$ are given by $\{f = 0\}$ and $\{H_1 = y\}, \{H_1 = y, H_2 = 0\}, \{H_1 = y, H_2 = 0, H_3 = 0\}$ respectively.

	$k = 0$	$k = 1$	$k = 2$	$k = 3$	$\mathrm{Eu}_{X^o}(\mathbf{p})$
$\mathbf{p}_0 = (3,2,1)$	3	10	10	3	0
$\mathbf{p}_1 = (3,3,2)$	3	10	10	2	1
$\mathbf{p}_2 = (1,1,2)$	3	10	10	1	2
$\mathbf{p}_3 = (1,1,1)$	3	10	9	1	1

4 Using the Package

The package computes the Euler obstruction function of a very affine variety $X^o \subset (\mathbb{C}^*)^n$ at the point $\mathbf{p}$. This is done by computing the k-th removal ML degrees of X^o with respect to the point $\mathbf{p}$ for $k = 0, 1, \ldots, \dim X^o + 1$. We define a new type of mutable hash table called the `RemovalMLDegree`, which is used to store the results of the computations of the removal ML degrees.

The package takes two approaches to computing removal ML degrees. The first approach uses symbolic computation like in the foundational paper [14]. The second approach uses homotopy continuation [1]. In each case, our algorithms are probabilistic and there exists a open Zariski dense set such that choices of random values will produce the true answer. How we generate random values can be changed in the `Configuration` when loading the package (see help `randomValue`), with the default producing random values by `random(1,30102)`. See the documentation for details about the keys of each new type of hash table.

4.1 Symbolic Computation

Preprocess. We assume the variety X^o is defined by an ideal I, but we store information about X^o in a new type of mutable hash table: `MLDegreeVariety`. To create this mutable hash table we use the method `newMLDegreeVariety`.

```
i1 :   loadPackage"MaximumlikelihoodObstructionFunction";
i2 :   R=QQ[p1,p2,p3];
i3 :   I=ideal((p-1)1^2-(p2-1)^2*(p3-1));
i4 :   L=newMLDegreeVariety(I)
o4 = MLDegreeVariety{...4...}
```

Solving. Let **p** denote a point in $(\mathbb{C}^*)^n$. If the point **p** is not specified by the user, then **p** is set to $(1, \ldots, 1)$. For $k = 0, 1, \ldots, \dim X^o + 1$, the package computes the k-th removal ML degree of X^o with respect to **p**. To store this information we introduce a type of mutable hash table called `RemovalMLDegree`. This hash table has three important keys `MLDegrees`, `ThePoint`, and `TheVariety`.

The method `solveRemovalMLDegree`, solves the likelihood equations (Definition 2) and stores the degree in `RemovalMLDegree` by appending k=>m to the list under the key `MLDegrees`, where k indexes the values of the removal ML degree m. This is the most difficult step in the computation.

```
i5 :   P={1,1,1};
i6 :   M=newRemovalMLDegree(L,P)
o6 = RemovalMLDegree{...3...}
i7 :   solveRemovalMLDegree M
o7 = {3, 10, 9, 1}
```

Extracting Information. Once all of the removal ML degrees of X^o are computed, we can extract the information using the methods `removalMLDegree` or `mlObstructionFunction`. The former return lists where the k-th element of the list is the k-th removal ML degree of X^o at **p**. The `mlObstructionFunction` returns the alternating sum of computed ML degrees.

```
i8 :   removalMLDegree(M)
o8 = {3, 10, 9, 1}
i9 :   mlObstructionFunction M
o9 = 1
```

4.2 Numeric Computation

In this subsection we compute removal ML degrees using homotopy continuation with the numerical algebraic geometry software BERTINI [3]. We use methods of the MACAULAY2 package BERTINI.M2 [2] to manipulate the input files.

Preprocess. Let **q** denote a general point in $(\mathbb{C}^*)^n$. In this step we compute the k-th removal ML degrees of X^o at **q** by solving the equations in Theorem 2 for a general choice of parameters. We denote by **q** the solution to this set of equations for a general choice of parameters. We assume the variety X^o is given to us by an ideal I, but we store information in a new type of mutable hash table called `MLDegreeWitnessCollection`. To create this mutable hash table we use the method `newMLDegreeWitnessCollection`.

```
i10 : loadPackage("MaximumlikelihoodObstructionFunction",
Reload=>true, Configuration=>{"RandomCoefficients"=>CC})
o10 = MaximumlikelihoodObstructionFunction
i11 :   s = temporaryFileName() | "/";
i12 :   mkdir s;
i13 :   R=CC[p1,p2,p3];
i14 :   I=ideal((p-1)1^2-(p2-1)^2*(p3-1))
i15 :   WC=newMLDegreeWitnessCollection(I,d,s)
o15 = MLDegreeWitnessCollection{...10...}
```

Computing Witness Sets. We solve Lagrange likelihood equations (Definition 3) to compute the removal ML degrees of X^o with respect to a generic point using the method `newMLDegreeWitnessSet`. We store this information in a directory determined by the key `Directory` of `MLDegreeWitnessCollection`. The most intensive part of the computation is this step were we compute witness sets. To avoid repeating this step, we save and load the collection of witness sets using the methods `saveWitnessCollectionConfiguration` and `getWitnessCollection`.

```
i16 :   newMLDegreeWitnessSet(WC)
o16 = {3, 10, 10, 3}
i17 :   saveWitnessCollectionConfiguration(WC,s)
```

Solving. With the method `homotopyRemovalMLDegree`, we use a parameter homotopy from Theorem 2 to determine the Euler obstruction at a point.

```
i18 :   P={1,1,1}
i19 :   M=newRemovalMLDegree(WC,P)
o19 = RemovalMLDegree{...4...}
i20 :   homotopyRemovalMLDegree M
o20 = {3, 10, 9, 1}
```

Extracting Information. Since we are working with floating point arithmetic, one must take care when classifying points in the coordinate hyperplanes. The method `reclassifyWitnessPoints` allows us to change the tolerances as we like.

5   Motivating Example

Matrices with rank constraints and their ML degrees have been studied in [12, 18, 20]. Moreover, their Euler obstructions have also been studied in [8, 23]. Consider $X \subset \mathbb{C}^5$ defined by the determinant of a 3×3 symmetric matrix with the bottom right entry set to one. We let $S_0 = \mathbb{C}^5 \setminus X$, $S_1 = X_{\text{reg}}$, and $S_2 = X_{\text{sing}}$. The kth column records $r_k(X, \mathbf{p})$ and the time to compute the kth witness set in

Sect. 4.2. The Sym and Num columns record the timing of the solving step in Sects. 4.1 and 4.2.

X_2	$k=0$	$k=1$	$k=2$	$k=3$	$k=4$	$k=5$	$\mathrm{Eu}_{X^\circ}(\mathbf{p})$	Sym	Num
$\mathbf{p}_0 = (1,2,3,5,7)$	0	16	47	49	21	3	0	1620 s	1 s
$\mathbf{p}_1 = (1,1,1,1,2)$	0	16	47	49	21	2	1	1991 s	1 s
$\mathbf{p}_2 = (1,1,1,1,1)$	0	16	47	49	19	1	0	2002 s	1 s
Time-Compute WS	8 s	197 s	321 s	90 s	68 s	9 s			

References

1. Allgower, E.L., Georg, K.: Continuation and path following. In: Acta Numerica 1993, pages 1–64. Cambridge University Press, Cambridge (1993)
2. Bates, D.J., Gross, E., Leykin, A., Rodriguez, J.I.: Bertini for Macaulay2. Preprint arXiv:1310.3297 (2013)
3. Bates, D.J., Hauenstein, J.D., Sommese, A.J., Wampler, C.W.: Bertini: software for numerical algebraic geometry. https://bertini.nd.edu/
4. Brasselet, J.-P., Trang, L.D., Seade, J.: Euler obstruction and indices of vector fields. Topology **39**(6), 1193–1208 (2000)
5. Catanese, F., Hoşten, S., Khetan, A., Sturmfels, B.: The maximum likelihood degree. Amer. J. Math. **128**(3), 671–697 (2006)
6. Dimca, A.: Sheaves in Topology. Universitext. Springer, Berlin (2004). https://doi.org/10.1007/978-3-642-18868-8
7. Franecki, J., Kapranov, M.: The Gauss map and a noncompact Riemann-Roch formula for constructible sheaves on semiabelian varieties. Duke Math. J. **104**(1), 171–180 (2000)
8. Gaffney, T., Grulha, Jr., N.G., Ruas, M.A.S.: The local Euler obstruction and topology of the stabilization of associated determinantal varieties. Preprint arXiv:1611.00749 (2017)
9. Grayson, D.R., Stillman, M.E.: Macaulay2, a software system for research in algebraic geometry. http://www.math.uiuc.edu/Macaulay2/
10. Gross, E., Drton, M., Petrović, S.: Maximum likelihood degree of variance component models. Electron. J. Stat. **6**, 993–1016 (2012)
11. Gross, E., Rodriguez, J.I.: Maximum likelihood geometry in the presence of data zeros. In: Proceedings of the 39th International Symposium on Symbolic and Algebraic Computation, ISSAC 2014, pp. 232–239. ACM, New York (2014)
12. Hauenstein, J.D., Rodriguez, J.I., Sturmfels, B.: Maximum likelihood for matrices with rank constraints. J. Algebr. Stat. **5**(1), 18–38 (2014)
13. Horobet, E., Rodriguez, J.I.: The maximum likelihood data singular locus. J. Symbolic Comput. **79**(part 1), 99–107 (2017)
14. Hoşten, S., Khetan, A., Sturmfels, B.: Solving the likelihood equations. Found. Comput. Math. **5**(4), 389–407 (2005)
15. Hoşten, S., Sullivant, S.: The algebraic complexity of maximum likelihood estimation for bivariate missing data. In: Gibilisco, P., Riccomagno, E., Rogantin, M.P., Wynn, H.P. (eds.) Algebraic and Geometric Methods in Statistics, pp. 123–134. Cambridge Books Online, Cambridge University Press, Cambridge (2009)
16. Huh, J.: The maximum likelihood degree of a very affine variety. Compos. Math. **149**(8), 1245–1266 (2013)

17. Huh, J., Sturmfels, B.: Likelihood geometry. In: Di Rocco, S., Sturmfels, B. (eds.) Combinatorial Algebraic Geometry. LNM, vol. 2108, pp. 63–117. Springer, Cham (2014). https://doi.org/10.1007/978-3-319-04870-3_3
18. Kubjas, K., Robeva, E., Sturmfels, B.: Fixed points of the EM algorithm and nonnegative rank boundaries. Ann. Statist. **43**(1), 422–461 (2015)
19. MacPherson, R.D.: Chern classes for singular algebraic varieties. Ann. Math. **2**(100), 423–432 (1974)
20. Rodriguez, J.I., Wang, B.: Computing Euler obstruction functions using maximum likelihood degrees. Arxiv:1710.04310 (2017)
21. Sommese, A.J., Wampler, II, C.W.: The Numerical Solution of Systems of Polynomials: Arising in Engineering and Science. World Scientific Publishing Co., Pte. Ltd., Hackensack (2005)
22. Uhler, C.: Geometry of maximum likelihood estimation in Gaussian graphical models. Ann. Stat. **40**(1), 238–261 (2012)
23. Zhang, X.: Local Euler Obstruction and Chern-Mather classes of Determinantal Varieties. Preprint arXiv:1706.02032 (2017)

IntegerSequences: A Package for Computing with k-Regular Sequences

Eric Rowland[(⊠)]

Department of Mathematics, Hofstra University, Hempstead, NY 11549, USA
eric.rowland@hofstra.edu

Abstract. INTEGERSEQUENCES is a *Mathematica* package for computing with integer sequences. Its support for k-regular sequences includes basic closure properties, guessing recurrences, and computing automata. Recent applications have included establishing the structure of extremal a/b-power-free words, obtaining a product formula for the generating function enumerating binomial coefficients by their p-adic valuations, and proving congruences for combinatorial sequences modulo prime powers.

Keywords: Integer sequences · Regular sequences
Automatic sequences

1 Introduction

INTEGERSEQUENCES [9] is a *Mathematica* package for identifying and computing with integer sequences from a variety of classes. It has a particular emphasis on the class of k-regular sequences, which arise widely in combinatorics, number theory, and theoretical computer science. The following code loads the package, assuming it is downloaded to one of the directories listed in $Path (the recommended location being the Applications subdirectory of $UserBaseDirectory).

```
In[1]:= << IntegerSequences`
```

A notebook version of this extended abstract containing executable code is available from the author's web site[1].

The following set of subsequences is central to the definition of a k-regular sequence.

Definition 1. *Let $k \geq 2$ be an integer. The k-kernel of a sequence $s(n)_{n \geq 0}$ is the set*

$$\{s(k^e n + i)_{n \geq 0} : e \geq 0 \text{ and } 0 \leq i \leq k^e - 1\}.$$

The k-kernel is the base-k analogue of the set of shifts $\{s(n + i)_{n \geq 0} : i \geq 0\}$. A sequence $s(n)_{n \geq 0}$ (such as the Fibonacci sequence) is *constant-recursive* if $\{s(n + i)_{n \geq 0} : i \geq 0\}$ is contained in a finite-dimensional vector space. We define k-regular (or k-constant-recursive) sequences analogously.

---

[1] https://wolfr.am/uZ4DJDth.

© Springer International Publishing AG, part of Springer Nature 2018
J. H. Davenport et al. (Eds.): ICMS 2018, LNCS 10931, pp. 414–421, 2018.
https://doi.org/10.1007/978-3-319-96418-8_49

Definition 2. *Let $k \geq 2$ be an integer. A sequence $s(n)_{n \geq 0}$ with entries in a field F is k-regular if its k-kernel is contained in a finite-dimensional F-vector space.*

For example, consider the *ruler sequence* [6, A007814]

$$0, 1, 0, 2, 0, 1, 0, 3, 0, 1, 0, 2, 0, 1, 0, 4, \ldots$$

whose nth term $s(n)$ is the exponent of 2 in the prime factorization of $n + 1$. The ruler sequence is 2-regular, since the recurrence

$$s(2n) = 0$$
$$s(4n + 1) = -s(n) + s(2n + 1) \tag{1}$$
$$s(4n + 3) = -s(n) + 2s(2n + 1)$$

establishes that the 2-kernel is contained in the $\mathbb{Q}$-vector space generated by $s(n)_{n \geq 0}$ and $s(2n + 1)_{n \geq 0}$.

The class of k-regular sequences was introduced by Allouche and Shallit [1], who established several equivalent characterizations and a number of fundamental properties. In particular, $s(n)_{n \geq 0}$ is k-regular if and only if there exists some integer $r \geq 0$ (the dimension of the associated vector space), $r \times r$ matrices $M(0), M(1), \ldots, M(k-1)$, a $1 \times r$ vector u, and an $r \times 1$ vector v such that

$$s(n) = u \, M(n_0) \, M(n_1) \cdots M(n_\ell) \, v$$

for all $n \geq 0$, where $n_\ell \cdots n_1 n_0$ is the standard base-k representation of n [1, Lemma 4.1]. For example, the ruler sequence can be represented by

$$u = \begin{bmatrix} 1 & 0 \end{bmatrix} \qquad M(0) = \begin{bmatrix} 0 & 0 \\ -1 & 1 \end{bmatrix} \qquad M(1) = \begin{bmatrix} 0 & 1 \\ -1 & 2 \end{bmatrix} \qquad v = \begin{bmatrix} 0 \\ 1 \end{bmatrix}.$$

The matrices $M(0)$ and $M(1)$ encode the recurrence (1). The vector v contains the 0th term of each generator sequence, namely $s(0) = 0$ and $s(2 \cdot 0 + 1) = s(1) = 1$. The vector u specifies which linear combination of the generators we are interested in, namely $s(n)_{n \geq 0} = 1 \cdot s(n)_{n \geq 0} + 0 \cdot s(2n + 1)_{n \geq 0}$.

INTEGERSEQUENCES uses the matrices $M(d)$ and the vectors u, v to represent a k-regular sequence. The syntax is as follows.

```
In[2]:= s = RegularSequence[{1, 0}, {{{0, 0}, {-1, 1}}, {{0, 1}, {-1, 2}}}, {0, 1}];
```

The design of `RegularSequence` parallels the built-in *Mathematica* symbol for representing a holonomic sequence, `DifferenceRoot`[2]. Passing an argument to a `RegularSequence` object computes a term of the sequence.

```
In[3]:= Table[s[n], {n, 0, 15}]
Out[3]= {0, 1, 0, 2, 0, 1, 0, 3, 0, 1, 0, 2, 0, 1, 0, 4}
```

```
In[4]:= Clear[s]
```

---

[2] http://reference.wolfram.com/language/ref/DifferenceRoot.html.

Basic closure properties for k-regular sequences established in [1] are implemented in the function `RegularSequenceReduce`, which attempts to reduce an expression to a single `RegularSequence` object. The following writes the Stern–Brocot and Thue–Morse sequences as 2-regular sequences and then computes their sum.

In[5]:= `RegularSequenceReduce[SternBrocot[n] + ThueMorse[n], n] //`
      `RegularSequenceMatrixForm`

Out[5]= $\text{RegularSequence}\left[\{1, 0, 1, 0\}, \left\{\begin{pmatrix} 1 & 0 & 0 & 0 \\ 0 & 1 & 0 & 0 \\ 0 & 0 & 1 & 0 \\ 0 & 0 & 1 & 1 \end{pmatrix}, \begin{pmatrix} 0 & 1 & 0 & 0 \\ 1 & 0 & 0 & 0 \\ 0 & 0 & 0 & 1 \\ 0 & 0 & -1 & 2 \end{pmatrix}\right\}, \{0, 1, 0, 1\}, 2\right][n]$

2   Guessing a k-Regular Sequence

Given the first N terms of a sequence, one is frequently interested in guessing a general form for the sequence. A procedure for guessing k-regular sequences was described by Shallit [14]. The implementation in INTEGERSEQUENCES works by maintaining a set B of generators, a set R of relations, and a set of k-kernel sequences S which have not yet been written as a linear combination of elements of B. Initialize $B = \{\}$, $R = \{\}$, and $S = \{s(n)_{n\geq 0}\}$. While $S \neq \{\}$, remove a sequence $t(n)_{n\geq 0}$ from S and determine, using the known terms, whether it is a linear combination of elements of B; if it is, add the linear relation to R; if it is not, add $t(n)_{n\geq 0}$ to B as a new generator and add its k subdivisions $t(kn + 0)_{n\geq 0}, \ldots, t(kn + (k-1))_{n\geq 0}$ to S. When S becomes empty, we have determined a conjectural basis B such that every element of the k-kernel can be written as a linear combination of the elements of B. The set of relations R, along with the initial term of each element of B, uniquely determines a sequence that agrees with $s(n)_{n\geq 0}$ on the known terms.

Since only finitely many terms of $s(n)_{n\geq 0}$ are known, it is possible that as we consider additional sequences from the k-kernel we will exhaust the known terms. If elements of B which were previously known to be linearly independent become linearly dependent due to truncating terms, then we do not have enough terms to confidently guess a recurrence.

This algorithm is implemented in `FindRegularSequenceFunction`. To our knowledge, this is the only publicly available guesser for k-regular sequences. The first argument is the list of terms, and the second argument is k. The following guesses a 2-regular representation for the number of 1s in the binary representation of n.

In[6]:= `terms = Table[DigitCount[n, 2, 1], {n, 0, 100}];`
      `FindRegularSequenceFunction[terms, 2] // RegularSequenceMatrixForm`

Out[7]= $\text{RegularSequence}\left[\{1, 0\}, \left\{\begin{pmatrix} 1 & 0 \\ 0 & 1 \end{pmatrix}, \begin{pmatrix} 0 & 1 \\ -1 & 2 \end{pmatrix}\right\}, \{0, 1\}, 2\right]$

A variant, `FindRegularSequenceRecurrence`, uses the same algorithm but outputs a recurrence rather than a `RegularSequence` object.

```
In[8]:= Column[FindRegularSequenceRecurrence[terms, 2, s[n]]]

        s[2 n] == s[n]
        s[1 + 4 n] == s[1 + 2 n]
Out[8]= s[3 + 4 n] == -s[n] + 2 s[1 + 2 n]
        s[0] == 0
        s[1] == 1
```

More generally, `FindRegularSequenceFunction` supports guessing multidimensional $(k_1, \ldots, k_d)$-regular sequences $s(n_1, \ldots, n_d)_{n_1 \geq 0, \ldots, n_d \geq 0}$. Let $\nu_p(n)$ be the p-adic valuation of n, that is, the exponent of the highest power of p dividing n. The 2-dimensional sequence consisting of the 2-adic valuations of $\binom{n}{m}$ (where we treat ∞ as a formal symbol) is 2-regular:

```
In[9]:= array = Table[IntegerExponent[Binomial[n, m], 2], {n, 0, 100}, {m, 0, 100}];
        FindRegularSequenceFunction[array /. ∞ → infinity, 2] //
        RegularSequenceMatrixForm
```

$$\text{Out[10]= RegularSequence}\Big[\{1, 0, 0, 0\},$$

$$\left\{\left\{\begin{pmatrix} 1 & 0 & 0 & 0 \\ 0 & 0 & 1 & 0 \\ 0 & -1 & 2 & 0 \\ 0 & 0 & 0 & 1 \end{pmatrix}, \begin{pmatrix} 0 & 1 & 0 & 0 \\ 0 & 0 & 1 & 0 \\ 0 & -1 & 2 & 0 \\ 0 & 0 & 1 & 0 \end{pmatrix}\right\}, \left\{\begin{pmatrix} 1 & 0 & 0 & 0 \\ 0 & 0 & 0 & 1 \\ -1 & 0 & 0 & 2 \\ 0 & 0 & 0 & 1 \end{pmatrix}, \begin{pmatrix} 1 & 0 & 0 & 0 \\ 0 & 0 & 1 & 0 \\ 0 & -1 & 2 & 0 \\ 0 & 0 & 0 & 1 \end{pmatrix}\right\}\right\},$$

$$\{0, \text{infinity}, \text{infinity}, 1\}, \{2, 2\}\Big]$$

We mention several applications where `FindRegularSequenceFunction` has been used to guess a sequence or family of sequences.

For a rational number $\frac{a}{b} > 1$, a word w is an $\frac{a}{b}$-*power* if it can be written $v^e x$ where e is a non-negative integer, x is a prefix of v, and $\frac{|w|}{|v|} = \frac{a}{b}$. For example, 011101 is the $\frac{3}{2}$-power $(0111)^{3/2}$ with $v = 0111$ and $x = 01$. The lexicographically least $\frac{3}{2}$-power-free infinite word on the non-negative integers [6, A269518] is

$$00110210011200110310011300110210 \cdots.$$

It is difficult *a priori* to guess whether the sequence of letters in such a word is k-regular and, if so, to guess the correct value of k. However, through experimentation, `FindRegularSequenceFunction` revealed that the letters in the lexicographically least $\frac{3}{2}$-power-free word form a 6-regular sequence [11].

```
In[11]:= FindRegularSequenceFunction[
         {0, 0, 1, 1, 0, 2, 1, 0, 0, 1, 1, 2, 0, 0, 1, 1, 0, 3, 1, 0, 0, 1, 1,
          3, 0, 0, 1, 1, 0, 2, 1, 0, 0, 1, 1, 4, 0, 0, 1, 1, 0, 3, 1, 0, 0, 1,
          1, 2, 0, 0, 1, 1, 0, 2, 1, 0, 0, 1, 1, 3, 0, 0, 1, 1, 0, 3, 1, 0, 0, 1}, 6] //
         RegularSequenceMatrixForm
```

$$\text{Out[11]= RegularSequence}\Big[\{1, 0, 0\}, \left\{\begin{pmatrix} 0 & 1 & 0 \\ 0 & 0 & 0 \\ 0 & 1 & 1 \end{pmatrix}, \begin{pmatrix} 0 & 0 & 0 \\ 0 & 1 & 1 \\ 0 & 0 & 0 \end{pmatrix},\right.$$

$$\left.\begin{pmatrix} 0 & 0 & 1 \\ 0 & 0 & 0 \\ 0 & 1 & 1 \end{pmatrix}, \begin{pmatrix} 0 & 1 & 1 \\ 0 & 1 & 1 \\ 0 & 0 & 0 \end{pmatrix}, \begin{pmatrix} 0 & 1 & 0 \\ 0 & 0 & 0 \\ 0 & 1 & 1 \end{pmatrix}, \begin{pmatrix} 1 & 2 & 2 \\ 0 & 1 & 1 \\ 0 & 0 & 0 \end{pmatrix}\right\}, \{0, 0, 1\}, 6\Big]$$

This discovery led to a large systematic study of the value of k for which the lexicographically least $\frac{a}{b}$-power-free word is k-regular, although it is an open question whether a k always exists [8].

Enumeration questions in combinatorics on words often turn out to have answers given by k-regular sequences for appropriate values of k. An explanation of this phenomenon in many cases was given by Charlier, Rampersad, and Shallit [2]. However, k-regular sequences also appear in enumeration questions not covered by their framework. For example, the ℓ-abelian complexity for many infinite words appears to be k-regular. The ℓ-*abelian complexity* of an infinite word counts factors up to ℓ-abelian equivalence — that is, two factors x and y are considered the same if $|x|_v = |y|_v$ for each word v of length $\leq \ell$. The 2-regularity of the 2-abelian complexities of two well-known words, the Thue–Morse word and the period-doubling word, were established by proving the 2-regularity of sequences satisfying a general reflection recurrence [7]. The 2-regular recurrences for such sequences were guessed by `FindRegularSequenceFunction`.

The intended use case of `FindRegularSequenceFunction` is a sequence of integers. However, the code is sufficiently general to support sequences of polynomials. Again, let $\nu_p(n)$ denote the p-adic valuation of n. Spiegelhofer and Wallner [15] considered the generating function counting binomial coefficients by their p-adic valuations $\nu_p(\binom{n}{m})$. For each prime p, `FindRegularSequenceFunction` is able to guess a p-regular recurrence for this generating function. For $p = 2$ we obtain the following. Note that the matrix entries are now polynomials in the formal variable x.

```
In[12]:= polynomials = Table[Sum[x^IntegerExponent[Binomial[n,m],2], {m, 0, n}], {n, 0, 100}];

        FindRegularSequenceFunction[polynomials, 2] // RegularSequenceMatrixForm

Out[13]= RegularSequence[{1, 0}, {({{0, 1}, {-2 x, 1 + 2 x}}), ({{2, 0}, {2, x}})}, {1, 1}, 2]
```

The basis chosen by `FindRegularSequenceFunction` may not be the most natural basis, since it (a) necessarily consists of k-kernel elements and (b) depends on the order in which the k-kernel is traversed. By performing a suitable change of basis for each prime p, the author conjectured the following, which can be proved by a bijective argument [10].

Theorem 1. *Let p be a prime. For each $d \in \{0, 1, \ldots, p-1\}$, let $M_p(d)$ be the 2×2 matrix*

$$M_p(d) = \begin{bmatrix} d+1 & p-d-1 \\ d\,x & (p-d)\,x \end{bmatrix}.$$

Let $n \geq 0$, and let $n_\ell \cdots n_1 n_0$ be the standard base-p representation of n. Then

$$\sum_{m=0}^{n} x^{\nu_p(\binom{n}{m})} = \begin{bmatrix} 1 & 0 \end{bmatrix} M_p(n_0)\, M_p(n_1) \cdots M_p(n_\ell) \begin{bmatrix} 1 \\ 0 \end{bmatrix}.$$

This theorem generalizes a well-known result of Fine [5] on the number of binomial coefficients not divisible by p. More generally, the analogous sequence of generating functions enumerating multinomial coefficients by their p-adic valuations is p-regular [10].

3   Computing Automata for Sequences Modulo p^α

A k-regular sequence whose terms take on finitely many distinct values is called k-*automatic*. This name derives from the characterization of k-automatic sequences as sequences whose nth term is the output of an automaton when fed the base-k digits of n.

Many integer sequences that arise in combinatorics have the property that reducing each term modulo p^α produces a p-automatic sequence. For algebraic sequences modulo p, this is explained by Christol's theorem, which states that a sequence over a finite field of characteristic p is p-automatic if and only if its generating function is algebraic [3]. Therefore, if $\sum_{n \geq 0} s(n)x^n \in \mathbb{Z}[\![x]\!]$ is algebraic (as it is for the Catalan numbers, for example), then $\sum_{n \geq 0} (s(n) \bmod p)x^n \in \mathbb{F}_p[\![x]\!]$ is algebraic, so $(s(n) \bmod p)_{n \geq 0}$ is p-automatic. In INTEGERSEQUENCES this is implemented in `AutomaticSequenceReduce`. The following computes a 3-automatic sequence, represented by an automaton, for the nth Catalan number modulo 3.

In[14]:= `AutomaticSequenceReduce[Mod[CatalanNumber[n], 3], n]`

Out[14]= AutomaticSequence[
    Automaton[{{1→2, 0}, {1→2, 1}, {1→3, 2}, {2→2, 0}, {2→4, 1},
      {2→5, 2}, {3→4, 0}, {3→5, 1}, {3→3, 2}, {4→4, 0}, {4→2, 1},
      {4→5, 2}, {5→5, 0}, {5→5, 1}, {5→5, 2}}, 1,
      {1→1, 2→1, 3→2, 4→2, 5→0}, InputAlphabet→{0, 1, 2}]][n]

The function `AutomatonGraph` produces the `Graph` object corresponding to an automaton.

In[15]:= `AutomatonGraph[AutomaticSequenceAutomaton[%]]`

Out[15]=

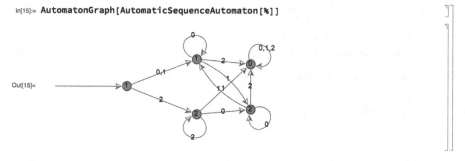

More generally, if $s(n)_{n \geq 0}$ is the diagonal of the power series of a multivariate rational expression $\frac{f}{g}$ whose denominator's constant term $g(0, \ldots, 0)$ is nonzero modulo p, then $(s(n) \bmod p^\alpha)_{n \geq 0}$ is p-automatic. An automaton for this sequence can be computed by embedding the p-kernel into the space of rational expressions with a certain fixed denominator [12]. The diagonal of a rational power series is represented in INTEGERSEQUENCES by `DiagonalSequence`.

In[16]:= `AutomaticSequenceReduce[Mod[DiagonalSequence[` $\frac{1-x}{1-(1+x)^2\,y}$ `, {x, y}][n], 4],`

   `n]`

   `AutomatonGraph[AutomaticSequenceAutomaton[%]]`

Out[16]= `AutomaticSequence[Automaton[{{1 → 2, 0}, {1 → 1, 1}, {2 → 2, 0},`
    `{2 → 3, 1}, {3 → 3, 0}, {3 → 4, 1}, {4 → 4, 0}, {4 → 4, 1}},`
    `1, {1 → 1, 2 → 1, 3 → 2, 4 → 0}, InputAlphabet → {0, 1}]][n]`

Out[17]=

Computing automata for sequences modulo p^α provides routine proofs of many congruences that were established in the literature by nontrivial case analyses. For example, Eu et al. [4] proved that no Motzkin number is divisible by 8. The following computation proves this in less than a second. The resulting automaton has 24 states.

In[18]:= `automaton = AutomaticSequenceAutomaton[`
       `AutomaticSequenceReduce[Mod[MotzkinNumber[n], 8], n]];`
   `AutomatonStateCount[automaton]`
   `Sort[AutomatonOutputAlphabet[automaton]]`

Out[19]= `24`

Out[20]= `{1, 2, 3, 4, 5, 6, 7}`

Closely related to diagonal sequences are *constant-term sequences*. Let f and g be (possibly multivariate) Laurent polynomials, and let $s(n)$ be the constant term of $f^n g$. An automaton for $(s(n) \bmod p^\alpha)_{n\geq 0}$ can be computed similarly [13]. `AutomaticSequenceReduce` also implements this algorithm. Constant-term sequences are represented by `ConstantTermSequence`, where the first argument is f and the second argument is g.

In[21]:= `automaton = AutomaticSequenceAutomaton[`
       `AutomaticSequenceReduce[`
       `Mod[ConstantTermSequence[1 +` $\frac{1}{x}$ `+ x, 1 - x², x][n], 25], n]];`
   `AutomatonStateCount[automaton]`
   `Complement[Range[0, 24], AutomatonOutputAlphabet[automaton]]`

Out[22]= `136`

Out[23]= `{0}`

In fact that constant-term sequence is the sequence of Motzkin numbers, so we have established that no Motzkin number is divisible by 25.

For many sequences, including the sequences of Catalan and Motzkin numbers, the constant-term representation is preferable to the diagonal representation since it uses polynomials in a single variable, whereas the diagonal representation requires at least two variables.

References

1. Allouche, J.-P., Shallit, J.: The ring of k-regular sequences. Theoret. Comput. Sci. **98**, 163–197 (1992)
2. Charlier, É., Rampersad, N., Shallit, J.: Enumeration and decidable properties of automatic sequences. Int. J. Found. Comput. Sci. **23**, 1035–1066 (2012)
3. Christol, G., Kamae, T., Mendès France, M., Rauzy, G.: Suites algébriques, automates et substitutions. Bulletin de la Société Mathématique de France **108**, 401–419 (1980)
4. Eu, S.-P., Liu, S.-C., Yeh, Y.-N.: Catalan and Motzkin numbers modulo 4 and 8. Eur. J. Comb. **29**, 1449–1466 (2008)
5. Fine, N.: Binomial coefficients modulo a prime. Amer. Math. Monthly **54**, 589–592 (1947)
6. The On-Line Encyclopedia of Integer Sequences. http://oeis.org
7. Parreau, A., Rigo, M., Rowland, E., Vandomme, É.: A new approach to the 2-regularity of the ℓ-abelian complexity of 2-automatic sequences. Electron. J. Comb. **22** (2015) #P1.27
8. Pudwell, L., Rowland, E.: Avoiding fractional powers over the natural numbers. Electron. J. Comb. **25** (2018) #P2.27
9. Rowland, E.: IntegerSequences. https://people.hofstra.edu/Eric_Rowland/packages.html#IntegerSequences
10. Rowland, E.: A matrix generalization of a theorem of Fine. Integers **18A**, A18 (2018)
11. Rowland, E., Shallit, J.: Avoiding 3/2-powers over the natural numbers. Discrete Math. **312**, 1282–1288 (2012)
12. Rowland, E., Yassawi, R.: Automatic congruences for diagonals of rational functions. Journal de Théorie des Nombres de Bordeaux **27**, 245–288 (2015)
13. Rowland, E., Zeilberger, D.: A case study in meta-automation: automatic generation of congruence automata for combinatorial sequences. J. Differ. Equ. Appl. **20**, 973–988 (2014)
14. Shallit, J.: Remarks on inferring integer sequences. https://cs.uwaterloo.ca/shallit/Talks/infer.ps
15. Spiegelhofer, L., Wallner, M.: An explicit generating function arising in counting binomial coefficients divisible by powers of primes. Acta Arithmetica **181**, 27–55 (2017)

A User-Friendly Hybrid
Sparse Matrix Class in C++

Conrad Sanderson[1,3,4]($\boxtimes$) and Ryan Curtin[2,4]

[1] Data61, CSIRO, Brisbane, Australia
conrad.sanderson@data61.csiro.au
[2] Symantec Corporation, Atlanta, USA
ryan@ratml.org
[3] University of Queensland, Brisbane, Australia
[4] Arroyo Consortium, San Francisco, USA

Abstract. When implementing functionality which requires sparse matrices, there are numerous storage formats to choose from, each with advantages and disadvantages. To achieve good performance, several formats may need to be used in one program, requiring explicit selection and conversion between the formats. This can be both tedious and error-prone, especially for non-expert users. Motivated by this issue, we present a user-friendly sparse matrix class for the C++ language, with a high-level application programming interface deliberately similar to the widely used MATLAB language. The class internally uses two main approaches to achieve efficient execution: (i) a hybrid storage framework, which automatically and seamlessly switches between three underlying storage formats (compressed sparse column, coordinate list, Red-Black tree) depending on which format is best suited for specific operations, and (ii) template-based meta-programming to automatically detect and optimise execution of common expression patterns. To facilitate relatively quick conversion of research code into production environments, the class and its associated functions provide a suite of essential sparse linear algebra functionality (eg., arithmetic operations, submatrix manipulation) as well as high-level functions for sparse eigendecompositions and linear equation solvers. The latter are achieved by providing easy-to-use abstractions of the low-level ARPACK and SuperLU libraries. The source code is open and provided under the permissive Apache 2.0 license, allowing unencumbered use in commercial products.

Keywords: Numerical linear algebra · Sparse matrix · C++ language

1 Introduction

Modern scientific computing often requires working with data so large it cannot fully fit in working memory. In many cases, the data can be represented as sparse, allowing users to work with matrices of extreme size with few nonzero elements.

© Springer International Publishing AG, part of Springer Nature 2018
J. H. Davenport et al. (Eds.): ICMS 2018, LNCS 10931, pp. 422–430, 2018.
https://doi.org/10.1007/978-3-319-96418-8_50

However, converting code from using dense matrices to using sparse matrices is not always straightforward.

Existing open-source frameworks may provide several separate sparse matrix classes, each with their own data storage format. For instance, SciPy [10] has 7 sparse matrix classes: `bsr_matrix`, `coo_matrix`, `csc_matrix`, `csr_matrix`, `dia_matrix`, `dok_matrix`, and `lil_matrix`. Each storage format is best suited for efficient execution of a specific set of operations (eg., matrix multiplication vs. incremental matrix construction). Other frameworks may provide only one sparse matrix class, with severe runtime penalties if it is not used in the right way. This can be challenging and bewildering for users who simply want to create and use sparse matrices, and do not have the expertise (or desire) to understand the advantages and disadvantages of each format. To achieve good performance, several formats may need to be used in one program, requiring explicit selection and conversion between the formats. This plurality of sparse matrix classes complicates the programming task, increases the likelihood of bugs, and adds to the maintenance burden.

Motivated by the above issues, we present a user-friendly sparse matrix class for the C++ language, with a high-level application programming interface (function syntax) that is deliberately similar to MATLAB. The sparse matrix class uses a hybrid storage framework, which *automatically* and *seamlessly* switches between three data storage formats, depending on which format is best suited for specific operations: (i) Compressed Sparse Column (CSC), used for efficient fundamental arithmetic operations such as matrix multiplication and addition, as well as efficient reading of individual elements; (ii) Co-Ordinate List (COO), used for facilitating operations involving bulk coordinate transformations; (iii) Red-Black Tree (RBT), used for both robust and efficient incremental construction of sparse matrices (i.e., construction via setting individual elements one-by-one, not necessarily in order). To further promote efficient execution, the class exploits C++ features such as template meta-programming to provide a compile-time expression evaluator, which can automatically detect and optimise common mathematical expression patterns.

The sparse matrix class provides an intuitive interface that is very close to a typical dense matrix API; this can help with rapid transition of dense-specific code to sparse-specific code. In addition, we demonstrate that the overhead of the hybrid format is minimal, and that the format is able to choose the optimal representation for a variety of sparse linear algebra tasks. This makes the format and implementation suitable for real-world prototyping and production usage.

Although there are many other sparse matrix implementations in existence, to our knowledge ours is the first to offer a unified interface with automatic format switching under the hood. Most toolkits are limited to either a single format or multiple formats the user must manually convert between. As mentioned earlier, SciPy contains no fewer than seven formats, and the comprehensive SPARSKIT package [12] contains 16. In these toolkits the user must manually convert between formats. On the other hand, both MATLAB and GNU Octave [5] contain sparse matrix implementations, but they supply only the CSC format, meaning that users must write their code in special ways to ensure its efficiency [9].

The source code for the sparse matrix class and its associated functions is included in recent releases of the cross-platform and open-source Armadillo library [13], available from http://arma.sourceforge.net. The code is provided under the permissive Apache 2.0 license [11], allowing unencumbered use in commercial products.

We continue the paper as follows. In Sect. 2 we overview the functionality provided by the sparse matrix class and its associated functions. In Sect. 3 we briefly describe the underlying storage formats used by the class, and the tasks that each of the formats is best suited for. Section 4 provides an empirical evaluation showing the performance of the hybrid storage framework in relation to the underlying storage formats. The salient points and avenues for further exploration are summarised in Sect. 5.

2   Functionality

To allow prototyping directly in C++ as well as to facilitate relatively quick conversion of research code into production environments, the sparse matrix class and its associated functions provide a user-friendly suite of essential sparse linear algebra functionality, including fundamental operations such as addition, matrix multiplication and submatrix manipulation. Various sparse eigendecompositions and linear equation solvers are also provided. C++ language features such as overloading of operators (eg., $*$ and $+$) [14] are exploited to allow mathematical operations with matrices to be expressed in a concise and easy-to-read manner. For instance, given sparse matrices A, B, and C, a mathematical expression such as

$$D = \frac{1}{2}(A + B) \cdot C^T$$

can be written directly in C++ as

$$\texttt{sp\_mat D = 0.5} * (A + B) * \texttt{C.t()};$$

Low-level details such as memory management are hidden, allowing the user to concentrate effort on mathematical details. Table 1 lists a subset of the available functionality for the sparse matrix class, sp_mat.

The sparse matrix class uses a delayed evaluation approach, allowing several operations to be combined to reduce the amount of computation and/or temporary objects. In contrast to brute-force evaluations, delayed evaluation can provide considerable performance improvements as well as reduced memory usage. The delayed evaluation machinery is accomplished through template meta-programming [15], where a type-based signature of a set of consecutive mathematical operations is automatically constructed. The C++ compiler is then induced to detect common expression subpatterns at compile time, and selects the corresponding optimised implementations. For example, in the expression trace(A.t() * B), the explicit transpose and time-consuming matrix multiplication are omitted; only the diagonal elements of A.t() * B are accumulated.

Sparse eigendecompositions and linear equation solutions are accomplished through integration with low-level routines in the de facto standard ARPACK [7] and SuperLU libraries [8]. The resultant high-level functions automatically take care of the cumbersome and error-prone low-level management required with these libraries.

3 Underlying Sparse Storage Formats

The three underlying storage formats (CSC, COO, RBT) were chosen so that the sparse matrix class can achieve overall efficient execution of the following five main use cases: (i) incremental construction of sparse matrices via quasi-ordered insertion of elements, where each new element is inserted at a location that is past all the previous elements according to column-major ordering; (ii) flexible ad-hoc construction or element-wise modification of sparse matrices via unordered insertion of elements, where each new element is inserted at a random location; (iii) operations involving bulk coordinate transformations; (iv) multiplication of dense vectors with sparse matrices; (v) multiplication of two sparse matrices.

Table 1. Selected functionality of the sparse matrix class, with brief descriptions. See http://arma.sourceforge.net/docs.html#SpMat for more detailed documentation. Several optional additional arguments have been omitted for brevity.

Function	Description
sp_mat X(100,200)	Declare sparse matrix with 100 rows and 200 columns
sp_cx_mat X(100,200)	As above, but use complex elements
X(1,2) = 3	Assign value 3 to element at location (1,2) of matrix X
X = 4.56 * A	Multiply matrix A by scalar
X = A + B	Add matrices A and B
X = A * B	Multiply matrices A and B
X = kron(A, B)	Kronecker tensor product of matrices A and B
X(span(1,2), span(3,4))	Provide read/write access to submatrix of X
X.diag(k)	Provide read/write access to diagonal k of X
X.print()	Print matrix X to terminal
X.save(filename, format)	Store matrix X as a file
speye(rows, cols)	Generate sparse matrix with values on diagonal set to one
sprandu(rows, cols, density)	Generate sparse matrix with random non-zero elements
sum(X, dim)	Sum of elements in each column ($dim = 0$) or row ($dim = 1$)
min(X, dim); max(X, dim)	Obtain extremum value in each col. ($dim = 0$) or row ($dim = 1$)
X.t() or trans(X)	Return transpose of matrix X
repmat(X, rows, cols)	Replicate matrix X in block-like fashion
norm(X, p)	Compute p-norm of vector or matrix X
normalise(X, p, dim)	Normalise each col. ($dim = 0$) or row ($dim = 1$) to unit p-norm
trace(A.t() * B)	Compute trace **omitting** explicit transpose and multiplication
eigs_gen(eigval, eigvec, X, k)	Compute k largest eigenvalues and eigenvectors of matrix X
svds(U, s, V, X, k)	Compute k singular values and singular vectors of matrix X
X = spsolve(A, b)	Solve sparse system $Ax = b$ for x

Below we briefly describe each storage format and its limitations. We use N to indicate the number of non-zero elements of the matrix, while n_rows and n_cols indicate the number of rows and columns, respectively.

3.1   Compressed Sparse Column

In the CSC format [12], three arrays are used: (i) the *values* array, which is a contiguous array of N floating point numbers holding the non-zero elements, (ii) the *row indices* array, which is a contiguous array of N integers holding the corresponding row indices (i.e., the n-th entry contains the row of the n-th element), and (iii) the *column offsets* array, which is a contiguous array of n_cols $+1$ integers holding offsets to the *values array*, with each offset indicating the start of elements belonging to each column. Let us denote the i-th entry in the column offsets array as $c[i]$, the j-th entry in the row indices array as $r[j]$, and the n-th entry in the values array as $v[n]$. All arrays use zero-based indexing, i.e., the initial position in each array is denoted by 0. Then, $v[c[i]]$ is the first element in column i, and $r[c[i]]$ is the corresponding row of the element. The number of elements in column i is determined using $c[i+1] - c[i]$, where, by definition, $c[0]$ is always 0 and $c[\text{n\_cols}]$ is equal to N.

The CSC format is well-suited for sparse linear algebra operations such as summation and vector-matrix multiplication. It is also suited for operations that do not change the structure of the matrix, such as element-wise operations on the nonzero elements. The format also affords relatively efficient random element access; to locate an element (or determine that it is not stored), a single lookup to the beginning of the desired column can be performed, followed by a binary search to find the element.

The main disadvantage of CSC is the effort required to insert a new element. In the worst-case scenario, memory for three new larger-sized arrays (containing the values and locations) must first be allocated, the position of the new element determined within the arrays, data from the old arrays copied to the new arrays, data for the new element placed in the new arrays, and finally the memory used by the old arrays deallocated. As the number of elements in the matrix grows, the entire process becomes slower.

There are opportunities for some optimisation, such as using oversized storage to reduce memory allocations, where a new element past all the previous elements can be readily inserted. It is also possible to perform batch insertions with some speedup by first sorting all the elements to be inserted and then merging with the existing data arrays. While the above approaches can be effective, they require the user to explicitly deal with low-level storage details instead of focusing on high-level functionality.

The CSC format was chosen over the related Compressed Sparse Row (CSR) format [12] for two main reasons: (i) to ensure compatibility with external libraries such as the SuperLU solver [8], and (ii) to ensure consistency with the surrounding infrastructure provided by the Armadillo library, which uses column-major dense matrix representation for compatibility with LAPACK [1].

3.2   Coordinate List Representation

The Coordinate List (COO) is a general concept where a list $L = (l_1, l_2, \cdots, l_N)$ of 3-tuples represents the non-zero elements in a matrix. Each 3-tuple contains the location indices and value of the element, i.e., $l = (\texttt{row}, \texttt{column}, \texttt{value})$. The format does not prescribe any ordering of the elements, and a linked list [2] can be used to represent L. However, in a computational implementation geared towards linear algebra operations [12], L is often represented as a set of three arrays: (i) the *values* array, which is a contiguous array of N floating point numbers holding the non-zero elements of the matrix, and the (ii) *rows* and (iii) *columns* arrays, which are contiguous arrays of N integers, holding the row and column indices of the corresponding values.

The array-based representation of COO is related to CSC, with the main difference that for each element the column indices are explicitly stored. As such, the COO format contains redundancy and is hence less efficient than CSC for representing sparse matrices. However, in the COO format the coordinates of all elements can be directly read and modified in a batch manner, which facilitates specialised/niche operations that involve bulk transformation of matrix coordinates (eg., circular shifts). In the CSC format such operations are more time-consuming and/or more difficult to implement, as the compressed structure must be taken into account. The general disadvantages of the array-based representation of COO are similar as for the CSC format, in that element insertion is typically a slow process.

3.3   Red-Black Tree

To address the problems with element insertion at arbitrary locations, we first represent each element as a 2-tuple, $l = (\texttt{index}, \texttt{value})$, where $\texttt{index}$ encodes the location of the element as $\texttt{index} = \texttt{row} + \texttt{column} \times \texttt{n\_rows}$. This encoding implicitly assumes column-major ordering of the elements. Secondly, rather than using a linked list or an array based representation, the list of the tuples is stored as a Red-Black Tree (RBT), a self-balancing binary search tree [2].

Briefly, an RBT is a collection of nodes, with each node containing the 2-tuple described above and links to two children nodes. There are two constraints: (i) each link points to a unique child node and (ii) there are no links to the root node. The ordering of the nodes and height of the tree is explicitly controlled so that searching for a specific index (i.e., retrieving an element at a specific location) has worst-case complexity of $\mathcal{O}(\log N)$. Insertion and removal of nodes (i.e., matrix elements), also has the worst-case complexity of $\mathcal{O}(\log N)$. If a node to be inserted is known to have the largest index so far (eg., during incremental matrix construction), the search for where to place the node can be omitted, thereby speeding up the insertion process close to $\mathcal{O}(1)$ complexity.

Traversing the tree in an ordered fashion (from the smallest to largest index) is equivalent to reading the elements in column-major ordering. This in turn allows the quick conversion of matrix data stored in RBT format into CSC format. Each element's location is simply decoded via $\texttt{row} = \texttt{index}$ mod $\texttt{n\_rows}$

and column $= \lfloor$index/n_rows$\rfloor$, with the operations accomplished via direct integer arithmetic on CPUs.

In our hybrid format, the RBT format is used for incremental construction of sparse matrices, either in an ordered or unordered fashion, and a subset of elementwise operations. This in turn enables users to construct sparse matrices in the same way they might construct dense matrices—for instance, a loop over elements to be inserted without regard to storage format.

4   Automatically Switching Between Storage Formats

To avoid the problems associated with selection and manual conversion between formats, our sparse matrix class uses a hybrid storage framework that *automatically* and *seamlessly* switches between the data storage formats described in Sect. 3.

By default, matrix elements are stored in CSC format. When required, data in CSC format is internally converted to either the RBT or COO format, on which an operation or set of operations is performed. The matrix is automatically converted ('synced') back to the CSC format the next time an operation requiring the CSC format is performed.

The actual underlying storage details and conversion operations are completely hidden from the user, who may not necessarily be knowledgeable about (or care to learn about) sparse matrix storage formats. This allows for simplified code, which in turn increases readability and lowers maintenance. In contrast, other toolkits without automatic format conversion can cause either slow execution (as a non-optimal storage format might be used), or require many manual conversions. As an example, Fig. 1 shows a short Python program using the SciPy toolkit and a corresponding C++ program using the sparse matrix class. Manually initiated format conversions are required for efficient execution in the SciPy version; this causes both development time and code size to increase.

To empirically demonstrate the usefulness of the hybrid storage framework we have performed several experiments: (i) quasi-ordered element insertion, i.e., incremental construction, (ii) unordered (random) insertion, and (iii) matrix multiplication. In all cases the sparse matrices have a size of 10,000 × 10,000, with four settings for the density of non-zero elements: 0.01%, 0.1%, 1%, 10%.

Figure 2(a) shows the time taken for unordered element insertion done directly using the underlying storage formats (i.e., CSC, COO, RBT, as per Sect. 3), as well as the hybrid approach which uses RBT followed by conversion to CSC. The CSC and COO formats use oversized storage as a form of optimisation. The RBT format is the quickest, generally by one or two orders of magnitude, with the conversion from RBT to CSC adding negligible overhead. The results for quasi-ordered insertion (not shown) follow a similar pattern.

Figure 2(b) shows the time taken to multiply two sparse matrices in either CSC or RBT format, with the matrix elements already stored in each format. The COO format was omitted due to its similarity with CSC. The hybrid storage format automatically uses CSC for matrix multiplication, which is faster than RBT by about two orders of magnitude.

```
X = scipy.sparse.rand(1000, 1000, 0.01)    |    sp_mat X = sprandu(1000, 1000, 0.01);

# manually convert to LIL format           |    // automatic conversion to RBT format
# to allow insertion of elements           |    // for fast insertion of elements
X = X.tolil()                              |
X[1,1]  = 1.23                             |    X(1,1)  = 1.23;
X[3,4] += 4.56                             |    X(3,4) += 4.56;

# random dense vector                       |    // random dense vector
V = numpy.random.rand((1000))              |    rowvec V(1000, fill::randu);

# manually convert X to CSC format          |    // automatic conversion of X to CSC
# for efficient multiplication              |    // prior to multiplication
X = X.tocsc()                              |
W = V * X                                  |    rowvec W = V * X;
```

Fig. 1. Left panel: a Python program using the SciPy toolkit, requiring explicit conversions between sparse format types to achieve efficient execution; if an unsuitable sparse format is used for a given operation, SciPy will emit *TypeError* or *SparseEfficiencyWarning*. Right panel: A corresponding C++ program using the sparse matrix class, with the format conversions automatically done by the class.

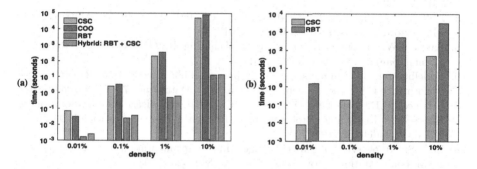

Fig. 2. Time taken to **(a)** insert elements at random locations into a sparse matrix to achieve various densities of non-zero elements, and **(b)** multiply two sparse matrices with elements at random locations and various densities. In both cases the sparse matrices have a size of $10{,}000 \times 10{,}000$.

5   Conclusion

Motivated by a lack of easy-to-use tools for sparse matrix development, we have proposed and implemented a sparse matrix class in C++ that internally uses a hybrid format. The hybrid format automatically converts between good representations for specific functionality, allowing the user to write sparse linear algebra without requiring to consider the underlying storage format. Internally, the hybrid format uses the CSC (compressed sparse column), COO (coordinate list), and RBT (red-black tree) formats. In addition, template meta-programming is used to optimise common expression patterns. We have made our implementation available as part of the open-source Armadillo C++ library [13].

The class has already been successfully used in open-source projects such as MLPACK, a C++ library for machine learning and pattern recognition [3]. It is used there to allow machine learning algorithms to be run on either sparse or dense datasets. Furthermore, bindings are provided to the R environment via RcppArmadillo [6].

Future avenues for exploration include integrating more specialised matrix formats in order to automatically speed up specific operations. For example, the Skyline formats [4] are useful for Cholesky factorisation and related operations.

References

1. Anderson, E., Bai, Z., Bischof, C., Blackford, S., Demmel, J., Dongarra, J., Du Croz, J., Greenbaum, A., Hammarling, S., et al.: LAPACK Users' Guide. SIAM, Philadelphia (1999)
2. Cormen, T.H., Leiserson, C.E., Rivest, R.L., Stein, C.: Introduction to Algorithms, 3rd edn. MIT Press, Cambridge (2009)
3. Curtin, R., Cline, J., Slagle, N., March, W., Ram, P., Mehta, N., Gray, A.: MLPACK: a scalable C++ machine learning library. J. Mach. Learn. Res. **14**, 801–805 (2013)
4. Duff, I.S., Erisman, A.M., Reid, J.K.: Direct Methods for Sparse Matrices, 2nd edn. Oxford University Press, Oxford (2017)
5. Eaton, J.W., Bateman, D., Hauberg, S., Wehbring, R.: GNU Octave 4.2 Reference Manual. Samurai Media Limited (2017)
6. Eddelbuettel, D., Sanderson, C.: RcppArmadillo: accelerating R with high-performance C++ linear algebra. Comput. Stat. Data Anal. **71**, 1054–1063 (2014)
7. Lehoucq, R.B., Sorensen, D.C., Yang, C.: ARPACK Users' Guide: Solution of Large-Scale Eigenvalue Problems with Implicitly Restarted Arnoldi Methods. SIAM, Philadelphia (1998)
8. Li, X.S.: An overview of SuperLU: algorithms, implementation, and user interface. ACM Trans. Mathe. Softw. (TOMS) **31**(3), 302–325 (2005)
9. MathWorks: MATLAB Documentation - Accessing Sparse Matrices (2018). https://www.mathworks.com/help/matlab/math/accessing-sparse-matrices.html
10. Nunez-Iglesias, J., van der Walt, S., Dashnow, H.: Elegant SciPy: The Art of Scientific Python. O'Reilly Media (2017)
11. Rosen, L.: Open Source Licensing. Prentice Hall, Upper Saddle River (2004)
12. Saad, Y.: SPARSKIT: A basic tool kit for sparse matrix computations. Technical report, NASA-CR-185876, NASA Ames Research Center (1990)
13. Sanderson, C., Curtin, R.: Armadillo: a template-based C++ library for linear algebra. J. Open Source Softw. **1**, 26 (2016)
14. Stroustrup, B.: The C++ Programming Language, 4th edn. Addison-Wesley, Boston (2013)
15. Vandevoorde, D., Josuttis, N.M.: C++ Templates: The Complete Guide, 2nd edn. Addison-Wesley, Boston (2017)

Intelligent Editor for Authoring Educational Materials in Mathematics e-Learning Systems

Shizuka Shirai[1]([✉]), Tetsuo Fukui[2], Kentaro Yoshitomi[3], Mitsuru Kawazoe[3],
Takahiro Nakahara[4], Yasuyuki Nakamura[5], Katsuya Kato[6],
and Tetsuya Taniguchi[7]

[1] Osaka University, Osaka, Japan
shirai@ime.cmc.osaka-u.ac.jp
[2] Mukogawa Women's University, Nishinomiya, Japan
fukui@mukogawa-u.ac.jp
[3] Osaka Prefecture University, Osaka, Japan
{yositomi,kawazoe}@las.osakafu-u.ac.jp
[4] Sangensha LLC., Chitose, Japan
nakahara@3strings.co.jp
[5] Nagoya University, Nagoya, Japan
nakamura@nagoya-u.jp
[6] Cybernet Systems Co., Ltd., Tokyo, Japan
katsuyak@cybernet.co.jp
[7] Nihon University, Tokyo, Japan
taniguchi.tetsuya@nihon-u.ac.jp

Abstract. E-learning systems for mathematics, such as STACK, Maple T.A., and MATH ON WEB that are able to assess answers using mathematical expressions, have been used for mathematics education at universities. The means for inputting mathematical expressions using current interfaces in these mathematics e-Learning systems are cumbersome not only for students entering their answers, but also for teachers authoring educational materials. In most editing software, teachers need to enter mathematical expressions according to LaTeX-style or computer algebra system-style. This exerts a heavy toll on teachers who have never used these systems. For general use of these systems, it is important to improve the means for entering mathematical expressions. In this study, we developed an intelligent editor for authoring educational materials in mathematics e-Learning systems by implementing a mathematical input interface, named MathTOUCH. This interface allows users to enter the desired mathematical expressions through predictive conversion that converts obscure linear strings presented in a colloquial-style into suitable formats. The results of our previous investigation show that MathTOUCH allows higher level of performance than the standard interfaces. Therefore, the proposed editor is expected to overcome the problem of inputting mathematical expressions in e-learning systems for mathematics education.

Keywords: Mathematics e-learning systems · Mathematics interfaces

© Springer International Publishing AG, part of Springer Nature 2018
J. H. Davenport et al. (Eds.): ICMS 2018, LNCS 10931, pp. 431–437, 2018.
https://doi.org/10.1007/978-3-319-96418-8_51

1   Introduction

In recent years, e-learning systems have gained popularity of use in higher-education institutions in accordance with the development of information and communications technologies. These systems offer multiple features such as providing teaching materials, message boards for discussion, and online testing. In particular, online testing is an important feature for self-directed study and measurement of students' abilities.

In online mathematics testing, several systems such as STACK [1], Maple T.A. [2] and MATH ON WEB [3] are used at several universities in JAPAN. These systems enable students to enter a mathematical expression directly as their answer. However, the current standard input interfaces for these systems are cumbersome for novice learners to enter their answer. To improve this issue, Fukui and Shirai have proposed a new mathematical input interface, named MathTOUCH [4–6]. This interface facilitates predictive conversion from a colloquial-style mathematical text to suitable two-dimensional mathematical expressions.

Meanwhile, we have proposed mathematics e-learning questions specification (MeLQS) for sharing questions among different systems [7]. We are also developing the system for authoring questions according to MeLQS. However, the input procedure for mathematical expressions is also troublesome for teachers authoring educational materials.

This study aims to address this shortcoming by introducing MathTOUCH, an intelligent-type mathematical input interface, as mentioned above. We present an intelligent editor for authoring educational materials in mathematics e-learning systems by implementing MathTOUCH.

2   MathTOUCH: Math Input Interface

2.1   Overview of MathTOUCH

Educational materials for mathematics e-learning systems are authored by an HTML editor that allows users to embed media or an equation into their editing text. Currently, there are two ways to enter mathematical expressions, namely text-based interfaces and structure-based interfaces.

Text-based interfaces such as LaTeX use only characters. These interfaces represent mathematical expressions with inline text. To represent relationships between mathematical elements, users need to input characters according to a command syntax explicitly. It is hard to use for novices [8] because these inline text notations for mathematical expressions are not as intuitive as desired.

Conversely, structure-based interfaces allow users to enter mathematical expressions using individual symbols and mathematical structures graphically from menu palettes. It is quite friendly for novices but they need to have previous understanding of the structures of the mathematical expressions. For instance, if users want to input the expression $\frac{x^2+3}{2}$, they need to choose the fraction symbol first, thereafter they insert $x^2 + 3$ and 2. This procedure is different from writing procedures on paper, so this type of interface also causes usability problems [9].

To address these issues, we propose an intelligent-type mathematical input interface, named MathTOUCH. This interface allows users to enter the desired mathematical expressions from obscure colloquial-style strings [4,5]. For instance, in the case of the expression $\frac{x^2+3}{2}$, the users first input the linear string "x2 + 3/2". The rules of colloquial-style linear string set the key letters (or words) linearly corresponding to the symbols for the elements of a mathematical expression in the order they are read or spoken [6]. It is unnecessary to enter signs that are not displayed, such as the power sign and the parentheses as a delimiter for the numerator. Thereafter, a list of candidates is displayed as system prediction proposals as in Fig. 1. After that, they simply choose the desired expression from the list. Finished mathematical expressions are output in formats such as LaTeX, MathML, PNG, JPEG, EPS, Maxima, Maple, and Mathematica.

MathTOUCH enables users to input almost any mathematical expression dealt with in the general categories of mathematics from junior high school level to university level without having to learn a complex language such as LaTeX. Some examples for linear strings for MathTOUCH and LaTeX-form are shown in Table 1. For example, the linear string for $\cos^2 \theta$ is denoted by "cos2t." However, the linear string of the expressions $\cos 2\theta$, $\cos^2 t$ and $\cos 2t$ are also denoted by "cos2t." Hence, there are some ambiguities in our linear string rules.

To address this shortcoming on such obscure rules, we have proposed a predictive algorithm to convert an linear string into the most suitable mathematical expressions using machine learning through a data set consisting of 4000 formulae [6].

Our prior research shows that MathTOUCH allows approximately 1.2–1.6 times faster task times than the standard interfaces. It shows higher satisfaction with regards to math input usability than the standard interfaces [5].

The results of our evaluation show that the prediction accuracy for the top ten ranking of our method is 85.2% [6].

2.2   Entering Mathematical Expressions

We explain the mathematical input process of MathTOUCH by using the case of the equation $y = x^2 \sin x$ which is illustrated in Fig. 2. First, users input a colloquial-style linear string for the desired mathematical expression. Then, a list of prediction proposals is displayed in a two-dimensional mathematical notation by using our proposed predictive algorithm through a machine learning. In this case, the linear string is "y = x2sinx" and the user then hits the top of prediction proposals in the list. After all the elements are interactively chosen, the desired expression is formed. Finally, the complete mathematical expression is outputted in the desired format.

MathTOUCH was developed using JavaScript and can be integrated into the other systems.

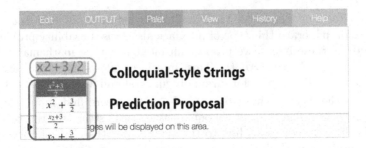

Fig. 1. MathTOUCH: math input interface.

Table 1. Examples of colloquial-style linear strings.

Mathematical expression	MathTOUCH	LaTeX-form
$x^2 + 3x + 2$	x2 + 3x + 2	x^{2} + 3x + 2
$\frac{2}{5}$	2/5	\frac{2}{5}
$\sqrt{3}$	root3	\sqrt {3}
$\cos^2 \theta$	cos2t	\cos^{2}\theta
$\log_{10} x$	log10x	\log_{10}x
$\sum_{k=1}^{n} a_k$	sumk = 1nak	\sum_{k=1}^{n}a_{k}
$\int_a^b f(x)dx$	intabf(x)dx	\int_{a}^{b}f\left(x\right)dx

3   Proposed Intelligent Editor

We have developed an editor for authoring educational materials in mathematics
e-learning systems that enables users to embed any mathematical expression
into the text using MathTOUCH. In this section, we describe a specification

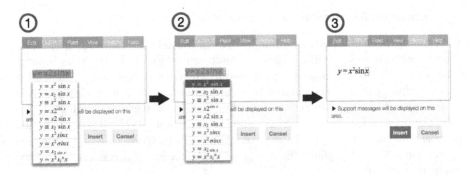

Fig. 2. Mathematical input process on MathTOUCH.

of a proposed intelligent editor and how to edit educational materials including mathematical expressions using our editor.

3.1   System Specification

This editor was created in JavaScript (HTML5) to make it compatible with other e-assessment systems.

Figure 3 represents our proposed intelligent editor window and their editing functions in the menu palettes. This editor has functions like other common HTML editors such as the ability to change font size, font color, and inserting images. All functions are available from buttons arranged at the top of the editor window. Users are able to insert any mathematical expression by calling MathTOUCH from the insert equation button (Fig. 3, No. 23). The documents inside of the entry area in Fig. 3 are an example of a calculus question for an e-assessment system.

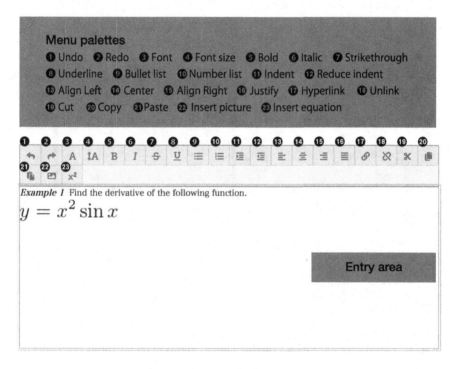

Fig. 3. Proposed intelligent editor.

3.2   Interaction Design

In this section, we explain the editing process of this editor. Figure 4 represents an example of authoring a question as in Fig. 3. First, the teacher inputs the quiz or question statement in the editing area. The MathTOUCH editor (the intelligent-type mathematical input interface), is available whenever it is called from the pop-up window using the insert equation button in the functional icon pallet (Fig. 4). In this case, in Fig. 4, the teacher inputted the text statement for a calculus question in the first line and called MathTOUCH from the insert equation button. After formatting the desired mathematical expression by Math-TOUCH as mentioned in Sect. 2, the two-dimensional mathematical expression is embedded into the editing text at the cursor point in the second line of the entry area.

Therefore, it is easy to imagine how the questions are displayed on the e-learning system. Moreover, the embedded mathematical expressions on this editor are amendable by calling the MathTOUCH window again.

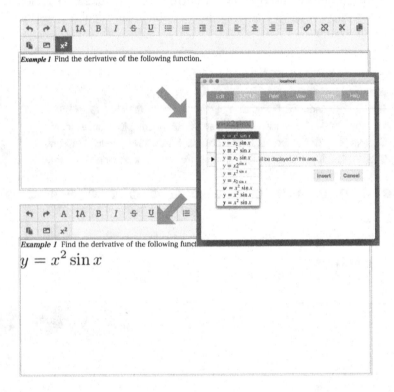

Fig. 4. Example of authoring a mathematical question on our proposed editor.

4   Conclusion and Future Work

In this paper, we proposed an intelligent editor for authoring educational materials in mathematics e-learning systems by implementing MathTOUCH. Math-TOUCH is an intelligent-type mathematical input interface that enables users to insert desired mathematical expressions into the text editor in a two-dimensional mathematical notation using predictive conversion from colloquial-style strings through a machine learning algorithm. The proposed intelligent editor enables teachers to embed their desired equations and/or formulae into any point of a mathematical materials. Especially, they are able to imagine how the authored materials consisting of mathematical expressions are displayed on e-learning systems and to amend all the embedded mathematical expressions. Therefore, the workload of authoring educational materials for teachers would be reduced.

The most important avenues for future research are evaluating the editor and implementing it in MeLQS systems that are created with Moodle.

Acknowledgments. This work was supported by JSPS KAKENHI Grant Numbers 16H03067, 16K16178, and 17K00501.

References

1. Sangwin, C.: Computer Aided Assessment of Mathematics. Oxford University Press, Oxford (2013)
2. Maple T.A.: Online Assessment System for STEM Courses 2013 Maplesoft. https://www.maplesoft.com/products/mapleta/
3. Osaka Prefecture University (2018) MATH ON WEB: Learning College Mathematics by web Mathematica. http://www.las.osakafu-u.ac.jp/lecture/math/MathOnWeb/
4. Fukui, T.: An intelligent method of interactive user interface for digitalized mathematical expressions (in Japanese). RIMS Kokyuroku **1780**, 160–171 (2012)
5. Shirai, S., Fukui, T.: MathTOUCH: mathematical input interface for e-assessment systems. MSOR Connect. **15**(2), 70–75 (2016)
6. Fukui, T., Shirai, S.: Predictive algorithm for converting linear strings to general mathematical formulae. In: Yamamoto, S. (ed.) HIMI 2017. LNCS, vol. 10274, pp. 15–28. Springer, Cham (2017). https://doi.org/10.1007/978-3-319-58524-6_2
7. Kawazoe, M., Yoshitomi, K., Nakahara, T., Nakamura, Y., Fukui, T., Shirai, S., Kato, K., Taniguchi, T.: MeLQS: mathematics e-learning questions specification - a common base for sharing questions among different systems. In: Proceedings of the International Workshop on Mathematical Education for Non-Mathematics Students Developing Advanced Mathematical Literacy, pp. 123–126 (2018)
8. Pollanen, M., Hooper, J., Cater, B., Kang, S.: A tablet-compatible web-interface for mathematical collaboration. In: Hong, H., Yap, C. (eds.) ICMS 2014. LNCS, vol. 8592, pp. 614–620. Springer, Heidelberg (2014). https://doi.org/10.1007/978-3-662-44199-2_92
9. Pollanen, M., Wisniewski, T., Yu, X.: XPRESS: a novice interface for the real-time communication of mathematical expressions. In: Proceedings of the Workshop on Mathematical User Interfaces (2007)

Recent Developments in Cayley Hash Functions

Bianca Sosnovski$^{(\boxtimes)}$ (iD)

Queensborough Community College - CUNY, Bayside, NY 11364, USA
`bsosnovski@qcc.cuny.edu`

Abstract. In 1994, Tillich and Zémor proposed a scheme for a family of hash functions. In the scheme, they used products of 2×2 matrices in special linear group over a field. Since then, other hash functions based on the Tillich and Zémor's design have been proposed. These cryptographic hash functions are called Cayley hash functions because of the correspondence between their constructions and Cayley graphs of (semi)groups. Most instances of Cayley hash functions have been proved insecure, but the algorithms used to break Cayley hash functions target specific vulnerabilities of each underlying (semi)group used. However, these algorithms don't seem to invalidate the generic scheme. An overview is presented of some of the latest proposals for Cayley hash functions and related open problems.

Keywords: Cryptography · Hash functions · Cayley hash functions

1 Introduction

With the increasing use of technology in communications, financial transactions and many other internet applications, cryptography is essential to the security of many online protocols. Most of the cryptosystems in use today are based on finite Abelian groups. Some of these cryptographic systems will be completely vulnerable to attacks once large quantum computers are made possible. Post-quantum cryptography is cryptography under the assumption that the attacker has a large quantum computer, and its objective is to provide cryptosystems that remain secure in such scenario [5].

Hash functions are an important tool for cryptography. They are fundamental blocks in the construction of several cryptographic primitives such as digital signature, encryption and key derivation systems. According to Bernstein [4], there are three classes of cryptographic systems that appear to be difficult to break even with a large quantum computer. One of these systems is the hash-based public-key signatures, which requires the use of a standard cryptographic hash function.

Provably secure hash functions are hash functions whose security are based on the difficulty of solving a known "hard" problem. Also in recent years, non-Abelian (semi)groups that are typically studied in combinatorial group theory

© Springer International Publishing AG, part of Springer Nature 2018
J. H. Davenport et al. (Eds.): ICMS 2018, LNCS 10931, pp. 438–447, 2018.
https://doi.org/10.1007/978-3-319-96418-8_52

and linear group theory have been considered with increasing interest in public key cryptography [1,3,16,18]. It seems reasonable to research replacements and enhancements of protocols that are based on finite Abelian groups.

Examples of provable-secure hash functions are the Cayley hash functions that are based on the Cayley graphs of (semi)groups. Cayley hash functions are designed so that their security would follow from the alleged hardness of a mathematical problem related to the Cayley graph of the underlying (semi)group [11].

The first cryptographic hash function using finite groups was proposed by Bosset in 1977 [6]. It was described as a signature scheme and uses multiplications of matrices in the group $GL_2(\mathbb{F}_p)$ where p is prime. It was broken by Camion in 1984 with a probabilistic factorization algorithm that uses a certain type of chain of subgroups with "small" subgroup indices to search for a text whose hash value is in a subgroup of the chain [9].

In 1991, Zémor introduced a hash function whose values correspond to products in $SL_2(\mathbb{F}_p)$ where p is prime [27]. It was broken by Tillich and Zémor, who then also provided the group $SL_2(\mathbb{F}_{2^n})$ as replacement to increase the security of the scheme [25,26].

The Tillich-Zémor hash function sustained attacks until 2009 when Grassl et al. [14] established a connection between the Tillich-Zémor function and maximal length chains in the Euclidean algorithm for polynomials over the field with two elements. Petit and Quisquater [22,23] suggested that security might be recovered by introducing new generators. The factorization, representation and balance problems in non-Abelian groups still are potentially hard problems for general parameters of Cayley hash functions. Other proposals for Cayley hash functions have been developed since then.

The goal of this paper is to describe some of the developments in Cayley hash functions and it is not intended to be a comprehensive description of the subject. It will provide an overview of some of the latest Cayley hash proposals and relevant open problems. For more details about these proposals and related results, we refer the reader to the references herein.

The remainder of the paper is organized as follows. In Sect. 2, we recall the properties of a cryptographic hash function. In Sect. 3, we present the Cayley hash function design and define the balance, representation and factorization problems. In Sect. 4, we briefly describe certain Cayley hash proposals and review some aspects related to their security, and we conclude the paper in Sect. 5.

2   Preliminaries

Hash functions are efficient compression functions that take a variable-length input and convert it to a fixed-length output.

Definition 1. *Let $h : X^* \longrightarrow X^n$ be a hash function. The alphabet X usually used is $\{0,1\}$. h is a cryptographic hash function if satisfy at least one of the following properties.*

- Preimage resistance: *Given a hash value y for which a corresponding input is not known, it is computationally infeasible (or "hard") to find any input x such that $y = h(x)$.*
- Second-preimage resistance: *Given an input x_1 it is computationally infeasible to find another input x_2 where $x_1 \neq x_2$ such that $h(x_1) = h(x_2)$.*
- Collision resistance: *It is computationally infeasible to find any two inputs x_1 and x_2 where $x_1 \neq x_2$ such that $h(x_1) = h(x_2)$.*

Our interest is in cryptographic hash functions whose constructions are based on directed Cayley graphs of (semi)groups.

3   Cayley Hash Functions

Cayley hash functions are families of cryptographic hash functions constructed from Cayley graphs. The initial idea of Cayley hash functions was to use groups whose Cayley graph are expander graphs to design hash functions that are collision-resistant.

Definition 2. *Let G be a finite (semi)group with a set of generators S that has the same size as the text alphabet $\mathcal{A}$. Choose a function: $\pi : \mathcal{A} \to S$ such that defines an one-to-one correspondence between $\mathcal{A}$ and S. The hash value of the text $x_1 x_2 \ldots x_k$ is the (semi)group element $\pi(x_1)\pi(x_2)\ldots\pi(x_k)$.*

One of the advantages of this design is that the computation of the hash value can be easily parallelized due to the concatenation property $\pi(xy) = \pi(x)\pi(y)$ for any texts x and y from $\mathcal{A}$. Unlike the SHA family of hash functions that hash blocks of input, this type of function hashes each bit individually.

Cayley hash functions have their security properties strongly related to mathematical problems.

Definition 3. *Let G be a (semi)group and $S = \{s_1, \ldots s_k\} \subset G$ be a generating set of G. Let L be of polylogarithmic (small) in the size of G.*

- Balance problem: *Find an efficient algorithm that returns two words $m_1 \ldots m_l$ and $m'_1 \ldots m'_{l'}$ with $l, l' < L$, $m_i, m'_i \in \{1, \ldots, k\}$ that yield equal products in G, that is,* $\prod_{i=1}^{l} s_{m_i} = \prod_{i=1}^{l'} s_{m'_i}$
- Representation problem: *Find and efficient algorithm that returns a word $m_1 \ldots m_l$ with $l < L$, $m_i \in \{1, \ldots, k\}$ such that* $\prod_{i=1}^{l} s_{m_i} = 1$.
- Factorization problem: *Find an efficient algorithm that given any element $g \in G$ returns a word $m_1 \ldots m_l$ with $l < L$, $m_i \in \{1, \ldots, k\}$ such that* $\prod_{i=1}^{l} s_{m_i} = g$.

A Cayley hash function is collision resistant if and only if the balance problem is hard in the underlying (semi)group. If the representation problem is hard in the (semi)group, the associated Cayley hash is second-preimage resistant and, it is preimage resistant if and only if the corresponding factorization problem is hard in (semi)group [19,23].

Other requirements considered by Tillich and Zémor in [25,27] in the construction of Cayley hash functions are that the Cayley graph of G with generator set S has large girth and small diameter. This is closely related to the Babai's conjecture.

Conjecture 1 (Conjecture 1.7 in [2]). If G is a non-Abelian finite simple group of order N, then $diam(G) < (\log N)^c$ for some absolute constant c.

The factorization problem in non-Abelian groups can also be seen as an constructive proof of Babai's conjecture and a constructive proof of Babai's conjecture would make all Cayley hash functions insecure [23].

In addition to the Tillich-Zémor hash function, we will provide a brief description of some of the latest developments in Cayley hashes.

4   Instances of Cayley Hash Functions

4.1   Tillich-Zémor Hash Function

The generators of this hash function are $A = \begin{pmatrix} \alpha & 1 \\ 1 & 0 \end{pmatrix}$ and $B = \begin{pmatrix} \alpha & \alpha+1 \\ 1 & 1 \end{pmatrix}$ with α as the root of an irreducible polynomial $p(x)$ of degree n in the ring of polynomials $\mathbf{F}_2[x]$, where $\mathbf{F}_2$ is the field with two elements. A and B are generators of the Cayley graph for the group $SL_2(\mathbb{F}_{2^n})$ with $\mathbb{F}_{2^n} \approx \mathbf{F}_2[x]/(p(x))$ where $(p(x))$ is the ideal generated by an irreducible polynomial $p(x)$.

To find collisions for the Tillich-Zémor hash functions one needs to find two distinct sequences of matrix generators such that the corresponding products coincide in the group $SL_2(\mathbb{F}_{2^n})$. The Tillich-Zémor hash function sustained early attacks. However, Grassl et al. [14] introduced an algorithm that finds collisions for this hash function. They discovered a pattern in the structure of hash values of palindromic messages (messages such that their representation in bit strings are the same backward as forward). The attack showed that the Tillich-Zémor hash function was not collision resistant.

Consequently, Petit and Quisquater [22] presented efficient algorithms that show that the Tillich-Zémor function is not preimage nor second-preimage resistant. Their two algorithms provide preimages of lengths $O(n^2)$ and $O(n^3)$. They also describe the following open problem.

Problem 1 (Petit and Quisquater, 2011). Since the size of $SL_2(\mathbb{F}_{2^n})$ is about 2^{3n}, it seems reasonable to conjecture that preimages of size $3n$ exist for any matrix. However, even if this conjecture is true, it is not clear that there exists an efficient algorithm computing preimages of this length.

Other instances of Cayley hashes based on Ramanujan expander graphs have been proposed after Tillich-Zémor functions. They are the *LPS hash function* and the *Morgenstern hash function*. The latter is a generalization of the former hash function. For information about these Cayley hash functions and their cryptanalyses, consult [20,21].

4.2   Bromberg-Shpilrain-Vdovina Hash Function (BSV)

In the paper [8], Bromberg et al. proposed specific cases of the following matrices as generators for the BSV Cayley hash function: $A(x) = \begin{pmatrix} 1 & x \\ 0 & 1 \end{pmatrix}$ and $B(y) = \begin{pmatrix} 1 & 0 \\ y & 1 \end{pmatrix}$ considered over $\mathbb{F}_p$ with p prime and $xy \geq 4$. Their choice is based on the fact that these matrices generate a free monoid over $\mathbb{Z}$ and that there cannot be any short relations over $\mathbb{F}_p$.

It is known that the Cayley graphs of groups generated by the pairs $(A(2), B(2))$ and $(A(3), B(3))$ over $\mathbb{F}_p$ are expander graphs. Bromberg et al. provide explicit lower bounds for the directed girth of the corresponding Cayley graphs. The semigroup generated by $A(2)$ and $B(2)$ over $\mathbb{F}_p$ has a Cayley graph with girth at least $\log_{b_2} p$ where $b_2 = \sqrt{3 + \sqrt{8}} \approx 2.4$, and the Cayley graph generated by $A(3)$ and $B(3)$ has girth at least $\log_{b_3} p$ where $b_3 = \sqrt{\frac{11 + \sqrt{117}}{2}} \approx 3.3$.

They also mention that the girth of expander graphs are not necessarily large in general and give explicit examples of that in the literature. In addition, they proved the following result.

Theorem 1 (Bromberg, Shpilrain and Vdovina, 2017). *There is an efficient heuristic algorithm that finds particular relations of the form $w(A(2), B(2)) = 1$, where w is a group word of length $O(\log p)$, and the matrices $A(2)$ and $B(2)$ are considered over $\mathbb{F}_p$.*

The algorithm combines Sanov's result (1947) about the form of all invertible matrices in the subgroup of $SL_2(\mathbb{Z})$ generated by $A(2)$ and $B(2)$ and the "lifting attack" by Tillich and Zémor [25] used to break the Zémor hash function [27]. Despite their algorithm, the security of the BSV Cayley hash function is not affected since only positive powers of $A(2)$ and $B(2)$ are used in hashing, and the group relations produced by the algorithm involve, with high probability, negative and positive powers. To the best of our knowledge, the BSV hash remains unbroken.

The following are related open problems listed in [7].

Problem 2 (Bromberg, Shpilrain and Vdovina, 2015). Find an analog of Sanov's form for the subgroup of $SL_2(\mathbb{Z})$ generated by $A(3)$ and $B(3)$.

Problem 3 (Bromberg, Shpilrain and Vdovina, 2015). Determine which words in the matrices $A(1)$ and $B(2)$ will have the fastest growth of their entries.

Problem 4 (Bromberg, Shpilrain and Vdovina, 2015). Find an efficient algorithm that finds relations of the form $w(A(3), B(3)) = 1$, where w is a group word of bounded length when the matrices are considered over $\mathbb{F}_p$.

In a paper by Han et al. [15], the authors answer the question in Problem 3 about the size of the entries of matrices in the monoid generated by the matrices of type $A(x)$ and $B(y)$ for $x, y \geq 1$, generalizing the results in [8].

Regarding Problems 2 and 4, Chorna, Geller and Shpilrain [12] provide results about the form of the matrices in $SL_2(\mathbb{Z})$ generated by $A(k)$ and $B(k)$ for $k \in \mathbb{Z}$ and an algorithm that decides whether or not a given matrix $M = (m_{ij})$ in $SL_2(\mathbb{Z})$ is in the subgroup of $SL_2(\mathbb{Z})$ generated by $A(k)$ and $B(k)$ for $k \in \mathbb{Z}, k \geq 2$ (and if it does, finds a presentation of the matrix as a group word in $A(k)$ and $B(k)$) in time $O(n \log n)$, where $n = \sum |m_{ij}|$.

4.3   Shpilrain-Sosnovski Hash Function

In the paper [24], Shpilrain and Sosnovski presented a Cayley hash function that uses linear functions in one variable over $\mathbb{F}_p$ with composition operation.

The semigroup generated by $f(x) = ax + b$ and $g(x) = cx + d$ under composition is isomorphic to the semigroup generated by $A = \begin{pmatrix} a & b \\ 0 & 1 \end{pmatrix}$ and $B = \begin{pmatrix} c & d \\ 0 & 1 \end{pmatrix}$ under matrix multiplication. Using results about the freeness of upper triangular matrices by Cassaigne et al. [10], they showed that the semigroup of linear functions over $\mathbb{Z}$ is free if the generators of the semigroup do not commute and $a, c \geq 2$.

The functions $f_0(x) = 2x + 1 \mod p$ and $f_1(x) = 3x + 1 \mod p$ with $p > 3$ are considered the generators of the proposed hash function. The hash value is obtained by first computing product $h(b_1 b_2 \cdots b_k) = f_{b_1} f_{b_2} \cdots f_{b_k} \pmod p$ where $b_i \in \{0, 1\}$ for $1 \leq i \leq k$. The corresponding product linear function is of the form $\ell(x) = rx + s$ where $s, r \in \mathbb{Z}_p$, and the hash value is defined as $H(b_1 b_2 \cdots b_k) = (r + s, s)$.

The corresponding hash functions are very efficient. A bit string of length n can be hashed by performing at most $2n$ multiplications and about $2n$ additions in $\mathbb{F}_p$.

Proposition 1 (*Spilrain and Sosnovski, 2016*).   *Let the "0" be hashed to $f_0(x) = 2x + 1$ and the "1" bit be hashed to $f_1(x) = 3x + 1$. If two bit strings U and V hash to the same value, then the length of either U or V is at least $\log_3 p$.*

An advantage of this hash function is that the output bit strings have length $2 \log p$, while the Tillich-Zémor hash function outputs bit strings of length $4 \log p$. With respect to the security of the hash function, the authors recommend that $p \approx 2^{512}$ or larger to prevent generic attacks. With this recommended value of parameter p there will be no collisions unless the length of at least one of the colliding strings is at least 323. If input text is short (323 bits or less), use padding to extend its length to 512 bits. Subgroup attacks and attacks using elements of small orders can be prevented by choosing p such that $p = 2q + 1$ where q is a "large" prime.

Shpilrain and Sosnovski, also noted the following problem.

Problem 5 (Shpilrain and Sosnovski, 2016). Is there a way of efficiently determining lifting functions of the form $\mathbb{L}(x) = Rx + S$ over $\mathbb{Z}$ where $R = r + k_1 p$ and $S = s + k_2 p$ for some $k_1, k_2 \in \mathbb{Z}$ to find a preimage?

Cryptanalysis of the Shpilrain-Sosnovski Cayley Hash

Though a version of the lifting attack is not known to exist for the Shpilrain-Sosnovski Cayley hash function, Monico [17] developed an attack that shows that the hash function is not second-preimage resistant for inputs larger than about 1.9 MB for parameter $p \approx 2^{256}$. Actually, in Monico's method the original bit string is not even required and it suffices to have only a bound on its length.

In Monico's attack, a hash value (x, y) in $\mathbb{F}_p$ of a bit string of known length L is given and inverted to $(r, s) = (x - y, y)$. Since $r = 2^a 3^b$ where a is the number of zeros in the original bit string and b is the number of ones (or vice-versa), then $L = a + b$. The values of a and b can be recovered with $O(L \log L)$ operations over $\mathbb{F}_p$ by precomputing L powers of 2, sorting them out and then computing and testing $r, 3^{-1}r, 3^{-2}r, \ldots$ until one of the values in the sequence matches one of the precomputed powers of 2.

Let $n = min\{a, b\}$, $Y = \begin{pmatrix} r & s \\ 0 & 1 \end{pmatrix}$ and $U = \begin{pmatrix} r & u \\ 0 & 1 \end{pmatrix}$, where U is a suitable matrix whose factorization in generators $A = \begin{pmatrix} 2 & 1 \\ 0 & 1 \end{pmatrix}$ and $B = \begin{pmatrix} 3 & 1 \\ 0 & 1 \end{pmatrix}$ is known and determined by the values of a and b found in the first step.

The attack's goal is to transform U into Y by replacing several of leading AB factors of U with BA. To do so, one must find $\mathbf{x} \in \{0, 1\}^n$ such that $\sum_{j=0}^{n-1} x_j 6^j \equiv t$ (mod p) where $t = s - u$ (mod p) (for more details, see [17]).

To provide a probabilistic algorithm to find such $\mathbf{x}$, Monico reduced the problem to a dense instance of the Random Modular Subset Sum Problem (RMSSP), which was considered by Lyubashevsky (2005). Heuristically, his algorithm is expected to succeed as long as the original bit string had at least n zeros and n ones for some $n \geq 2\sqrt{2 \log_2 p}$. According to Monico, the algorithm's expected running time is $O(n^2 \log n)$ with an implied constant small enough to keep the attack practical for $p \approx 2^{256}$.

In the paper [17], the following open problems are listed.

Problem 6 (Monico, 2018). If n is small compared to $\log_2 p$, there need not exist a solution. Prove that a solution $\mathbf{x}$ exists when is n sufficient large.

Problem 7 (Monico, 2018). Is there a modification which would avoid Monaco's attack while still retaining the algebraic advantage?

We add the following to the list.

Problem 8. If we increase the prime parameter p of the hash function without affecting too much efficiency, is Monico's algorithm still practical?

4.4   Ghaffari-Mostaghim Linear Hash Function Variation

A modification of the Shpilrain-Sosnovski hash function is proposed by Ghaffari and Mostaghim [13].

As discussed in [24], preimages can be easily computed for short messages in the Shpilrain-Sosnovski hash and one option to avoid this is to use padding.

In this modification, a similar idea introduced in [19] is suggested. The functions $f_0(x) = 2x + 1 \mod p$ and $f_1(x) = 3x + 1 \mod p$, where $p > 3$ is a prime, are also considered as generators in this hash variation. Let $H(m_1 m_2 \cdots m_l) = f_{m_1} f_{m_2} \cdots f_{m_l} \pmod{p}$ for $m = m_1 m_2 \cdots m_l \in \{0,1\}^*$.

To make the factorization problem harder, the following is suggested. Let S the group generated by f_0 and f_1 over $\mathbb{Z}_p$, $t > 1$ an integer and $g \in S \setminus \{e, f_0, f_1\}$, where e is the identity element of S. Define $\widehat{H} : \{0,1\}^* \to S$ by

$$\widehat{H}(m) = \prod_{i=1}^{l} C_i$$

where

$$C_i = \begin{cases} f_{m_i} & \text{if } t \nmid i \\ f_{m_i} g & \text{if } t \mid i \end{cases}.$$

Now define $\widehat{H}_2(m) = \widehat{H}(m)\widehat{H}((\widehat{H}(m) \oplus c_{rand}))$, where c_{rand} is a constant bit string whose bits look like random.

Ghaffari and Mostaghim, showed that $\widehat{H}$ is at least as secure as H, and consequently so is $\widehat{H}_2$. Because Proposition 1, choosing $t < \log_3 p$ would make $\widehat{H}$ safer than H since a collision for H cannot be directly used to find a collision for $\widehat{H}$.

Problem 9. Can a variation of Monico's algorithm produce a preimage for the Ghaffari-Mostaghim hash function?

For an input bit string of length l, the computation of $\widehat{H}$ requires $\lfloor l/t \rfloor$ multiplications more than the original Cayley hash function proposed by Shpilrain and Sosnovski, thus not affecting too much the performance of the hash.

5   Conclusion

Switching to cryptosystems that remain secure against attacks by quantum computers is needed and hash-based cryptography is one of the classes of cryptographic systems that are considered resistant to quantum attacks, if large quantum computers are built.

As non-Abelian (semi)groups are involved in the design of Cayley hash functions, this category of cryptographic hash functions may be resistant to quantum attacks. It seems reasonable to research improvements and new designs for Cayley hash functions with the goal of sustaining quantum attacks. With this in mind, we provided an overview of instances of Cayley hash functions and some relevant open problems.

References

1. Anshel, I., Atkins, D., Goldfeld, D., Gunnells, P.E.: Post quantum group theoretic cryptography. Tech. Rep., SecureRF Corporation, Shelton, CT, USA (2016)
2. Babai, L., Seress, Á.: On the diameter of permutation groups. Eur. J. Comb. **13**(4), 231–243 (1992). https://doi.org/10.1016/S0195-6698(05)80029-0
3. Baumslag, G., Fine, B., Xu, X.: Cryptosystems using linear groups. Appl. Algebra Eng. Commun. Comput. **17**(3), 205–217 (2006). https://doi.org/10.1007/s00200-006-0003-z
4. Bernstein, D.J.: Introduction to Post-Quantum Cryptography, pp. 1–14. Springer, Heidelberg (2009). https://doi.org/10.1007/978-3-540-88702-7_1
5. Bernstein, D.J., Lange, T.: Post-quantum cryptography. Nature **549**(7671), 188–194 (2017). https://doi.org/10.1038/nature23461
6. Bosset, J.: Contre les risques d'altération, un systeme de certification des informations. 01 Informatique **107**, (1977)
7. Bromberg, L.: Some applications of noncommutative groups and semigroups to information security. Ph.D. thesis, Graduate Center, the City University of New York (2015)
8. Bromberg, L., Shpilrain, V., Vdovina, A.: Navigating in the Cayley graph of $SL_2(\mathbb{F}_p)$ and applications to hashing. Semigroup Forum **94**(2), 314–324 (2017). https://doi.org/10.1007/s00233-015-9766-5
9. Camion, P.: Can a fast signature scheme without secret key be secure. In: Poli, A. (ed.) AAECC 1984. LNCS, vol. 228, pp. 215–241. Springer, Heidelberg (1986). https://doi.org/10.1007/3-540-16767-6_67
10. Cassaigne, J., Harju, T., Karhumäki, J.: On the undecidability of freeness of matrix semigroups. Int. J. Algebra Comput. **09**(03n04), 295–305 (1999). https://doi.org/10.1142/S0218196799000199
11. Charles, D.X., Lauter, K.E., Goren, E.Z.: Cryptographic hash functions from expander graphs. J. Cryptol. **22**(1), 93–113 (2009). https://doi.org/10.1007/s00145-007-9002-x
12. Chorna, A., Geller, K., Shpilrain, V.: On two-generator subgroups in $SL_2(\mathbb{Z})$, $SL_2(\mathbb{Q})$ and $SL_2(\mathbb{R})$. arxiv.org/abs/1605.05226 (2017)
13. Ghaffari, M.H., Mostaghim, Z.: More secure version of a Cayley hash function. Groups Complex. Cryptol. **10**(1) (2018). https://doi.org/10.1515/gcc-2018-0002
14. Grassl, M., Ilić, I., Magliveras, S., Steinwandt, R.: Cryptanalysis of the Tillich-Zémor hash function. J. Cryptol. **24**(1), 148–156 (2011). https://doi.org/10.1007/s00145-010-9063-0
15. Han, S., Masuda, A.M., Singh, S., Thiel, J.: Collision-free bounds for the BSV hash. arxiv.org/abs/1703.02388v2 (2017)
16. Kahrobaei, D., Cavallo, B., Garber, D. (eds.): Algebra and Computer Science, vol. 677. American Mathematical Society (2016)
17. Monico, C.: Cryptanalysis of a hash function, and the modular subset sum problem (2018). http://www.math.ttu.edu/~cmonico/research/linearhash.pdf
18. Myasnikov, A., Shpilrain, V., Ushakov, A.: Group-Based Cryptography. Springer, Heidelberg (2008). https://doi.org/10.1007/978-3-7643-8827-0
19. Petit, C.: On graph-based cryptographic hash functions. Ph.D. thesis, Universit Catholique de Louvain (2009)
20. Petit, C., Lauter, K., Quisquater, J.-J.: Full cryptanalysis of LPS and morgenstern hash functions. In: Ostrovsky, R., De Prisco, R., Visconti, I. (eds.) SCN 2008. LNCS, vol. 5229, pp. 263–277. Springer, Heidelberg (2008). https://doi.org/10.1007/978-3-540-85855-3_18

21. Petit, C., Lauter, K.E., Quisquater, J.-J.: Cayley hashes: A class of efficient graph-based hash functions (2007). http://perso.uclouvain.be/christophe.petit/files/Cayley.pdf
22. Petit, C., Quisquater, J.-J.: Preimages for the Tillich-Zémor hash function. In: Biryukov, A., Gong, G., Stinson, D.R. (eds.) SAC 2010. LNCS, vol. 6544, pp. 282–301. Springer, Heidelberg (2011). https://doi.org/10.1007/978-3-642-19574-7_20
23. Petit, C., Quisquater, J.-J.: Rubik's for cryptographers. Not. Am. Math. Soc. **60**(6), 733–739 (2013)
24. Shpilrain, V., Sosnovski, B.: Compositions of linear functions and applications to hashing. Groups Complex. Cryptol. **8**(2), (2016). https://doi.org/10.1515/gcc-2016-0016
25. Tillich, J.-P., Zémor, G.: Group-theoretic hash functions. In: Cohen, G., Litsyn, S., Lobstein, A., Zémor, G. (eds.) Algebraic Coding 1993. LNCS, vol. 781, pp. 90–110. Springer, Heidelberg (1994). https://doi.org/10.1007/3-540-57843-9_12
26. Tillich, J.-P., Zémor, G.: Hashing with SL_2. In: Desmedt, Y.G. (ed.) CRYPTO 1994. LNCS, vol. 839, pp. 40–49. Springer, Heidelberg (1994). https://doi.org/10.1007/3-540-48658-5_5
27. Zémor, G.: Hash functions and graphs with large girths. In: Davies, D.W. (ed.) EUROCRYPT 1991. LNCS, vol. 547, pp. 508–511. Springer, Heidelberg (1991). https://doi.org/10.1007/3-540-46416-6_44

Mathematical Research Data, Software, Models, and the Publication-Based Approach

Wolfram Sperber$^{(\boxtimes)}$

FIZ Karlsruhe/zbMATH, Franklinstr. 11, 10587 Berlin, Germany
wolfram@zbmath.org

Abstract. Scientific publications are still the most important medium for publishing mathematical research results. They serve as a container for different types of mathematical research data, especially mathematical models, theories, theorems, conjectures, proofs, algorithms, etc. They also link to mathematical software and simulations which has became more and more important for mathematics and applications. Therefore it seems to be natural to use publications for a more sophisticated analysis of mathematical research data, especially software. Mathematical publications are well-structured and use a more or less standard terminology for content, e.g., theorems, proofs, etc, and the formal structure. Nevertheless, publications could be used as a starting point to develop information services for mathematical research data. In the talk, the publication-based approach for mathematical software and a possible extension to mathematical models are discussed.

Keywords: Mathematical research data · Software
Mathematical models · Information services · Heuristics

1 Introduction: Mathematical Knowledge, Mathematical Research Data, and Publishing Formats

Mathematical research data is a not exactly defined term. The subject of mathematical knowledge management is mathematical research data. The OMDoc [1] approach distinguishes between different levels

- the object level (atomic or formulae level) (In some sense there is an analogy to atoms and elementary particles. Formulae are complex structures compounded by elementary mathematical objects, operations, relations. Moreover, the object level covers also other mathematics-relevant objects, e.g., diagrams or graphs.)
- the statement level (axioms, definitions, mathematical models, theorems, conjectures, examples, ...),
- the theory level (top or global level, e.g., proofs).

Remark: The most elements on the different levels defined above are used in the same way by the mathematical community. But some terms are more difficult, especially the term "mathematical model". Mathematical modeling arises in applications (mathematical modeling is used to describe and steer the whole spectrum of real-world problems) and also in mathematics. Mathematical models differ in the level of abstraction, modeling of real world problems bases on laws and entities. In a next step, these models are often transformed to abstract mathematical entities. In other words, the communities and terminologies used by them have an influence on the presentation of the model. Moreover, different real-world problems can be transformed to the same abstract mathematical model. Last but not least, the complexity of mathematical models is increasing, e.g., instead of a single equation there is a system of equations, which allows to consider also different influencing factors. Thus, the presentation of mathematical models in books and articles is very heterogeneous.

Mathematical knowledge can be published in several formats. The first publishing format in modern mathematics was books. A famous example is "Euclids elements". Later journal articles, allowing to publish a new research result in short time, became the dominant publishing format. The development of computational mathematics was the birth of new mathematical object classes and publication formats, especially software, visualizations, and simulations. It is useful to distinguish between mathematical research data and publishing formats.

Mathematical publication formats transport mathematical knowledge and serve as containers for mathematical research data.

Mathematical books, articles, preprints, etc., in the following called "mathematical documents", are written in natural mathematical language: text in natural language, often English, plus mathematical expressions. Mathematical documents are written for the human reader. They have a more or less standardized formal structure (abstract, introduction, sections, subsections, references) and metadata. A standardized semantic markup of mathematical research data in mathematical documents would be an important step in improving knowledge-based analysis of mathematical documents and citation of research data. Semantic mathematical markup languages like OMDoc [1] and MathML [2] provide semantic tagging.

Mathematical software and simulations are different from mathematical documents. Mathematical software, more precisely the software code, is written in a special language, the programming language. Programming languages are formalized languages specifying a set of instructions which can be executed by a machine. There is a great variety of mathematical programming languages, universal programming languages or those which are specialized in a certain class of problems. Software bases on algorithms which describe unambiguous specifications to solve a problem. Algorithms are translated into software by using a programming language. Typically, software code is not part of publications (sometimes publications contain pseudo-code of a software). But there is an increasing number of mathematical documents which describe mathematical software or cite it. Mathematical research data and software are integrated and closely linked with mathematical publications. Mathematical software has dramatically pushed forward the use of mathematical methods especially in applications. Furthermore, it has changed the way mathematical knowledge is applied. Mathematical knowledge and tools can be experimentally used, simulated with the help of software.

In the following, we discuss how publications can be used to improve access to and the information about mathematical research data, especially mathematical software and models.

2   The Publication-Based Approach for Mathematical Software

Until the end of the 20th century mathematical publications were the most important resource for mathematical knowledge. Its importance has led to the development of a powerful information infrastructure for the mathematical literature. It covers services which provide the access to publications and tools for mathematical knowledge management: libraries, bibliographic reviewing media, especially zbMATH [3] and MathSciNet [4], encyclopedias, the Mathematical Subject Classification (MSC) [5], etc.

As said above, mathematical software is becoming more and more important not only in mathematical research but there is no counterpart to zbMATH and MathSciNet for mathematical software for some reasons

- The missing tradition of software information
  The first reviewing journal in mathematics, the "Jahrbuch über die Fortschritte der Mathematik" [6] was founded by mathematicians in the mid of the 19th century. For more than 100 years the reviewing journals have developed in the course of time methods for content analysis (especially reviewing and classification) and the presentation of mathematical literature in compact form. The switching to digital formats has led to some extensions, e.g., by adding citations, author disambiguation, and author profiles. The current mathematics reviewing databases zbMATH and MathSciNet have an institutional basis and permanent editorial staff. Software information is relatively new and the capabilities for software information services are limited.
- The dynamic character of software
  Typically, a software name stands for a series of versions. Of course, also mathematical documents undergo a development process, but only the final version is published. For software, the situation is different. Often, different development stages of a software are published and are in use. The reasons for versioning of software are different, performance, adaption to changed hard- and software environments or programming languages, licenses, etc. A resulting problem is the persistence of software information. Archiving of software is non-trivial and requires a specific technical and organizational infrastructure.

- Usability of software
  Unlike mathematical knowledge in mathematical publications, the use of the mathematical software is dependent on its form, whether it is a service or can be used via an API or by integrating, adapting, or further developing the source code. Mathematical software can be used also as a blackbox, e.g., in cloud computing.
- Software information has different features
  The core of software is code but software is more than software code. But software information also includes documentations, installation guides, tutorials, license and usability information, information about hard- and software dependencies, programming languages, etc.
- Software development as collaborative work
  Software development is a complex task and cannot be reduced to translate an algorithm into a programming language. Also an intuitive user interface, or tools for visualization are necessary. Often software development is widely distributed, also the developers may have different roles.

We can postulate that software information and an information infrastructure for software is a more complex challenge for developing a powerful information infrastructure than for documents. The discussion on a software citation standard should illustrate this. Software citations are done in different ways. A common praxis for software citations is to refer to a publication instead of a software due to the advantage of persistence and human-addressed information. Other citations are references to websites or documentations. Last but not least, a big number of references cite software only by name. Summarizing, today software citations refer to different objects and documents and are incomplete. Especially the citations often contain no information about the version which was used. The increasing role of software has led to intensified discussions on how to cite a software, see the Software Citation Principles [7] of the FORCE11 working group. The principles require among others persistent identifiers for software. These identifiers should point to so-called landing pages which provide at least persistent meta-information about a software. The CodeMeta initiative [8] has developed a metadata scheme for software. This scheme contains a series of URLs which refer to different objects, e.g., the source code (of the actual version and the archive), the documentation, license information, etc.

The swMATH project, see [9,10], uses the information about mathematical publications to develop an information service for mathematical software. This results in two essential advantages. Mainly machine-based methods can be used to realize the approach and the obtained information also covers data about the use cases of software. In detail:

- **zbMATH as starting point:** The reviewing database zbMATH ia used as starting point.
- **Identification of software:** The identification of software citations starts with heuristic methods which search for special term/phrase combinations in connection with a pattern search and is completed by manual support by zbMATH editorial office.

- **Standard and user publications:** The publications citing a software are divided in standard and user publications. Standard publications are such ones which are focused on the description of the software. User publications are focused on research results which were achieved by software. User publications and benchmarks often cite different software and compare the results which have been achieved by different software.
- **Analysis of the information from the publication-based approach:** The classification corresponding the MSC, a mandatory feature of the zbMATH data, provides valuable context information which characterize both the mathematical subjects and the algorithms behind the software as well as the application areas.
- **Aggregation and ranking of publication-based information:** The zbMATH data can be used to aggregate, summarize and rank the information given in all zbMATH entries citing a software: profiles for mathematical subjects and applications, for the acceptance of a software, and for listing similar software. On average, each software is cited more than 10 times in zbMATH but the number of citations varies strongly. Broadly accepted software packages have thousands of citations, small software packages only a few. This allows statements about the acceptance and the dissemination of a software. The documents of zbMATH are peer-reviewed which is also an indirect quality measure for all tools which were used within the publication.

But the publication-based approach has also some limitations

- **Completeness:** There are several reasons which influence the completeness of software information in swMATH. The zbMATH data of mathematical documents contain possibly a lower set of software citations than the full texts. zbMATH is the most comprehensive bibliographic database in mathematics. But this does not mean, that all documents which are application-focused but use mathematical methods, are listed in the database zbMATH. This is particularly true for mathematical software. Physicists, biologists, and chemists also have specialized journals for scientific software for their subjects which are out of the scope of zbMATH. For more completeness we additionally analyze further sources, especially arXiv [12], journals with focus on mathematical software as TOMS [11], and software repositories as CRAN [13]. Moreover there is a time delay between publication of the document and its inclusion in the database zbMATH.
- **Type vagueness:** The distinction between software and software-related objects as algorithms, programming languages and environments, or benchmarks is sometimes difficult. Moreover, software is provided in different forms, e.g., as service, package, or library. There is used a very primitive scheme for typing of the swMATH entries, see the "Browse software by type" in swMATH.

The publication-based approach basing on the zbMATH data seems to be an efficient way to identify software and extract information about software and to build a portal for mathematical software. It has also led to a nearly comprehensive portal of mathematical software and software-relevant research data. The swMATH portal is a search engine for software and provides separate Web pages for each software (product) containing a persistent identifier which can be used for citation. The swMATH pages combine the information about software from publications (indirect information), namely:

- name of the software product
- description
- keyword cloud
- list of publications citing a software
- list of related (similar) software products
- a MSC profile of the software product and a ranked lists of mathematical subjects and application areas
- a timeline of the number of publications citing a software

with Web information (direct information), especially from the website of a software, repositories, and developer platforms as github, or specialized archives for software code, especially Software Heritage [14]. For more details for extracting Web information see [15]. The publication-based approach extends the information about software especially by information about its use and dissemination. This information is helpful for searching suitable software to solve a problem. Such metadata are not contained in the CodeMeta metadata set, see [8], which addresses especially the direct information about software. A citation standard for software would be very helpful for a secure and precise machine-based identification of software, a standardized metadata scheme would improve the description and the search facilities.

3   Mathematical Models

3.1   Mathematical Models as Research Data

Is the publication-based approach also applicable to other mathematical research data? It is evident that the publications have the potential to detect detailed information about research data. In principle the problems are the same as for software: at first identification of research data in publications and at second analysis of information about mathematical research data. But its realization is challenging. Research data in mathematical documents appear in diverse forms: as explicitly defined objects or as citations. Up to now, the use of semantic markup of mathematical research data is low. This would require that the mathematical documents are offered in a semantic markup language und not only as textual terms in PDF files. But typographic characteristics as bold fonts etc. in combination with controlled vocabulary as proof, theorem, conjecture, etc. can be used for identification. Citations of mathematical research data which are defined as hyperlinks can be easily detected. The publication-based approach for other mathematical research data is confronted with a further challenge: Mathematical research data are explicitly described in full texts not in the zbMATH data. But the increasing number of Open Access journals and repositories like arXiv [12] opens up the possibility to extend the publication-based approach also to full texts. Each class of mathematical research data requires the development of its own concepts.

In the following, we will discuss the possibilities of a publication-based approach for mathematical models. Mathematical models are mathematical objects, this means they are described by a definition in mathematical language. But this covers a broad spectrum for the presentation of a mathematical model. Mathematical models are heterogeneous in content and form. The presentations are influenced by different factors,

- **User communities and their scientific background:** Different user communities prefer different terminologies and notations for presentations. Different real-world problems can lead to the same mathematical models.
- **Research aims and the degree of abstractions:** The research topics influence the selected presentation of mathematical models, e.g., for the numerical solution of a model specification, generalization, approximation. For instance, the numerical solution of a model can require the approximation of the original model, qualitative properties as stability or regularity are valid only for subclasses of the original model, etc. The degree of abstraction ranges from applications-related formulation by domain-specific laws and quantities containing dimensioned variables to abstract mathematical formulations, e.g., general equations, operators, or algebraic structures.

The missing infrastructure for mathematical models hampers the communication between different communities and leads to duplication and reinventing the wheel. Mathematical models are an important and independent class of mathematical research data. Especially the mathematical modeling of the whole spectrum of real-world problems has increased the importance of mathematical modeling. An indicator for this is the growing number of documents with the subject mathematical modeling. Therefore a concept for a uniform presentation, structuring, and semantification of mathematical models should be developed, containing

- **A compact and in some sense unified presentation of the investigated mathematical model(s):** Mathematical models are structured and often complex mathematical objects having inputs and outputs, parameters which are combined by operations and relations. For more transparency, mathematical models should be formalized in a unified way. A possible approach for a standardized formalization are the Model Path Diagrams [16].
- **Relations to other models:** This requires a specification (formalization) of the different relations between mathematical models, e.g., transformations, specializations, generalizations, approximations, etc.

3.2   The Publication-Based Approach for Mathematical Models

The first major problem is the (automatic) identification of mathematical models. Up to now, a special markup for mathematical models is missing and therefore the identification of model information in documents is challenging for the following reasons. Typically, the presentation of mathematical models in documents is given by mathematical expressions (formulae, diagrams, ...) which are embedded in text. Often complex mathematical models are not presented in a self-contained form. The components of a model can be positioned at different places of the document. The descriptions of a lot of mathematical models are incomplete and ambiguous. Names are used for citations but different names are used for the same model a lot of times. Also if the name is a named entity, a unique resolution is difficult because the names often describe classes of mathematical models and not a unique model. Mathematical documents contain more than one model which will be derived from the basic model by transformation, specification, generalization, approximation etc.

If we compare the situation between mathematical models and software we detect some analogies but also essential differences.

Software	Models
Software is given by a name (names can be used as identifier) (software code versus definition)	Models are given by a definition or names (but only prominent models have a name)
References: links to information about software products, e.g., documentations, websites strong growth	*References:* links to publications low growth
"Standard publications" for software: (Publication describing a software)	*"Standard publications" for models:* (Publication with focus on modeling or with sections about models)

The table shows that the features for software and models are only partially the same. For this reason the methods for software cannot be used without changes. As said above only named mathematical models appear in the zbMATH data. The zbMATH data are important to detect documents which contain significant information about mathematical models, especially in the title, keywords, review/abstract, and the MSC codes.

The most promising method to identify mathematical models is to search for explicit definitions of models which are given in fulltexts. Especially TEX-encoded documents as in [12] are the preferred object, because such files use a standardized markup for the document structure and mathematical expressions.

- **Phrase-based analysis:** Characteristic word phrases (containing the term "model") in zbMATH could also be used for model search. But phrase-based identification of mathematical models is for models less important than for software. Not all models are characterized by such a phrase, models are also termed as objects or defined only by mathematical expressions. This has to be investigated in detail by a statistical analysis.

- **Document structure:** We assume that the structural analysis of documents is more important than a phrase-based one. The document structure is very informative for the identification of mathematical models. All mathematical documents contain at least the problem which contains the mathematical object (model) which is investigated, the mathematical treatment, and results.

  Typically, research documents which originate from real-world problems start with the description of the background and the mathematical model. Often the mathematical model is described in a section of its own. It seems to be possible to identify these sections in documents by heuristic means. But this is only a first step, these sections can contain both too much (e.g., similar models) or too less information (e.g., missing notations which are described in another section). So, we have to develop further methods to extract a complete and compact form of the investigated mathematical model. A first attempt could be to extract all mathematical expressions in the relevant sections of a document which can be extended by a formulae analysis.

  Documents with the focus on mathematical modeling are particularly relevant for scientists working on the interface between real-world problems and mathematics. In these publications mathematical models are research results. This suggests that models used up to now are listed in the introduction and the new mathematical model is highlighted as the main result of the publication.

All features for the identification of mathematical models should be combined and tested in a planned project. A result could be a database of mathematical models which contains the original information given in the documents plus some metadata, e.g., keywords, classification codes, links to similar models, and a list of documents citing a model. The database also allows to cite a model. In a next step, the original model presentations could be enhanced by formalized presentations that facilitate understanding of the models for the human reader and express semantic and structural information in a machine-readable way which is significant for increasing the reusability mathematical models.

4   Summary and Outlook

Mathematical documents are containers for mathematical research data. The publication-based approach includes concepts to analyze the mathematical content by automatic means and extract specific information about mathematical research data. In the article, the publication-based approach for two types of mathematical research data is discussed, for mathematical software which was the starting point for the swMATH project and its potential use for mathematical models. The publication-based approach can be used to create new services, e.g., specialized databases for mathematical research data and for linking mathematical research data with each other. The publication-based approach opens the opportunity to extend the document-based information services by specialized services of mathematical research data and to enrich the search. But it has been shown that the transfer of the concept to other classes of research data is non-trivial and requires a rethinking by authors and publishers: open access to full texts in markup formats and not only in PDF. Currently, the existing services, especially arXiv, can be used to develop, evaluate and demonstrate concepts and prototypes for a sophisticated management of mathematical research data.

References

1. Open Mathematical Documents (OMDoc). https://kwarc.info/systems/omdoc/
2. Mathemathics Markup Language (MathML). https://www.w3.org/TR/MathML3/
3. The database zbMATH (1868–). https://www.zbmath.org
4. MathSciNet. http://www.ams.org/mathscinet
5. The Mathematics Subject Classification (2010). http://www.msc2010.org
6. Jahrbuch über die Fortschritte der Mathematik. https://www.emis.de/MATH/JFM/JFM.html
7. Smith, A.M., Katz, D.S., Niemeyer, K.E.: Software citation principles. PeerJ Comput. Sci. **2**, e86 (2016). https://doi.org/10.7717/peerj-cs.86. https://www.force11.org/software-citation-principles
8. CodeMeta. https://github.com/codemeta/
9. The database swMATH (2018–). http://www.swmath.org
10. Greuel, G.-M., Sperber, W.: swMATH – an information service for mathematical software. In: Hong, H., Yap, C. (eds.) ICMS 2014. LNCS, vol. 8592, pp. 691–701. Springer, Heidelberg (2014). https://doi.org/10.1007/978-3-662-44199-2_103. http://www.mathematik.uni-kl.de/~greuel/Paper/GeneralType/2014-ICMS_swmath.pdf
11. ACM Transactions on Mathematical Software (TOMS) (1975–). http://toms.acm.org/
12. arXiv.org e-Print archive. http://arxiv.org/
13. CRAN. https://cran.r-project.org/web/packages/
14. Software Heritage. https://www.softwareheritage.org/
15. Chrapary, H., Dalitz, W.: Software products, software versions, archiving of software, and swMATH. In: Davenport, J.H., Kauers, M., Labahn, G., Urban, J. (eds.) ICMS 2018. LNCS, vol. 10931, pp. 123–127. Springer, Cham (2018)
16. Koprucki, T., Kohlhase, M., Tabelow, K., Mller, D., Rabe, F.: Model pathway diagrams for the representation of mathematical models. Opt. Quant. Electron. **50** (2018). https://doi.org/10.1007/s11082-018-1321-7

HomotopyContinuation.jl: A Package for Homotopy Continuation in Julia

Paul Breiding[1](✉) and Sascha Timme[2](✉)

[1] Max Planck Institute for Mathematics in the Sciences, Leipzig, Germany
breiding@mis.mpg.de
[2] Technische Universität Berlin, Berlin, Germany
timme@math.tu-berlin.de
http://personal-homepages.mis.mpg.de/breiding/,
http://page.math.tu-berlin.de/~timme/

Abstract. We present the Julia package HomotopyContinuation.jl, which provides an algorithmic framework for solving polynomial systems by numerical homotopy continuation. We introduce the basic capabilities of the package and demonstrate the software on an illustrative example. We motivate our choice of Julia and how its features allow us to improve upon existing software packages with respect to usability, modularity and performance. Furthermore, we compare the performance of HomotopyContinuation.jl to the existing packages Bertini and PHCpack.

Keywords: Numerical algebraic geometry
Solving polynomial equations · Homotopy continuation · Julia

1 Introduction

Numerical algebraic geometry is concerned with the study of algebraic varieties by using numerical methods. The main computational building block therein is *homotopy continuation* which is a technique to approximate zero-dimensional solution sets of polynomial systems $F : \mathbb{C}^n \to \mathbb{C}^n$. The idea is that one first forms another polynomial system G related to F in a prescribed way, which has known or easily computable solutions. Then the systems G and F can be connected by setting up a homotopy $H : \mathbb{C}^n \times [0,1] \to \mathbb{C}^n$. An example for this would be the linear homotopy $H(x,t) = (1-t)F + tG$. For a properly formed homotopy, there are continuous solution paths leading from the solutions of G to those of F which may be followed using predictor-corrector methods. Singular solutions of F cause numerical difficulties, so singular endgames [18] are typically employed.

There are several software packages publicly available to make computations with homotopy continuation such as Bertini [3] and PHCpack [15]. We add the new and actively developed package HomotopyContinuation.jl[1] to that list. The

---

[1] www.JuliaHomotopyContinuation.org.

© Springer International Publishing AG, part of Springer Nature 2018
J. H. Davenport et al. (Eds.): ICMS 2018, LNCS 10931, pp. 458–465, 2018.
https://doi.org/10.1007/978-3-319-96418-8_54

package is programmed in Julia [4], which has recently gained much popularity in the numerical mathematics community. HomotopyContinuation.jl offers new and innovative features as well as a flexible design, which allows the user to adapt the code to the structure of their specific polynomial systems with little effort.

2   Functionality

HomotopyContinuation.jl aims at having an intuitive user interface. Assume we are interested in the solution set of the polynomial system

$$F := \begin{bmatrix} x^2 + y^2 - 1 \\ 3x - 2y \end{bmatrix} \tag{1}$$

which is the intersection of a quadric with a line. The code to solve this system is as follows:

```
using HomotopyContinuation # load package

@polyvar x y # we define variables x and y
solve([x^2+y^2-1, 3x-2y]) # define F and solve the system
```

In the background the software first constructs the total degree start system

$$G := \begin{bmatrix} x^2 - 1 \\ y - 1 \end{bmatrix} \tag{2}$$

and then defines the homotopy $H(x,t) := (1 - t)F + \gamma tG$ where $\gamma \in \mathbb{C}$ is choosen randomly. The two solutions $(-1, 1)$ and $(1, 1)$ of G are tracked towards the solutions of F. By default, we use the classical Runge-Kutta predictor and Newton's method for correction. Internally all computations are executed in the complex projective plane $\mathbb{P}^2$ on a (local) affine coordinate patch. In general, envoking the solve() command on any square system of polynomials will let HomotopyContinuation.jl generate a total degree starting system like (2).

HomotopyContinuation.jl also features a predictor-corrector scheme for overdetermined systems of polynomials $F : \mathbb{C}^N \to \mathbb{C}^n$ with $N < n$. However, in the overdetermined case there is no way to automatically generate a suitable starting system, but the user has to provide it. Furthermore, the input to HomotopyContinuation.jl is not limited to explicitly defined polynomial systems. Custom-defined homotopies are allowed. An example for a custom-defined homotopy for a family of overdetermined systems is given in Sect. 3.1.

In order to deal with singular solutions, an endgame strategy which combines the power series [8,18] and Cauchy endgame [18] is implemented. The solution can be computed in serial-processing as well as in parallel on a single machine by multiple threads.

3   Technical Contribution

Existing software packages are, as most scientific software, written in a fast, statically compiled language like C or C++. They then have to rely on files as input and output format, which can be cumbersome to write and parse, or they build a wrapper in a dynamic language like Python to allow the user to interact with the core software. While such a wrapper is preferable to a file based user interface it also has disadvantages. It puts an additional development and maintenance burden on the software authors and ultimately limits the flexibility of possible user input.

By contrast, HomotopyContinuation.jl is completely written in Julia, a high-level, dynamic programming language. There is no separation between the computational core and a wrapper with which the user interacts, everything is pure Julia. Julia programs are organized around *multiple dispatch*, which allows built-in and user-defined functions to be overloaded for different combinations of argument types. With its modular design HomotopyContinuation.jl exploits Julia's architecture. It is easy for users to extend and modify the capabilities of the package and to adapt the program to specific applications. An illustration of this is given in the following section, where we explain how to use the modular design for creating a homotopy that computes singular points on *symmetroids*.

Julia's LLVM-based just-in-time (JIT) compiler combined with the language's design allows to approach and often match the performance of C. For specific applications one can even surpass the performance of conventional C programs by making use of Julia's metaprogramming capabilities and its JIT compiler. One of these specific applications, which is of particular interest in the context of homotopy continuation, is the evaluation of polynomials. Let f be a polynomial with support $A \subset \mathbb{N}^n$. Generating optimal source code to evaluate polynomials with support A moves work from runtime to compile time, a tradeoff well worth if the same polynomials are evaluated very often, as it is the case during homotopy continuation. Horner's method for polynomials over the reals or a Goertzel-like method for complex polynomials [10, Sect. 4.6.2] may be employed to reduce the number of operations. Processor instructions like fused multiply-add (FMA) improve the performance and numerical accuracy. An experimental implementation of this idea by the second author is available under https://github.com/JuliaAlgebra/StaticPolynomials.jl. It also possible to use this optionally with HomotopyContinuation.jl.

3.1   Implementing Custom Homotopies – An Example

Above, we emphasized the *modular design* of HomotopyContinuation.jl and claimed that it is useful for creating homotopies for specific problems. A generic homotopy like the straight-line homotopy built from the total degree starting system (2) is not suited for highly structured problems. In fact, treating structured problems with structured homotopies may be decisive in making a computation feasible. The following example illustrates this.

Let $\mathcal{A} = (A_0, A_1, A_2, A_3) \in \mathrm{Sym}(\mathbb{R}^{n \times n})^{\times 4}$ be a 4-tuple of real symmetric matrices. The associated *symmetroid* $S_\mathcal{A}$ is the hypersurface in complex projective 3-space $\mathbb{P}^3$ given by the polynomial

$$f_\mathcal{A}(x_0, x_1, x_2, x_3) := \det(x_0 A_0 + x_1 A_1 + x_2 A_2 + x_3 A_3).$$

Already studied by Cayley [6], symmetroids are objects of interest at the intersection between algebraic geometry and optimization. Let us explain the connection to the latter. For a point $x = (x_0, \ldots, x_3) \in \mathbb{P}^3_\mathbb{R}$ with $x_0 \neq 0$ let us write $z_i = \frac{x_i}{x_0}$ for affine coordinates. The set of real points $x = (x_0, \ldots, x_3) \in \mathbb{P}^3_\mathbb{R}$ such that $A_0 + z_1 A_1 + z_2 A_2 + z_3 A_3$ is positive semi-definite is called a *spectrahedron* [16] and we denote it by $\Sigma_\mathcal{A}$. Spectrahedra are feasible sets in semi-definite programming, which is a generalization of linear programming [1,12]. For instance, problems as finding the smallest eigenvalue of a symmetric matrix or optimizing a polynomial function on the sphere can be formulated as a semidefinite programme. Because a linear function on a spectrahedron attains its maximum in a real singular point of the boundary with a positive probability, the number of singularities on the boundary of $\Sigma_\mathcal{A}$ matters. The boundary of $\Sigma_\mathcal{A}$ is $\Sigma_\mathcal{A} \cap S_\mathcal{A}$.

If A_1, A_2, A_3, A_4 are generic, the singular locus of the symmetroid $S_\mathcal{A}$ consists of $\binom{n+1}{3}$ isolated points. It is known how to construct a tuple $\mathcal{B} = (B_0, B_1, B_2, B_3)$ together with all of the associated $\binom{n+1}{3}$ singular points on $S_\mathcal{B}$. Moreover, the construction is such that the $\binom{n+1}{3}$ singular points of $S_\mathcal{B}$ are all real; see, e.g., [13, Theorem 1.1]. By contrast, a tuple $\mathcal{B} = (B_0, B_1, B_2, B_3)$ with $\Sigma_\mathcal{B} \cap S_\mathcal{B} = S_\mathcal{B}$, i.e., a tuple $\mathcal{B}$ for which all the associated singular points are at the same time points on the spectrahedron is only known for $n = 4$. This is due to work by Degtyarev and Itenberg [7]. In [14] Sturmfels poses the question:

How many of the $\binom{n+1}{3}$ singular points of $S_\mathcal{A}$ can lie on the boundary of $\Sigma_\mathcal{A}$?

By using homotopy continuation we can compute all the $\binom{n+1}{3}$ singular points on a symmetroid $S_\mathcal{A}$, from which we can check how many of them actually lie on the boundary of the spectrahedron. This way we advance in answering Sturmfels' question. We are currently working on a full featured implementation of the symmetroid-homotopy and will publish it in the near future. For the rest of this subsection let us explain the idea and sketch how an implementation of a symmetroid-homotopy in HomotopyContinuation.jl could look like.

To study the singularities of $S_\mathcal{A}$ we are interested in the zeros of the system

$$F_\mathcal{A}(x_0, x_1, x_2, x_3) := \left(f_\mathcal{A}, \frac{\partial f_\mathcal{A}}{\partial x_0}, \frac{\partial f_\mathcal{A}}{\partial x_1}, \frac{\partial f_\mathcal{A}}{\partial x_2}, \frac{\partial f_\mathcal{A}}{\partial x_3} \right). \tag{3}$$

A homotopy from a symmetroid $S_\mathcal{B}$ to a symmetroid $S_\mathcal{A}$ is then defined as

$$H_{\mathcal{A},\mathcal{B}}(x, t) := F_{(1-t)\mathcal{A}+t\mathcal{B}}(x). \tag{4}$$

Note that the number of monomials in $H_{\mathcal{A},\mathcal{B}}(x, t)$ in (x_0, x_1, x_2, x_3, t) for the generic choice of symmetric matrices is $(n+1)\binom{n+3}{n} + 4(n+1)\binom{n+2}{n-1}$. For $n = 20$

this number is 166551. The size of the polynomials prevents us from working with explicit expressions in the monomial basis. Already evaluating F_A and its Jacobian by considering the representation of f_A in the monomial basis becomes prohibitively expensive. On the other hand, the number of solutions of the system (3) is $\binom{21}{3} = 1330$, which is reasonably small.

Nevertheless, homotopy continuation algorithms never require to have the polynomial written down explicitly. What is needed for tracking the solution paths of a homotopy $H(x,t)$ is a function to evaluate $H(x,t)$ for all x and t and functions for evaluating the derivatives $\frac{\partial H(x,t)}{\partial x}$ and $\frac{\partial H(x,t)}{\partial t}$. Using matrix calculus and linear algebra, we find that the evaluation of $H_{A,B}$ and its Jacobian matrix at x are given by the first and second order derivatives of f_A at x. Denoting $A(x) := x_0 A_0 + x_1 A_1 + x_2 A_2 + x_3 A_3$ and $P_i(x) := A(x)^{-1} A_i$ they can be written in the following compact form:

$$\tfrac{\partial f_A}{\partial x_i}(x) = \det(A(x))\mathrm{tr}(P_i(x))$$

$$\tfrac{\partial^2 f_A}{\partial x_i \partial x_j}(x) = \det(A(x))\mathrm{tr}(P_i(x))\mathrm{tr}(P_j(x)) - \det(A(x))\mathrm{tr}(P_i(x)P_j(x))$$

where we used the fact that $\frac{\partial A(x)^{-1}}{\partial x_i} = -A(x)^{-1}A_i A(x)^{-1}$. The derivative of $H(x,t)$ with respect to t is obtained by a similar computation. Hence, the evaluation of F_A and its partial derivative can be done efficiently, because evaluating determinants can be done efficiently.

We use the aforementioned construction from [13, Theorem 1.1] for building a start system F_B. The Runge-Kutta predictor scheme and the overdetermined Newton corrector are employed for tracking the solutions from F_B to F_A.

Implementing this homotopy in existing software packages is very onerous and slow since the predefined interfaces can only handle the polynomial representation of $H_{A,B}$. By contrast, in HomotopyContinuation.jl the homotopy can be implemented in an efficient way. Since everything is defined in Julia, we have a full-fledged programming language at our hand to evaluate $H_{A,B}$. An illustrative example of the subset of the code necessary to handle $H_{A,B}$ in HomotopyContinuation.jl is depicted in Fig. 1.

4   Comparison

We compare HomotopyContinuation.jl against the established software packages Bertini and PHCpack. For this we pick a range of real-world polynomial systems of different type, presented in Table 1[2], and solve each polynomial system 10 times.

In particular, we take the perspective of a non-expert user and solve every system without any modification to the default parameters of the respective software packages. The only excemption is that for Bertini we distinguish between

---

[2] The authors discovered the examples in the excellent database of Jan Verschelde available at http://homepages.math.uic.edu/~jan/.

```
import HomotopyContinuation.Homotopies:
  AbstractHomotopy, evaluate!, jacobian!

struct SymmetroidHomotopy{T} <: AbstractHomotopy
  A_tuple::NTuple{4, Symmetric{T, Matrix{T}}}
  B_tuple::NTuple{4, Symmetric{T, Matrix{T}}}
end


# The SymmetroidHomotopy is a homotopy of 5 polynomials
# in 4 variables
Base.size(::SymmetroidHomotopy) = (5, 4)

# This computes H(x,t) and stores the result in u
function evaluate!(u, H::SymmetroidHomotopy, x, t, cache)
  tuple_at_t = (1-t) .* H.A_tuple .+ t .* H.B_tuple
  A = sum(x[i] * tuple_at_t[i] for i in 1:4)
  # Compute the inverse and determinant of the matrix A
  inv_A, det_A = inv(A), det(A)
  u[1] = det_A
  for i=1:4
    u[i+1] = det_A + trace(inv_A * tuple_at_t[i])
  end
end


# This computes the Jacobian of H at (x,t) and stores
# the result in U
function jacobian!(U, H::SymmetroidHomotopy, x, t, cache)
  tuple_at_t = (1-t) .* H.A_tuple .+ t .* H.B_tuple
  A = sum(x[i] * tuple_at_t[i] for i in 1:4)
  inv_A, det_A = inv(A), det(A)
  P = [inv_A * tuple_at_t[i] for i in 1:4]
  traces = trace.(P)
  U[1,:] = det_A .* traces
  for i=1:4, j=i:4
    U[1+i,j] = U[1+j,i] = det_A * traces[i] * traces[j] -
                          det_A * trace(P[i] * P[j])
  end
end
```

Fig. 1. Subset of the code necessary to track solutions of the homotopy $H_{A,B}$. In addition it is necessary to define a function dt!, which evaluates $\frac{\partial}{\partial t} H(x, t)$. Furthermore, it is possible to define a function evaluate_and_jacobian! that evaluates $H(x, t)$ and computes its Jacobian simultaneously. This is in particular useful here due to the shared structure of the derivatives. Although this code is able to solve the problem, it is written in an illustrative style. In a full featured implementation we would define an additional cache object to precallocate structures to avoid unnecessary temporary allocations.

Table 1. Overview of the polynomial systems choosen for the comparison. In the characteristics n is the number of unknowns, D is the Bézout number of the system and MV is the mixed volume. The system were taken from the database by Jan Verschelde.

| Polynomial systems | | Characteristics | | | | # Roots | |
Name	Description	Ref	n	D	MV	$\mathbb{C}$	$\mathbb{R}$
cyclic7	The cyclic 7-roots problems	[5]	7	5,024	924	924	56
ipp2	The 6R inverse position problem	[17]	11	1,024	288	16	0
heart	The heart-dipole problem	[11]	8	576	121	4	2
katsura11	A problem of magnetism in physics	[9]	12	2,048	2,048	2,048	326

Table 2. The results obtained for the systems in Table 1 using serial processing.

| | | # Solutions | | # Failed paths | | Runtime |
Systems	Package	Correct	Avg.	Avg.	Med.	Avg.
cyclic7	Bertini	8/10	923.3	1196.5	1300	48.93 s
	Bertini (adaptive precision)	10/10	924.0	0	0	1028.21 s
	PHCpack	0/10	918.4	5.6	5.5	6.48 s
	HomotopyContinuation.jl	10/10	924.0	0	0	8.38 s
heart	Bertini	10/10	4.0	66.0	73.5	4.88 s
	Bertini (adaptive precision)	10/10	4.0	0	0	30.63 s
	PHCpack	10/10	4.0	16.5	16	1.33 s
	HomotopyContinuation.jl	10/10	4.0	0	0	1.39 s
ipp2	Bertini	10/10	16.0	0.5	0	10.03 s
	Bertini (adaptive precision)	10/10	16.0	0	0	13.15 s
	PHCpack	10/10	16.0	272	272	6.67 s
	HomotopyContinuation.jl	10/10	16.0	0	0	3.07 s
katsura11	Bertini	8/10	2047.7	0.2	0	28.97 s
	Bertini (adaptive precision)	10/10	2048.0	0	0	28.88 s
	PHCpack	0/10	2043.7	2.3	2.0	179.13 s
	HomotopyContinuation.jl	10/10	2048.0	0	0	9.30 s

a version which uses adaptive precision [2] and one which uses standard 64 bit floating point arithmetic since HomotopyContinuation.jl as well as PHCpack also only compute by default with standard 64 bit floating point arithmetic.

We compare the packages with respect to outside observation. This is the number of times the correct number of solutions, the average number of solutions found, the average and median number of reported path failures (since these introduce uncertainity about the correctness of the result) and the average runtime. The results of the comparison are presented in Table 2. They were run on a MacBook Pro with a 2 GHz Intel i5-6360U CPU. We used MacOS 10.13.4 and Julia 0.6.2, Bertini v1.5.1 and PHCpack v2.4.52 and HomotopyContinuation.jl v0.2.0-alpha.2.

References

1. Alizadeh, F.: Interior point methods in semidefinite programming with applications to combinatorial optimization. SIAM J. Optim. **5**(1), 13–51 (1995)
2. Bates, D.J., Hauenstein, J.D., Sommese, A.J., Wampler, C.W.: Adaptive multi-precision path tracking. SIAM J. Numer. Anal. **46**(2), 722–746 (2008)
3. Bates, D.J., Hauenstein, J.D., Sommese, A.J., Wampler, C.W.: Bertini: Software for Numerical Algebraic Geometry. bertini.nd.edu, https://doi.org/10.7274/R0H41PB5
4. Bezanson, J., Edelman, A., Karpinski, S., Shah, V.B.: Julia: a fresh approach to numerical computing. SIAM Rev. **59**(1), 65–98 (2017)
5. Björck, G., Fröberg, R.: A faster way to count the solutions of inhomogeneous systems of algebraic equations, with applications to cyclic n-roots. J. Symbolic Comput. **12**(3), 329–336 (1991)
6. Cayley, A.: A memoir on quartic surfaces. Proc. London Math. Soc. **3**, 19–69 (1869/1871). (Collected Papers, VII, 133–181; see also the sequels on pages 256–260, 264–297)
7. Degtyarev, A., Itenberg, I.: On real determinantal quartics. In: Proceedings of the Gökova Geometry Topology Conference 2010 (2011)
8. Huber, B., Verschelde, J.: Polyhedral end games for polynomial continuation. Numer. Algorithms **18**(1), 91–108 (1998)
9. Katsura, S.: Spin glass problem by the method of integral equation of the effective field. In: New Trends in Magnetism, pp. 110–121 (1990)
10. Knuth, D.E.: The Art of Computer Programming, 3rd edn. Seminumerical Algorithms, vol. 2. Addison-Wesley Longman Publishing Co. (1997)
11. Nelson, C.V., Hodgkin, B.C.: Determination of magnitudes, directions, and locations of two independent dipoles in a circular conducting region from boundary potential measurements. IEEE Trans. Biomed. Eng. **12**, 817–823 (1981)
12. Ramana, M., Goldman, A.J.: Some geometric results in semidefinite programming. J. Global Optim. **7**, 33–50 (1995)
13. Sanyal, R.: On the derivative cones of polyhedral cones. Adv. Geometry **13**(2), 315–321 (2011)
14. Sturmfels, B.: Spectrahedra and their shadows. Talk at the Simons Institute Workshop on Semidefinite Optimization, Approximation and Applications (2014). https://simons.berkeley.edu/sites/default/files/docs/2039/slidessturmfels.pdf
15. Verschelde, J.: Algorithm 795: PHCpack: a general-purpose solver for polynomial systems by homotopy continuation. ACM Trans. Math. Software (TOMS) **25**(2), 251–276 (1999)
16. Vinzant, C.: What is... a Spectrahedron? Not. AMS **61**(5), 492–494 (2014)
17. Wampler, C., Morgan, A.: Solving the 6R inverse position problem using a generic-case solution methodology. Mech. Mach. Theory **26**(1), 91–106 (1991)
18. Wampler, I.C.W.: The Numerical Solution of Systems of Polynomials Arising in Engineering and Science. World Scientific (2005)

Polynomial Constraints and Unsat Cores in TARSKI

Fernando Vale-Enriquez and Christopher W. Brown[(✉)]

United States Naval Academy, Annapolis, USA
Fernvale25@gmail.com, wcbrown@usna.edu

Abstract. This paper gives a brief overview of TARSKI, a system for computing with Tarski formulas, which are boolean combinations of non-linear polynomial constraints over the reals. It gives an overview of TARSKI's basic functionality, then goes into more detail on facilities TARSKI provides for checking the satisfiability of conjunctions of constraints that are able to produce "unsat cores" for unsatisfiable inputs.

Keywords: Non-linear polynomial constraints · Unsat cores · SMT

1   Introduction

TARSKI is a system for computing with Tarski formulas/semi-algebraic sets[1]. The purpose of this paper is the give a broad overview of TARSKI and its functionality, followed by a more detailed description of tools it provides for computing "Unsat Cores" of conjunctions of polynomial constraints. This functionality is useful on its own, but particularly relevant in the context of Satisfiability Modulo Theory (SMT) solving, which requires theory solvers to provide unsat cores as "explanations" for sets of constraints that are not mutually satisfiable in the theory.

1.1   Related Work

Exact computations involving real, non-linear polynomial constraints is an area with a long history. Tarski's celebrated result from the 1930's [20] showed that the theory admits quantifier elimination by providing an algorithm to do it. This algorithm has a running time that is not even primitive recursive, so it was the beginning rather than the end of research in algorithms and programs to solve the quantifier elimination problem. Today, there are, for example, general quantifier elimination algorithms implemented in Mathematica, in Maple directly

---

[1] TARSKI is available at https://www.usna.edu/Users/cs/wcbrown/tarski/. It is open source, distributed under an ISC-style license.

This is a U.S. government work and its text is not subject to copyright protection in the United States; however, its text may be subject to foreign copyright protection 2018
J. H. Davenport et al. (Eds.): ICMS 2018, LNCS 10931, pp. 466–474, 2018.
https://doi.org/10.1007/978-3-319-96418-8_55

and through the Regular Chains and SyNRAC projects, in Reduce through the Redlog library, and as a stand-alone application in QEPCAD B. Quantifier elimination is a general, and computationally very difficult problem [5,10,21]. Research and software implementations for many other, usually more specific, problems involving exact computations with non-linear polynomial constraints has been a part of research in computation for as long as there have been computers. For example, for the problem of solving systems of equations there are packages like RAGlib and the Regular Chains library [8]; for sum-of-squares decomposition to prove positivity there is SOSTOOLS [18]; and for SAT-solving for real non-linear polynomial constraints there are systems like SMT-RAT [9], Z3 [15] and Yices [12]. This is just a small sample of the many problems researchers have considered and the many software packages they have produced.

The primary long-term focus of the TARSKI system is to compute well with formulas that are either automatically generated or come from users without the background to phrase problems to fit algorithmic tools well. It aims to provide tools for quantifier elimination, formula simplification, and satisfiability solving.

1.2 Background

The typical object we compute with in TARSKI is a *Tarski formula*, a first-order formula over the real numbers with relational operators as predicates and multiplication, addition and subtraction as functions. These may or may not have free variables. For example

$$\exists y[x^2+y^2 < 1 \wedge x+y > 0] , \quad xyz < 1 \wedge [x^2-2yz+y-2z-1 = 0 \vee x+2y+z > 2]$$

are Tarski formulas. For Tarski formula $F(x_1, \ldots, x_n)$, where $x_1, \ldots, x_n$ are the free variables, we naturally have an associated geometric object in $\mathbb{R}^n$ defined as $\{(\alpha_1, \ldots, \alpha_n) \in \mathbb{R}^n \,|F(\alpha_1, \ldots, \alpha_n)\}$. A subset of $\mathbb{R}^n$ that can be defined this way is called a *semi-algebraic set*.

Some basic computational problems for Tarski formulas include:

- *satisfiability*, are there assignments of real values to variables for which the formula evaluates to true;
- *quantifier elimination*, give an equivalent quantifier-free formula in the unquantified variables; and
- *simplification*, given a quantifier-free formula, provide an equivalent formula that is, in some sense, simpler.

These are the problems we are primarily interested in solving with TARSKI, and as stated previously, we would like, to the greatest extent possible, for TARSKI to be robust with respect to problem formulation.

For simplification and satisfiability solving for conjunctions of constraints, TARSKI also is able to produce "Unsat Cores" or "Explanations". An "Unsat Core" for an unsatisfiable conjunction of constraints is a subset of the original constraints that is unsatisfiable. For a human user, unsat cores are useful for understanding why a given input in unsatisfiable. In the context of automated SMT solving, a theory solver needs to be able to produce unsat cores in order to fit the general SMT framework [17].

1.3 Organization

The remainder of this paper is organized as follows. First we describe input, output and the general nature of interactions with TARSKI. We then describe TARSKI's facilities for calling QEPCAD B to perform quantifier elimination or formula simplification. This and some related functionality take steps to deal with problem formulation issues to which QEPCAD B is very sensitive. TARSKI also includes commands for constructing and querying Non-uniform Cylindrical Algebraic Decompositions (NuCADs), which are described. Finally, we describe recently developed functionality that provides unsat cores for unsatisfiable conjunctions of constraints.

2 Input, Output and General Interaction

TARSKI is an interpreter whose syntax is a (limited) variant of Scheme. The primary addition to Scheme syntax are built-in types *algebraic* and *Tarski formula*. Literals of these types are defined inside [ ]'s using a syntax similar to QEPCAD B's syntax for polynomials and formulas.

```
$ tarski
> (def F [ x^2 + y^2 < 1 /\ x + y > 0 ])
:void
> (def G [ex y[ $F ]])
:void
> G
ex y[y^2 + x^2 - 1 < 0 /\ y + x > 0]:tar
```

TARSKI is also able to read formulas written in the SMT-LIB version 2 format [1].

```
> (smtlib-load "Lyapunov1a-chunk-0015.smt2")
[17200 skoZ - 493 skoY - 2540 skoX <= 0 /\ -1(7200 skoZ^2 - 413 skoY skoZ
- 2130 skoX skoZ + 14100 skoY^2 + 10500 skoX skoY + 26100 skoX^2 - 1000)
> 0 /\ [-1(17200 skoZ - 493 skoY - 2540 skoX) > 0 \/ 17200 skoZ
- 493 skoY - 2540 skoX > 0] /\ [skoX /= 0 \/ skoY /= 0 \/ skoZ /= 0]]:tar
```

TARSKI is also able to output formulas in SMB-LIB format. Additionally, it can output formulas in the syntax of a number of other systems, e.g.

```
> (syntax 'mathematica G)
"Exists[{y} , y^2 + x^2 - 1 < 0 && y + x > 0]":str
> (syntax 'synrac G)
"Ex([y] , And(y^2 + x^2 - 1 < 0, y + x > 0))":str
```

3 Tarski as a Front-End to Qepcad b

QEPCAD B is a well-known system for quantifier elimination and formula simplification based on Cylindrical Algebraic Decomposition (CAD) [2]. There many options for how problems are phrased or input into the system, and these can hugely impact performance. Some of these are system-specific, but many are common to any CAD implementation. In general, problem phrasing for CAD-based methods is very important (see for example [11,13,14]). TARSKI can either be used to call QEPCAD B, making these decisions for the user, or can suggest problem formulation to the user. TARSKI chooses a variable order, declares equational constraints, makes use of QEPCAD B-specific features like special quantifiers, and assumptions.

```
> (def F [ex x,y [ a > 0 /\ b > 0 /\ b^2 (x - c)^2 + a^2 y^2 - a^2 b^2
= 0 /\ x^2 + y^2 > 1]])
:void
> (suggest-qepcad F)
"[]
(b,a,c,x,y)
3
(E x)(E y)[a^2 y^2 + b^2 x^2 - 2 b^2 c x + b^2 c^2 - a^2 b^2 = 0 /\
y^2 + x^2 - 1 > 0].
assume [a > 0 /\ b > 0]
prop-eqn-const
go
eqn-const-poly a^2 y^2 + b^2 x^2 - 2 b^2 c x + b^2 c^2 - a^2 b^2.
go
go
sol T
quit
":str
> (qepcad-qe F)
[a > 0 /\ b > 0 /\ [c + a - 1 > 0 \/ c - a + 1 < 0 \/ [-1(b^2 - a) < 0
/\ b^2 c^2 + b^4 - a^2 b^2 - b^2 + a^2 > 0]]]:tar
```

TARSKI also offers facilities for breaking up problems prior to calling QEPCAD B or simplifying formulas prior to calling QEPCAD B.

4   Non-uniform Cylindrical Algebraic Decomposition

Cylindrical Algebraic Decomposition (CAD) is a data-structure that provides an explicit representation of the semi-algebraic set defined by a Tarski formula. It is a computational tool used by a number of systems, including QEPCAD B, Mathematica and Redlog, to solve problems involving Tarski formulas. Non-uniform Cylindrical Algebraic Decomposition (NuCAD) relaxes some of the requirements of CAD, which allows semi-algebraic sets to be represented with smaller data structures. "Open" NuCAD construction is implemented in TARSKI and the OpenNuCAD data type supports a number of operations. Note: an "Open" NuCAD is not a true decomposition of real space, but rather a collection of disjoint open cells whose closures cover all of real space.

```
> (def D (make-NuCADConjunction [ x > 0 /\ y > 0 /\ x + y > 2/3 /\ x^2
+ y^2 < 1 ]))
:void
> (msg D 'get-var-order)
( y:sym x:sym )
> (msg D 'print-t-cells)
"Cell C1U2U2L1U TRUE (27/32 1/4) [ x > _root_1 x /\ x < _root_2 y^2 +
x^2 - 1  /\  y > _root_1 3 y - 2 /\ y· < _root_1 y - 1 ]
Cell C1U2U2LX TRUE (1/2 3/8) [ x > _root_1 3 y + 3 x - 2 /\ x <
_root_2 y^2 + x^2 - 1  /\  y > _root_1 y /\ y < _root_1 3 y - 2 ]
":str
> (msg D 'plot-leaves "-2 2 -2 2 600 600" "C"  "tmp.svg")
:void
```

Development of algorithms for NuCAD and implementations of those algorithms in TARSKI is ongoing work. One piece of functionality based on NuCAD that is quite useful is SAT solving for conjunctions of strict inequalities. "Open" NuCAD (and CAD, for that matter) suffices for SAT-solving when formulas contain only strict inequalities, i.e. $<, >, \neq$, and no logical negation [16,19]. TARSKI's SAT-NuCADConjunction command constructs an Open NuCAD for an input conjunction, terminating early as soon as it finds a satisfying point. In the following example, SAT-NuCADConjunction quickly finds a satisfying point for a problem that a number of other systems find challenging:

```
> (def F20 [   72 x - 99 y - 18 z x + 71 w - 59 > 0 /\ -68 x + 43 y +
68 z + 80 w^2 - 19 > 0 /\ -48 x - 47 y - 76 z^2 - 92 w - 95 > 0 /\ 17
x y + 78 y - 8 z - 70 w + 20 > 0 /\ -83 x - 20 y + 72 z x + 80 w + 25
> 0 /\ -9 x - 64 y^2 + 99 z + 39 w + 93 > 0 ] )
:void
> (head (SAT-NuCADConjunction F20))
( ( y:sym z:sym w:sym x:sym ) ( -41/16:num 25353/16384:num 3:num
-37825/4096:num ) )
```

5   Unsat Cores

In general, the SMT paradigm (see for example [17]) requires theory solvers to produce "explanations" for unsatisfiable inputs. More specifically, theory solvers are presented with a set of theory constraints, and the theory solver is supposed to determine whether these constraints are simultaneously satisfiable and, if so, to produce a satisfying assignment and, if not, to produce a (hopefully small) subset of the constraints that is unsatisfiable. The subset of constraints that is unsatisfiable, the *unsat core*, serves as the explanation. TARSKI includes implementations of algorithms that adapt the "fast simplification" algorithms of [3,4,6] to produce fast partial-solvers — theory solvers that can deduce "false" or "unknown", but never "true". These algorithms are always fast, and produce unsat cores for unsatisfiable instances.

5.1 bbsat

"BlackBox" simplification abstracts a formula by treating each factor as a separate variable. So a formula like $[(x - y)(xz - 1)^2(y^2 + z) > 0 \land (x - y)z < 0]$ is abstracted to $[ab^2c > 0 \land ad < 0]$. We can actually decide satisfiability in this abstraction quite quickly [3], moreover we can deduce new facts about the signs of the variables and products of the variables. The bbsat command determines satisfiability in the BlackBox abstraction, and returns unsat cores for unsatisfiable input.

```
(def F [ (x + 2) (x y - z)^2 (y - x) > 0 /\ (x + 2)  z^2 < 0 /\ y^2 z
(z + x) > 0 /\ (y - x) > 0 ])
:void
> (bbsat F)
( UNSAT:sym [(x + 2)(y - x)(z - x y)^2 > 0 /\ (x + 2)(z)^2 < 0 /\ y -
x > 0]:tar )
```

5.2 wbsat

"WhiteBox" simplification has a slightly more complex description, but it essentially allows for deductions on the signs of factors appearing in the formula based on known sign information on the individual variables and other factors. Once again when deductions lead to "unsat", TARSKI is able to produce unsat cores.

```
> (wbsat [ y x^2 + 1 < 0 /\ x + y < 0 /\ x > 0 /\ x - y < 0 /\ y > 0 ])
( UNSAT:sym [y - x > 0 /\ y > 0 /\ y + x < 0 /\ x > 0 /\ x^2 y + 1 < 0
]:tar )
```

The kinds of deductions allowed in WhiteBox are described in [6]. The augmentation of WhiteBox (and BlackBox) simplification algorithm to produce unsat cores is a new contribution due to the first author.

5.3 bbwb

BlackBox and WhiteBox methods can be combined to produce a more powerful fast partial-solver. The TARSKI command bbwb does this, alternating between BlackBox deductions and WhiteBox deductions, each potentially allowing the other to deduce still more facts. If unsat is deduced, bbwb traces back through the deductions to find a subset of the initial constraints that suffices to deduce unsat. In the following example, both BlackBox and WhiteBox deductions are used to deduce a contradiction:

```
> (def G [q0 < 0 /\ q1 > 0 /\ 2 B q2 + A eps^2 B + 2 q3 eps B - 2 x2 B
+ A^2 eps^2 + 2 A q3 eps + q3^2 < 0 /\ (2 B q2 + A eps^2 B + 2 q3 eps
B - 2 x2 B + A^2 eps^2 + 2 A q3 eps + q3^2)(B) < 0 /\ B > 0 /\ (B)(2 B
q2 - 2 x2 B + q3^2) < 0 /\ (B)(v1^2 - 2 x2 B + 2 x1 B) < 0 /\ (B)(2 B
q2 + 2 q1 q3 B + q1^2 A B - 2 x2 B + q3^2 + 2 q1 A q3 + q1^2 A^2) >= 0
/\ 2 B q2 - 2 x2 B + q3^2 < 0 /\ q2 - x2 < 0 /\ q3 >= 0 /\ x2 - x1 > 0
/\ v1 >= 0 /\ eps > 0 /\ A > 0 /\ 2 B q2 + 2 q1 q3 B + q1^2 A B - 2 x2
B + q3^2 + 2 q1 A q3 + q1^2 A^2 >= 0 /\ v1^2 - 2 x2 B + 2 x1 B - eps >
0])
:void
> (bbwb G)
( UNSAT:sym [v1^2 - 2 x2 B + 2 x1 B - eps > 0  /\ eps > 0  /\ B > 0 /\
(B)(v1^2 - 2 x2 B + 2 x1 B) < 0]:tar )
```

This result is deduced in two steps. First, $v_1^2 - 2x_2B + 2x_1B < 0$ is deduced from $B > 0$ and $(B)(v_1^2 - 2x_2B + 2x_1B) < 0$. This is a BlackBox deduction. Then a contradiction is derived from $v_1^2 - 2x_2B + 2x_1B < 0$, $v_1^2 - 2x_2B + 2x_1B - \epsilon > 0$ and $\epsilon > 0$. This is a WhiteBox deduction. Of course, an interesting feature of this example is that there are many constraints in the input conjunction, but the unsat core is quite small.

5.4   Unsat Cores in Qepcad b

Recent versions of QEPCAD B (version 1.72 and above) have also included facilities for computing unsat cores for fully existentially quantified conjunctions of constraints. This functionality is accessible through TARSKI.

```
> (def G [x > 0 /\ y > 0 /\ x + y < -1 /\ x^2 + y^2 < 1 /\ x + y > 5/2])
:void
> (qepcad-sat G)
[x > 0 /\ y^2 + x^2 - 1 < 0 /\ 2 y + 2 x - 5 > 0]:tar
```

Since QEPCAD B is a complete theory solver, we are able to do quick-checks for satisfiability using bbwb and then, for input that cannot be quickly determined to be unsat, fall back to QEPCAD B. In both cases, when the input is found to be unsatisfiable, TARSKI provides unsat cores.

Acknowledgments. The first author would like to thank the Office of Naval Research for its support of the United States Naval Academy's Trident Scholar program. He would also like to acknowledge H2020-FETOPEN-2016-2017-CSA project SC^2 (712689) for the SC^2 Summer School 2017, which he attended and which impacted and inspired work that he has incorporated into TARSKI. Both authors have also received support in this work from National Science Foundation Grant 1525896.

References

1. Barrett, C., Stump, A., Tinelli, C.: The SMT-LIB standard: version 2.0. In: Gupta, A., Kroening, D. (eds.) Proceedings of the 8th International Workshop on Satisfiability Modulo Theories, Edinburgh, UK (2010)
2. Brown, C.W.: QEPCAD B: a program for computing with semi-algebraic sets using CADs. ACM SIGSAM Bull. **37**(4), 97–108 (2003)
3. Brown, C.W.: Fast simplifications for Tarski formulas. In: Proceedings of the 2009 International Symposium on Symbolic and Algebraic Computation, ISSAC 2009, New York, NY, USA, pp. 63–70. ACM (2009)
4. Brown, C.W.: Fast simplifications for Tarski formulas based on monomial inequalities. J. Symb. Comput. **47**(7), 859–882 (2012)
5. Brown, C.W., Davenport, J.H.: The complexity of quantifier elimination and cylindrical algebraic decomposition. In: Proceedings of the 2007 International Symposium on Symbolic and Algebraic Computation, ISSAC 2007, New York, NY, USA, pp. 54–60. ACM (2007)
6. Brown, C.W., Strzeboński, A.: Black-box/white-box simplification and applications to quantifier elimination. In: Proceedings of the 2010 International Symposium on Symbolic and Algebraic Computation, ISSAC 2010, New York, NY, USA, pp. 69–76. ACM (2010)
7. Caviness, B.F., Johnson, J.R. (eds.): Quantifier Elimination and Cylindrical Algebraic Decomposition. Texts and Monographs in Symbolic Computation. Springer, Heidelberg (1998). https://doi.org/10.1007/978-3-7091-9459-1
8. Chen, C., Davenport, J.H., Lemaire, F., Maza, M.M., Xia, B., Xiao, R., Xie, Y.: Computing the real solutions of polynomial systems with the RegularChains library in Maple. ACM Commun. Comput. Algebra **45**, 166–168 (2011)
9. Corzilius, F., Loup, U., Junges, S., Ábrahám, E.: SMT-RAT: an SMT-compliant nonlinear real arithmetic toolbox. In: Cimatti, A., Sebastiani, R. (eds.) SAT 2012. LNCS, vol. 7317, pp. 442–448. Springer, Heidelberg (2012). https://doi.org/10.1007/978-3-642-31612-8_35
10. Davenport, J.H., Heintz, J.: Real quantifier elimination is doubly exponential. J. Symb. Comput. **5**, 29–35 (1997)
11. Dolzmann, A., Seidl, A., Sturm, T.: Efficient projection orders for CAD. In: Gutierrez, J. (ed.) Proceedings of the 2004 International Symposium on Symbolic and Algebraic Computation (ISSAC 2004), Santander, Spain, July 2004. ACM (2004)
12. Dutertre, B.: Yices 2.2. In: Biere, A., Bloem, R. (eds.) CAV 2014. LNCS, vol. 8559, pp. 737–744. Springer, Cham (2014). https://doi.org/10.1007/978-3-319-08867-9_49
13. England, M., et al.: Problem formulation for truth-table invariant cylindrical algebraic decomposition by incremental triangular decomposition. In: Watt, S.M., Davenport, J.H., Sexton, A.P., Sojka, P., Urban, J. (eds.) CICM 2014. LNCS (LNAI), vol. 8543, pp. 45–60. Springer, Cham (2014). https://doi.org/10.1007/978-3-319-08434-3_5
14. Huang, Z., England, M., Davenport, J.M., Paulson, L.C.: Using machine learning to decide when to precondition cylindrical algebraic decomposition with Groebner bases. In: 2016 18th International Symposium on Symbolic and Numeric Algorithms for Scientific Computing (SYNASC), pp. 45–52, September 2016
15. Jovanović, D., de Moura, L.: Solving non-linear arithmetic. In: Gramlich, B., Miller, D., Sattler, U. (eds.) IJCAR 2012. LNCS (LNAI), vol. 7364, pp. 339–354. Springer, Heidelberg (2012). https://doi.org/10.1007/978-3-642-31365-3_27

16. McCallum, S.: Solving polynomial strict inequalities using cylindrical algebraic decomposition. Comput. J. **36**(5), 432–438 (1993)
17. Nieuwenhuis, R., Oliveras, A., Tinelli, C.: Solving SAT and SAT modulo theories: from an abstract Davis-Putnam-Logemann-Loveland procedure to DPLL(T). J. ACM **53**, 2006 (2006)
18. Papachristodoulou, A., Anderson, J., Valmorbida, G., Prajna, S., Seiler, P., Parrilo, P.A.: SOSTOOLS: sum of squares optimization toolbox for MATLAB (2013). http://www.eng.ox.ac.uk/control/sostools. http://arxiv.org/abs/1310.4716
19. Strzebonski, A.: Solving systems of strict polynomial inequalities. J. Symb. Comput. **29**, 471–480 (2000)
20. Tarski, A.: A Decision Method for Elementary Algebra and Geometry, Second edn. University of California Press, Berkeley (1951). rev. Reprinted in [7]
21. Weispfenning, V.: The complexity of linear problems in fields. J. Symb. Comput. **5**(1–2), 3–27 (1988)

Private-Key Fully Homomorphic
Encryption for Private Classification

Alexander Wood[1,2,3,4]([✉]), Vladimir Shpilrain[5,6], Kayvan Najarian[2,3,4],
Ali Mostashari[7], and Delaram Kahrobaei[1,8]

[1] Department of Computer Science, The Graduate Center, CUNY, New York, USA
awood@gradcenter.cuny.edu
[2] Department of Computational Medicine and Bioinformatics,
University of Michigan, Ann Arbor, USA
[3] Center for Integrative Research in Critical Care, University of Michigan,
Ann Arbor, USA
[4] Emergency Medicine Department, University of Michigan, Ann Arbor, USA
[5] Department of Mathematics, The Graduate Center, CUNY, New York, USA
[6] Department of Mathematics, The City College of New York, New York, USA
[7] LifeNome Inc., New York, NY, USA
[8] Department of Computer Science, Tandon School of Engineering,
New York University, New York, USA

Abstract. Fully homomophic encryption enables private computation
over sensitive data, such as medical data, via potentially quantum-safe
primitives. In this extended abstract we provide an overview of an implementation of a private-key fully homomorphic encryption scheme in a
protocol for private Naive Bayes classification. This protocol allows a
data owner to privately classify her data point without direct access to
the learned model. We implement this protocol by performing privacy-
preserving classification of breast cancer data as benign or malignant.

Keywords: Fully homomorphic encryption · Data privacy
Machine learning

1 Introduction

Fully-homomorphic encryption (FHE) encompasses potentially quantum-safe
primitives which allows for computation of arbitrary functions over encrypted
data. The majority of current FHE research is public-key. In contrast, private
key cryptosystems require prior knowledge of the encryption/decryption key(s).
While this is considered a disadvantage when the goal is purely communication,
these cryptosytems are in fact excellent for applications involving sensitive data
[20].

A number of papers have approached the problem of private computation
over medical data. Some of these applications focus specifically on genomic computation, including edit distance [10], string matching [3,28], genomic tests such

© Springer International Publishing AG, part of Springer Nature 2018
J. H. Davenport et al. (Eds.): ICMS 2018, LNCS 10931, pp. 475–481, 2018.
https://doi.org/10.1007/978-3-319-96418-8_56

as ancestry and paternity [9], and other genomic tests [22,23]. Other research focuses on the task of private classification, including neural networks [17,27], decision trees [5], and Fisher's linear discriminant classifier [19]. All of these applications take place in the public-key setting.

In this extended abstract we propose a fully homomorphic private-key protocol for private Naive Bayes classification, a private argmax protocol, and a new fully homomorphic method of encoding floating point values. We test the protocol on publicly available breast cancer data [24] using the GKS FHE scheme [20] to achieve private classification in under one second.

2   Fully Homomorphic Encryption

A scheme is called additively or multiplicatively homomorphic, respectively, if $[\![x+y]\!] = [\![x]\!] \oplus [\![y]\!]$ and $[\![x \cdot y]\!] = [\![x]\!] \otimes [\![y]\!]$ for operations $\oplus$ and $\otimes$ in the ciphertext space and $[\![\alpha]\!]$ denotes the encryption of a value α. Because a boolean circuit can describe arbitrary computation, a scheme is called fully homomorphic if it is both additively and multiplicatively homomorphic [26].

The first FHE scheme was a lattice-based public-key encryption scheme introduced by Gentry which was theoretically revolutionary but impractical implementation-wise [12]. Improvements on this method led quickly to the second generation of fully homomorphic encryption schemes [6,7,11] including the BGV scheme [6–8] and YASHE [4]. More recent improvements have built upon this foundation to yield even faster schemes [2,13–16,21]. These schemes are all lattice-based, which is broadly considered a potentially quantum-resistant primitive.

More recently, interest has grown in private-key fully homomoprhic encryption. The GKS scheme [20] is a ring- and group-based private-key FHE scheme. It avoids some of the computational overhead required for fully homomorphic public key encryption and joins other schemes as a potentially quantum-resistant primitive. The GKS scheme is secure against a ciphertext-only attack.

3   Privacy-Preserving Classification

The main utility of fully homomorphic encryption lies in privacy-preserving classification, where a user classifies her datapoint using a data owner's learned model and neither party learns information about the other party's data [5]. A major application of PPC lies in the medical field. With PPC, a patient could use her medical data to perform medical analyses without worrying about revealing any of her personal information.

Leveled homomorphic encryption schemes (LHE), a variant on fully homomorphic encryption which allows for computation up to a predefined depth, such as YASHE have been used in an application of neural networks to encrypted data called CryptoNets [17]. ML Confidential [19] uses LHE to run classification using Linear Means and Fisher's Linear Discriminant classifiers. The authors in [5] construct protocols for privacy-preserving classification via hyperplane detection, Naive Bayes, and decision trees using two additively homomorphic encryption schemes, Quadratic Reciprocity (QR) [18] and Paillier [25], and one leveled homomorphic encryption scheme, HELib [21].

4   Proposed Method for Fully Homomorphic Private Classification with Naive Bayes

We use private-key fully homomorphic encryption in order to provide private classification using a learned Naive Bayes model. Our method follows from the method of Bost et al. [5] but varies in several important ways. Bost et al. re-encode ciphertexts between two additively homomorphic, public-key encryption schemes. Our implementation uses one fully homomorphic, private-key encryption scheme.

Assume that a Data Owner, D, wishes to classify her vector X which contains q features based off of a learned model w owned by a Classification Model Owner, C. The group G contains r distinct classes, $G_1, \ldots, G_r$. During this protocol C should learn no unnecessary information about the input provided by D, and D should learn nothing but the predicted class index of X.

C prepares tables P represented as a column vector of degree r where $P_i = \Pr(G = G_i)$, the prior probability on class G_i, and T, an $r \times q$ matrix where entry T_{ij} represents $\Pr(X = X_j | G = G_i)$. Private classification proceeds as follows:

1: C prepares the tables P and T and sends $[\![P]\!]$ and $[\![T]\!]$ to D.
2: For each class G_i for i from 1 to r, D computes

$$[\![P_i]\!] \cdot \prod_{j=1}^{p} [\![T_{ij}]\!] = \left[\!\!\left[P_i \cdot \prod_{j=1}^{p} T_{ij} \right]\!\!\right] = [\![\Pr(G_i | X)]\!].$$

3: D computes $i = \operatorname*{argmax}_{1 \leq i \leq r} [\![\Pr(G_i | X)]\!]$ using a private `argmax` protocol.

The privacy of the learned model is derived from the FHE scheme used during the protocol. The Data Owner's privacy depends on the argmax protocol in step 3. Let $\mathcal{F}$ denote a family of monotone, continuous, additively homomorphic functions that commute with encryption. Our protocol for computing private `argmax` is as follows:

1: Set $I = \{1, 2, \ldots, r\}$.
2: **while** $|I| > 1$ **do**
3:     D computes a random permutation π on I and randomly chooses $f \leftarrow \mathcal{F}$
4:     D computes $v = f([\![p_{\pi(1)}]\!]) - f([\![p_{\pi(2)}]\!]) = [\![f(p_{\pi(1)} - p_{\pi(2)})]\!]$
5:     D sends v to C.
6:     C decrypts v and recovers $f(p_{\pi(1)} - p_{\pi(2)})$. If this value is negative, C sends the bit $b = 0$ to D, otherwise send $b = 1$.
7:     If $b = 0$, remove $\pi(1)$ from I. Otherwise remove $\pi(2)$.
8: **end while**
9: D returns I.

During this protocol, C collects r values representing the result of a monotone function applied to the difference between random pairs of the posterior probabilities. The application of an unknown monotone function to this difference prevents C from learning partial information from the decrypted value.

5   Implementation

5.1   Fully Homomorphic Encoding

We must encode values in a way which preserves the fully homomorphic properties of the scheme. In the GKS scheme, elements are encoded in the ring

$$S_n = \langle x_1, x_2, \ldots, x_n | p \cdot 1 = 0, x_i^2 = x_i,$$
$$\text{and } x_i x_j = x_j x_i \text{ for all } i, j \rangle.$$

The plaintext ring P and a public ciphertext ring C are rings of the above form, where $P \subset C$. Elements have coefficients which lie in the field $\mathbb{Z}_p$. The authors of [20] provide a straightforward method of implementing a fully homomorphic encoding of integer values by mapping the element 1 of $\mathbb{Z}_p$ to any idempotent element of P. In order to encode floating point values, the most straightforward approach is to scale the floating point values to integer values with a fixed precision. Decoding requires the user to keep track of the number of times the protocol performs multiplication.

5.2   Experiments

To test the above protocols, we implemented a Naive Bayes algorithm to create a learned model and encrypted this learned model using the GKS Encryption Scheme [20]. The size of the ciphertext ring in the experiments was $2^8 = 256$. The family of functions used during private `argmax` is given by

$$\mathcal{F} = \{f : R \to R : f(m) = km\}$$

for sufficiently small $k \in \mathbb{Z}$ to avoid overflow over the 198-bit prime modulus p. Our protocols were implemented in C++ using GNU Multiple Precision Library (GMP) [1] and run on a MacBook Pro using El Capitan, a 2.3 GHz Intel Core i7, and with 16 GB memory.

Additive smoothing was performed on the prior probability tables before encryption. Specifically, each probability was increased by 0.1. Any value which was greater than or equal to 1 after smoothing was reset to $0.\overline{9}$, truncated at 20 digits.

Data from the UCI Machine Learning Repository was used to test the performance of the protocols [24]. Specifically, we looked at the Breast Cancer Wisconsin (Original) Data Set which contains 683 complete data points each containing an ID along with 9 attributes and a binary classification. The data gives measurements taken from fine-needle aspirate (FNA) biopsies of benign and malignant

Table 1. Unencrypted versus encrypted experimental results

	Time (s)	Accuracy	Sensitivity	Specificity
Unencrypted	0.00001	0.96003	0.93389	0.97410
Encrypted	0.52506	0.96003	0.93389	0.97410

breast tumors. Each of the nine attributes was measured by a clinician on a scale of 1 to 10 at the time it was collected. Previous research found that while each measurement holds clinical significance in diagnosing a breast tumor as benign or as malignant, a single attribute is not enough to distinguish between the two cases [29]. In the statistical analysis provided, a positive classification denotes a malignant classification and a negative classification denotes a benign classification. We used 10 by 10-fold cross validation to evaluate the performance of the algorithm.

The time data in the Table 1 represents the number of seconds it takes to classify a single data point. The time increase between encrypted and unencrypted classification is quite steep. However, this is to be expected and occurs to varying degrees in all fully homomorphic implementations. Classification of a single data point occurs in less than one second, a practical amount of time for clinical applications.

References

1. The GNU MP Bignum Library. https://gmplib.org/
2. Alperin-Sheriff, J., Peikert, C.: Faster bootstrapping with polynomial error. In: Garay, J.A., Gennaro, R. (eds.) CRYPTO 2014. LNCS, vol. 8616, pp. 297–314. Springer, Heidelberg (2014). https://doi.org/10.1007/978-3-662-44371-2_17
3. Ayday, E., Raisaro, J.L., Hengartner, U., Molyneaux, A., Hubaux, J.-P.: Privacy-preserving processing of raw genomic data. In: Garcia-Alfaro, J., Lioudakis, G., Cuppens-Boulahia, N., Foley, S., Fitzgerald, W.M. (eds.) DPM/SETOP -2013. LNCS, vol. 8247, pp. 133–147. Springer, Heidelberg (2014). https://doi.org/10.1007/978-3-642-54568-9_9
4. Bos, J.W., Lauter, K., Loftus, J., Naehrig, M.: Improved security for a ring-based fully homomorphic encryption scheme. In: Stam, M. (ed.) IMACC 2013. LNCS, vol. 8308, pp. 45–64. Springer, Heidelberg (2013). https://doi.org/10.1007/978-3-642-45239-0_4
5. Bost, R., Ada Popa, R., Tu, S., Goldwasser, S.: Machine learning classification over encrypted data. In: Symposium on Network and Distributed System Security (NDSS), February 2015
6. Brakerski, Z., Vaikuntanathan, V.: Efficient fully homomorphic encryption from (Standard) LWE. In: Proceedings of the 2011 IEEE 52nd Annual Symposium on Foundations of Computer Science, FOCS 2011, pp. 97–106. IEEE Computer Society, Washington, DC (2011)
7. Brakerski, Z., Vaikuntanathan, V.: Fully homomorphic encryption from ring-LWE and security for key dependent messages. In: Rogaway, P. (ed.) CRYPTO 2011. LNCS, vol. 6841, pp. 505–524. Springer, Heidelberg (2011). https://doi.org/10.1007/978-3-642-22792-9_29

8. Brakerski, Z., Vaikuntanathan, V., Gentry, C.: Fully homomorphic encryption without bootstrapping. In: Innovations in Theoretical Computer Science (2012)

9. Bruekers, F., Katzenbeisser, S., Kursawe, K., Tuyls, P.: Privacy-preserving matching of DNA profiles. Technical report (2008)

10. Cheon, J.H., Kim, M., Lauter, K.: Homomorphic computation of edit distance. In: Brenner, M., Christin, N., Johnson, B., Rohloff, K. (eds.) FC 2015. LNCS, vol. 8976, pp. 194–212. Springer, Heidelberg (2015). https://doi.org/10.1007/978-3-662-48051-9_15

11. van Dijk, M., Gentry, C., Halevi, S., Vaikuntanathan, V.: Fully homomorphic encryption over the integers. In: Gilbert, H. (ed.) EUROCRYPT 2010. LNCS, vol. 6110, pp. 24–43. Springer, Heidelberg (2010). https://doi.org/10.1007/978-3-642-13190-5_2

12. Gentry, C.: Fully homomorphic encryption using ideal lattices. In: Proceedings of the Forty-first Annual ACM Symposium on Theory of Computing, STOC 2009, pp. 169–178. ACM (2009)

13. Gentry, C., Halevi, S., Peikert, C., Smart, N.P.: Ring switching in BGV-style homomorphic encryption. In: Visconti, I., De Prisco, R. (eds.) SCN 2012. LNCS, vol. 7485, pp. 19–37. Springer, Heidelberg (2012). https://doi.org/10.1007/978-3-642-32928-9_2

14. Gentry, C., Halevi, S., Smart, N.P.: Better bootstrapping in fully homomorphic encryption. In: Fischlin, M., Buchmann, J., Manulis, M. (eds.) PKC 2012. LNCS, vol. 7293, pp. 1–16. Springer, Heidelberg (2012). https://doi.org/10.1007/978-3-642-30057-8_1

15. Gentry, C., Halevi, S., Smart, N.P.: Fully homomorphic encryption with polylog overhead. In: Pointcheval, D., Johansson, T. (eds.) EUROCRYPT 2012. LNCS, vol. 7237, pp. 465–482. Springer, Heidelberg (2012). https://doi.org/10.1007/978-3-642-29011-4_28

16. Gentry, C., Sahai, A., Waters, B.: Homomorphic encryption from learning with errors: conceptually-simpler, asymptotically-faster, attribute-based. In: Canetti, R., Garay, J.A. (eds.) CRYPTO 2013. LNCS, vol. 8042, pp. 75–92. Springer, Heidelberg (2013). https://doi.org/10.1007/978-3-642-40041-4_5

17. Gilad-Bachrach, R., Dowlin, N., Laine, K., Lauter, K., Naehrig, M., Wernsing, J.: CryptoNets: applying neural networks to encrypted data with high throughput and accuracy. In: PMLR, pp. 201–210, June 2016

18. Goldwasser, S., Micali, S.: Probabilistic encryption and how to play mental poker keeping secret all partial information. In: Proceedings of the Fourteenth Annual ACM Symposium on Theory of Computing, STOC 1982, pp. 365–377. ACM, New York (1982)

19. Graepel, T., Lauter, K., Naehrig, M.: ML confidential: machine learning on encrypted data. In: Kwon, T., Lee, M.-K., Kwon, D. (eds.) ICISC 2012. LNCS, vol. 7839, pp. 1–21. Springer, Heidelberg (2013). https://doi.org/10.1007/978-3-642-37682-5_1

20. Gribov, A., Kahrobaei, D., Shpilrain, V.: Private-key fully homomorphic encryption in rings. Groups, Complexity, Cryptology 10 (2018)

21. Halevi, S.: HElib: an implementation of homomorphic encryption (2013). https://github.com/shaih/HElib

22. Kim, M., Lauter, K.: Private genome analysis through homomorphic encryption. Cryptology ePrint Archive, Report 2015/965 (2015). http://eprint.iacr.org/2015/965

23. Lauter, K., López-Alt, A., Naehrig, M.: Private computation on encrypted genomic data. In: Aranha, D.F., Menezes, A. (eds.) LATINCRYPT 2014. LNCS, vol. 8895, pp. 3–27. Springer, Cham (2015). https://doi.org/10.1007/978-3-319-16295-9_1
24. Lichman, M.: UCI machine learning repository (2013). http://archive.ics.uci.edu/ml
25. Paillier, P.: Public-key cryptosystems based on composite degree residuosity classes. In: Stern, J. (ed.) EUROCRYPT 1999. LNCS, vol. 1592, pp. 223–238. Springer, Heidelberg (1999). https://doi.org/10.1007/3-540-48910-X_16
26. Peikert, C.: A decade of lattice cryptography. Found. Trends Theor. Comput. Sci. 10(4), 283–424 (2016)
27. Shokri, R., Shmatikov, V.: Privacy-preserving deep learning. In: Proceedings of the 22nd ACM SIGSAC Conference on Computer and Communications Security, CCS 2015, pp. 1310–1321. ACM, New York (2015)
28. Troncoso-Pastoriza, J.R., Katzenbeisser, S., Celik, M.: Privacy preserving error resilient DNA searching through oblivious automata. In: ACM Conference on Computer and Communications Security (CCS), pp. 519–528. ACM Press, Alexandria, October 29–Nov 2 2007 (2007)
29. Wolberg, W.H., Mangasarian, O.L.: Multisurface method of pattern separation for medical diagnosis applied to breast cytology. Proc. Nat. Acad. Sci. U.S.A 87, 9193–9196 (1990)

On μ-Symmetric Polynomials and D-Plus

Jing Yang[1]([✉]) and Chee K. Yap[2]([✉])

[1] SMS-HCIC, Guangxi University for Nationalities, Nanning, China
yangjing0930@gmail.com
[2] Courant Institute of Mathematical Sciences,
New York University, New York, USA
yap@cs.nyu.edu

Abstract. We study functions of the roots of a univariate polynomial of degree $n \geq 1$ in which the roots have a given multiplicity structure μ, denoted by a partition of n. For this purpose, we introduce a theory of μ-symmetric polynomials which generalizes the classic theory of symmetric polynomials. We designed three algorithms for checking if a given root function is μ-symmetric: one based on Gröbner bases, another based on preprocessing and reduction, and the third based on solving linear equations. Experiments show that the latter two algorithms are significantly faster. We were originally motivated by a conjecture about the μ-symmetry of a certain root function $D^+(\mu)$ called D-plus. This conjecture is proved to be true. But prior to the proof, we studied the conjecture experimentally using our algorithms.

1 Introduction

Suppose $P(x) \in \mathbb{Z}[x]$ is a polynomial with m distinct complex roots $r_1, \ldots, r_m$ where r_i has multiplicity μ_i. Write $\boldsymbol{\mu} = (\mu_1, \ldots, \mu_m)$ where we may assume $\mu_1 \geq \mu_2 \geq \cdots \geq \mu_m$. Thus $n = \sum_{i=1}^{m} \mu_i$ is the degree of $P(x)$. Consider the following function of the roots

$$D^+(P(x)) := \prod_{1 \leq i < j \leq m} (r_i - r_j)^{\mu_i + \mu_j}.$$

Call this the **D-plus** root function. This root function[1] was introduced by Becker et al. [2] in their complexity analysis of a root clustering algorithm. The original motivation of this paper was to try to prove that $D^+(P(x))$ is a rational function in the coefficients of $P(x)$.

Jing's work is supported by the Special Fund for Guangxi Bagui Scholars (WBS 2014-01) and the Startup Foundation for Advanced Talents in Guangxi University for Nationalities (2015MDQD018).

Chee's work is supported by Guangxi University for Nationalities and by NSF Grant # CCF-1564132.

[1] In [2], the D-plus function was called a generalized discriminant.

J. H. Davenport et al. (Eds.): ICMS 2018, LNCS 10931, pp. 482–491, 2018.
https://doi.org/10.1007/978-3-319-96418-8_57

We may write "$D^+(\boldsymbol{\mu})$" instead of $D^+(P(x))$ since the expression in terms of roots $\mathbf{r} = (r_1, \ldots, r_m)$ depends only on the multiplicity structure $\boldsymbol{\mu}$. For example, if $\boldsymbol{\mu} = (2, 1)$ then $D^+(\boldsymbol{\mu}) = (r_1 - r_2)^3$ and this turns out to be $[a_1^3 - (9/2)a_0a_1a_2 + (27/2)a_0^2a_3]/a_0^3$ where $P(x) = \sum_{i=0}^3 a_{3-i}x^i$. More generally, for any function $F(\mathbf{r}) = F(r_1, \ldots, r_m)$, we ask whether evaluating F at the m distinct roots of a polynomial $P(x)$ with multiplicity structure $\boldsymbol{\mu}$ is rational in the coefficients of $P(x)$. In case $P(x)$ has only simple roots, the Fundamental Theorem of Symmetric Functions tells us the complete answer: $F(\mathbf{r})$ is rational iff $F(\mathbf{r})$ is a symmetric polynomial. We extend this theorem to the case of non-simple roots: if the roots of $P(x)$ have multiplicity structure $\boldsymbol{\mu}$, then we define what it means for $F(\mathbf{r})$ to be $\boldsymbol{\mu}$-symmetric. As expected, this characterizes when $F(\mathbf{r})$ is rational in the coefficients of $P(x)$. It is non-trivial to check if any given root function F (in particular $F = D^+(\boldsymbol{\mu})$) is $\boldsymbol{\mu}$-symmetric. We will design three algorithms for this task. Although we feel that $\boldsymbol{\mu}$-symmetry is a natural concept, to our knowledge, this has not been systematically studied before.

Overview of Paper. In Sect. 2, we define $\boldsymbol{\mu}$-symmetric polynomials and show some preliminary properties of such polynomials. Then three algorithms for checking $\boldsymbol{\mu}$-symmetry are given in Sects. 3–5. Section 6 proves the $\boldsymbol{\mu}$-symmetry of $D^+(\boldsymbol{\mu})$. In Sect. 7, we show experimental results from our Maple implementation of the three algorithms. We conclude in Sect. 8.

The full version of this paper includes 3 appendices: A: Maple source code, B: Description of benchmark polynomials, and C: All the proofs. This full version may be downloaded from http://cs.nyu.edu/exact/papers/.

2 μ-Symmetric Polynomials

Throughout, assume K is a field of characteristic 0. For our purposes, $K = \mathbb{Q}$ will do. We fix three sequences of variables $\mathbf{x} = (x_1, \ldots, x_n)$, $\mathbf{z} = (z_1, \ldots, z_n)$ and $\mathbf{r} = (r_1, \ldots, r_m)$ where $n \geq m \geq 1$.

Let $\boldsymbol{\mu} = (\mu_1, \ldots, \mu_m)$ be a partition of n where $\mu_1 \geq \mu_2 \geq \cdots \geq \mu_m \geq 1$. We may denote this relation by $\boldsymbol{\mu} \vdash n$. We call $\boldsymbol{\mu}$ an m-**partition** if it has exactly m parts. A **specialization** σ is any function of the form $\sigma : \{x_1, \ldots, x_n\} \rightarrow \{r_1, \ldots, r_m\}$. We say σ is of **type** $\boldsymbol{\mu}$ if $|\sigma^{-1}(r_i)| = \mu_i$ for $i = 1, \ldots, m$. We say σ is **canonical** if $\sigma(x_i) = r_j$ and $\sigma(x_{i+1}) = r_k$ implies $j \leq k$. Clearly the canonical specialization of type $\boldsymbol{\mu}$ is unique, and we may denote it by $\sigma_{\boldsymbol{\mu}}$.

Consider the polynomial rings $K[\mathbf{x}]$ and $K[\mathbf{r}]$. Any specialization

$$\sigma : \{x_1, \ldots, x_r\} \rightarrow \{r_1, \ldots, r_m\}$$

can be extended naturally into a K-homomorphism $\sigma : K[\mathbf{x}] \rightarrow K[\mathbf{r}]$ where $P = P(\mathbf{x}) \in K[\mathbf{x}]$ is mapped to $\sigma(P) = P(\sigma(x_1), \ldots, \sigma(x_n))$. When σ is understood, we may write "$\overline{P}$" for the homomorphic image $\sigma(P)$.

We denote the i-**th elementary symmetric functions** ($i = 1, \ldots, n$) in $K[\mathbf{x}]$ by $e_i = e_i(\mathbf{x})$. E.g., $e_1 := \sum_{i=1}^n x_i$, $e_2 := \sum_{1 \leq i < j \leq n} x_i x_j$, ..., $e_n := \prod_{i=1}^n x_i$. Also define $e_0 := 1$. Typically, we write $\overline{e}_i$ for $\sigma_{\boldsymbol{\mu}}(e_i)$ when $\boldsymbol{\mu}$ is understood.

The key definition is the following: a polynomial $F \in K[\mathbf{r}]$ is said to be **μ-symmetric** if there is a symmetric polynomial $\widehat{F} \in K[\mathbf{x}]$ such that $\sigma_\mu(\widehat{F}) = F$ where $n = \sum_{i=1}^m \mu_i$. We call $\widehat{F}$ the **μ-lift** (or simply "lift") of F. If $\mathring{F} \in K[\mathbf{z}]$ satisfies $\mathring{F}(e_1, \ldots, e_n) = \widehat{F}(\mathbf{x})$ then we call $\mathring{F}$ the **μ-kernel** of F.

Remarks. 1. Note that the μ-lift of F is defined if and only if F is μ-symmetric.
2. We view the z_i's as symbolic representation of $e_i(\mathbf{x})$'s.
3. Although $\widehat{F}$ and $\mathring{F}$ are mathematically equivalent, the kernel concept lends itself to direct evaluation based on coefficients of $P(x)$.

The Fundamental Theorem on Symmetric Functions implies the following:

Lemma 1. *If $f(\mathbf{r}) \in K[\mathbf{r}]$ is μ-symmetric, then for any $P(x) = \sum_{i=1}^n c_i x^i \in K[x]$ of degree n, if P has m distinct roots $\rho_1, \ldots, \rho_m$ with multiplicity $\mu = (\mu_1, \ldots, \mu_m)$, $F(\rho_1, \ldots, \rho_m) \in K$.*

We want to study the lift $\widehat{F} \in K[\mathbf{x}]$ of a μ-symmetric polynomial $F \in K[\mathbf{r}]$ of total degree δ. If we write F as the sum of its homogeneous parts, $F = F_1 + \cdots + F_\delta$, then $\widehat{F} = \widehat{F}_1 + \cdots + \widehat{F}_\delta$. Hence, we may restrict F to be homogeneous.

Next consider a polynomial $G(\mathbf{z}) \in K[\mathbf{z}]$. Suppose there is a **weight function**

$$\omega : \{z_1, \ldots, z_n\} \to \mathbb{N} = \{1, 2, \ldots\}$$

then for any term $t = \prod_{i=1}^n z_i^{e_i}$, its ω-**degree** is $\sum_{i=1}^n e_i \omega(z_i)$. Normally, $\omega(z_i) = 1$ for all i; but in this paper, we are also interested in the weight function where $\omega(z_i) = i$. For short, we simply call this ω-degree of t its **weighted degree**. E.g., the weighted degree of $z_1^2 z_3$ is 5. The weighted degree of a polynomial $G(\mathbf{z})$ is just the maximum weighted degree of terms in its support. A polynomial $G(\mathbf{z})$ is said to be **weight homogeneous** if all of its terms have the same weighted degree. Note that the kernel $\mathring{F}$ of F is not unique: for any kernel $\mathring{F}$, we can decompose it as $\mathring{F} = \mathring{F}_0 + \mathring{F}_1$ where $\mathring{F}_0$ is the weight homogeneous part of $\mathring{F}$ of weighted degree δ, and $\mathring{F}_1 := \mathring{F} - \mathring{F}_0$. Then $\mathring{F}(\overline{e}_1, \ldots, \overline{e}_n) = F$ implies that $\mathring{F}_0(\overline{e}_1, \ldots, \overline{e}_n) = F$ and $\mathring{F}_1(\overline{e}_1, \ldots, \overline{e}_n) = 0$. We can always omit $\mathring{F}_1$ from the kernel of F. We shall call any polynomial $G(\mathbf{z}) \in K[\mathbf{z}]$ a **μ-constraint** if $G(\overline{e}_1, \ldots, \overline{e}_n) = 0$. Thus, $\mathring{F}_1$ is a μ-constraint. We may check that the set of μ-constraints forms an ideal in $K[\mathbf{z}]$ which we call the **μ-ideal**.

3 Computing Kernels via Gröbner Bases

In this section, we consider a Gröbner basis algorithm to compute the μ-kernel of a given polynomial $F \in K[\mathbf{r}]$, or detect that it is not μ-symmetric. For this purpose, define the following ideal:

$$\mathcal{I}_\mu := \langle v_1, \ldots, v_n \rangle \tag{1}$$

where $v_i := z_i - \overline{e}_i$ $(i = 1, \ldots, n)$. Note that $\mathcal{I}$ is an ideal in $K[\mathbf{z}, \mathbf{r}]$. Moreover, we define $\mathcal{G}_\mu$ to be the Gröbner basis of $\mathcal{I}_\mu$ relative to the term ordering where

$z_i \prec r_j$ for all i and j. The following is a generalization of Proposition 4 in [3, Chap. 7, Sect. 1].

Theorem 1. *Let $R \in K[\mathbf{r}, \mathbf{z}]$ be the normal form of $F \in K[\mathbf{r}]$ relative to $\mathcal{G}_\mu \subseteq K[\mathbf{r}, \mathbf{z}]$. Then F is μ-symmetric iff $R \in K[\mathbf{z}]$. Moreover, if $R \in K[\mathbf{z}]$ then R is the μ-kernel of F.*

Theorem 1 leads to the algorithm of Fig. 1.

G-kern(F, μ):
    Input: $F \in K[\mathbf{r}]$ and $\mu = (\mu_1, \ldots, \mu_m)$.
    Output: the μ-kernel of F or say "$\mathring{F}$ does not exist"
        $\mathcal{B} \leftarrow \{z_1 - \bar{e}_1(\mathbf{r}), \ldots, z_n - \bar{e}_n(\mathbf{r})\}$
        $ord \leftarrow plex(r_m, \ldots, r_1, z_n, \ldots, z_1)$
        $\mathcal{G} \leftarrow GroebnerBasis(\mathcal{B}, ord)$
        $R \leftarrow NormalForm(F, \mathcal{G}, ord)$
        If $\deg(R, \mathbf{r}) > 0$ then
            Return "$\mathring{F}$ does not exist"
        Return R

Fig. 1. Kernel Algorithm based on Gröbner Bases

4 Checking μ-Symmetry via Preprocessing and Reduction

In the previous section, we show how to compute μ-kernels using Gröbner bases. This algorithm is quite slow especially when $\mu \neq (1, 1, \ldots, 1)$; this may be seen from the timings in Table 1 below. In this and the next section, we will design two alternative methods based on an analysis of the following two K-vector spaces:

- $K^\delta_{\text{sym}}[\mathbf{x}]$: the set of symmetric homogeneous polynomials of degree δ in $K[\mathbf{x}]$
- $K^\delta_\mu[\mathbf{r}]$: the set of μ-symmetric polynomials of degree δ in $K[\mathbf{r}]$.

The first method is based on preprocessing and reduction: we first compute a basis for $K^\delta_\mu[\mathbf{r}]$, and then use the basis to reduce $F(\mathbf{r})$. The second method directly computes the μ-kernel of $F(\mathbf{r})$ by solving linear equations.

First consider $K^\delta_{\text{sym}}[\mathbf{x}]$. By a **weak partition** of δ, we mean

$$\alpha = (\alpha(1), \alpha(2), \ldots, \alpha(\delta))$$

where $\alpha(1) \geq \alpha(2) \geq \cdots \geq \alpha(\delta) \geq 0$ and $\alpha = \sum \alpha(i) = \delta$. Note that $\alpha(i)$ can be 0 in weak partitions. If α is a weak partition of δ with no part $\alpha(i)$ larger than n, we will write $\alpha \vdash (\delta, n)$. Let $e_\alpha := \prod_{i=1}^\delta e_{\alpha(i)}$. E.g., if $\delta = 4, n = 2, \alpha = (2, 1, 1, 0)$ then $e_\alpha = e_2 e_1^2 e_0 = e_2 e_1^2$.

Let $T(\mathbf{x})$ denote the set of terms of $\mathbf{x}$, and $T^\delta(\mathbf{x})$ denote those terms of degree δ. A typical element of $T^\delta(\mathbf{x})$ is $\prod_{i=1}^n x_i^{e_i}$ where $e_1 + \cdots + e_n = \delta$. We totally order the terms in $T^\delta(\mathbf{x})$ using the lexicographic ordering in which $x_1 \prec x_2 \prec \cdots \prec x_n$.

Given any $F \in K(\mathbf{x})$, its **support** is $\mathrm{Supp}(F) \subseteq T(\mathbf{x})$ such that F can be uniquely written as

$$F = \sum_{p \in \mathrm{Supp}(F)} c(p)p \qquad (2)$$

where $c : \mathrm{Supp}(F) \to K \setminus \{0\}$ denote the coefficients of F. Let the **leading term** $\mathrm{Lt}(F)$ be equal to the $p \in \mathrm{Supp}(F)$ which is the largest under the lexicographic ordering. For instance, $\mathrm{Supp}(e_1) = \{x_1, \ldots, x_n\}$ and $\mathrm{Lt}(e_1) = x_n$. Also $\mathrm{Supp}(e_1 e_2) = \{x_i x_j x_k : 1 \le i \ne j \le n, 1 \le k \le n\}$ and $\mathrm{Lt}(e_1 e_2) = x_n^2 x_{n-1}$. The coefficient of $\mathrm{Lt}(F)$ in F is the **leading coefficient** of F, denoted by $\mathrm{Lc}(F)$. Call $\mathrm{Lm}(F) := \mathrm{Lc}(F)\mathrm{Lt}(F)$ the **leading monomial** of F. This is well-known:

Proposition 2. *The set $\mathcal{B}_n^\delta := \{e_\alpha : \alpha \vdash (\delta, n)\}$ is a K-basis for $K_{\mathrm{sym}}^\delta[\mathbf{x}]$.*

Next consider the set $K_\mu^\delta[\mathbf{r}]$ comprising the μ-symmetric functions of degree δ. The map $\sigma_\mu : K_{\mathrm{sym}}^\delta[\mathbf{x}] \to K_\mu^\delta[\mathbf{r}]$ is an onto K-homomorphism. Thus $K_\mu^\delta[\mathbf{r}]$ is a vector space which is generated by the set $\sigma_\mu(\mathcal{B}_n^\delta) := \{\sigma_\mu(G) : G \in \mathcal{B}_n^\delta\}$. It follows that there is a maximal independent set $\overline{\mathcal{B}}_n^\delta \subseteq \sigma_\mu(\mathcal{B}_n^\delta)$ that is a basis for $K_\mu^\delta[\mathbf{r}]$. The set $\overline{\mathcal{B}}_n^\delta$ may be a proper subset of $\sigma_\mu(\mathcal{B}_n^\delta)$.

Now we generate the basis of the vector space $K_\mu^\delta[\mathbf{r}]$ with which one could easily check whether a given polynomial is in this vector space or not. For this purpose, we introduce a reduction procedure and its applications. A set $\mathcal{B} \subseteq K[\mathbf{r}]$ is **linearly independent** if any non-trivial K-linear combination over $\mathcal{B}$ is nonzero; otherwise, $\mathcal{B}$ is **linearly dependent**. We say $\mathcal{B}$ is **canonical** if $\mathcal{B}$ is linearly independent and ordered as $\mathcal{B} = (B_1, \ldots, B_\ell)$ with $\mathrm{Lt}(B_i) \prec \mathrm{Lt}(B_j)$ for all $i < j$.

Given a polynomial $F \in K[\mathbf{r}]$, we say it is **reduced** relative to $\mathcal{B} = (B_1, \ldots, B_\ell)$ if $\mathrm{Lt}(B_i) \notin \mathrm{Supp}(F)$ for each $i = 1, \ldots, \ell$. We can reduce F relative to $\mathcal{B}$ by subtracting from F a linear combination of elements in $\mathcal{B}$ as shown in Fig. 2.

The termination of the **reduce** algorithm is guaranteed by the following:

Lemma 3. *The number of loops in the algorithm* $\mathrm{reduce}(F, \mathcal{B})$ *is bounded by*

$$\#\mathrm{Supp}(F) + \sum_{i=1}^{\ell} \#\mathrm{Supp}(B_i) - 1.$$

Moreover, this bound is tight in the worst case.

It is easy to see that $\mathrm{reduce}(F, \mathcal{B}) = 0$ iff $\mathcal{B} \cup \{F\}$ is linearly dependent. This gives rise to the **canonize** algorithm in Fig. 3 for constructing a canonical set from any set $\mathcal{B} \subseteq K[\mathbf{r}]$. Clearly $\mathrm{canonize}(\mathcal{B})$ terminates in $|\mathcal{B}|$ loops. Finally, we use **reduce** and **canonize** algorithms to construct the isMuSymmetric algorithm for checking the μ-symmetry of a polynomial.

Lemma 4. *The algorithm* isMuSymmetric *halts. Moreover, it outputs "Yes" iff F is μ-symmetric.*

```
reduce(F, B):
    Input:   F ∈ Kδ[r], B = (B₁, ..., Bℓ) is canonical and each Bᵢ ∈ Kδ[r]
    Output:  R such that F = Σℓᵢ₌₁ cᵢBᵢ + R with cᵢ ∈ K and
             R is reduced relative to B.
        Let R ← 0, i ← ℓ
        If B = () then
            Return F
        While (F ≠ 0)
            p ← Lt(F)
            If p = Lt(Bᵢ) then
                F ← F − Lc(F)/Lc(Bᵢ) Bᵢ; i ← i − 1
            elseif p ≻ Lt(Bᵢ) then
                R ← R + Lc(F) · p; F ← F − Lc(F) · p
            else
                i ← i − 1
            If i = 0 then Return R + F
        Return R
```

Fig. 2. The `reduce` algorithm.

```
canonize(B):
    Input:   B ⊆ K[r].
    Output:  a maximal canonical C ⊆ B
        Let C ← () (empty sequence)
        While B ≠ ∅
            B ← pop(B)
            B' ← reduce(B, C)
            If B' ≠ 0 then
                C ← prepend(B', C)
                C ← sort(C)
        Return C
```

```
isMuSymmetric(F, μ):
    Input:   F ∈ K[r], μ = (μ₁, ..., μₘ)
    Output:  Yes if F is μ-symmetric;
             otherwise return No.
        δ ← deg(F, r)
        n ← Σᵐᵢ₌₁ μᵢ
        B ← {ēα : α ⊢ (δ, n)}
        C ← canonize(B)
        If reduce(F, C) = 0 then
            Return "Yes"
        Return "No"
```

Fig. 3. The `canonize` and `isMuSymmetric` algorithms.

5   Computing Kernels via Solving Linear Systems

We now outline a method to compute the kernel of $F(\mathbf{r})$ by solving a linear system of equations.

Recall that $F \in K[\mathbf{r}]$ is μ-symmetric iff there is a $\mathring{F} \in K[\mathbf{z}]$ such that $\mathring{F}(\bar{e}_1, \ldots, \bar{e}_n) = F$. We propose to first write $\mathring{F}(\mathbf{z})$ as an indeterminate polynomial $G(\mathbf{k}; \mathbf{z}) \in K[\mathbf{k}][\mathbf{z}]$ which has homogeneous weighted degree δ with indeterminate coefficients $\mathbf{k}$. Each term of weighted degree δ has the form $\mathbf{z}_\alpha := \prod_{i=1}^{\delta} z_{\alpha(i)}$ where $\alpha = (\alpha(1), \ldots, \alpha(\delta))$ is a weak partition of δ with parts at most n, i.e., $\alpha \vdash (\delta, n)$. Let the set of all such partitions be denoted $I_n^\delta := \{\alpha : \alpha \vdash (\delta, n)\}$ Then $G(\mathbf{k}; \mathbf{z})$ can be written as $G(\mathbf{k}; \mathbf{z}) := \sum_{\alpha \in I_n^\delta} k_\alpha \mathbf{z}_\alpha$ where each k_α is an

E-kern(F, μ):
    Input: $F \in K[\mathbf{r}]$ and $\mu = (\mu_1, \ldots, \mu_m)$
    Output: the kernel of F if F is μ-symmetric;
            otherwise, return "F is not μ-symmetric".
        $\delta \leftarrow \deg(F, \mathbf{r})$; $n \leftarrow \sum_{i=1}^m \mu_i$
    Construct the index set I_n^δ.
    $G \leftarrow \sum_{\alpha \in I_n^\delta} k_\alpha z_\alpha$
    $H \leftarrow G(\mathbf{k}; \bar{e}_1, \ldots, \bar{e}_n)$
    Extract Coeffs$(H, \mathbf{r})$ and Coeffs$(F, \mathbf{r})$.
    Find a solution $\mathbf{k} = \mathbf{k}_0$ of the linear system
            Coeffs$(H, \mathbf{r}) = $ Coeffs$(F, \mathbf{r})$.
    If $\mathbf{k}_0$ is nondefined
            Return "F is not μ-symmetric"
    Return $G(\mathbf{k}_0; \mathbf{z})$

Fig. 4. Kernel Algorithm via Linear System Solving

indeterminate. Here, $\mathbf{k} := (k_\alpha : \alpha \in I_n^\delta)$. Next, we plug in $\bar{e}_i$'s for the z_i's to get $H(\mathbf{k}; \mathbf{r}) := G(\mathbf{k}; \bar{e}_1, \ldots, \bar{e}_n)$ which we view as a polynomial in $K[\mathbf{k}][\mathbf{r}]$. We then set up the equation

$$H(\mathbf{k}; \mathbf{r}) = F(\mathbf{r}) \tag{3}$$

to solve for the values of $\mathbf{k}$. Note that total degree of G in $\mathbf{k}$ is 1, i.e., $\deg(G, \mathbf{k}) = 1$. Therefore, $\deg(H, \mathbf{k}) = 1$. Thus (3) amounts to solving a linear system of equations in $\mathbf{k}$. The above procedure can be summarized as the E-kern algorithm in Fig. 4.

6   The μ-Symmetry of $D^+(\mu)$

The conjecture that motivated this paper is that the root function $D^+(\mu) = \prod_{1 \le i < j \le m}(r_i - r_j)^{\mu_i + \mu_j}$ is μ-symmetric. This conjecture is proved in following theorem:

Theorem 2. $D^+(\mu)$ is μ-symmetric, with kernel $\overset{\circ}{D}{}^+(\mu) = \frac{1}{c}H$ where

- $c = c(\mu) = (-1)^{mn + \frac{n(n-1)}{2} + \sum_{i=1}^m i\mu_i} \cdot (n - m)! \prod_{i=1}^m \mu_i^{\mu_i}$
- $H := \frac{\partial^{n-m} D}{\partial c_n^{n-m}}\big|_{c_i = (-1)^i z_i c_0} \big/ c_0^{m+n-2}$.
- Here, $D = D(P)$ is the discriminant of $P(x) = \sum_{i=0}^n c_{n-i}x^i$ with multiplicity structure μ.

This theorem further tells us that the kernel polynomial of $D^+(\mu)$ is, up to a constant, determined by the number m of distinct roots. In other words, if $\mu = (\mu_1, \ldots, \mu_m)$ and $\mu' = (\mu'_1, \ldots, \mu'_{m'})$ with $m = m'$, then there exists a constant a such that $\overset{\circ}{D}{}^+(\mu) = a\overset{\circ}{D}{}^+(\mu')$.

Since the lift/kernel of D^+ is non-unique, it may be possible to rewrite it in simpler forms, especially when μ has some special structure. The rest of this section gives two such theorems. First, we introduce a μ-symmetric function Δ that is useful in deriving simple expressions for D^+. Δ is closely related to the notion of subdiscriminants [1, Sect. 4.1]. To define them, we need some notations: let $[n] := \{1, \ldots, n\}$, and $\binom{[n]}{k}$ denote the set of k-subsets of $[n]$. For $k = 0, \ldots, n-2$, we may define the function

$$S_k^n = S_k^n(\mathbf{x}) := \sum_{I \in \binom{[n]}{n-k}} \prod_{i \neq j \in I} \left(x_i - x_j\right)^2 \tag{4}$$

called the kth **subdiscriminant** in n variables. We may also define $S_{n-1}^n := 1$. When $k = 0$, we have $S_0^n = \prod_{i \neq j \in [n]} \left(x_i - x_j\right)^2$. If the x_i's are roots of a polynomial $P(x)$ of degree n, then S_0^n is the standard discriminant of $P(x)$. Clearly S_k^n is a symmetric polynomial in $\mathbf{x}$.

Lemma 5. *Define $\Delta := \prod_{1 \leq i < j \leq m} (r_i - r_j)^2$.*

(a) Δ is μ-symmetric with lift given by $\widehat{\Delta} = \frac{1}{\prod_{i=1}^m \mu_i} \cdot S_{n-m}^n$ where S_{n-m}^n is the $(n-m)$-th subdiscriminant.

(b) When $m = 2$, we have an explicit formula for the lift of Δ: with $n = \mu_1 + \mu_2$,

$$\widehat{\Delta} = \frac{(n-1)e_1^2 - 2ne_2}{\mu_1 \mu_2}.$$

The following results provide explicit formulas for the lift $\widehat{D^+}$ or kernel $\overset{\circ}{D^+}$ in two special cases of μ. First, consider the case where $\mu = (a, a, \ldots, a)$.

Theorem 3. *If all μ_i's are equal to a, then $D^+(\mu)$ is μ-symmetric with lift given by $\widehat{F}_n(\mathbf{x}) = \left(\frac{1}{a^m} \cdot S_{n-m}^n\right)^a$ where S_{n-m}^n is given by Lemma 5(a).*

Another special case of μ is when $m = 2$:

Theorem 4. *For all $\mu = (\mu_1, \mu_2)$, $D^+(\mu)$ has a μ-kernel $\overset{\circ}{F}_n$ where $\mu \vdash n$:*

– n is even: $\overset{\circ}{F}_n = \left(\frac{(n-1)z_1^2 - 2nz_2}{\mu_1 \mu_2}\right)^{n/2}$

– n is odd: $\overset{\circ}{F}_n = \left(\frac{(n-1)z_1^2 - 2nz_2}{\mu_1 \mu_2}\right)^{\frac{n-3}{2}} \left(k_1 z_1^3 + k_2 z_1 z_2 + k_3 z_3\right)$

where $(k_1, k_2, k_3) = \left(\frac{-(n-1)(n-2)}{d}, \frac{3n(n-2)}{d}, \frac{-3n^2}{d}\right)$ and $d = \mu_1 \mu_2 (\mu_1 - \mu_2)$.

7 Software and Experiments

Table 1 shows timings of our algorithms (G-kern, E-kern and isMuSymmetric) for checking the existence of the μ-kernel of $F \in K[\mathbf{r}]$, or reporting "No" otherwise. They are implemented in Maple (see code in Appendix A in full version). These experiments use Maple 2017 on a Windows laptop with an Intel(R) Core(TM) i7-7660U CPU (2.50 GHz, 8 GB RAM). We use a test suite of 12 polynomials of degrees ranging from 6–20 (see Appendix B in full version), with corresponding μ with $n = \sum_{i=1}^{m} \mu_i$ ranging from 4–6. These polynomials are either D^+ polynomials or subdiscriminants, or their variants to create non-μ-symmetric polynomials.

Table 1 shows that E-kern is significantly faster than G-kern on all but in this case, $\mu = (1, \ldots, 1)$, i.e., the ideal $\mathcal{I}_\mu = \langle v_1, \ldots, v_n \rangle$ is symmetric in $\mathbf{r}$. Possibly, the Gröbner basis algorithm in Maple is highly optimized for such ideals. One may also see that isMuSymmetric is also a very efficient method for checking the μ-symmetry of a polynomial. In particular, the preprocessing procedure canonize is independent on F, so one can compute the canonical set first and store it in a database. The actual time to reduce a given F using a canonical set is relatively small. The speedup of G-kern/isMuSymmetric may be partly attributed to the fact that G-kern outputs more information than isMuSymmetric. In the full paper, we will extend isMuSymmetric into an algorithm to actually compute the kernel.

Table 1. Timing for computing μ-kernel of F of degree δ. Here $n = \sum_{i=1}^{m} \mu_i$, canonize is a preprocessing step in isMuSymmetric and total= the sum of canonize time and reduce time.

F	δ	μ	n	Y/N	G-kern Time (sec)	E-kern Time (sec)	Speedup (G-kern/E-kern)	isMuSymmetric			speedup (G-kern /isMuSymmetric)
								canonize (sec)	reduce (sec)	Total (sec)	
F1	12	[1, 1, 1, 1]	4	Y	0.453	0.235	1.9	0.094	0.000	0.094	4.8
F2	8	[2, 1, 1]	4	Y	0.328	0.015	21.9	0.016	0.015	0.031	10.6
F3	20	[1, 1, 1, 1, 1]	5	Y	34.141	187.703	0.2	3.766	0.031	3.797	9.0
F4	15	[2, 1, 1, 1]	5	Y	>600.000	1.875	>320.0	0.391	0.015	0.406	>1477.8
F4x	6	[2, 1, 1, 1]	5	N	>600.000	0.015	>40000.0	0.000	0.016	0.016	>37500.0
F5	6	[2, 2, 1]	5	Y	68.031	0.032	2126.0	0.000	0.000	0.000	Inf
F5x	6	[2, 2, 1]	5	N	0.078	0.000	Inf	0.000	0.016	0.016	4.9
F6	10	[2, 2, 1]	5	Y	0.438	0.078	5.6	0.031	0.000	0.031	14.1
F6x	10	[2, 2, 1]	5	N	0.406	0.047	8.6	0.031	0.016	0.047	8.6
F7	18	[3, 1, 1, 1]	6	Y	>600.000	9.000	>66.7	3.390	0.063	3.453	>173.8
F8	12	[3, 2, 1]	6	Y	>600.000	0.360	>1666.7	0.187	0.000	0.187	>3208.6
F9	6	[2, 2, 2]	6	Y	8.734	0.000	Inf	0.000	0.000	0.000	Inf

8   Conclusion

We introduced the concept of μ-symmetric polynomial as a generalization of the classical symmetric polynomial. We further designed some efficient algorithms to compute μ-kernel of such polynomials. The notion of a "μ-kernel" could be generalized by considering other bases for the symmetric polynomials. In this paper, we used the elementary symmetric functions as basis. This was a natural choice for showing rationality of root functions. But all our algorithms could be extended to other bases, for instance, the power basis.

This paper proved that the root function $D^+(\boldsymbol{\mu})$ is $\boldsymbol{\mu}$-symmetric. This implies that we can now obtain explicit constants on the bit complexity of the root clustering algorithm in [2] when applied to polynomials with integer or algebraic coefficients. We plan to explore this application in future work.

References

1. Basu, S., Pollack, R., Roy, M.-F.: Algorithms in Real Algebraic Geometry (Algorithms and Computation in Mathematics), 2nd edn. Springer, New York (2016). https://doi.org/10.1007/3-540-33099-2
2. Becker, R., Sagraloff, M., Sharma, V., Xu, J., Yap, C.: Complexity analysis of root clustering for a complex polynomial. In: 41st Proceedings of ISSAC, Waterloo, Canada, 19–22 July, pp. 71–78 (2016)
3. Cox, D., Little, J., O'Shea, D.: Ideals, Varieties and Algorithms: An Introduction to Computational Algebraic Geometry and Commutative Algebra, 3rd edn. Springer, New York (2007). https://doi.org/10.1007/978-0-387-35651-8

Generation of Abundant Multi-choice or STACK Type Questions Using CAS for Random Assignments

Kentaro Yoshitomi[✉]

Osaka Prefecture University, Sakai, Japan
yositomi@las.osakafu-u.ac.jp

Abstract. In the past four years, the author has developed more than 120 video lectures for a linear algebra course, each of which is of about 10 min duration, and has tried to conduct flipped class. The author set tasks for the students using Moodle (LMS) questions, which are mainly of STACK type, in order to check if they have really studied the video materials and how deeply they understand them. According to the questionnaire result, in the case that the questions have too many input fields or require some CAS-formatted texts, the students may have some difficulty to answer the questions, especially with mobile phones. Multiple choice type or true-false questions are very simple to answer with such devices. In the latter half of the last academic year, the author developed CAS programs to automatically generate multiples of 100 mutually different matrices for some problems. For example, one of the programs generates the question data to select diagonalizable matrices from 10 matrices of degree 3. In order to prove that the students actually solved the questions by themselves, they were instructed to submit papers describing the process of the solution. We are planning to develop more generating programs to cover the entire course, including non-computational tasks.

Keywords: Flipped class · LMS · STACK · Linear algebra

1 Introduction

Recently, the flipped class or flipped learning has become a major blended learning strategy, which is usually conducted with video lectures for the preparation activities of students. Linear algebra courses are quite important in college mathematics. However, in Japan, there have been only a few examples of flipped learning/class trials for linear algebra courses. In the second semester of the 2014 academic year, the author developed video lectures on some topics for the students to review after class. In addition, in the second semester of the following two years, the author was in charge of a class of non-engineering students. More than half of the students in the class were not good at mathematics. The course in the 2nd semester includes some abstract topics related to abstract vector spaces, direct sums, linear maps, and the concept of matrix representation.

© Springer International Publishing AG, part of Springer Nature 2018
J. H. Davenport et al. (Eds.): ICMS 2018, LNCS 10931, pp. 492–497, 2018.
https://doi.org/10.1007/978-3-319-96418-8_58

Thus, some effective active learning strategies were required. Meanwhile, taking note of the usability of smartphones, the author found it necessary to have a "smartphone-friendly" learning strategy.

2 Video Lectures and Questions

The author developed more than 50 short video lectures for the preparation, all of which have a duration of about 5–10 min. In the last academic year, the author tried to conduct flipped learning/classes for the linear algebra course of the engineering school of our university (Fig. 1).

The video lectures were uploaded to YouTube site and/or were exported to local files. The local files were uploaded to the LMS of our university, which is based on Moodle [1]. The slides used in the video lectures were authored with LaTeX and Beamer class using the Tikz library. More animation effects were added using the authoring tool Camtasia Studio [2]. Since last autumn, it is possible to record screenshots with voice on iOS devices i.e. iPad. With this function, it seems that the video lectures have become more visible and effective, especially in the case of videos used to explain examples.

Since 2005, the web-based learning and testing system MATH ON WEB [5], which use *Mathematica* and web*Mathematica*, has been used in our university. The site has more than 1000 questions, and is therefore quite useful. However, the system has not been linked to our Moodle based LMS, so it is not easy to monitor and analyze the learning activities of the students. Hence, the author converted the question data of MATH ON WEB to STACK [3] data, and developed more new STACK questions.

Even if video materials are prepared abundantly and each video lecture is short enough for the students to view easily, there are not many cases where students actively use them. What matters is whether they under-

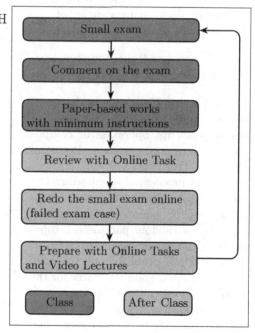

Fig. 1. Learning flow

stand the content, rather than just see it. To evaluate the preparation of the students, the author first used the STACK type questions mentioned above along with a few additional multiple-choice type questions and/or true/false type questions. Problems with these question types have gradually become clear, as seen from the comments in the last academic year's questionnaire.

3 Question Examples and Problems

The question shown in Fig. 2 is a STACK type question to check whether or not the students understand the elementary transformations. To use the random selection feature of Moodle, about fifty variations were prepared (Fig. 3).

CAS-based systems, for example, STACK, Maple T.A. [4], or MATH ON WEB, are useful for checking students' answers algebraically by using an algorithm. Especially, they are useful for quizzes that have infinite correct answers, for example, questions such as:

- "Find a basis of the null space of..."
- "Find a basis of the image of the linear map defined by following matrices:..."
- "Find the linear dependencies of the vectors if not linearly independent."

On the other hand, these systems have some problems:

1. Authoring problem: it is hard to author question data because of heavy javascript operation or complicated programming such as Potential Response Tree (PRT).
2. Parameter problem: it is important but subtly difficult to control the appropriateness of question data, which depends on random parameters, especially in generating matrices.
3. User interface problem: only recently have there been good interfaces for the students to enter the CAS formula in the input forms, such as MathTOUCH [6] and FlickMath [7]. However, there are still difficulties in doing this as described in the following paragraph.

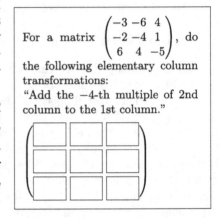

For a matrix $\begin{pmatrix} -3 & -6 & 4 \\ -2 & -4 & 1 \\ 6 & 4 & -5 \end{pmatrix}$, do the following elementary column transformations:
"Add the -4-th multiple of 2nd column to the 1st column."

Fig. 2. Sample of a quiz: elementary column transformations (3×3/C1+$=$C2*a/Gen)

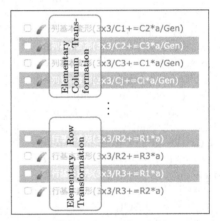

Fig. 3. More than 40 patterns of elementary column/row transformations (mathbank)

If the questions require the students to enter a 4 by 4 matrix, they should put 16 values into the input fields. This is somewhat annoying for them, especially in the case of putting them on smartphone. In the questionnaire, they had actually commented that it is troublesome to do them. In addition, there were the following cases: some students come up with complaints that they were told that their answers were not correct. The answers were actually correct, but they were put in double-byte characters. Even if there is difficulty in providing the correct answer, it is meaningless if it is too laborious to enter the answer.

True/false questions are some of the most simple questions. Here is a sample small online test consisting of such questions:

- A square matrix has an inverse matrix if it has non-zero entry.
- A square matrix has an inverse matrix if all of the entries are not zero.
- For a square matrix A, if there exists a square matrix B such that $AB = BA = E_n$, then A has another matrix B', which has the same property.
- If AB has the inverse matrix, then both A and B have inverse matrices and $(AB)^{-1} = B^{-1}A^{-1}$.
- If A has an inverse matrix A^{-1}, then the transpose of A also has an inverse matrix and $({}^tA)^{-1} = {}^t(A^{-1})$.

Such quizzes are easy in terms of inputting the answer, but are not necessarily suggestive. Students remember the answer and after some trials they would answer the questions correctly. This is somewhat different from what we aim to achieve. More random elements are required so that the students cannot answer correctly without having a deep understanding. For example, "Select all the invertible matrices from $\begin{pmatrix} 2 & 3 \\ 0 & 0 \end{pmatrix}$, $\begin{pmatrix} 1 & 1 \\ 1 & 1 \end{pmatrix}$, $\begin{pmatrix} 0 & 0 \\ 0 & 0 \end{pmatrix}$, $\begin{pmatrix} 1 & 1 \\ 0 & 1 \end{pmatrix}$,...". Appropriate feedback will result in better understanding for the students.

It seems that some kind of suggestive but easy-to-input tasks are necessary.

In the Moodle case, the quizzes are of multiple choice type, or STACK [3] type with a minimum input form (if required) to check the correctness algebraically by using an algorithm.

Finally, authoring the Moodle question data also has the following problems in general:

- When there are multiple analogous questions and some corrections are required, it is hard to edit them on Moodle or edit the exported XML codes directly.
- All the questions cannot be viewed at once, therefore, one cannot tell which of the questions have already been fixed. If the number of questions are in the hundreds, this is almost impossible. Systematic procedures using CAS to check the appropriateness or the validy of question data may be required.

Considering the above problems, we thought that it is reasonable to generate appropriate and large amounts of question data automatically by using a CAS.

4   Question Data Generator

For the reasons described in the preceding section, it is reasonable and natural to generate large amounts of question data using a CAS such as Maxima [8], *Mathematica* [9], or Maple [10]. In general, a CAS has the facility to handle matrices, strings, and lists which are required to construct XML question data.

Fortunately, Osaka Prefecture University, to which the author is affiliated, has a site license. This is the reason for using *Mathematica* to generate question data. As a matter of course, the **ideas** of the questions are important, and not which CAS one uses.

An excerpt of the flow of "Find eigen and diagonalizability" question data is shown in Fig. 4. This generates 100, 200, or 300 question data, each having 10 matrices, from 2 to 8 matrices of which are diagonalizable. The actual program and sample data including the cases of degree 4 and 5 are available from the author's website [11].

1. `TransMatToMaxima` is a procedure to Transform matrix Mathematica format to Maxima format.
2. `ChkMatrix` check the entries in the viewpoint of size of each entries and count of zero entries.
3. `GenerateFindEigen` generates from given Jordan standard form matrices satisfying some condistions using `ChkMatrix`.
4. `GenerateFindEigen` generates from given Jordan standard form matrices satisfying some condistions using `ChkMatrix`.
5. `LD3` and `LJ3` are lists of diagonalizable and non-diagonalizable matrices which have originally length of more than 5000 and 3000 per each and using DeleteDuplicates function, each has length of about 1600 and 1200.
6. Finally `MakeXML` generates from template XML exported from moodle. Output XML consists of 100,200, or 300 questions and can be imported by moodle.

Fig. 4. Excerpt of flow of generating of "Find diagonalizable matrix" question data

Actually, the author used these questions as a task for the students and some of them responded that "it was a great learning opportunity".

5   Future Works

As for generating the XML question data of various computational tasks automatically, it requires time and labor; however, it is not so difficult for users familiar with CAS. Nagasaka developed a *Mathematica* program that generates question data in Moodle XML format [12]. The author is planning a joint work with Nagasaka with the support of JSPS KAKENHI (cf. Acknowledgements).

In the near future, a complete set of XML data for generic linear algebra courses will be generated, and its effectiveness in flipped learning will be verified. Learning analytics or LA is expected.

Here are some issues remaining in this project.

One of them is that the student will be kept waiting until all the listed formulae (e.g. matrices) are displayed, because MathJax rendering is slightly slow. In this regard, some measures are required. We expect Moodle's official support for KaTeX or something similar.

Another one of the problems is that this method is not friendly for CAS beginners. Authoring tools for the automatic generators of question data that have a user-friendly interface for CAS beginners should be developed based on our CAS scripts. Since the data needs to be customized according to the requirements of any lecturer or the course of any school, the specifications of the generators are important. This could be integrated with the MeLQS [13] project.

Acknowledgments. This work was supported by JSPS KAKENHI Grant Number 15K00926 and is now supported by JSPS KAKENHI Grant Number 18K02941.

References

1. The Moodle project. https://moodle.org/. Accessed 27 Apr 2018
2. Camtasia Studio. https://www.techsmith.com/video-editor.html. Accessed 29 Apr 29 2018
3. moodle-qtype_STACK: https://github.com/maths/moodle-qtype_stack. Accessed 29 Apr 2018. (cf. Sangwin, C.: Computer Aided Assessment of Mathematics. Oxford University Press (2013))
4. Online Assessment System for STEM Courses 2013 Maplesoft. https://www.maplesoft.com/products/mapleta/. Accessed 27 Apr 2018
5. MATH ON WEB: Learning College Mathematics by web Mathematica. Osaka Prefecture University (2018). http://www.las.osakafu-u.ac.jp/lecture/math/MathOnWeb/. Accessed 27 Apr 2018
6. Shirai, S., Fukui, T.: MathTOUCH: mathematical input interface for e-assessment systems. MSOR Connect. **15**(2), 70–75 (2016)
7. Nakamura, Y., Nakahara, T.: Development of a math input interface with flick operation for mobile devices. In: Proceedings of 12th International Conference on Mobile Learning, pp. 113–116 (2016)
8. Maxima, A Computer Algebra System. http://maxima.sourceforge.net/. Accessed 28 Apr 2018
9. Wolfram Mathematica. https://www.wolfram.com/mathematica/. Accessed 28 Apr 2018
10. Mathematics-based software & services for education, engineering, and research. https://maplesoft.com/. Accessed 28 Apr 2018
11. http://www.las.osakafu-u.ac.jp/~yositomi/moodle_xml/. Accessed 28 Apr 2018
12. http://wwwmain.h.kobe-u.ac.jp/~nagasaka/research/xml_quiz/. Accessed 28 Apr 2018
13. Kawazoe, M., Yoshitomi, K., Nakahara, T., Nakamura, Y., Fukui, T., Shirai, S., Kato, K., Taniguchi, T.: MeLQS: mathematics e-learning questions Specification - a common base for sharing questions among different systems. In: Proceedings of the International Workshop on Mathematical Education for Non-Mathematics Students Developing Advanced Mathematical Literacy, pp. 123–126 (2018)

Intuitive Interface for Solving Linear and Nonlinear System of Equations

Zhonggang Zeng$^{(\boxtimes)}$

Northeastern Illinois University, Chicago, USA
zzeng@neiu.edu
http://www.homepages.neiu.edu/~zzeng

Abstract. An innovative approach is proposed in designing intuitive interfaces for solving systems of linear and nonlinear equations over Cartesian products of general vector spaces. The interfaces enable scientific computing practitioners and learners to enter equations in WYSIWYG manner, lets the software generate vector representations of equations/variables internally and outputs the solutions in the desired forms automatically. Such interfaces save more time than algorithmic improvement for one-time users and students.

Keywords: Interface · System of equations

1 Introduction

Solving linear and nonlinear systems of equations is one of the fundamental tasks in scientific computing. Existing software packages, however, require systems to be in matrix-vector form $A\mathbf{x} = \mathbf{b}$, or represented using multivariate functions in the form of

$$\begin{cases} f_1(x_1, \ldots, x_n) & = & 0 \\ \vdots & & \vdots & \vdots \\ f_m(x_1, \ldots, x_n) & = & 0 \end{cases} \quad . \tag{1}$$

Variables are allowed only as arrays of real/complex numbers. More common in practical computation and classroom teaching, linear equations are formulated in the form of

$$L(\mathbf{x}_1, \ldots, \mathbf{x}_n) \;\; = \;\; (\mathbf{b}_1, \ldots, \mathbf{b}_m)$$

where L is a linear transformation, and nonlinear equations are given as

$$\mathbf{f}(\mathbf{x}_1, \ldots, \mathbf{x}_n) \;\; = \;\; (0, \ldots, 0)$$

where $\mathbf{f}$ is a differentiable mapping between certain spaces. Furthermore, the variables $\mathbf{x}_1, \ldots, \mathbf{x}_n$ can be arrays of numbers, vectors, matrices, polynomials,

Research is supported in part by NSF under grant DMS-1620337.

functions of certain forms, etc. It is a tedious, time-consuming and error-prone process to transform the equations into the function representations before being solved by existing software and, after obtaining results, to make backward representations. Such back-and-forth representation processes are particularly daunting for beginners and students. We propose an innovative approach in designing intuitive interfaces for solving equations over Cartesian products of general vector spaces in our Matlab toolbox NAClab[1] for numerical algebraic computation [3]. The interfaces enable computing practitioners and learners to enter equations in WYSIWYG manner, let the software generate vector representations of equations/variables internally and output the solutions in the desired forms automatically. Such interfaces save more time than algorithmic improvement for one-time users and students.

2   Interface for Solving Linear Systems of Equations

In practical scientific computing, a general system of linear equations is in the form of

$$L(\mathbf{x}_1, \ldots, \mathbf{x}_n) \;=\; (\mathbf{b}_1, \ldots, \mathbf{b}_m) \tag{2}$$

where

$$L : X_1 \times \cdots \times X_n \longrightarrow Y_1 \times \cdots \times Y_m$$

is a linear transformation between Cartesian products of general vector spaces such as $\mathbb{C}^n$ of n-dimensional vectors, $\mathbb{P}_n$ of univariate or multivariate polynomials of degree up to n, $\mathbb{C}^{m \times n}$ of $m \times n$ matrices, etc. Furthermore, the linear transformation L may be under-determined, overdetermined or rank-deficient. There are two drawbacks in current software for numerical solutions of linear system of equations:

- Systems can only be solved in the matrix-vector form $A\mathbf{x} = \mathbf{b}$ to which the general linear system (2) must be transformed by users.
- Rank-deficient linear systems may not be solvable accurately from empirical data.

Our Matlab toolbox NACLAB is developed with an attempt to fill these two gaps. The interface module LinearSolve for solving general linear systems requires users to provide the following input items:

- The Matlab function or Matlab anonymous function for carrying out the evaluation of the linear transformation L.
- The domain $X_1 \times \cdots \times X_n$ in the form of a typical vector $(\mathbf{x}_1, \ldots, \mathbf{x}_n)$ with all the relevant components as a Matlab cell array.
- The parameters of the linear transformation as a cell array.
- The right-hand side vector $(\mathbf{b}_1, \ldots, \mathbf{b}_m)$.
- The error tolerance ε.

---

[1] http://www.homepages.neiu.edu/~zzeng.

We illustrate the interface $\texttt{LinearSolve}$ in the following example.

Example 1. Suppose, with the data error bound 10^{-4}, the polynomials

$$\tilde{f}(x) = 5.99999 - 9\,x + 3\,x^3 - 4\,x^4 + 6\,x^5 - 2\,x^7$$
$$\tilde{g}(x) = -2 + 5\,x - 3\,x^2 - x^3 + x^4 + 2\,x^5 - 3.00002\,x^6 + x^8$$

are empirical data of a polynomial pair f and g with a greatest common divisor

$$u \;=\; gcd\,(f, g) \;\approx\; 2 - 2.99999\,x + 1.00001\,x^3.$$

given approximately. It is known that the greatest common divisor is in the range of the linear transformation

$$
L \;:\; \begin{array}{ccc}
\mathbb{P}_5 \times \mathbb{P}_4 & \longrightarrow & \mathbb{P}_{12} \\
(p, q) & \longmapsto & p\,f + q\,g
\end{array}
\tag{3}
$$

and the kernel

$$\mathcal{K}ernel\,(L) \;=\; span\left\{\left(\frac{f}{u},\,\frac{g}{u}\right)\right\}$$

where the notation $\mathbb{P}_n$ denotes the vector space of polynomials with degrees up to n. The question: *Can we use the empirical data of f, g and u to solve the linear equation*

$$p\tilde{f} + q\tilde{g} \;=\; \tilde{u}$$

for $(p, q) \in \mathbb{P}_5 \times \mathbb{P}_4$ approximately with an accuracy comparable to the data accuracy? There appear to be no available software for finding numerical solution of such a rank-deficient linear system, and all software systems require users to go through a tedious process of transforming the equation into a matrix-vector form.

NACLAB provides a comprehensive linear equation solver $\texttt{LinearSolve}$ for such problems directly. Instead of constructing the representation matrix for the linear transformation, write a simple Matlab anonymous function implementing the linear transformation (3) with parameters f and g as it is:

```
>> L = @(p,q,f,g) ...          % Matlab anonymous function for the linear transformation
   PolynomialPlus(PolynomialTimes(p,f),PolynomialTimes(q,g));   % L : (p,q) |--> p*f+q*g
```

Here the syntax rules require the input items to start with variables p and q followed by parameters f and g. The modules $\texttt{PolynomialPlus}$ and $\texttt{PolynomialTimes}$ are interface functionalities in NACLAB to perform polynomial additions and polynomial multiplications with polynomials entered as character strings in WYSIWYG style:

```
>> f = '5.99999 - 9*x + 3*x^3 - 4*x^4 + 6*x^5 - 2*x^7';          % data for polyn. f
>> g = '-2 + 5*x - 3*x^2 - x^3 + x^4 + 2*x^5 - 3.00002*x^6 + x^8';   % data for polyn. g
>> u = '2-2.99999*x+1.00001*x^3';                                % data for polyn. u
```

Define the domain of the linear transformation by providing two polynomials in $\mathbb{P}_5$ and $\mathbb{P}_4$ with all the relevant monomials, along with the parameter:

```
>> domain = {'1+x+x^2+x^3+x^4+x^5','1+x+x^2+x^3+x^4'};   % domain of variable (p,q) of L
>> parameter = {f,g};                        % parameter cell array of the linear transf. L
>> error = 1e-4;                             % error tolerance
```

Call the NACLAB interface **LinearSolve** with input items consist of the linear transformation array $\{L,\ \texttt{domain},\ \texttt{parameter}\}$, the right-hand side u, and error tolerance 10^{-4}:

```
>> [Z,K,lcond,res] =   % solve L(p,q) = u, in the 'domain' with 'parameter' within 'error'
   LinearSolve({L,domain,parameter}, u, error);
```

The cell array Z contains the numerical minimum-norm solution in $\mathbb{P}_5 \times \mathbb{P}_4$ in WYSIWYG style

```
>> Z                              % display the numerical minumum-norm solution
Z =
  '0.31521363813893 + 0.032589573209552*x + 0.02170375447856*x^2 + 0.054260211908843*x^3
   + 0.13564951660988*x^4 + 0.02390813281878*x^5'  '-0.05435802677693 + 0.04340953379911*x
   + 0.10852114641191*x^2 + 0.27129848252721*x^3 + 0.047816465413613*x^4'
```

The cell array K contains the 1-dimensional numerical Kernel in $\mathbb{P}_5 \times \mathbb{P}_4$ in WYSIWYG style

```
>> K{:}                              % display the basis for numerical kernel
ans =
     '0.24999938925534 - 0.25000118965734*x - 0.25000215206104*x^5'   '0.74999762022995
   - 0.500002204088944*x^4'
```

The numerical solution can be verified using the linear transformation function L.

```
>> p = Z{1};  q = Z{2};                          % extract p and q
>> h = L(p,q,f,g);                    % evaluate h = L(p,q) with parameter f and g
>> PolynomialClear(h,1e-5)           % clear numerical tiny coefficients below 1e-5
ans =
1.99999473025102 - 2.9999948313716*x + 1.00000604534541*x^3
```

The output **lcond** and **res** show the condition number for the linear equation is healthy at 52.1 and the residual is below the error tolerance at about 5.27×10^{-6}.

The user can study the linear equation further by investigating the representation matrix of the linear equation by calling **LinearTransformMatrix** in NACLAB using the above defined input items

```
>> A = LinearTransformMatrix(L,domain,parameter);         % representation matrix of L
```

obtaining the 13×11 matrix with respect to natural bases for the domain and codomain

$$
\begin{bmatrix}
5.99999 & 0 & 0 & 0 & 0 & 0 & -2.0 & 0 & 0 & 0 & 0 \\
-9.0 & 5.99999 & 0 & 0 & 0 & 0 & 5.0 & -2.0 & 0 & 0 & 0 \\
0 & -9.0 & 5.99999 & 0 & 0 & 0 & -3.0 & 5.0 & -2.0 & 0 & 0 \\
3.0 & 0 & -9.0 & 5.99999 & 0 & 0 & -1.0 & -3.0 & 5.0 & -2.0 & 0 \\
-4.0 & 3.0 & 0 & -9.0 & 5.99999 & 0 & 1.0 & -1.0 & -3.0 & 5.0 & -2.0 \\
6.0 & -4.0 & 3.0 & 0 & -9.0 & 5.99999 & 2.0 & 1.0 & -1.0 & -3.0 & 5.0 \\
0 & 6.0 & -4.0 & 3.0 & 0 & -9.0 & -3.00002 & 2.0 & 1.0 & -1.0 & -3.0 \\
-2.0 & 0 & 6.0 & -4.0 & 3.0 & 0 & 0 & -3.00002 & 2.0 & 1.0 & -1.0 \\
0 & -2.0 & 0 & 6.0 & -4.0 & 3.0 & 1.0 & 0 & -3.00002 & 2.0 & 1.0 \\
0 & 0 & -2.0 & 0 & 6.0 & -4.0 & 0 & 1.0 & 0 & -3.00002 & 2.0 \\
0 & 0 & 0 & -2.0 & 0 & 6.0 & 0 & 0 & 1.0 & 0 & -3.00002 \\
0 & 0 & 0 & 0 & -2.0 & 0 & 0 & 0 & 0 & 1.0 & 0 \\
0 & 0 & 0 & 0 & 0 & -2.0 & 0 & 0 & 0 & 0 & 1.0
\end{bmatrix}
$$

This is the matrix representation of the linear system the user would have to construct without using the interface `LinearSolve`.   □

In fact, the module `LinearTransformMatrix` is the most crucial component of the interface `LinearSolve` that automates the process of solving linear systems and frees users from the tedious representation process.

Notice that there is a profound difference in `LinearSolve` compared to currently standard linear system solving implementations: The representing matrix shown above is considered numerically rank-deficient by virtue of a singular value 0.000003318035559 that is below the error tolerance. A number of magnitude below error tolerance can be considered a zero.

Standard linear system solvers treat the matrix as full-ranked but the condition number 6.6×10^6 suggests a questionable solution due to an error tolerance 10^{-4}. Numerical kernel is not returned in standard implementations.

3   Interface for Solving Nonlinear System of Equations

Similar to linear cases, a general system of nonlinear equations in scientific computing is usually in the form of

$$
\mathbf{f}(\mathbf{x}_1, \ldots, \mathbf{x}_n) = (0, \ldots, 0)
$$

where $\mathbf{f} : X_1 \times \cdots \times X_n \longrightarrow Y_1 \times \cdots \times Y_m$ is a differentiable mapping between Cartesian products of vector spaces where $\mathbf{f}$ may be square (i.e. $m = n$) or overdetermined ($m > n$). The most commonly used method for solving such a nonlinear system is Newton's iteration or, in overdetermined case, the Gauss-Newton iteration. For beginners and those who needs to solve such a nonlinear system only once or twice, the most daunting and time consuming part of computation is to transform the system into its multivariate function form (1) and to construct the corresponding Jacobian matrix. Our interface `GaussNewton` in NACLAB is an attempt to simplify the representation into a few WYSIWYG steps. We shall use the following example to illustrate the process.

Example 2. A defective eigenvalue λ_* of a matrix $A \in \mathbb{C}^{n \times n}$ can be calculated accurately by solving the equation

$$
\mathbf{g}(\lambda, X) = (O, O)
$$

where, knowing the multiplicity support is $m \times k$, the holomorphic mapping **g** is given as

$$\mathbf{g} \;:\; \begin{array}{l} \mathbb{C} \times \mathbb{C}^{n \times k} \;\longrightarrow\; \mathbb{C}^{n \times k} \times \mathbb{C}^{m \times k} \\ (\lambda, X) \;\longmapsto\; \left(A X - \lambda X - X S, \; C^{\mathsf{H}} X - T \right). \end{array}$$

with constant parameters A, S, C and T. Detailed theoretical and computational issues on such equations can be found in [4]. For the purpose of illustration, we use the following parameters:

$$n = 9, \quad m = 3, \quad k = 2,$$

$$A = \begin{bmatrix} 3 & 3 & 3 & 3 & -2 & 2 & 1 & -1 & -1 \\ 1 & 4 & 4 & 3 & 0 & 2 & 0 & -1 & 0 \\ 0 & 0 & 2 & 0 & -2 & 0 & 0 & 0 & -1 \\ -2 & -3 & -4 & -2 & 6 & -2 & -2 & 1 & 3 \\ 0 & 0 & 0 & 0 & 5 & 0 & 0 & 0 & 3 \\ 2 & 4 & 4 & 5 & -3 & 5 & 3 & -3 & -1 \\ 1 & 0 & 1 & 1 & -2 & 0 & 3 & 0 & -1 \\ 1 & 6 & 7 & 7 & 2 & 6 & 1 & -4 & 2 \\ 0 & 0 & 0 & 0 & -6 & 0 & 0 & 0 & -4 \end{bmatrix},$$

$$S = \begin{bmatrix} 0 & 1 & & \\ & \ddots & \ddots & \\ & & \ddots & 1 \\ & & & 0 \end{bmatrix}, \quad T = \begin{bmatrix} 1 & 0 & \cdots & 0 \\ 0 & 0 & \cdots & 0 \\ \vdots & \vdots & \ddots & \vdots \\ 0 & 0 & \cdots & 0 \end{bmatrix},$$

and C is a 9×3 random matrix. The exact eigenvalues is $\lambda_* = 2$. Using a standard numerical eigenvalue solver, one can only obtain a few accurate digits of the eigenvalue, say $\lambda_0 = 1.9999$ due to the defectiveness of the eigenvalue.

On top of the time-consuming task of transforming the system of equations $\mathbf{g}(\lambda, X) = (O, O)$ into multivariate function form, the main difficulties for a one-time user include formulating the Jacobian matrix. As a key aspect of our interface, it is much more convenient to consider the Jacobian of the mapping **g** at any (λ_0, X_0) as a linear transformation that is also known as Fréchet derivative

$$J(\lambda_0, X_0) \;:\; \begin{array}{l} \mathbb{C} \times \mathbb{C}^{n \times k} \;\longrightarrow\; \mathbb{C}^{n \times k} \times \mathbb{C}^{m \times k} \\ (\lambda, X) \;\longmapsto\; \left(-\lambda X_0 + (A - \lambda_0 I) X - X S, \; C^{\mathsf{H}} X \right). \end{array}$$

with the same domain/codomain of the mapping **g** and additional parameters λ_0, X_0. The main innovation of our interface development is to enable users to enter both the mapping **g** and the Jacobian $J(\lambda_0, X_0)$ into Matlab as either function m-files or in-line anonymous functions exactly as they are:

```
>> g = @(lambda, X, A, S, C, T) ...        % Matlab anonymous function for the mapping g
      {A*X - lambda*X - X*S, C'*X - T}; % g : (lambda,X) |--> (A*X-lambda*X-X*S, C'*X-T)
>> J = @(lambda, X, lambda0, X0, A, S, C, T) ... % Matlab anonymous function for Jacobian
      {-lambda*X0 + A*X - lambda0*X - X*S, C'*X};
```

The syntax rules stipulate that the input items start with the common variables `lambda`, `X` followed by the common parameters `A`, `S`, `C`, `T`, and the additional parameters `lambda0`, `X0` for J are in between. The transformation of the system into the multivariate function form (1) is not needed either by the user or in the internal implementation since what is required is the evaluation of the mapping. The construction of the Jacobian matrix is carried out internally by the interface `LinearSolve` described in the previous section.

After using `LinearSolve` to solve the overdetermined linear system of equations

$$\mathbf{g}(\lambda_0, X) = (O, O) \quad \text{for} \quad X \in \mathbb{C}^{n \times k}$$

for its least squares solution X_0, we can apply the Gauss-Newton iteration from the initial iterate (λ_0, X_0) and solve the nonlinear equation $\mathbf{g}(\lambda, X) = (O, O)$ by the NACLAB interface `GaussNewton` with a few intuitive statements:

```
>> S = [0 1; 0 0]; T = [1 0; 0 0; 0 0]; C = rand(9,3);      % enter parameters S, T, C
>> domain = {1, ones(9,2)}                         % domain of (lambda,X) for mappings g and J
>> parameters = {A, S, C, T};                      % common parameters for both g and J
>> z0 = {1.9999,X0}                                % initial iterate
>> [z,res,fcond] = ...                             % call GaussNewton with input items
   GaussNewton({g,domain,parameters},J,z0,2,5e-10);   % and display the 1st component
     Step   1:     residual =    2.89e-03
   1.989243561668298
     Step   2:     residual =    6.89e-04
   2.000311883756314
     Step   3:     residual =    6.13e-08
   2.000000057498796
     Step   4:     residual =    4.94e-15
   1.999999999999999
   ...
```

The result includes an accurate approximation of the eigenvalue $\lambda_* = 2$.

As shown in the example, the users do not need to transform the system into multivariate form, nor do they need to construct the representation matrix of the Jacobian. Instead, the system is entered into Matlab directly as it is defined and the Jacobian is conveniently entered as a linear transformation in WYSIWYG style.

4   Underlying Theory and Technical Contribution

The underlying theory of the interface is quite simple in basic linear algebra. Let $\mathcal{V}$ and $\mathcal{W}$ be general vector spaces, possibly Cartesian products of vector spaces over a number field, say $\mathbb{C}$. Vectors in $\mathcal{V}$ and $\mathcal{W}$ can be represented as column vectors in $\mathbb{C}^n$ and $\mathbb{C}^m$ respectively via isomorphisms that can be denoted by $\phi : \mathcal{V} \to \mathbb{C}^n$ and $\psi : \mathcal{W} \to \mathbb{C}^m$ respectively. A linear transformation $L : \mathcal{V} \to \mathcal{W}$ can thus be represented as a matrix T in $\mathbb{C}^{m \times n}$ that makes the following diagram commute:

$$
\begin{array}{ccc}
\mathcal{V} & \xrightarrow{\ L\ } & \mathcal{W} \\
\phi \downarrow & & \uparrow \psi^{-1} \\
\mathbb{C}^n & \xrightarrow{\ T\ } & \mathbb{C}^m
\end{array}
$$

Namely, we have the identity

$$L(\mathbf{v}) \;\equiv\; \psi^{-1}\big(T\,\phi(\mathbf{v})\big).$$

Let $\{\mathbf{v}_1, \ldots \mathbf{v}_n\}$ and $\{\mathbf{w}_1, \ldots \mathbf{w}_m\}$ be some kind of standard bases for $\mathcal{V}$ and $\mathcal{W}$ respectively. Then the representation matrix can be constructed column-by-column according to

$$T \;=\; \big[\psi(L(\mathbf{v}_1)), \ldots, \psi(L(\mathbf{v}_n))\big]$$

where the j-th column is $\psi(L(\mathbf{v}_j)) \in \mathbb{C}^m$ for $j = 1, 2, \ldots, n$.

At the current stage of development, we assume the domain $\mathcal{V}$ and the codomain $\mathcal{W}$ are Cartesian products of matrix spaces and polynomial spaces. A standard basis $\{\mathbf{v}_1, \ldots, \mathbf{v}_n\}$ for the domain $\mathcal{V}$ can be extracted automatically from a typical vector as input item for the domain, so can a standard basis $\{\mathbf{w}_1, \ldots, \mathbf{w}_m\}$ for the codomain $\mathcal{W}$. The aforementioned isomorphisms ϕ and ψ can be implemented in a straightforward process due to the simplicity of the standard bases for the spaces of matrices and polynomials.

The theoretical foundation for the numerical solution of rank-deficient linear systems of equations is related to the subject of numerical rank-revealing [1,2] and beyond the scope of this abstract.

For the nonlinear system of equations $\mathbf{f}(\mathbf{x}_1, \ldots, \mathbf{x}_n) = (0, \ldots, 0)$ where $\mathbf{f} \;:\; X_1 \times \cdots \times X_n \to Y_1 \times \cdots \times Y_m$ is a differentiable mapping between Cartesian products of general vector spaces, the Gauss-Newton iteration

$$(\mathbf{x}_1^{(k+1)}, \ldots, \mathbf{x}_n^{(k+1)}) \;=\; (\mathbf{x}_1^{(k)}, \ldots, \mathbf{x}_n^{(k)}) - J(\mathbf{x}_1^{(k)}, \ldots, \mathbf{x}_n^{(k)})^\dagger \, \mathbf{f}(\mathbf{x}_1^{(k)}, \ldots, \mathbf{x}_n^{(k)})$$
$$k = 0, 1, \ldots$$

where

$$J(\mathbf{x}_1^{(k)}, \ldots, \mathbf{x}_n^{(k)}) \;:\; X_1 \times \cdots \times X_n \;\longrightarrow\; Y_1 \times \cdots \times Y_m$$

is the Jacobian (linear transformation) of $\mathbf{f}$ at $(\mathbf{x}_1^{(k)}, \ldots, \mathbf{x}_n^{(k)})$ and $(\cdot)^\dagger$ denote the pseudo-inverse. At every step of the Gauss-Newton iteration, the linear system of equations

$$J(\mathbf{x}_1^{(k)}, \ldots, \mathbf{x}_n^{(k)})\,(\Delta\mathbf{x}_1, \ldots, \Delta\mathbf{x}_n) \;=\; -\,\mathbf{f}(\mathbf{x}_1^{(k)}, \ldots, \mathbf{x}_n^{(k)}) \tag{4}$$
$$\text{for} \quad (\Delta\mathbf{x}_1, \ldots, \Delta\mathbf{x}_n) \in X_1 \times \cdots \times X_n$$

is to be solved for its least squares solution so that

$$\mathbf{x}_j^{(k+1)} \;=\; \mathbf{x}_j^{(k)} + \Delta\mathbf{x}_j \quad \text{for} \quad j = 1, 2, \ldots, n.$$

As a result, an interface for solving the system $\mathbf{f}(\mathbf{x}_1, \ldots, \mathbf{x}_n) = (0, \ldots, 0)$ of general nonlinear systems of equations can be built on top of the interface for solving general linear systems so that the system (4) can be solved at every step of the Gauss-Newton iteration. The key is to treat the Jacobian J at every point $(\mathbf{x}_1^{(k)}, \ldots, \mathbf{x}_n^{(k)})$ as a linear transformation with the same pair of domain/codomain as the mapping $\mathbf{f}$ and can be implemented as Matlab function in WYSIWYG style. Most importantly, the Jacobian as a linear transformation can be derived from elementary differentiation rules. In contrast, the Jacobian as a matrix is tedious to construct.

The interfaces LinearSolve and GaussNewton in NACLAB are particularly useful for research in numerical algebraic computation and classroom teaching, in which we often need to experiment with various formulations of computational models and each system of equations needs to be solved mostly only once or twice. In such settings, the efficiency and complexity of the underlying algorithm are secondary since the sizes of testing problems are usually small. Substantial portion of time and effort are spent on setting up the systems into a format acceptable to the existing software, particularly in representing the system into matrix-vector form or multivariate function form. For students and beginners, such time-consuming and error-prone processes are daunting and may even be intimidating. The equation-solving interfaces make it possible for our graduate students to conduct computing projects involving sophisticated algebraic equations.

References

1. Li, T.-Y., Zeng, Z.: A rank-revealing method with updating, downdating and applications. SIAM J. Matrix Anal. Appl. **26**, 918–946 (2005)
2. Lee, T.-L., Li, T.-Y., Zeng, Z.: A rank-revealing method with updating, downdating and applications. Part II. SIAM J. Matrix Anal. Appl. **31**, 503–525 (2009)
3. Zeng, Z.: NAClab: a Matlab toolbox for numerical algebraic computation. ACM Commun. Comput. Algebra **47**(3/4), 170–173 (2013)
4. Zeng, Z.: Sensitivity and computation of a defective eigenvalue. SIAM J. Matrix Anal. Appl. **37**, 798–817 (2016)

Author Index

Printed in the United States
By Bookmasters